ESTUARINE PROCESSES

Volume II

Circulation, Sediments, and Transfer of Material in the Estuary

Produced by the
Estuarine Research Federation
with support from the

Bureau of Land Management, Department of Interior
Fish and Wildlife Service, Department of Interior
U. S. Environmental Protection Agency
National Marine Fisheries Service, National Oceanographic and Atmospheric Administration
Marine Ecosystems Analysis Program, NOAA
Office of Applications, National Aeronautics and Space Administration
Council on Environmental Quality
Energy Research and Development Administration

ESTUARINE PROCESSES

Volume II
Circulation,
Sediments,
and Transfer of Material in the Estuary

Edited by

MARTIN WILEY

Chesapeake Biological Laboratory
University of Maryland
Center for Environmental
and Estuarine Studies
Solomons, Maryland

ACADEMIC PRESS New York San Francisco London 1976
A Subsidiary of Harcourt Brace Jovanovich, Publishers

ACADEMIC PRESS, INC.
111 Fifth Avenue, New York, New York 10003

United Kingdom Edition published by
ACADEMIC PRESS, INC. (LONDON) LTD.
24/28 Oval Road, London NW1

Library of Congress Cataloging in Publication Data

International Estuarine Research Conference, 3d, Galveston, 1975.
Circulation, sediments, and transfer of material in the estuary.

(Its Estuarine processes ; v. 2)
1. Estuarine oceanography–Congresses.
2. Oceanographic circulation–Congresses.
3. Estuarine sediments–Congresses. 4. Sediment transport–Congresses. I. Wiley, Martin L.
II. Title.
GC96.I57 1975 vol. 2 [GC97] 551.4′609s
ISBN 0–12–751802–9 [551.4′609] 76-49578

PRINTED IN THE UNITED STATES OF AMERICA

CONTENTS

LIST OF CONVENERS

Numbers in parentheses indicate the pages on which the conveners' sessions begin.

A. J. ELLIOTT (321), *Chesapeake Bay Institute, The Johns Hopkins University, Baltimore, Maryland 21218*

RAY B. KRONE (1), *Department of Civil Engineering, University of California, Davis, Davis, California 95616*

MAYNARD M. NICHOLS (33, 107), *Virginia Institute of Marine Science, Gloucester Point, Virginia 23062*

SCOTT W. NIXON (217), *Graduate School of Oceanography, University of Rhode Island, Kingston, Rhode Island 02881*

WILLIAM E. ODUM (217, 280), *Department of Environmental Sciences, University of Virginia, Charlottesville, Virginia 22903*

JEROME WILLIAMS (379), *Environmental Sciences Department, U.S. Naval Academy, Annapolis, Maryland 21402*

LIST OF CONTRIBUTORS

Numbers in parentheses indicate the pages on which the authors' contributions begin.

G. P. ALLEN (63), *Centre Océanologique de Bretagne–CNEXO, B.P. 337, 29273 Brest Cedex, France*

RANJAN ARIATHURAI (98), *Department of Civil Engineering, University of California at Davis, Davis, California 95616*

KEITH BANCROFT (270), *Institute of Ecology, Marine Institute, The University of Georgia, Athens, Georgia 30602*

JOHN BREED (270), *Institute of Ecology, Marine Institute, The University of Georgia, Athens, Georgia 30602*

W. F. BOHLEN (109), *Marine Sciences Institute and Department of Geology, University of Connecticut, Avery Point, Groton, Connecticut 06340*

KEVIN T. BIDDLE (167), *Department of Geology, Rice University, Houston, Texas 77001*

ALAN FRED BLUMBERG[1] (321), *Department of Earth and Planetary Sciences, The Johns Hopkins University, Baltimore, Maryland 21218*

THOMAS J. BUTLER (255), *Center for Wetland Resources, Louisiana State University, Baton Rouge, Louisiana 70803*

ENRIQUE A. CAPONI[2] (332), *Institute for Fluid Dynamics and Applied Mathematics, University of Maryland, College Park, Maryland 20742*

HARRY H. CARTER (48), *Marine Sciences Research Center, State University of New York, Stony Brook, New York 11794*

[1]Present address: Geophysical Fluid Dynamics Program, Princeton University, Princeton, New Jersey 08540

[2]Present address: Laboratorio de Hidráulica Aplicada/INCYTH, Casilla de Correos 21, Aeropuerto Ezeiza, Provincia de Buenos Aires, Argentina

P. CASTAING (63), *Institut de Géologie du Bassin d'Aquitaine, 351, Cours de la Libération, 33405 Talence, France*

A. CHALMERS (241), *University of Georgia Marine Institute, Sapelo Island, Georgia 31327*

R. R. CHRISTIAN (270), *Institute of Ecology, Marine Institute, The University of Georgia, Athens, Georgia 30602*

DAVID R. COLBY (416), *Atlantic Estuarine Fisheries Center, National Marine Fisheries Service, Beaufort, North Carolina 28516*

WILLIAM H. CONNER (255), *Center for Wetland Resources, Louisiana State University, Baton Rouge, Louisiana 70803*

T. J. CONOMOS (82), *U.S. Geological Survey, 345 Middlefield Road, Menlo Park, California 94025*

L. EUGENE CRONIN (18), *Center for Environmental and Estuarine Studies, University of Maryland, Box 775, Cambridge, Maryland 21613*

JOHN W. DAY, Jr. (255), *Center for Wetland Resources, Louisiana State University, Baton Rouge, Louisiana 70803*

K. R. DYER (124), *Institute of Oceanographic Sciences, Crossway, Taunton, United Kingdom*

RICHARD W. FAAS (136), *Lafayette College, Easton, Pennsylvania 10842*

DAVID A. FLEMER[3] (219, 309), *University of Maryland Center for Environmental and Estuarine Studies, Chesapeake Biological Laboratory, Solomons, Maryland 20688*

DIRK FRANKENBERG[4] (270), *Department of Zoology, The University of Georgia, Athens, Georgia 30602*

RONALD J. GIBBS (35), *College of Marine Studies, University of Delaware, Lewes, Delaware 19958*

E. HAINES (241), *University of Georgia Marine Institute, Sapelo Island, Georgia 31327*

J. R. HALL (270), *Institute of Ecology, Marine Institute, The University of*

[3]Present address: Office of Biological Studies, U.S. Fish and Wildlife Service, Department of the Interior, Washington, D.C. 20240

[4]Present address: Curriculum in Marine Science, University of North Carolina, Chapel Hill, North Carolina

Georgia, Athens, Georgia 30602

PETER HAMILTON (347), *Department of Oceanography, University of Washington, Seattle, Washington 98195*

R. HANSON (241), *University of Georgia Marine Institute, Sapelo Island, Georgia 31327*

DONALD R. HEINLE (219, 309), *University of Maryland, CEES, Chesapeake Biological Lab, Box 38, Solomons, Maryland 20688*

J.M. JOUANNEAU (48), *Institut de Géologie du Bassin d'Aquitaine, 351 Cours de la Libération 33405 Talence, France*

MARTIN A. KJELSON (416), *Atlantic Estuarine Fisheries Center, National Marine Fisheries Service, Beaufort, North Carolina 28516*

VYTAUTAS KLEMAS (381), *College of Marine Studies, University of Delaware, Newark, Delaware 19711*

TED S. Y. KOO (18), *Chesapeake Biological Laboratory, Center for Environmental and Estuarine Studies, University of Maryland, Box 38, Solomons, Maryland 20688*

RAY B. KRONE (98), *Department of Civil Engineering, University of California at Davis, Davis, California 95616*

ROBERT R. LANKFORD (182), *UNESCO Marine Geologist, Centro de Ciencias del Mar y Limnolgia, UNAM, Apartado Postal 70-305, Mexico 20, D.F., Mexico*

VICTOR LOTRICH (18), *College of Marine Studies, University of Delaware, Newark, Delaware 19958*

JAMES H. McKAY, Jr. (404), *U.S. Army Corps of Engineers, P.O. Box 1715, Baltimore, Maryland 21203*

L. G. MAURER[5] (270), *Department of Zoology, The University of Georgia, Athens, Georgia 30602*

GERALD T. ORLOB (3), *Resource Management Associates, 3706 Mt. Diablo Blvd., Suite 200, Lafayette, California 94549*

D. H. PETERSON (82), *U.S. Geological Survey, 345 Middlefield Road, Menlo Park, California 94025*

[5]Present address: Environmental Quality Lab, General Development Engineering Co., Port Charlotte, Florida 33950

JAMES C. PICKRAL (280), *Department of Environmental Sciences, University of Virginia, Charlottesville, Virginia 22903*

L. R. POMEROY (270), *Department of Zoology, The University of Georgia, Athens, Georgia 30602*

DONALD W. PRITCHARD (18), *Chesapeake Bay Institute, The Johns Hopkins University, Baltimore, Maryland 21218*

RANDY A. ROWLAND (219), *Botany Department, University of Maryland, College Park, Maryland 20742*

LUIS A. SANCHEZ-BARREDA (167), *Department of Geology, Rice University, Houston, Texas 77001*

G. SAUZAY (63), *S.A.P.R.A., Centre d'Etudes Nucléaires–C.E.A., B.P. 2, 91190 Gif-Sur-Yvette, France*

J. R. SCHUBEL (48), *Marine Sciences Research Center, State University of New York, Stony Brook, New York 11794*

B. SHERR (241), *University of Georgia Marine Institute, Sapelo Island, Georgia 31327*

ROBERT J. SMALL (219), *Botany Department, University of Maryland, College Park, Maryland 20742*

DAVID D. SMITH (150), *David D. Smith and Associates, Evironmental Consultants, P.O. Box 929-E, San Diego, California 92109*

J. COURT STEVENSON (219), *University of Maryland, CEES, Box 775, Cambridge, Maryland 21613*

JOSEPH F. USTACH[6] (219, 309), *University of Maryland Center for Environmental and Estuarine Studies, Chesapeake Biological Laboratory, Solomons, Maryland 20688*

GEORGE H. WARD, Jr. (365), *Manager, Engineering Programs, Espey, Huston and Associates, Inc., 3010 South Lamar, Austin, Texas 78704*

JOHN E. WARME (167), *Department of Geology, Rice University, Houston, Texas 77001*

STANISLAS I. WARTEL (136), *Royal Belgian Institute for Natural Sciences, Brussels, Belgium*

[6]Present address: Department of Zoology, North Carolina State University, Raleigh, North Carolina 27607

R. L. WETZEL[7] (270, 293), *Institute of Ecology, Marine Institute, The University of Georgia, Athens, Georgia 30602*

W. J. WIEBE (270), *Institute of Ecology, Marine Institute, The University of Georgia, Athens, Georgia 30602*

R. G. WIEGERT (270), *Institute of Ecology, Marine Institute, The University of Georgia, Athens, Georgia 30602*

[7]Present address: Virginia Institute of Marine Science, Gloucester Point, Virginia 23062

PREFACE

Planning for the Third International Estuarine Research Conference began at Myrtle Beach, while the second conference was still in session. Before moving ahead on details, the Estuarine Research Federation had several questions to answer: "Does the rate of *Recent Advances* in our field warrant another invited symposium two-years hence; if so, is a unifying theme readily apparent; how can we keep expenses down for students and the established investigators of tomorrow; and where should we meet to recognize surging interests of the Federation's new affiliates in the Southern and Gulf states?"

The Governing Board decided that the next *Recent Advances* would emphasize estuarine processes–an attempt to focus on dynamic interactions at several levels of organization. As it worked out, more than 70 papers were presented before 550 registrants in Galveston, Texas, October 7-9, 1975. Most of the papers appear in these two volumes. Success and failure reflect not only the Board's wisdom, but also instabilities that started at a global level with an oil embargo and came to bear on the outlook of all.

On the success side of the balance sheet, special thanks go to the convenors. They were given authority to organize sessions and follow through in the peer review process. The Federation also thanks the eight Federal agencies which provided support for the Conference and associated editorial work. The comments we received on the proposals submitted to the agencies were inevitably coupled with our direct experience, yielding a uniquely comprehensive view on the role of big scientific meetings enriched with opportunities for environmental management.

I believe that our shortcomings–judged from scientific value in the Conference's written record–are intertwined with positive attitudes emanating from the Federation's affiliated societies. Their history is a commitment to excellence in a field that on occasion has been as variable as the very structure of the environment studied. If certain sessions appear less comprehensive and sophisticated than usual, the result may well reflect traditions deeply embedded in these societies. The societies exist to challenge and redirect the results of research, and they rely upon an especially open system of appraisal. It follows that the negative side of the balance sheet is more indicative of needs–either for encouragement or for curtailment in specific areas–than it is an enumeration of wasted effort.

These volumes and the reviews they attract should serve well in the development of critical perspective, the foremost requirement today in shallow-water oceanography.

For the Governing Board, 1973-75

H. Perry Jeffries, President
Estuarine Research Federation

EFFECTS OF PHYSICAL ALTERATIONS

Convened by:
Ray B. Krone
Department of Civil Engineering
University of California, Davis
Davis, California 95616

Estuaries are the transition regions from the fresh water unidirectional flows of streams to the tidal, saline ocean. The flows in an estuary are affected by the conditions at both ends of this transition zone and are modified by the configuration of the estuary, by winds, and by waste discharges.

The oscillating character of ocean tides produce reversing flows in estuaries, and the difference in salinities causes the denser ocean waters to intrude along the bed of an estuary. Vertical mixing, resulting from the bed friction and tidal flows, produces vertical and longitudinal salinity gradients in the mixing region. The augmentation of seaward flowing surface waters by the intruding sea water enhances the capacity of estuarial waters for transporting wastes.

Man modifies the configurations of estuaries, the fresh water inflows, and waste discharges. These modifications affect currents, suspended solids concentrations, salinities, tides, and dissolved materials in the estuary. All biota are subject to these effects. Are these effects beneficial? How do we determine their desirability? This session is devoted to a description of effects of physical alterations on estuarial hydraulics, dissolved and particulate material transport, and on aquatic biota.

IMPACT OF UPSTREAM STORAGE AND DIVERSIONS ON SALINITY BALANCE IN ESTUARIES

Gerald T. Orlob[1]
Resource Management Associates
3706 Mt. Diablo Blvd., Suite 200
Lafayette, CA 94549

ABSTRACT: Impoundment of runoff and its diversion for purposes of power production, flood control, irrigation, or other beneficial uses usually results in significant alteration in the hydrologic regimen of the river system. When diversions are made for in-basin consumptive use or to extra-basin uses, net flows may be reduced to levels not experienced historically. The consequences to water quality, when these diversions occur during periods of low downstream flows, may be such as to impact adversely on the use of water for agriculture.

Cases of special interest are the lower San Joaquin Valley and the southern Sacramento-San Joaquin Delta. Since the advent of the Central Valley Project in California in the mid-40's, the quality of inflows to the agriculturally rich southern Delta has progressively declined. Today the area faces a critical shortage of water of adequate quality to sustain agricultural production.

Historic changes in salinity balance in the estuarial zone of the Sacramento-San Joaquin Delta are reviewed. Selected examples of upstream flow manipulation are presented to show how the salinity balance of the estuarial system can be most effectively managed.

INTRODUCTION

California's Central Valley includes approximately 38 percent of the State's land area, 70 percent of its irrigated land area, and is said to produce nearly one-fifth of the food served daily on American tables. The valley's river system, the Sacramento in the north and the San Joaquin in the south, produces roughly

[1] Professor of Civil Engineering, University of California, Davis, and Principal, Resource Management Associates, Lafayette, California

40 percent of the annual natural runoff of the state, an estimated 33,000,000 acre feet.

The state's enormous requirements for water, largely agricultural, have been met over the years by construction of dams, reservoirs, pumping facilities and conveyance systems to redistribute this resource from areas of surplus, usually in the northern portion of the state to areas of demand, predominantly in the south. The Bureau of Reclamation's Central Valley Project which went into

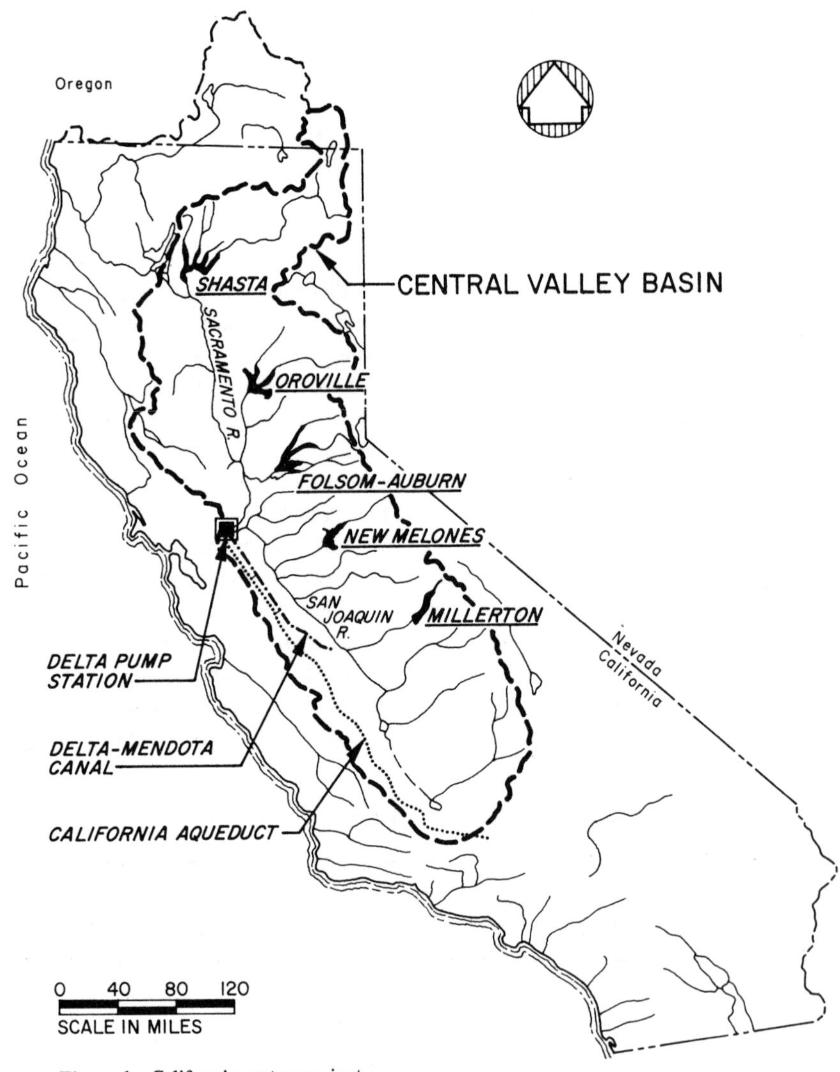

Figure 1. California water projects.

operation in the mid-40's and the ambitious California Water Project, implemented in the last decade, are outstanding examples of engineering works designed to manage the state's water resources. Between these two enormous systems there is an active storage capacity of some 16,000,000 acre feet and conveyance facilities capable of transporting 11,000,000-12,000,000 acre feet annually. The principal features of these systems are shown in Fig. 1.

At the heart of California's water resources is the Sacramento-San Joaquin Delta, the Central Valley's unique estuary, shown in Fig. 2. Comprised of some 700 miles of meandering channels, the Delta receives the residual runoff of undeveloped surface waters and the return flows of man's activities upstream. These flows, modified in quantity and quality, are passed through the Delta, thence through Carquinez Straits and San Francisco Bay, finally passing through the Golden Gate into the Pacific Ocean. The spatial and temporal distributions of salinity in this complex system are governed primarily by the transient hydrologic flux, the quality of flows delivered at the landward boundary of the estuarial system, and the actions of tides. Activities of man, however, play an important role in determining salinity balance in this system because of the relatively large redistributions of water accomplished by the Central Valley and State Water Projects and the attendant changes in water quality attributable to operation of these systems.

This paper seeks to illustrate some of the influences on estuarial salinity balance of major alterations in the hydrologic regimen of the Sacramento-San Joaquin Delta. In particular, two dramatic examples of the impact of flow regulation are described and used to point up some consequences, both positive and negative, of upstream flow regulations and diversion. The first of these episodes was Andrus-Brannan Island levee break, which occurred in June 1972, and occassioned an abrupt change in the operating policies of both the state and federal projects to avoid a disastrous impact on Delta water quality. The second was an abnormal release in September 1972 of water from New Exchequer Dam in the San Joaquin drainage which produced a dramatic response in the salinity balance in the extreme Southern Delta.

GENERAL BACKGROUND–HYDROLOGY AND WATER BALANCE

Both events occurred during the irrigation season. The first occurred in the early part of the season when the Delta was receiving water of generally high quality throughout. The second event occurred in the latter part of the season when quality deterioration was more notable, due both to the prior event and the generally lower quality of inflowing water, especially in the southern Delta. The water balance picture for this entire period depicts the significant flow changes that occurred as each episode was imposed on the otherwise rather steady hydrologic regimen.

Fig. 3 is a schematic diagram of the water balance of the Delta identifying each significant component used in determining the so-called Delta Outflow

Index, a measure of the net balance of water of the system. Of primary interest in the following discussion are the two major inflows from the Sacramento and San Joaquin rivers, the total export, and the Delta Outflow Index itself. Each of

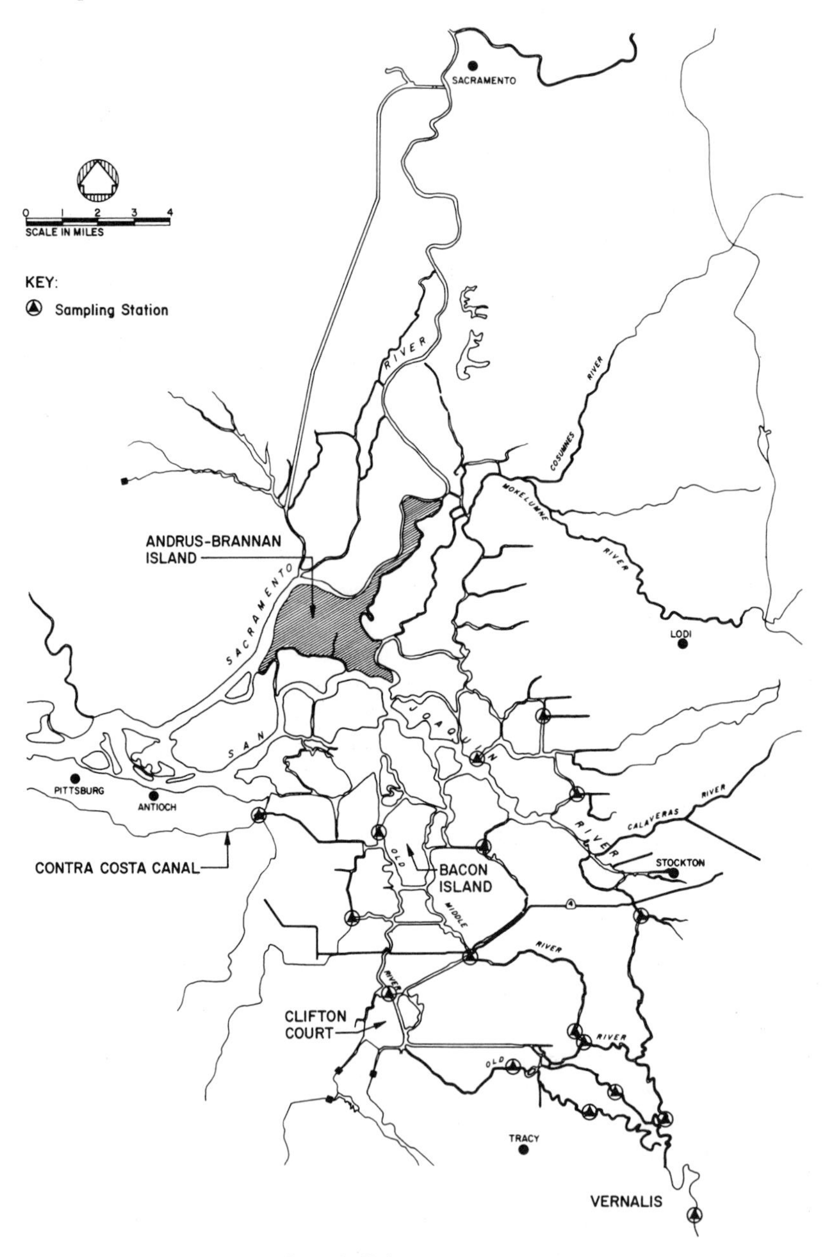

Figure 2. Sacramento-San Joaquin Delta.

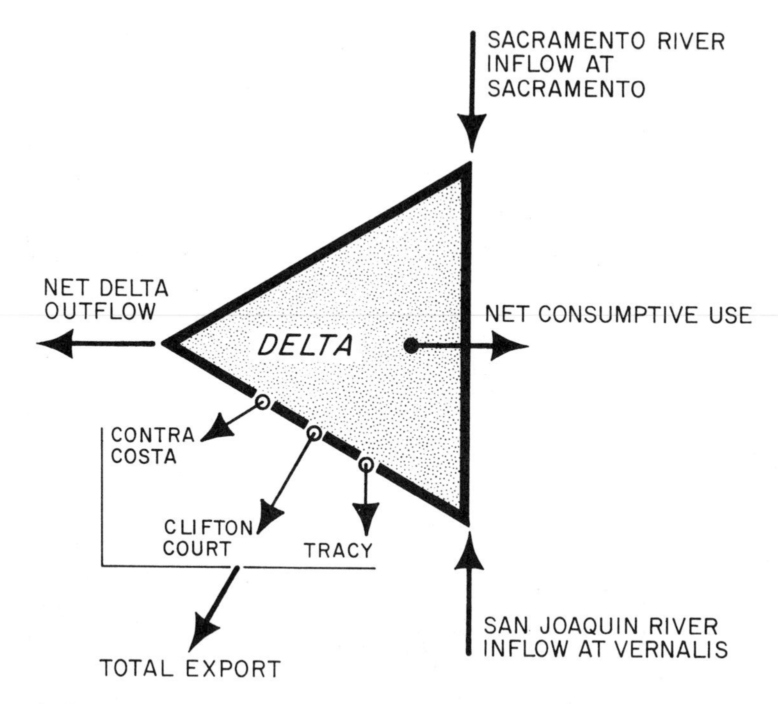

Figure 3. Schematic diagram of Delta water balance.

these varied markedly during the summer of 1972. Net consumptive use, corresponding primarily to the water requirement of Delta agriculture, followed a normal seasonal pattern.

Fig. 4 presents a history of the principal water balance components over the period May through October 1972. The singular features of this history are the abrupt changes in Delta export and outflow that occurred immediately following the Andrus-Brannan Island break on June 21 and the sharp rise in San Joaquin River inflow that succeeded the release of water from New Exchequer Reservoir on September 5. Each of these events gave rise to dramatic changes in salinity within the Delta; these are discussed below.

ANDRUS-BRANNAN ISLAND BREAK

In the early morning hours of June 21, 1972, 500 feet of levee on lower Andrus Island in the northern Delta failed (see Fig. 2 for location) and allowed about 150,000 acre feet of water to flood 11,000 acres of prime agricultural land on Andrus and Brannan islands.

Prior to this catastrophic event the net Delta outflow had been maintained at roughly 4300 cfs, sufficient to control salinity intrusion into the western Delta. Chloride levels at Antioch, for example, were steady at about 1680 mg/l while in

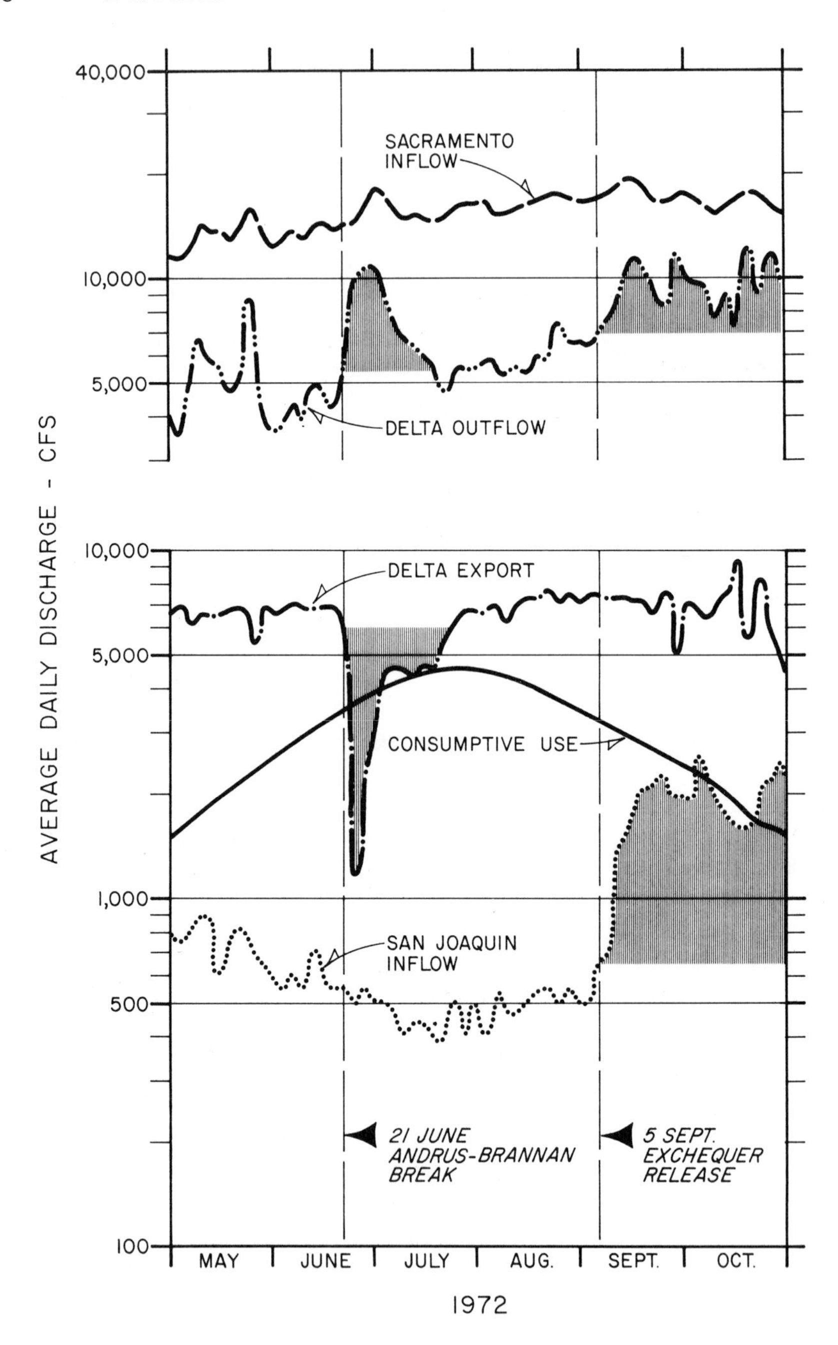

Figure 4. Delta water balance, May-Oct. 1972.

the interior of the Delta concentrations ranged from 70 to 150 mg/l (Clifton Court and Bacon Island). At the point of diversion for the Contra Costa Canal in the western Delta the chloride concentration was at 150 mg/l, corresponding roughly to the upper limit for industrial water users served by the canal. In the southwestern Delta near Tracy, where diversions totaling about 6500 cfs were

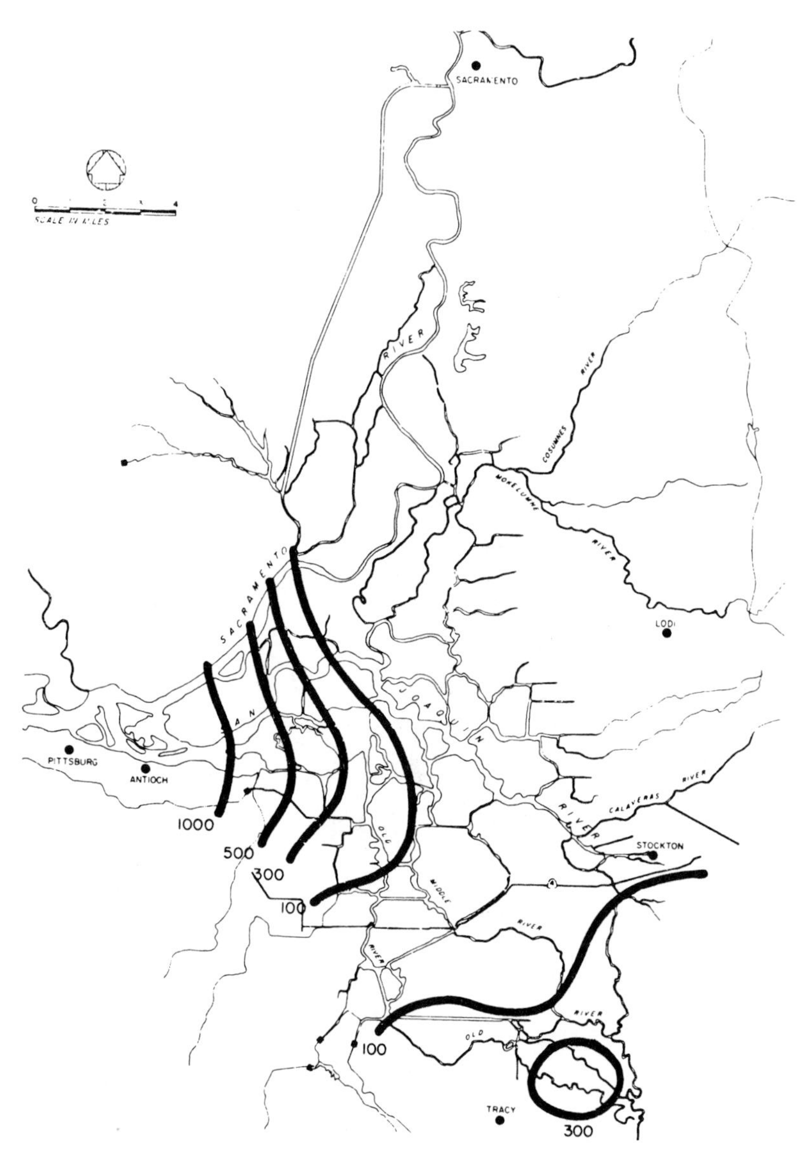

Figure 5. Pre-break conditions mean daily chlorides June 21, 1972.

being made to the Delta Mendota Canal and the California Aqueduct, the chloride concentration corresponded closely to that of the Sacramento River entering the Delta from the north.

Immediately following the break the flood of water moving into the island reversed the prevailing direction of outflow causing saline waters lower in the

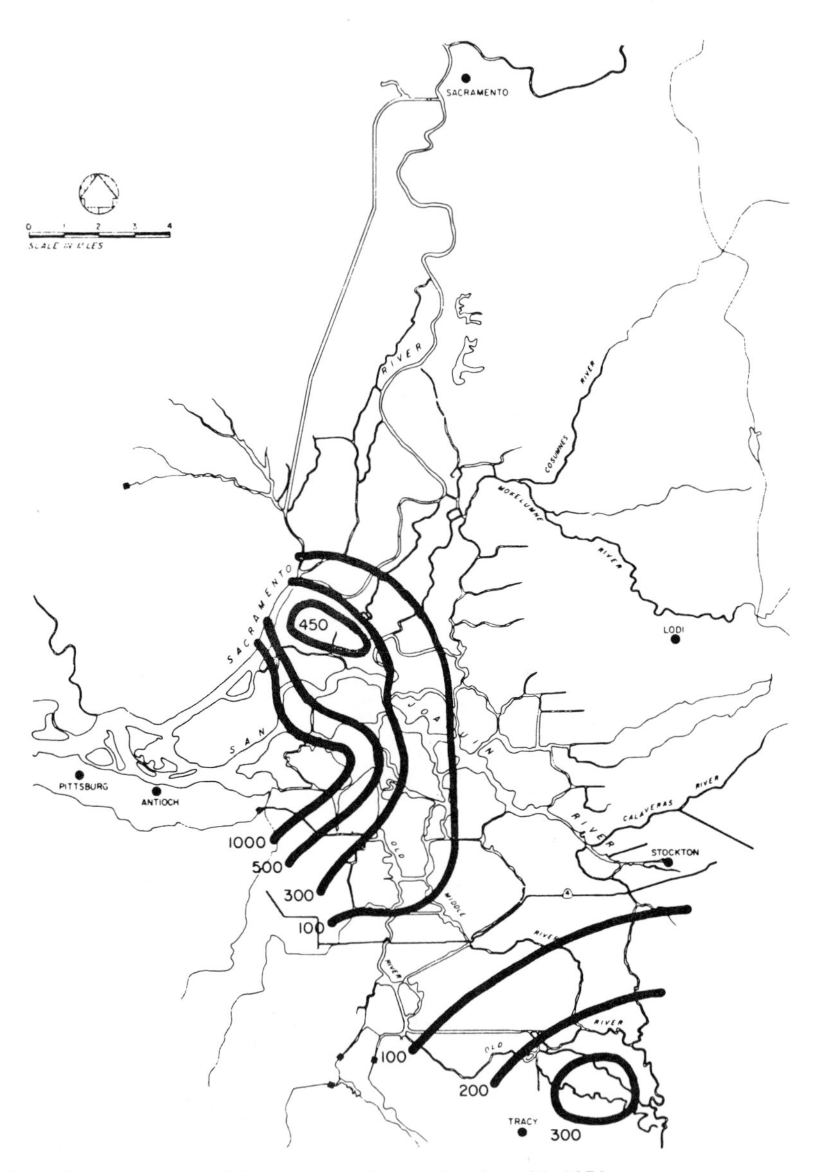

Figure 6. Post-break conditions mean daily chlorides June 26, 1972.

estuary to intrude into the western Delta. The hydraulic barrier, normally maintained by state and federal projects upstream, was temporarily disrupted. To offset this effect and to restore a favorable water balance in the Delta, two actions were taken by the project operators: releases from upstream projects were increased by about 4000 cfs and export pumping was dropped in stages by

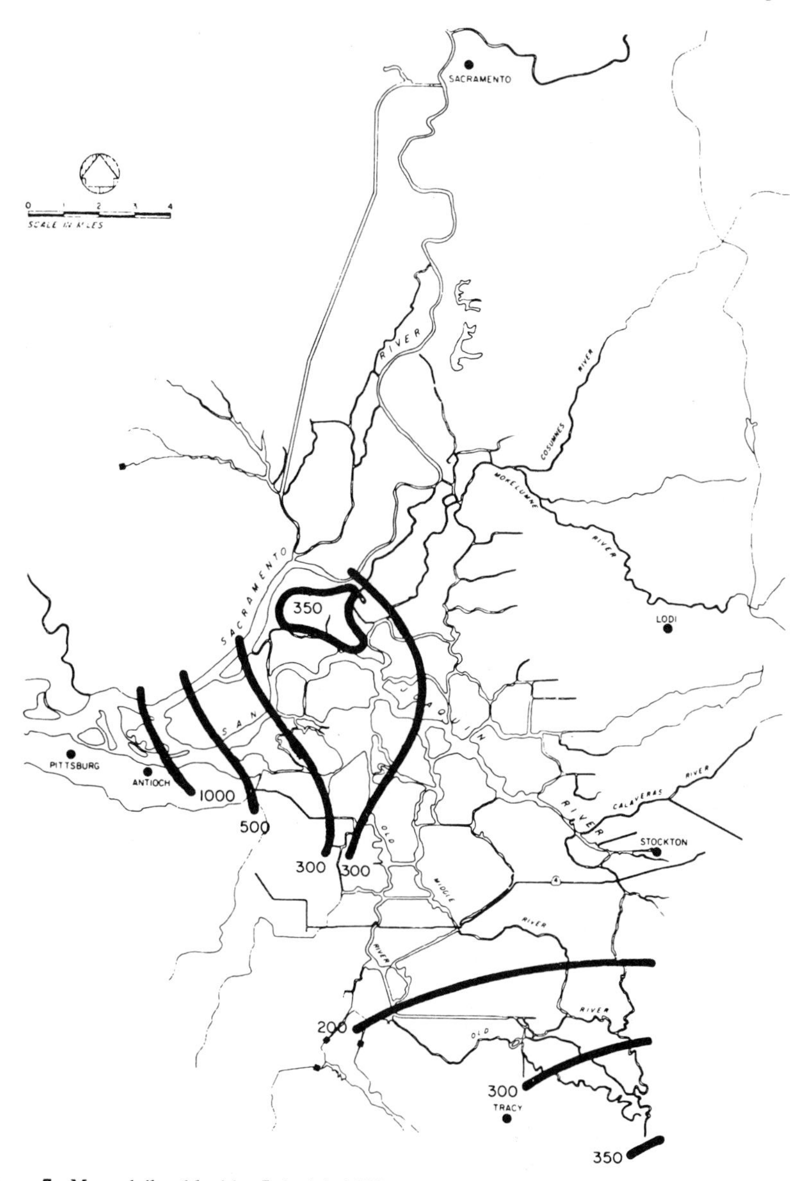

Figure 7. Mean daily chlorides July 16, 1972.

Table 1. Mean daily chlorides (PPM) estimated by DWR during levee break period June 21-July 16, 1972.[1]

	Antioch	Blind Point	Jersey Point	Emmaton	False R. at Webb Pump	Intake C.C. Canal	Old R. at Holland Tract	Old R. at Bacon Is.	Clifton Ct. Ferry at Old R.	Frank's Tract
June 21	1,680	870	590	--	190	150	330	150	70	--
22	3,020	1,970	1,370	--	440	150	380	160	70	--
23	3,380	2,500	1,410	--	650	170	390	160	70	--
24	3,080	2,030	1,310	--	640	140	410	180	70	--
25	2,930	1,940	1,250	--	620	160	440	200	70	810
26	2,810	1,960	1,140	--	600	160	430	230	70	870
27	2,640	1,730	1,010	--	540	190	400	250	70	840
28	2,370	1,430	810	--	530	240	400	280	70	770
29	2,070	1,270	750	--	500	270	400	380	70	730
30	1,880	1,090	620	--	460	310	--	400	90	650
July 1	1,860	1,050	620	570	420	340	--	420	110	--
2	1,790	960	590	570	360	380	--	430	163	--
3	1,500	750	510	370	410	420	--	400	162	620
4	1,320	640	470	310	390	440	--	400	215	--
10	1,120	460	370	360	270	370	310	310	270	--
16	1,310	450	380	--	210	250	220	240	200	--

[1] Within the flooded areas of Andrus-Brannan Islands an average of 1500 EC was recorded on June 26 based on measurements at seven sites. On July 7 the average EC for seven sites was 1200 μmhos/cm. This corresponds roughly to 450 ppm Cl^- on June 26 and 360 ppm Cl^- on July 7.

about 6000 cfs. As may be seen in Fig. 4, these actions resulted in an increase of about 7000 cfs in net Delta outflow to repulse salinity while at the same time slowing the movement of water across the Delta toward the pumps. The effect of shutting down the export pumps was felt within a few hours while the impact of upstream releases was not seen until about four days after the break.

The history of this accident and the consequent control actions is summarized in Table 1, a compilation of mean daily chloride concentrations observed at key stations in the Delta over the period from June 21 through July 16. Figs. 5, 6, and 7 illustrate the impact of the massive intrusion of saline water into the western and central Delta and its gradual dissipation as a result of remedial measures and the resumption of normal project operation about four weeks later.

Fig. 5 shows the pre-break pattern of chlorides, more or less typical of summertime operation during the height of the agricultural season. The cross-Delta movement of water to the pumps near Tracy is graphically illustrated by the "corridor" of high quality water enveloped between the 100 mg/l Cl^- contours.

In Fig. 6, corresponding to conditions about 5 days after the break, salinity is seen to have intruded well into the western Delta, surrounding Andrus-Brannan Island. A vestigal remnant of the highly saline water that intruded through the break is seen as the closed contour of 450 mg/l lying wholly within the island. The corridor of high quality water is seen as being pinched somewhat due to encroachment of saline water from the northwest.

In Fig. 7, conditions in the western Delta have returned to "normal," that is, salinities are at or below pre-break conditions. However, in the southwestern Delta, near the point of export salinities are substantially higher, by roughly a factor of 3, than prior to the break. The trapped saline water is gradually being exported from the system by the pumps at Tracy and by local diversions throughout the interior Delta.

NEW EXCHEQUER RELEASES

A second event that illustrates the response of the Delta to upstream flow regulation to control salinity was an unscheduled release of about 57,000 acre feet of stored water into the southern Delta.

In order to complete repairs on hydraulic structures at New Exchequer Dam on the Merced River, the Merced Irrigation District found it necessary to draw down the reservoir at an average release flow of 1360 cfs, beginning on September 5, 1972. Releases continued for a period of about three weeks, through September 25. At the commencement of releases the mean daily flow in the San Joaquin River near Vernalis was about 500-600 cfs. During the following three weeks the flow gradually increased to about 2250 cfs. Thereafter the flow was sustained generally between about 1500 and 2500 cfs, due in part to a gradual decline in upstream consumptive use of irrigation diversions, i.e., an increase in the volume of return flows.

The consequences to water quality in the southern Delta to releases of high quality Merced River water are illustrated graphically in Figs. 8 and 9. Fig. 8 shows the pattern of total dissolved solids (TDS) in the Delta as it was just prior

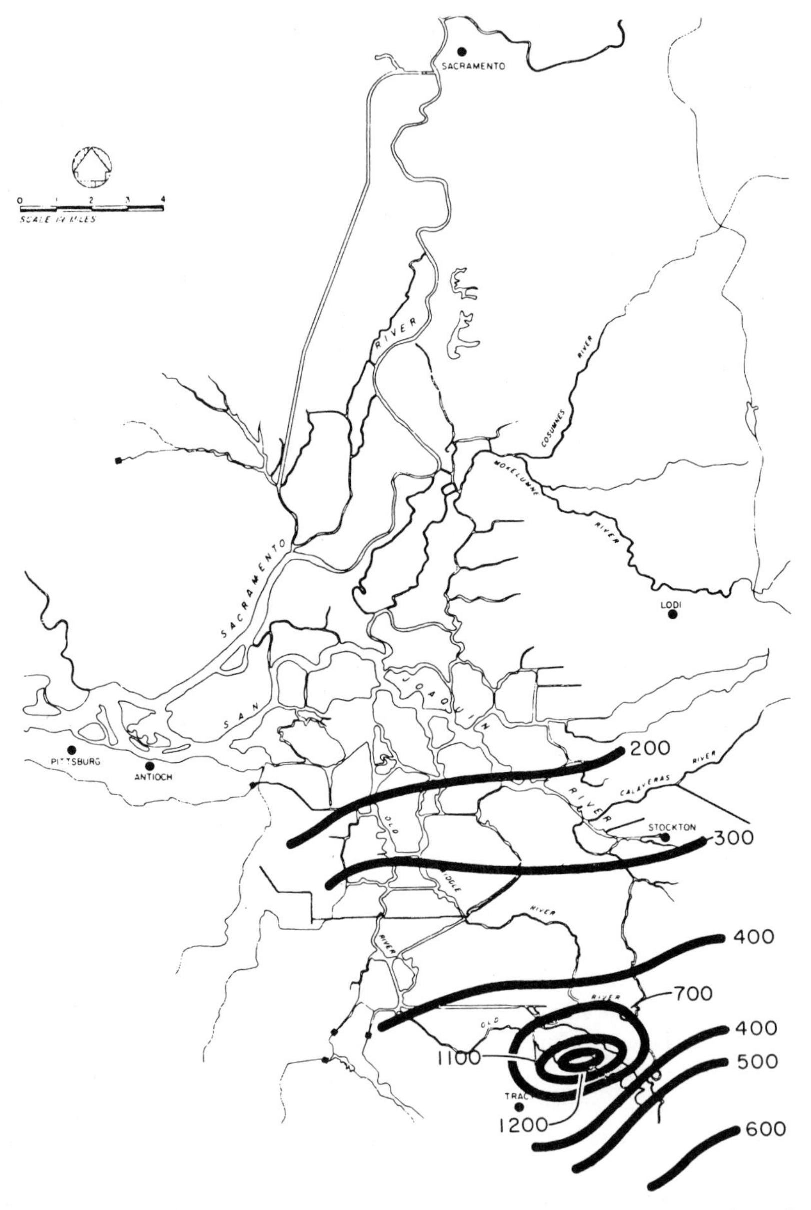

Figure 8. TDS conditions on September 7 before arrival of New Exchequer release (base flow in San Joaquin River = 612 cfs).

to arrival of the released water on September 7, 1972. Especially notable is the TDS level of 600 mg/l in the extreme southern Delta and the "pocket" of high TDS water in the interior channels (closed contours).

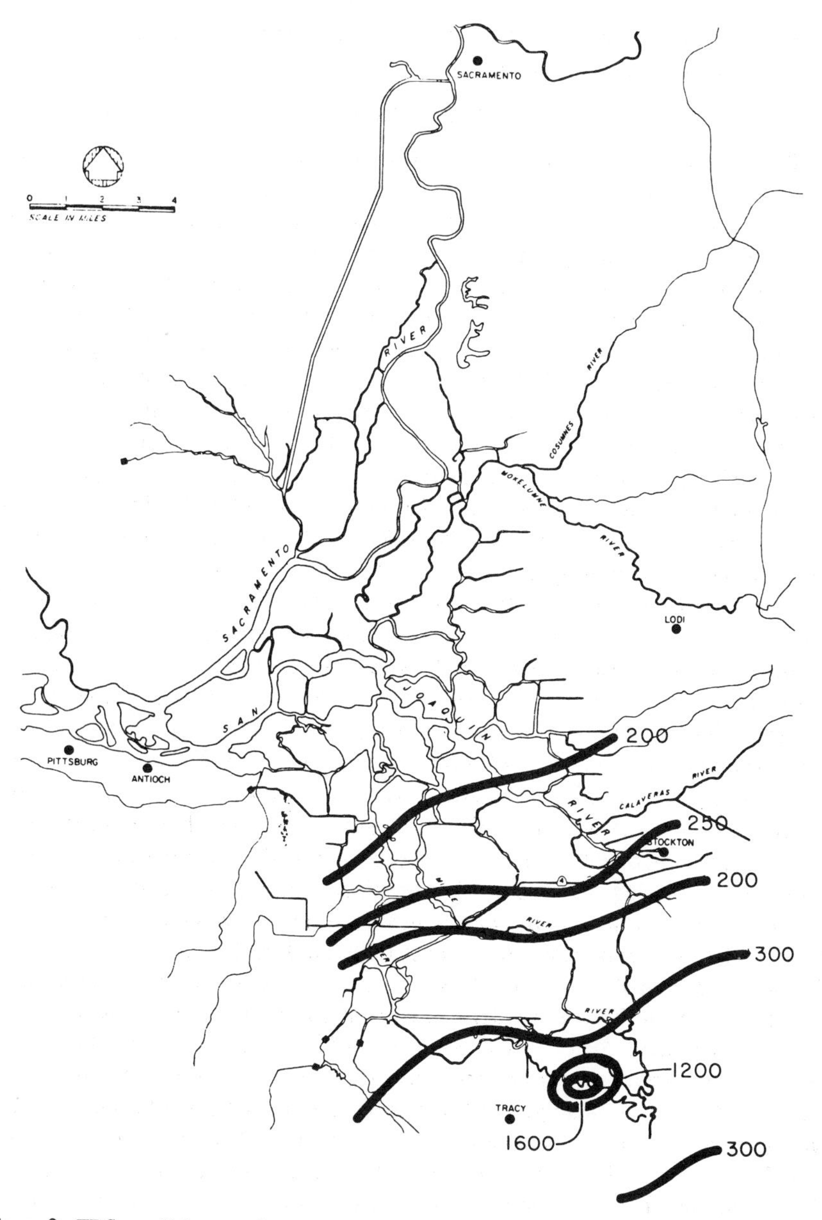

Figure 9. TDS conditions on September 25 after arrival of 57,000 acre-feet (1360 cfs daily) from New Exchequer.

Fig. 8 corresponds to TDS conditions near the conclusion of the release period on September 25. With the exception of the pocket of high TDS water, which has been noticeably diminished in areal extent, conditions are substantially improved. TDS concentrations over most of the area are below 300 mg/l, a reflection of the dilution effects of the Merced water whose TDS concentration was of the order of 50 mg/l or less.

DISCUSSION

The Sacramento-San Joaquin Delta is an integral link in the California Water Resources System. It is not only an element in the conveyance facilities of the state and federal water projects, but serves as a source of water for some 550,000 acres of prime agricultural land. Maintenance of water quality, in particular control of salinity, is crucial to agriculture, to fisheries, and to the many other beneficial uses of water derived from or transported through the Delta.

With the aid of the state and federal projects, it is possible to regulate the flow of water through the Delta and thereby facilitate quality control. However, these same projects are designed to serve the water needs of lands outside the Delta, sometimes with priorities that may be detrimental to in-channel quality control. The degradation of quality in the southern Delta, for example, is primarily a consequence of depletion of natural outflow of the San Joaquin system to serve water demands in the Southern Central Valley coupled with the burden of highly saline return irrigation flows.

There are needs to regulate flows and quality, both in times of emergency and during the course of normal water use activities in the Delta. The advantages of having available the capacity to supplement flows is apparent in both the cases illustrated in this paper. In the case of the Andrus-Brannan Island levee break, however, the most immediate effect in repulsing the saline surge moving toward the break was the cessation of export pumping. The effect of releases from Shasta Reservoir higher in the system were not felt for 5 to 7 days after the break and when they arrived at the Delta they served primarily to replace waters that had to be pumped for export. It would appear that such releases are not so valuable in repulsing salinity intrusion resulting from such accidents as they may be in merely maintaining an equilibrium condition of salinity in the estuary. Incidentally, the experience of Andrus-Brannon Island indicates that a Delta Outflow Index in the order of 4000 cfs is necessary to maintain a reasonable salinity balance in the western Delta, say about 1500 mg/l Cl^- at Antioch.

In the southern Delta, where salinity problems become acute during the height of the irrigation season, the only solution currently available is supplemental water of high quality. The benefits of upstream releases of such water are clearly evident in the New Exchequer episode in which flows of 1500 to 200 cfs, including about 1300 cfs from the Merced River, restored a salinity balance of about 300 mg/l TDS throughout most of the southern Delta. Currently a target for maintaining quality for agricultural use in the area is about 400 mg/l. A steady flow of 1200 to 1500 cfs of the average quality delivered at Vernalis during September 1972 would assure achievement of this objective.

REFERENCE

Kibler, David F. "Summer Sampling Program, 1972," report to Delta Water Users Association by WRE, Inc. Jan. 1973. 53 p.

EFFECTS OF ENLARGEMENT OF THE CHESAPEAKE AND DELAWARE CANAL[1]

L. Eugene Cronin
Center for Environmental and Estuarine Studies
University of Maryland
Box 775, Cambridge, Md. 21613

Donald W. Pritchard
Chesapeake Bay Institute
The Johns Hopkins University
Baltimore, Maryland 21218

Ted S. Y. Koo
Chesapeake Biological Laboratory
Center for Environmental and Estuarine Studies
University of Maryland
Solomons, Md. 20688

Victor Lotrich
College of Marine Studies
University of Delaware
Newark, Delaware 19958

ABSTRACT: Recent enlargement of this artificial waterway has resulted in substantial alterations of the physical hydrography, chemical environment and biotic populations of the canal and its approaches. Net transport from the Chesapeake to the Delaware has been increased with enlarged tidal velocities and excursion, and the hydrographic pattern will remain complex. Salinity distribu-

[1] Contribution No. 656, Center for Environmental and Estuarine Studies, University of Maryland; No. 224, Chesapeake Bay Institute, The Johns Hopkins University; and No. 106, College of Marine Studies, University of Delaware. The work was performed under Contract No. DACW 61-71-C-0062 with the Philadelphia District, U.S. Army Corps of Engineers.

tion has been slightly altered. The probability of transport of fish eggs and larvae into the Delaware is increased. The canal contains an abundant and diverse population of fish from freshwater, estuarine and marine sources and is used as a migration pathway by many species. Exceptionally dense concentrations of eggs and larvae of striped bass occur along with these stages of other species. Experimental observations on the effects of alterations in salinity and suspended sediment similar to those observed did not indicate detriment to eggs and larvae. Excessive transport to unfavorable water and minor damage from shear forces may be detrimental. Benthic populations are moderately abundant and no substantial damage to these has been observed. Because the results of environmental alterations are often subtle, and because uses of the canal are intensifying, continuing monitoring is strongly recommended.

INTRODUCTION

The Chesapeake and Delaware Canal is an artificial 22.5 km (14 miles) waterway, between the waters of a tributary of the head of the Chesapeake Bay, with fresh water in spring and low salinities at all seasons, and the Delaware River at a site of somewhat higher average salinity (Fig. 1). In calendar year 1974, nearly 11,000 vessels transited the canal, transporting about 12,400,000 tons of commerce. The canal was suggested as early as 1661, completed as a four-lock canal in 1829, changed to sea level without locks by 1927, and enlarged to 8.2 m (27 ft) deep and 76.2 m (250 ft) wide by 1938 (3, 9). Studies were not made of the environmental effects of these substantial alterations except for technical analyses of tides and currents (31). Further enlargement of the canal and its approaches to 10.7 m (35 ft) deep and 147.1 m (450 ft) wide was authorized in 1954 and initiated in 1958 (6). In 1970, when enlargement was about 80% completed, substantial concerns were expressed before the Congress about the effects of enlargement on the net transport of water from the Chesapeake to the Delaware, on velocities and turbulence, on the salinity regime, on the biological population of the canal regions and on the environment of the upper Chesapeake Bay (6). As a result, the Corps of Engineers proposed and implemented a program of field measurements, hydraulic model testing, mathematical model analyses, preliminary design of flow control structures, and ecological studies. The principal hydrographic investigations and all of the biological studies were arranged by contract through the present authors and are reported here, based on reports to the sponsors (2, 3), 16 Appendices to those reports, and publications on some of the specific studies (8, 11, 12, 13, 14, 15, 16, 17, 18, 19, 20, 21, 24, 25, 26, 27, 29, 30).

The event in the history of the canal which had the most significant environmental impact was almost certainly conversion to a sea-level, unobstructed waterway linking two somewhat different estuarine areas of low salinity. This permitted both pelagic and planktonic organisms to move from one system to the other, substantially altered salinity distribution and the net transport of

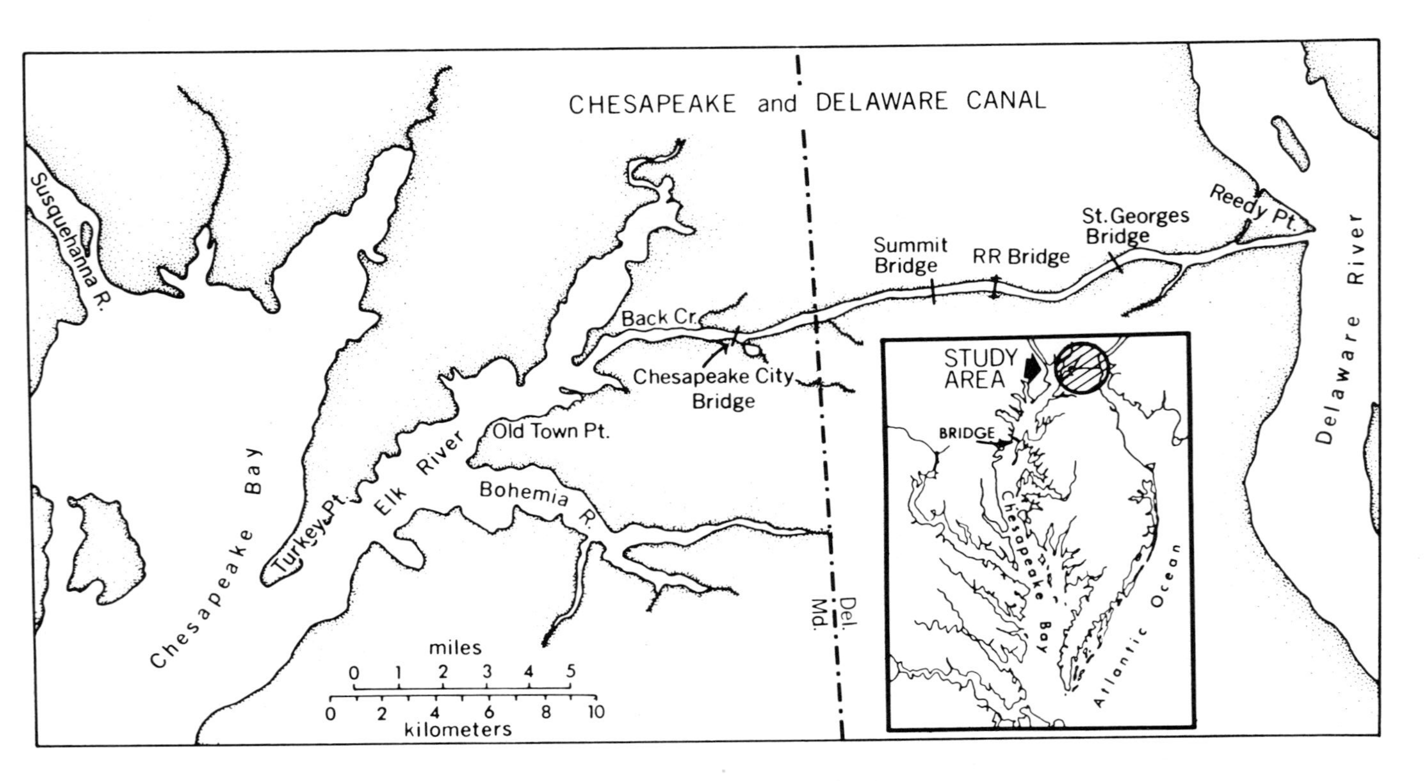

Figure 1. Chesapeake and Delaware Canal area.

water, and, because tidal amplitude and timing are different at the ends, established a turbulent new biological environment. The changes were not closely observed, but the inadvertent creation of a valuable spawning and nursery ground for the striped bass, *Morone saxatilis*, one of the most important recreational and commercial species of the Atlantic Coast, was discovered by Dovel (4, 7). He suggested that this waterway may have become the most important nursery ground for the species. Pritchard and Cronin (23) estimated that enlargement of the canal would increase net Chesapeake-to-Delaware transport from about 1000 cfs to 2700 cfs, increase velocities, lengthen tidal and non-tidal excursions, modify salinity in the canal and slightly alter salinities in the Chesapeake above the Chesapeake Bay Bridge near Annapolis. Biologically, they predicted increased transport of organisms from the Chesapeake to the Delaware, change in the biological success of striped bass in the canal, small shifts in the distribution of some Chesapeake species and additional biological responses which could not be specified from the limited data available.

The biological consequences of this engineering activity can be economically and ecologically important in the Maryland and Delaware regions and the far larger region which utilizes fish nurtured in their valuable low-salinity systems. Such consequences are not, however, likely to have as much significance as connecting two water masses which are inherently different and have been separated for millenia, as in the creation of the Suez, Panama, and Welland canals. These have been shown to have effects ranging from economic disruption to biological alteration, and from insignificant to profound (1, 22, 28).

The present studies were designed to obtain the best possible estimates of the ecological effects of enlargement of the canal. Serious constraints exist in that observations were initiated late in the process of enlargement, and the period of observation has been too short for detection and measurement of all of the effects. Specific goals were to learn:

(1) the patterns of salinity and flow in the canal;
(2) the value of the canal as nursery area for fish;
(3) the effects of observed changes on eggs and larvae of fish;
(4) the movements of fish through the canal, and
(5) the populations of other organisms in the canal and its approaches.

In each case, the effects of enlargement have been sought and the most advantageous operation of the canal, from various ecological points of view, has been considered.

It is appropriate to note that only brief description is provided here of methods and of results. Original references should be consulted for these. In addition to the material summarized below, they contain much valuable physical, chemical and biological information on the region studied and the concepts considered.

HYDROGRAPHIC EFFECTS

Four geographic features of the canal strongly influence its hydrographic characteristics and processes (Fig. 1). First, the Chesapeake end is much farther from the ocean than the Delaware end, so that any one tidal phase reaches the western end about 11 hours later than the eastern end, but appears to lead by about 1.4 hours. Second, the Chesapeake is large and irregular in shape, whereas the Delaware is funnel-shaped so that the tidal amplitude is about 5 feet at the Delaware end and 2 feet at the Chesapeake end. Third, the western end of the canal system is near the mouth of the Susquehanna River, the largest source of fresh water on the Atlantic U. S. Coast, so that salinity is about 2 °/oo lower at that end of the canal. Fourth, the western end of the canal has a higher mean elevation, by an estimated 2.5 cm, than the Delaware end. These features combine to create highly variable patterns of flow dominated by the relative elevation of water at the two ends. Meteorological conditions can cause very large temporary deviations and variations. Net transport over any extended period is eastward because of the difference in the mean water level at the ends.

Estimates of pre-enlargement transport had been made by Wicker (31) and by Pritchard and Cronin (23). These indicated that net eastward transport was 950-1000 cfs. The present hydrographic studies were made from May 1969 to January 1974, and are reported with detailed analysis by Pritchard and Gardner (3, 8, 24). They involved four periods of intensive observation with deployed current meters, 57 cruises through the canal at approximately slack water to sample salinity and temperature, and the development and verification of a numerical model of the flow in the canal. This portion of the studies was conducted by the Chesapeake Bay Institute.

Results

- Flow in the canal involves rapid and substantial changes in direction and in the volume of flow. Net transport is relatively small in comparison with the large short-term movements of water.
- The long-term average net non-tidal flow will increase from about 900 cfs eastward for the pre-enlargement canal to about 2450 cfs eastward for the post-enlargement canal. Average flows may be somewhat larger in spring and somewhat smaller in late summer and fall.
- Increased eastward flow will tend to lower salinity in the Delaware and to increase salinity in the upper Chesapeake Bay. These changes will be small compared to natural variations.
- Salinity in the canal will be little changed, except that short-term fluctuations will be increased and the region of fresh water will tend to extend farther to the east.
- Tidal velocities and tidal excursion in the post-enlargement canal will be about 15% higher than in the pre-enlargement canal.

- Any one water mass, at Chesapeake City, for example, will move out of the canal in about 15% less time than prior to enlargement. The probability that fish eggs and larvae, or other planktonic material, will be carried into the Delaware is increased.
- The standard deviation of the long-term average, including about two-thirds of the cases, for individual tidal cycle values of the net non-tidal flow will increase from about ± 5980 cfs for the pre-enlargement canal to about ± 14,830 cfs after enlargement.
- About 60% of the tidal cycles will have net eastward flow in the post-enlargement canal and 40% will have net westward flow. This is about the same partition as in the pre-enlargement flow.
- The average net non-tidal flow in the eastward cycles will increase from about 4425 cfs for the pre-enlargement canal with a standard deviation of about ± 4315 cfs to about 10,965 with a standard deviation of about ± 10,695 cfs after enlargement.
- The average net non-tidal flow in the westward cycles will increase from about 4130 cfs (± 4095 standard deviation) to about 10,240 cfs (± 10,155 standard deviation).
- The expected maximum net non-tidal flows for one tidal cycle will increase in both directions as the result of enlargement. Eastward maximum flow will rise from about 20,100 cfs to about 48,800. Westward maximum flow will increase from about 15,300 cfs to about 37,900 cfs.
- Model analysis indicates that the maximum salinity increase in the upper Chesapeake Bay will be about 3 ‰, occurring during the fall. During the rest of the year the change is less, nearly zero in spring. The effect decreases with distance south from a point 30 km below the mouth of the Susquehanna and is expected to be equivalent to a longitudinal shift of about 4 km during the fall.

BIOLOGICAL EFFECTS

Previous limited observations of the biota of the canal and approaches (4, 6) had shown large populations of fish eggs and larvae, especially striped bass, but provided little data on any other biota. In the present investigation, five parallel studies were conducted.

Benthos

Survey and analysis were made of the benthic community for species composition and biomass, both spatially and temporally. Replicate samples were obtained quarterly by 0.1 m Van Veen grab for sediment analysis and quantitative biological analysis by aliquot of material passing a 0.7 mm mesh screen. Supplementary samples from a Menzies trawl, Peterson grab, biological dredge, and modified Ockelman detritus sledge were also examined. Salinity, temperatures, dissolved oxygen and turbidity were observed at each station. These studies were conducted and reported by Pfitzenmeyer, and Miller et al. for the Chesapeake Biological Laboratory at the Maryland end (21, 14), and Taylor et al. for the College of Marine Studies in Delaware waters (29).

Results

- Macroscopic benthic animals are varied and moderately abundant in the canal and its approaches. In the Chesapeake approach, 25 species were noted, with a density of at least 865 individuals per square meter. Biomass decreases into the canal. In Delaware, 22 species occur, with lower density of individuals and of biomass per square meter than in the western canal or the Chesapeake approach.
- The benthic animals are of considerable importance as food for resident and migratory fish. Usually, the most abundant benthic species were the most abundant food items in fish stomachs.
- Blue crabs occur in the canal and its approaches during the warm months. Most of them are small or medium-size males, and some of them move through the canal.

Fish

Field collections and analyses were made to learn the distribution and abundance of fish in the canal and approaches. Twenty-five to thirty foot semi-balloon otter trawls were operated monthly in paired sampling at 11 stations in Md. and 8 in Delaware, with the total sampling area extending from Chesapeake Bay into the Delaware River. Beach seines were also employed during the summer and data from other sources were incorporated into the analyses. Salinity, temperatures and other environmental conditions were noted. Complementary phases were conducted and reported by Taylor et al. (30) for the College of Marine Studies and Ritchie and Koo (25) of the Chesapeake Biological Laboratory.

Results

- The canal contains an abundant and diverse population of fishes. At least 62 species occur in the canal and its approaches. They include freshwater fish, estuarine species and marine fish, and both catadromous and anadromous migrants.
- The canal is used by various species as a permanent residence, as a nursery area for juveniles, as a feeding ground, or as a path for migrations or other large-scale movements. Seasonal variation in the catch is large, with the smallest number of species taken in winter and the largest in summer and fall.
- The Chesapeake approaches and the western portion of the canal had a more

diverse fish population (45 species) than the Delaware approach and eastern portion (33 species).

- White perch, weakfish, bay anchovy, catfishes, spot and clupeids were the most abundant fish in the catches.

Fish Movements

Available species, especially the white perch, *Morone americanus*, and the striped bass, *Morone saxatilis*, were tagged with Carlin tags, the CBL modification of the Carlin, or Peterson disc tags; 12,533 tagged fish of 22 species yielded 1035 returns. The results were summarized by Ritchie and Koo (26) for the Chesapeake Biological Laboratory and Smith et al. for the College of Marine Studies (27). In addition, sonic tags were fabricated and placed by Koo and Wilson in five adult striped bass to track local movements for periods of 4-90 hours (13).

Results

- White perch and striped bass move extensively within the canal and in both directions through the system.
- The canal appears to now be the principal route for departure of striped bass from the Chesapeake Bay system.
- The upper Chesapeake Bay and the western portion of the Chesapeake and Delaware Canal is the major spawning area for striped bass found in Delaware Bay.
- The canal is an important avenue in the migratory movements of American shad, *Alosa sapidissima*, hickory shad, *Alosa mediocris*, and other anadromous species.
- Fish which use the canal region enter the commercial and recreational fisheries from Virginia to Maine.

Fish Eggs and Larvae

Survey and analysis were completed for the distribution and abundance of fish eggs and larvae in the canal and approaches. In 1971, 28 stations were sampled from near the mouth of the Susquehanna through the canal to the Delaware. Two 61 cm conical nets equipped with flow meters were towed for 5 minutes at each station, one near the surface and one near mid-depth. Sampling was usually repeated every other day during April and May and less frequently through December. In 1972, paired nets were employed, fewer stations were sampled, and intensive 30-hour sampling was performed each month from March through June. This project was conducted by the Chesapeake Biological Laboratory and details of sampling and analyses have been presented by Johnson (11) and Johnson and Koo (12).

Results

- The canal is exceptionally important as a hatchery and nursery for striped bass, with most eggs and larvae in the western portion.
- Larval and juvenile stages of more than 20 species of fishes were caught in the canal.
- Freshwater spawning species included striped bass, white perch, alewife and blueback herring. Estuarine spawners were represented by bay anchovy, naked goby and silversides. Marine spawners contributing larvae and juveniles were the American eel, Atlantic menhaden, spot, weakfish and Atlantic croaker.
- April, May and June are the months of abundance for eggs and larvae, but young fishes of various species are present in the canal area throughout the year.

Effects of Specific Environmental Changes on Eggs and Larvae.

Experimental studies were undertaken to determine the effects of enlargement-caused changes in salinity, water movement and suspended sediment on the eggs and larvae of striped bass and white perch. Additional observations were made on the biological adequacy of water from several relevant locations. These studies were made and reported by Morgan and his associates (15, 16, 17, 18, 19, 20). Eggs and larvae were experimentally exposed to a range of each environmental variable exceeding that which was known to occur in the canal.

Results

- Salinity between 0 and 10 °/oo did not affect the development rate of white perch, the percentage of hatch and survival of striped bass, or the egg size and larval length of striped bass. It had some effect on the egg size of white perch, and the rate of development for striped bass was faster at 0.5 °/oo than at 2.5 °/oo or higher.
- Striped bass eggs hatched best between 19 and 22 C, and survival was highest between 16 and 23 C.
- The optimum temperature for white perch egg development was 11 to 16 C, for striped bass eggs, 16 to 22 C.
- Deformed striped bass (pugheaded) were observed at temperatures above 24.5 C.
- The percentage of eggs hatching was not affected by sediment levels from 50 to 5250 ppm of suspended sediment for white perch or 20 to 2300 ppm for striped bass.
- The rate of development of eggs of white perch and striped bass was retarded in concentrations above 1500 ppm of suspended sediments.

- White perch eggs were resistant to sediment blanketing of 0.45 mm or less, but development rates were lowered with sediment thickness above 0.8 mm.
- Larval mortality LD_{50} for white perch was 11,600 ppm suspended sediment for one-day exposure and 2700 ppm for two days of exposure. For striped bass, LD_{50} for larval mortality was 7800 ppm for one day, or 3400 ppm for two days. Both species suffered some mortalities at one-day exposure to 1600 ppm.
- Striped bass eggs were resistant to low-intensity shear forces below 0.88 dynes/cm^2, but killed at that level. White perch were affected by 0.2, 0.5 and 0.88 dynes/cm^2, but with less than total mortality.
- For high-intensity, low-duration shear force exposure, white perch showed LD_{50}'s for eggs from 1.7 dynes/cm^2 for 1 minute to 0.88 dynes/cm^2 for 5 minutes, and for larvae from 1.63 dynes/cm^2 for 1 minute to 0.90 dynes/cm^2 for 4 minutes. Striped bass had LD_{50}'s for eggs from 2.1 dynes/cm^2 for 1 minute to 1.04 dynes/cm^2 for 4 minutes, and for larvae from 3.4 dynes/cm^2 for 1 minute to 1.25 dynes/cm^2 for 4 minutes.
- Acute static bioassays with eggs and larvae of white perch and striped bass showed no significant differences in mortality rates for two-day exposure in water from sites along the length of the canal. Water from one station, located in the Delaware River immediately upstream from the eastern end of the canal, consistently caused higher mortality rates than the sources in the canal.

DISCUSSION

Adequate assessment of the effects of environmental alterations is notoriously difficult, and the Chesapeake and Delaware Canal enlargement is no exception. The creation or modification of any waterway between established aquatic systems holds potentials for changing the environment for various biological resources, and the effects range from the immediate and obvious to the long-term and subtle (1, 5, 22, 28).

Ten environmental modifications are known to have occurred during enlargement of the canal, and each merits discussion and as much evaluation as present information will permit.

Dredging and Removal of Bottom Materials

No direct observations were made. Such actions obviously remove the substrate and organisms inescapably associated with it, and the dredging operation may produce sedimentation and other changes. The effects of such dredging were probably of relatively short duration as indicated by the substantial populations of benthic animals re-established by the time of this study and as observed in other studies (4, 5).

Increased Gross Transport of Water

The combined effect of 15% increase in tidal velocities and tidal excursion with enlarged section of the canal is a 2.5-fold increase in the quantity of water moved by tidal action. There is no present indication of biological significance in this rather complex movement of water back and forth in the system, except that it may increase the opportunity for planktonic and pelagic species to move into or out of the canal from the approach areas.

Increased Net Flow Eastward

Long-term average net non-tidal flow increase from about 900 cfs to about 2450 cfs may have adverse effects if it causes the transport of eggs and larvae to unfavorable locations, as observed at one Delaware River site. The probability appears to be low during average conditions of flow. Deleterious transport is most likely to occur when meteorological events cause an exceptional period of eastward flow, so that eggs and larvae are flushed into the Delaware system.

Salinity Alteration in the Canal

Available data are not adequate for measurement of the shifts in salinity resulting from enlargement, and inherent variation is normally great. It is probable, however, that short-term fluctuation will increase and that the region of fresh water will be extended farther to the east. This would expand the available area of the canal for freshwater spawners and shift eastward the volume of very low salinity for species like the striped bass which use it for spawning.

Salinity Change in the Chesapeake Bay System

Withdrawal of Susquehanna-source water from the head of the bay will superimpose a longitudinal up-bay shift of a few kilometers on the fluctuating salinity patterns of the Chesapeake above the Bay Bridge near Annapolis. The change will be trivial during high spring flows, but maximal following the low river input of fall. The species of animals and plants which are most strongly affected by spring conditions (upper limit of oysters and clams, of some of their predators and parasites, of sea nettle wintering stages, etc.) are likely to show no change. Species affected by fall conditions (some molluscan parasites, sea nettle medusae, intruding late summer and fall marine fish, the condition of upper bay oysters, etc.) will respond to increased salinities.

Fresh Water Release into the Delaware River

Delaware waters, which still carry a substantial load of pollutants from domestic and industrial sources, seem likely to be improved by dilution because of the additional 1500 cfs net inflow (ca 10% of the mean flow from the Delaware River). In addition, there are untested possibilities that the canal may attract anadromous migrating fish in the Delaware and direct them to the Chesapeake. This could affect spawning success and the future migratory inclinations of such species.

Increased Velocity

No biological effect is apparent.

Increased turbulence

Striped bass eggs are semi-bouyant and dependent upon turbulence to prevent unfavorable settlement onto bottom sediments. The inherently dynamic hydraulic system in the canal is very likely to be the reason for the success of this species. There is insufficient understanding of the requirements and tolerances of eggs and larvae of the many species presently available to permit identification of species for which the new environment is advantageous or disadvantageous.

Increased Shear Force

Experimental observation indicates that the shear forces injurious to the eggs and larvae of two principal species are limited in the canal to those in the immediate vicinity of ship propellers. Except for possible localized damage from larger or more rapidly rotating propellers, no environmental effect is known.

Increased Suspended Sediments

Sediment load is likely to be increased by greater hydraulic vigor. Laboratory studies and field observations indicate that the quantity of suspended materials is unlikely to exceed the tolerance of eggs and larvae of striped bass and white perch during normal conditions and that enlargement will not increase the probability of injury.

We conclude that enlargement has probably had little or no predictable detrimental effect on the resources of the region except that Delaware waters will be enhanced for biological use, the upper Chesapeake will be somewhat saltier in late summer, and benthic and pelagic population in the canal will shift somewhat to the east. There is, however, increased opportunity for unusual meteorological circumstances during the spawning period for striped bass to produce mass transport of eggs and larvae into the Delaware estuarine system where water quality is sometimes low.

We have recommended that structures capable of controlling flow in the canal at strategic times be designed, tested as models, and environmentally evaluated. We have further recommended that they not be constructed unless there is substantial evidence that they would be of value in protecting and enhancing the biota of the region.

The responses of an ecosystem to significant alterations involve both immediate and long-term adjustments. Adjustment of the physical hydrographic regime can be measured at an early date whereas biological adjustment involves all seasons and changes which are difficult to measure and to associate with a cause. The full effects of widening and deepening this artificial waterway are not yet known. We have therefore also urged that continuing monitoring of water movements and biotic responses be carefully designed and conducted for an extended period and that the accumulating understanding of this notable environmental alteration be continuously assessed. The need for such assessment grows annually. Consideration is already being given to siting of two nuclear electric generating plants along the canal and to the possible placement of wastes inthese waters. There may be future requests for further widening and deepening. It is increasingly imperative that a sufficient basis exist for reasonable selection of the best balance among possible alternatives.

REFERENCES

1. Aron, W. I., and S. H. Smith. 1971. Ship canals and aquatic ecosystems. Science 174:13-20.
2. Chesapeake Biological Laboratory (Univ. Md.), Chesapeake Bay Institute (The Johns Hopkins Univ.), and College of Marine Studies (Univ. Del.). 1972. Hydrographic and ecological effects of enlargement of the Chesapeake and Delaware Canal. Summary of interim findings. Report to Phila. Dist., U. S. Army Corps of Engineers. 36 p. (unpubl. MS).
3. ____, ____, and ____. 1973. Hydrographic and ecological effects of enlargement of the Chesapeake and Delaware Canal. Final Report. Summary of research findings. Report to the Phila. Dist., U. S. Army Corps of Engineers. 137 p. plus 15 Appendices. (unpubl. MS).
4. Cronin, L. E., R.B. Biggs, D. A. Flemer, H.T. Pfitzenmeyer, F. Goodwyn, Jr., W. L. Dovel and D. E. Ritchie, Jr. Gross physical and biological effects of overboard disposal in upper Chesapeake Bay. Special Report No. 3, Nat. Res. Inst., Univ. Md. 63 p. plus Appendix.
5. ____, G. Gunter and S. H. Hopkins. 1971. Effects of engineering activities on coastal ecology. Report to: The Office of the Chief of Engineers, Corps of Engineers, U. S. Army. 48 p.
6. Committee on Public Works, House of Representatives, 91st Congress, 2nd Session: The Chesapeake and Delaware Canal. Hearings of April 7, 8 and May 20, 1970. 322 p.
7. Dovel, W. L., and J. R. Edmunds, IV. 1971. Recent changes in striped bass (*Morone saxatilis*) spawning sites and commercial fishing areas in upper Chesapeake Bay; possible influencing factors. Chesapeake Science 12(1): 33-39.
8. Gardner, G. B., and D. W. Pritchard. 1971. Verification and use of a numerical model of the Chesapeake and Delaware Canal. CBI Tech. Rept. 87, Ches. Bay Inst. The Johns Hopkins Univ. 44 p. plus Appendices.
9. Gray, R. D. 1967. The National Waterway. A history of the Chesapeake and Delaware Canal, 1769-1965. University of Illinois Press, Urbana. 229 p.
10. Groves, R. H. 1970. A study program for evaluating the effects upon the environment of the Chesapeake and Delaware Canal. Testimony *in* Committee on Public Works, 1970. (6 above) 299-306.
11. Johnson, R. K. 1973. Production and distribution of fish eggs and larvae in the Chesapeake and Delaware Canal. App. I. to Ches. Biol. Lab.; Ches. Bay Inst. and Coll. Mar. Stud., 1973. NRI Ref. 73-31. 137 p.

12. ——, and T. S. Y. Koo. 1975. Production and distribution of striped bass (*Morone saxatilis*) eggs in the Chesapeake and Delaware Canal. Chesapeake Science 16(1): 39-55.
13. Koo, T. S. Y., and J. S. Wilson. 1972. Sonic tracking striped bass in the Chesapeake and Delaware Canal. Trans. Amer. Fish. Soc. 101:453-462.
14. Miller, R. E., S. D. Sulkin and R. L. Lippson. 1975. Composition and seasonal abundance of the blue crab, *Callinectes sapidus*, in the Chesapeake and Delaware Canal and adjacent waters. Chesapeake Science 16 (1):27-31.
15. Morgan, R. P., II. 1973. Marking fish eggs with biological stains. Chesapeake Science 14(4):303-305.
16. ——. 1975. Distinguishing larval white perch and striped bass by electrophoresis. Chesapeake Science 16(1):68-70.
17. ——, and V. J. Rasin, Jr. 1973. Effects of salinity and temperature on the development of eggs and larvae of striped bass and white perch. App. X *to* Ches. Biol. Lab., Ches. Bay Inst., Coll. Mar. Stud., 1973. NRI Ref. 73-109. 37 p.
18. ——, ——, and Linda Noe. 1973. Effects of suspended sediments on the development of eggs and larvae of striped bass and white perch. App. XI *to* Ches. Biol. Lab., Ches. Bay Inst., Coll. Mar. Stud., 1973. NRI Ref. 73-110. 22 p.
19. ——, ——, ——, and G. B. Gray. 1973. Effects of water quality in the Chesapeake and Delaware Canal region on the survival of eggs and larvae of striped bass and white perch. App. XIII *to* Ches. Biol. Lab., Ches. Bay Inst., Coll. Mar. Stud. 1973. NRI Ref. 73-112. 38 p.
20. ——, R. E. Ulanowicz, V. J. Rasin, Jr., L. A. Noe and G. B. Gray. 1973. Effects of water movement on eggs and larvae of striped bass and white perch. App. XII *to* Ches. Biol. Lab., Ches. Bay Inst., Coll. Mar. Stud., 1973. NRI Ref. 73-111. 29 p.
21. Pfitzenmeyer, H. T. 1973. Benthos of Maryland waters in and near Chesapeake and Delaware Canal. App. III *to* Ches. Biol. Lab., Ches. Bay Inst., and Coll. Mar. Stud., 1973. NRI Ref. 73-113. 41 p.
22. Por, F. D. 1971. One hundred years of Suez Canal—a century of Lessepsian migration: Restrospect and view points. Syst. Zool. 20(2) :138-159.
23. Pritchard, D. W., and L. E. Cronin. 1971. Chesapeake and Delaware Canal affects environment. Am. Soc. Civil Eng., Nat. Water Res. Eng. Meeting. 25 p.
24. ——, and G. B. Gardner. 1974. Hydrography of the Chesapeake and Delaware Canal. Tech. Rept. 85, Ref. 74-1, Chesapeake Bay Inst., The Johns Hopkins Univ. 77 p. plus Appendix.
25. Ritchie, D. E., Jr., and T. S. Y. Koo. 1973. Fish Survey in Maryland portion of the Chesapeake and Delaware Canal. App. VI *to* Ches. Biol. Lab., Ches. Bay Inst. and Coll. Mar. Stud., 1973. NRI Ref. 74-41. 28 p. plus 32 tables and 20 figures.
26. ——, and ——. 1973. Fish movements—Maryland study. App. VIII-A *to* Ches. Biol. Lab., Ches. Bay Inst., and Coll. Mar. Stud., 1973. NRI Ref. 74-42. 56 p.
27. Smith, R. W., M. H. Taylor, L. M. Katz, F. C. Daiber and V. Lotrich. Delaware fish migration. App. IX *to* Ches. Biol. Lab., Ches. Bay Inst., and Coll. Mar. Stud., 1973. 45 p. (unpubl. MS).
28. Steinitz, H. 1968. Remarks on the Suez Canal as a pathway and habitat. Rapp. Comm. Int. Mer. Medit. 19:139-141.

29. Taylor, M. H., W. R. Hall, R. W. Smith, L. M. Katz, F. C. Daiber and V. Lotrich. 1973. Delaware benthos. App. IV *to* Ches. Biol. Lab., Ches. Bay Inst., Coll. Mar. Stud., 1973. 51 p. (unpubl. MS).
30. ——, R. W. Smith, L. M. Katz, F. C. Daiber and V. Lotrich. Delaware fish survey. App. VII *to* Ches. Biol. Lab., Ches. Bay. Inst. and Coll. Mar. Stud., 1973. 75 p. (unpubl. MS).
31. Wicker, C. F. 1939. Tides and currents in the Chesapeake and Delaware Canal. Tech. Report to Phila. Dist., U. S. Army Corps of Engineers. Testimony *in* Comm. on Public Works, 1970. (6 above) p. 138-180.

INTRODUCTION TO SEDIMENTARY PROCESSES I

Convened by:
Maynard M. Nichols
Virginia Institute of Marine Science
Gloucester Point, Virginia 23062

Transport processes in estuaries operate by the dissipation of energy from the tide, waves, wind, density gradient and river inflow. Such processes, which include advection, diffusion, aggregation, deposition and resuspension, result in a change in bottom geometry or a transfer of sediment. Although transport varies widely in scale and intensity in different estuaries, the processes are similar in kind and they are often organized into coherent patterns. By learning the behavior of processes in different estuaries it should be possible to learn how transport is affected by various amounts of energy and to recognize routes of transport common to many estuaries. These papers bring together a series of mainly field studies, from different estuaries of the world, that span a range of estuarine mixing types, from the salt-wedge structure in the Amazon to the well-mixed tidal reaches of southern San Francisco Bay. These studies should offer students an introduction to the subject and stimulate further research.

DISTRIBUTION AND TRANSPORT OF SUSPENDED PARTICULATE MATERIAL OF THE AMAZON RIVER IN THE OCEAN

Ronald J. Gibbs
College of Marine Studies
University of Delaware
Lewes, Delaware 19958

ABSTRACT: The transport of suspended material of the Amazon River was determined based on data from seven cruises in the Amazon River/Atlantic Ocean area with measurements of currents, suspended material concentration, temperature, and salinity made at three anchored stations extending over complete tidal cycles. The suspended material is thrust out the river mouth onto the shelf where it encounters a two-layer flow with entrainment and mixing. It is also carried westward along shore by a strong ocean current, as well as by prevailing longshore currents. Off the river mouth, transport is oceanward at all depths with the majority being transported in the lower half of the water column. There is a transition oceanward until the upper third of the water column on the outer shelf has negligible suspended material transport, the middle third is transported oceanward, and the bottom third of the column is transported landward. A state of equilibrium existing between the bottom transports produces a turbidity maximum out on the shelf oceanward of the river mouth and extending northwestward parralleling the coast. The sedimentation patterns are in opposition to the classic pattern with sands and silts on the shelf and winnowed mud being deposited on the coast.

INTRODUCTION

Rivers supply the majority of the suspended material to the world's oceans. It is, therefore, important to understand the transport processes of this material after it is delivered to the ocean. The objective of this study is the understanding of the transport mechanisms of the suspended material discharged by the world's largest river, the Amazon River, into the Atlantic Ocean.

The previous research on the suspended material of estuaries was reviewed by Meade (4). A regional study of the suspended material distribution of the surface water of the Amazon River in the Atlantic Ocean was published by Gibbs (2). Milliman, et al. (5) also reported the distribution of surface suspended material in the area just off the Amazon River.

STUDY AREA AND METHODOLOGY

This study draws on the data obtained on seven cruises into this area with 3 anchored stations extending through a tidal cycle. The study area and ship track showing sampled stations used for this synoptic view are shown on Fig. 1. On each of the non-anchored stations an *in situ* profile of salinity, temperature, and light scattering and transmission was obtained. Samples were taken based on the data of this *in situ* profile. From the surface to about 90 meters depth, a pump was utilized to pump samples up to 200 liters directly into a pressure filtering tank. This process virtually eliminated the possibility of contamination since the sample was never exposed to open air. Further, the large volume of the sample makes any possible contamination negligible. Samples obtained at greater depths using a 30 1 Niskin bottle were drained directly into a pressure filtering tank. Filters with 0.45μ pore size were used. They were rinsed four times with distilled water to remove sea water and then stored for weighing in the laboratory. The filters were weighed before and after use at the same humidity with blanks used for standards. Calibration curves for the optical data were prepared utilizing samples taken at the same position as the optical reading. This procedure provided detailed data on concentration of suspended material between the

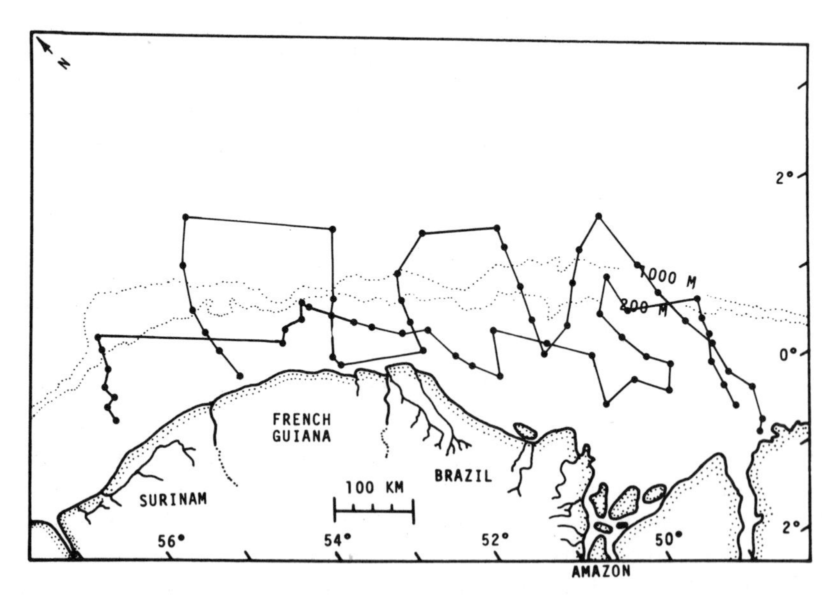

Figure 1. The cruise track and stations used for synoptic studies for the Amazon River/ Atlantic Ocean for June-July 1971.

sampling points. Bottom samples at each station were obtained utilizing a Shipek sampler or pipe dredge. On the anchored stations an *in situ* profile of salinity, temperature, and light scattering and transmission along with measurements of current velocity and direction were obtained at one- or two-hour intervals through at least one tidal cycle.

PHYSICAL OCEANOGRAPHY

The understanding of the physical circulation of the water is crucial to an understanding of the transport of the suspended material. A review of the physical circulation is therefore presented, with more details given in other papers by Gibbs (1, 2, 3). An idea of the overall circulation is best shown in Fig. 2. This three-dimensional drawing shows the fresh water thrust out over the sea water with some entrainment and mixing similar to a two-layer estuary model, although it actually occurs out on the open shelf. The strong Guiana current on the outer shelf and the longshore transport developed by the trade winds turn this brackish plume northwestward where it is carried along the shelf as a detectable plume for over 1000 km. As the currents move northward away from the Equator, the Coriolis effect becomes greater, producing an upwelling which comes up the continental slope and across the shelf as a bottom current.

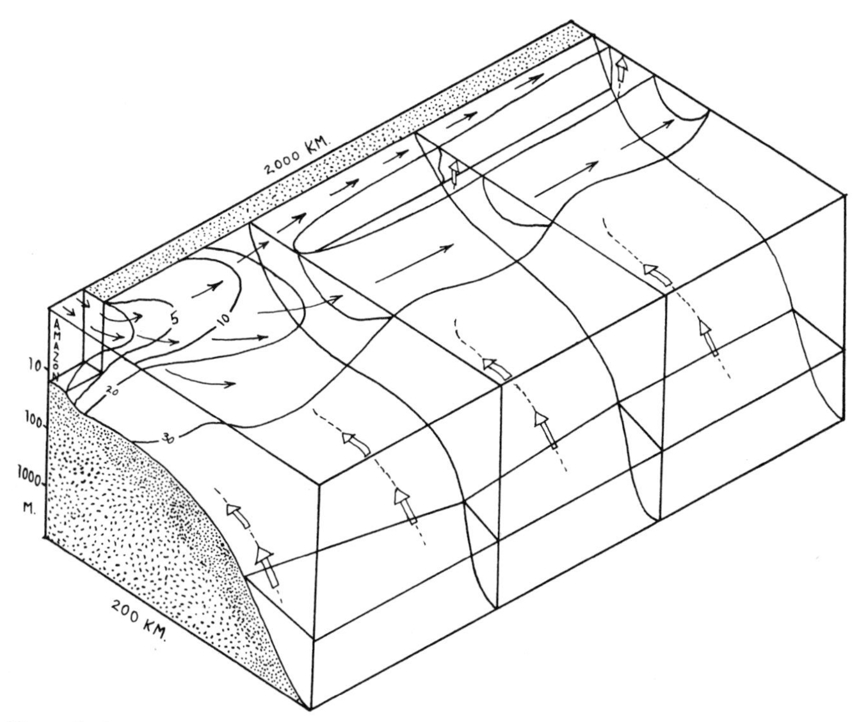

Figure 2. Three-dimensional drawing of the salinity distribution (ppt), Amazon River/Atlantic Ocean for June-July 1971.

SUSPENDED MATERIAL TRANSPORT

The surface distribution of suspended material shows that there is a turbidity maximum on the continental shelf in front of the Amazon River. The material is then carried off to the northwest in a plume on the shelf and within a zone along

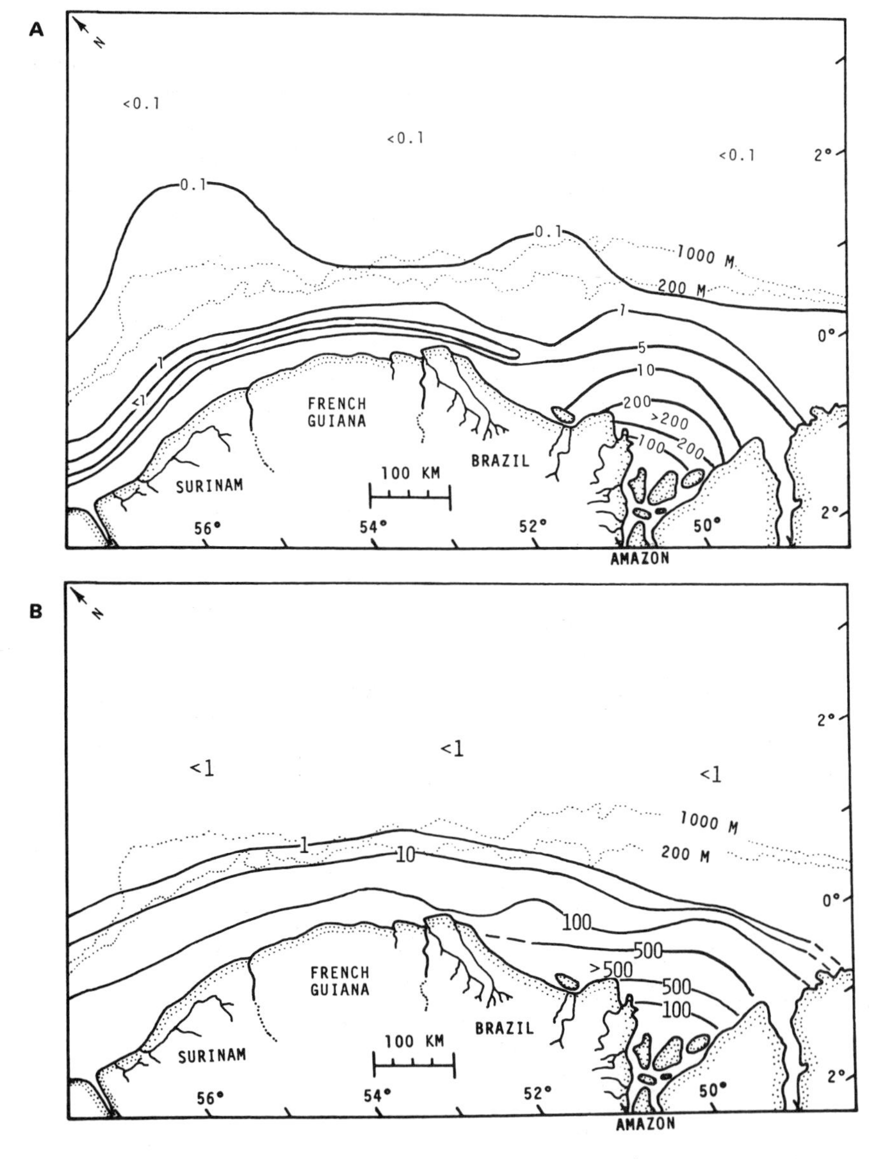

Figure 3. Distribution of the suspended particulate material in mg/1 for June-July 1971: A, surface distribution; B, bottom distribution.

the shore (see Fig. 3). This pattern shifts only slightly with the tides, and shifts about 10 to 15% between seasons. The surface plume undulates with time, as observed from the air on several occasions.

The turbidity maximum for suspended material near the bottom is located on the shelf in front of the Amazon River and extends northwestward along the coast (see Fig. 3). It should be noted that the major difference between the surface and bottom distributions of suspended material is the absence of the offshore plume along the bottom. The shoreward-moving bottom water continually moves toward the shore any material settling from the surface plume.

To understand the three-dimensional distribution of the suspended material in greater detail, a series of profiles (locations in Fig. 4) is given in Fig. 5. A log depth axis has been used on these profiles to facilitate presentation of details on the shelf region as well as to show the continental slope and rise data. Profile F of Fig. 5 shows the case to the south of the Amazon River. Since the ocean current carries most of the suspended material to the northwest, this profile shows very little suspended material.

In the next two profiles northwestward (Fig. 5D and E), the turbidity maximum can be seen off the mouth of the Amazon River. The landward-flowing bottom current can be seen, with the plume of turbid water moving oceanward at intermediate depths.

The remaining profiles of Fig. 5A, B, and C show the extension of the turbidity maximum along the shore, the surface plume over the shelf, and the

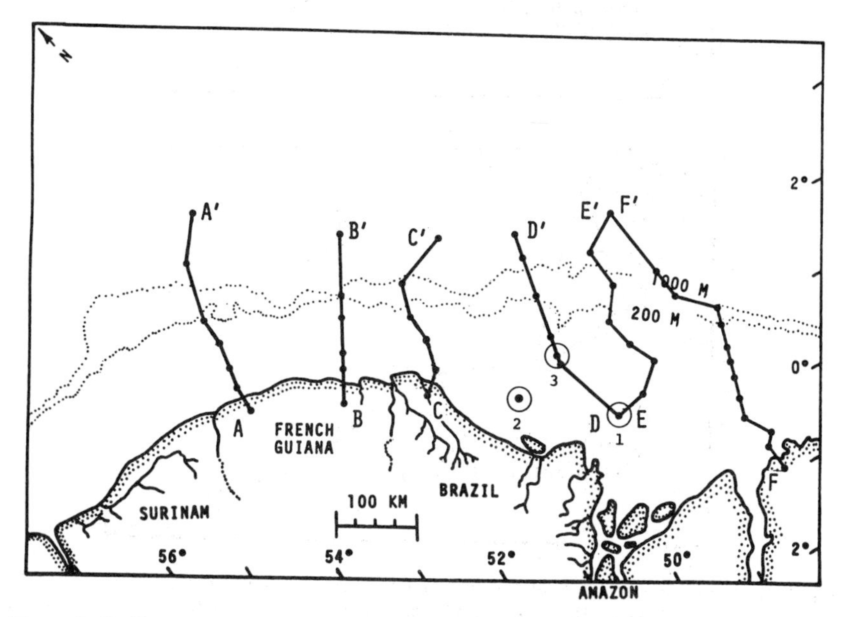

Figure 4. Profile and anchored station location map for the Amazon River/Atlantic Ocean for June-July 1971. Circles = anchored stations.

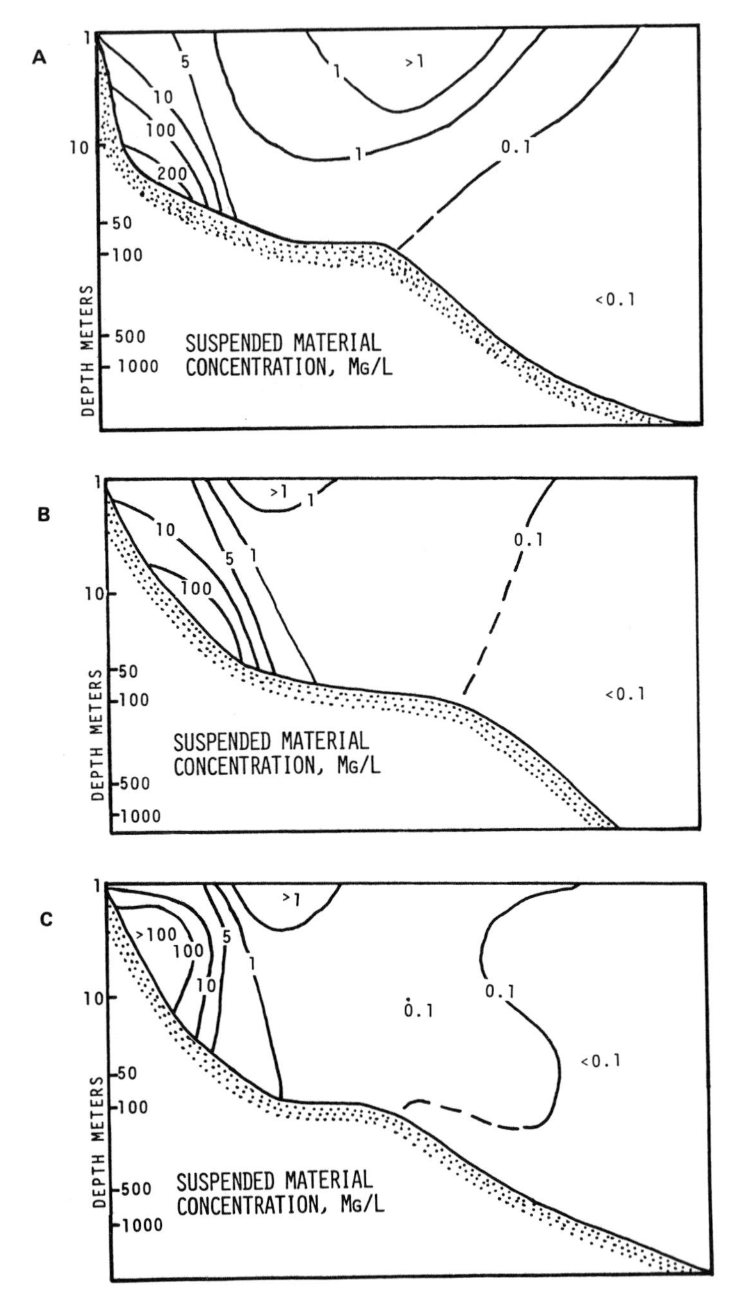

Figure 5. Profiles A through F of the suspended particulate material concentration (mg/1), for the Amazon River/Atlantic Ocean for June-July 1971 (see Fig. 4 for profile locations and horizontal scales).

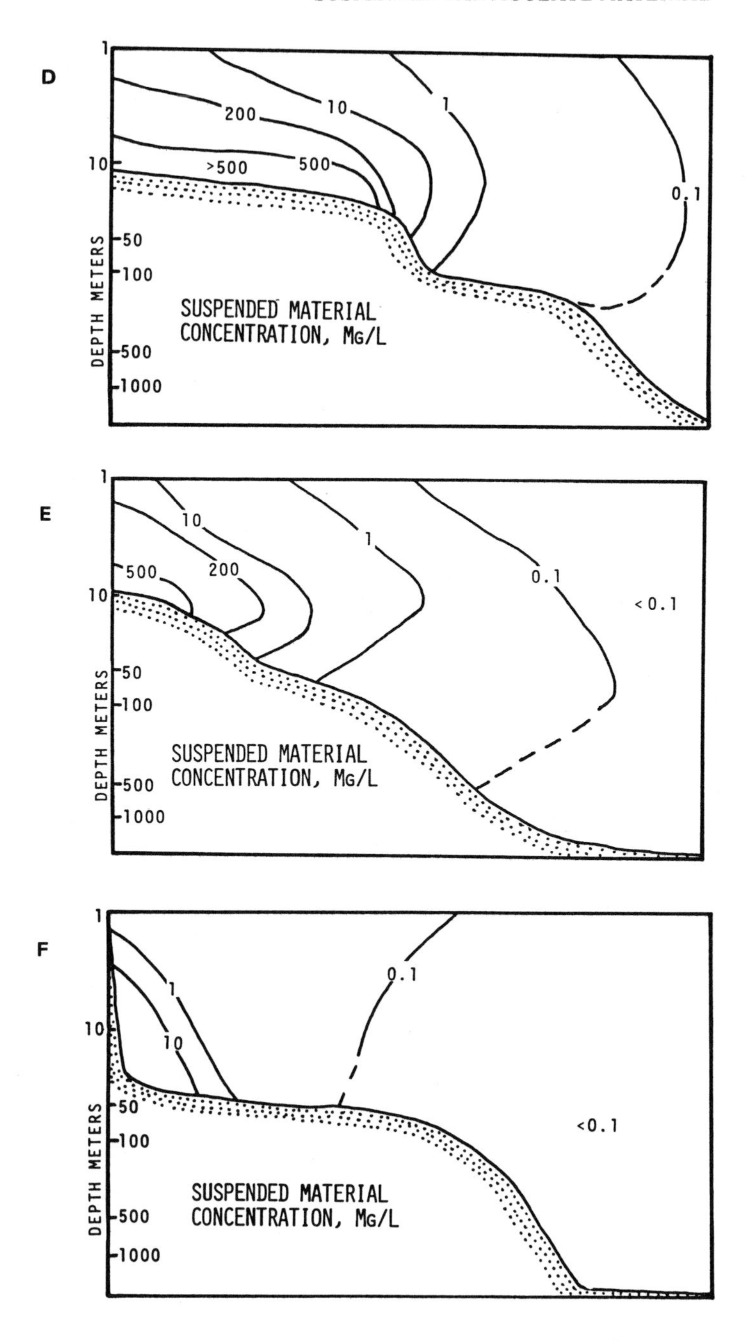

Figure 5 (Continued)

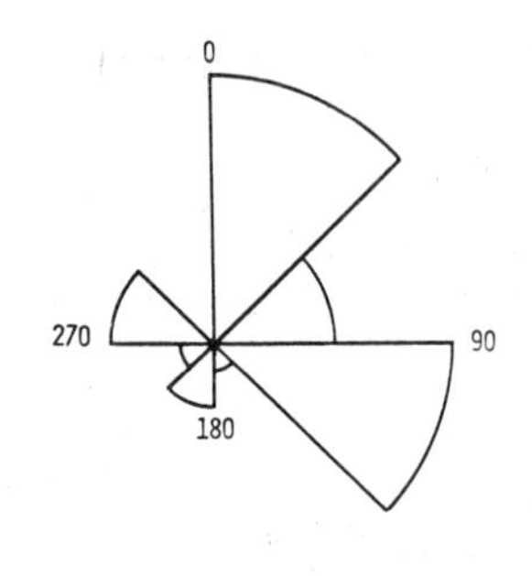

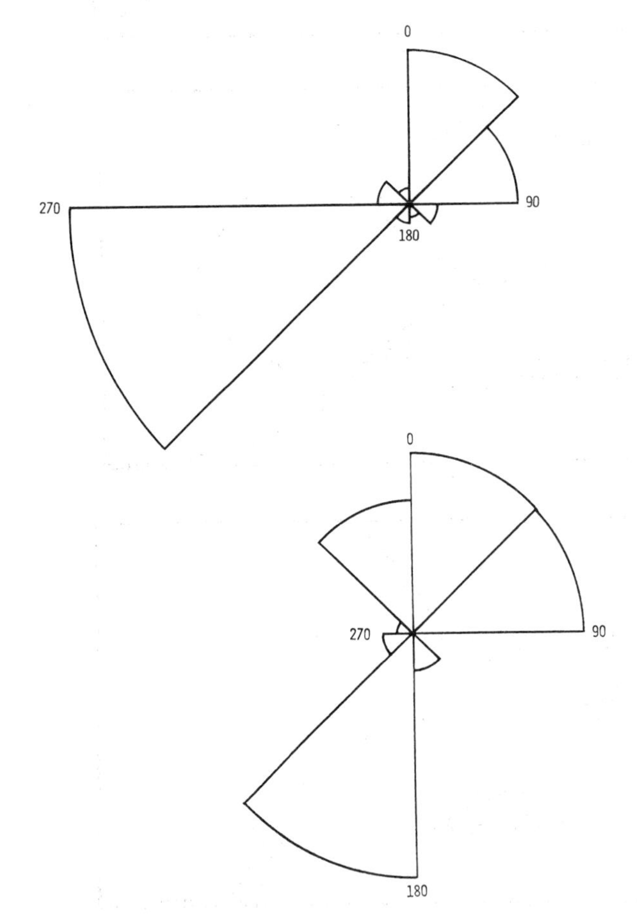

Figure 6. Vector diagrams of total transport of suspended particulate material for all depths for Stations 1(top), 2(middle), and 3(lower) for the Amazon River/Atlantic Ocean for June-July 1971 (see Fig. 4 for station locations).

slow disappearance of the plume of turbid water flowing oceanward at intermediate depths.

These maps and profiles give an instant picture of the distribution, but do not give the actual movement of the suspended material. For the actual transport of the suspended material, let us examine the data from three anchored stations, positions of which are shown on Fig. 4. These stations were occupied for 13 to 14 hours during which the current velocity and direction and the suspended material concentration were measured at a number of depths every one to two hours. Combining these data, we obtain the suspended material transport at various depths. If the suspended material transport is summed up for the various depths to see the total material transported in various directions, we obtain the vector diagrams (Fig. 6) for the three anchored stations. The diagrams of the river shows two large offshore modes north and east and very little shoreward transport. The diagram for the outer shelf station, on the other hand, shows one mode offshore and one onshore, both of about the same magnitude indicating that the net transport in either direction is very small. The third diagram is the station on the inner shelf and about 125 km northwest of the river mouth. This diagram has large onshore and offshore components, but it has an offshore net transport. So we change from a case near the river of almost complete offshore transport to a case up the coast of both onshore and offshore (but still net offshore) to the outer shelf with onshore and offshore canceling each other out to give almost zero net transport. To look in greater detail at this transition in transport, Fig. 7 shows the transport by direction and depth for the same three anchored stations.

For example, at the station closest to the mouth of the Amazon River the data have been divided into eight directional sections (45° each) and into two-meter depth intervals and are shown in Fig. 7 as positive numbers for landward transport and negative numbers for offshore transport with the summation of the onshore-offshore component of the eight sections shown as a dotted line. It should be noted that the major transport is offshore and that this is mainly in the bottom half of the water column with the surface transport negligible.

A look at the station up the coast about 125 km and in approximately the same depth of water shows a more complex condition. A few generalities can, however, be concluded: (1) there is negligible transport evident in the top five meters of the water column; (2) the offshore transport is dominated by that of the intermediate depths; and (3) landward transport is evident in the bottom three meters.

An interesting comparison with these two stations is a station located on the outer continental shelf. Again the top third of the water column transports a negligible amount of sediment, the middle third of the water column is characterized by offshore transport, and the bottom third is characterized by landward transport. The onshore and offshore transport are of about the same magnitude, whereas the previous profile closer to shore evidenced greater offshore

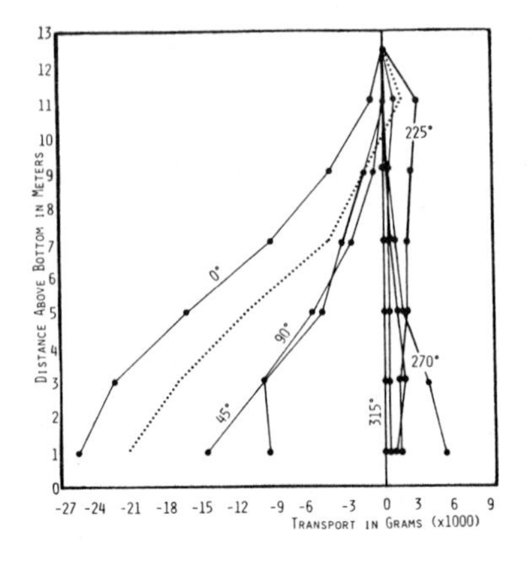

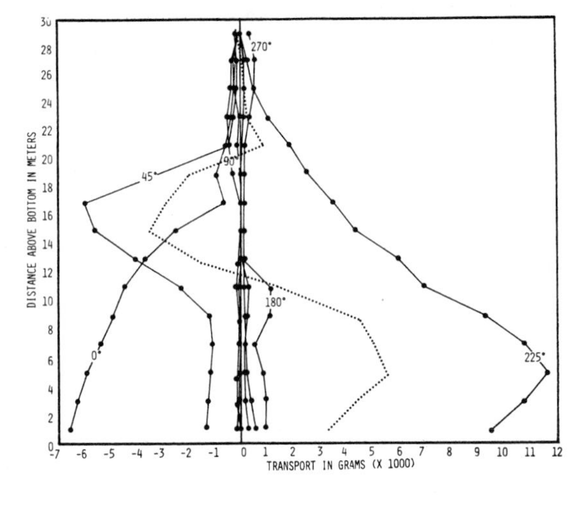

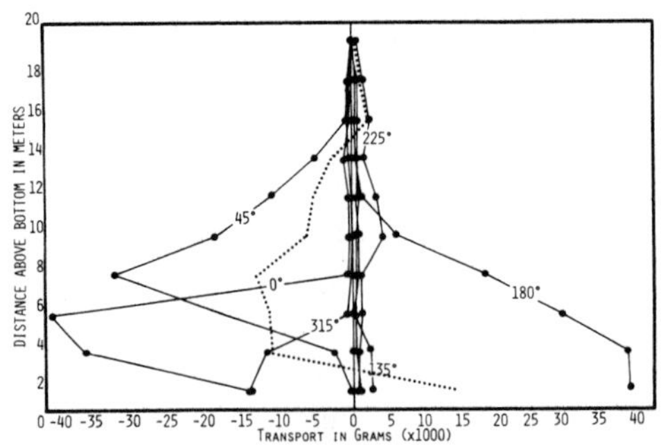

Figure 7. Suspended particulate material transport onshore and offshore at various depths and directions for the anchored Stations 1(top), 2(middle), and 3 (lower) for the Amazon River/Atlantic Ocean for June-July 1971. Dotted line is summation of all directions (see Fig. 4 for anchored station locations).

than onshore transport. Undoubtedly some of the suspended material which is transported offshore at intermediate depths is finding its way later into the onshore bottom transport and is being cycled back toward shore.

The present study is in general agreement with the work of Milliman, et al. (5); however, using only ocean surface suspended material, they concluded that most suspended matter transported by the Amazon River appears to be deposited within the river mouth, but they were unaware of the oceanward transport a few meters below the surface.

The organic material of the suspended particulate material is composed of two components: one, the land-derived material carried by the river, and two, the biologically-produced material being formed in the ocean area. In Fig. 8 can be seen the distribution of the organic carbon in the suspended particulate material. To a first approximation, the organic carbon can be seen to parallel the distribution of the total suspended particulate material. The percentage of carbon in the river samples ranges from 1 to 2%. Organic material comprises up to 90% of the suspended material in the open-ocean surface samples. Of the suspended particulate material of biological origin that possesses hard parts capable of adding to the sediments, the diatoms dominate. The diatoms become a significant source of suspended particulate material at the surface, but they are rare in the bottom water and in the bottom sediments. Milliman, et al. (5) found similar diatom distribution in this area. Apparently the diatoms dissolve upon settling and the released silica is used by more diatoms, but the eventual fate of the silica is not evident. The diatom population adds to the suspended particulate material over a wide area on the shelf, but since it is related to a different source (biological utilization of a dissolved nutrient) it appears to act as an "overlay" on the physical transport of the river-derived suspended particulate material.

CONCLUSIONS

The basic model is then a thrusting out onto the shelf of the suspended material which is being carried seaward at all depths. Farther out on the shelf, a landward-moving bottom current develops and, at the landward side of this bottom current, the turbidity maximum is developed. Moving directly oceanward from the river at about the edge of the continental shelf the landward bottom transport of suspended material reaches a point of equaling the seaward transport. Further up the coast, there is landward transport of the suspended material transport along the bottom, and offshore transport at intermediate depths. At all three anchored stations there is negligible transport of suspended material in the surface waters.

ACKNOWLEDGMENTS

Sincere thanks are extended to the captains and crews of Woods Hole Oceanographic Institute's *Chain,* the *Oregon* of the U.S. Bureau of Commercial Fisheries, the Brazilian Navy vessels, *Bertioga* and *Almirante Saldanha*, and the tugboat of the Companhia de Industria e Comercio de Minas Gerais (ICOMI) and to the various organizations and institutions that support them. Appreciation for support for this study is extended to the Office of Naval Research under contracts N00014-67-A-0356-0011 and N00014-75-C-0355.

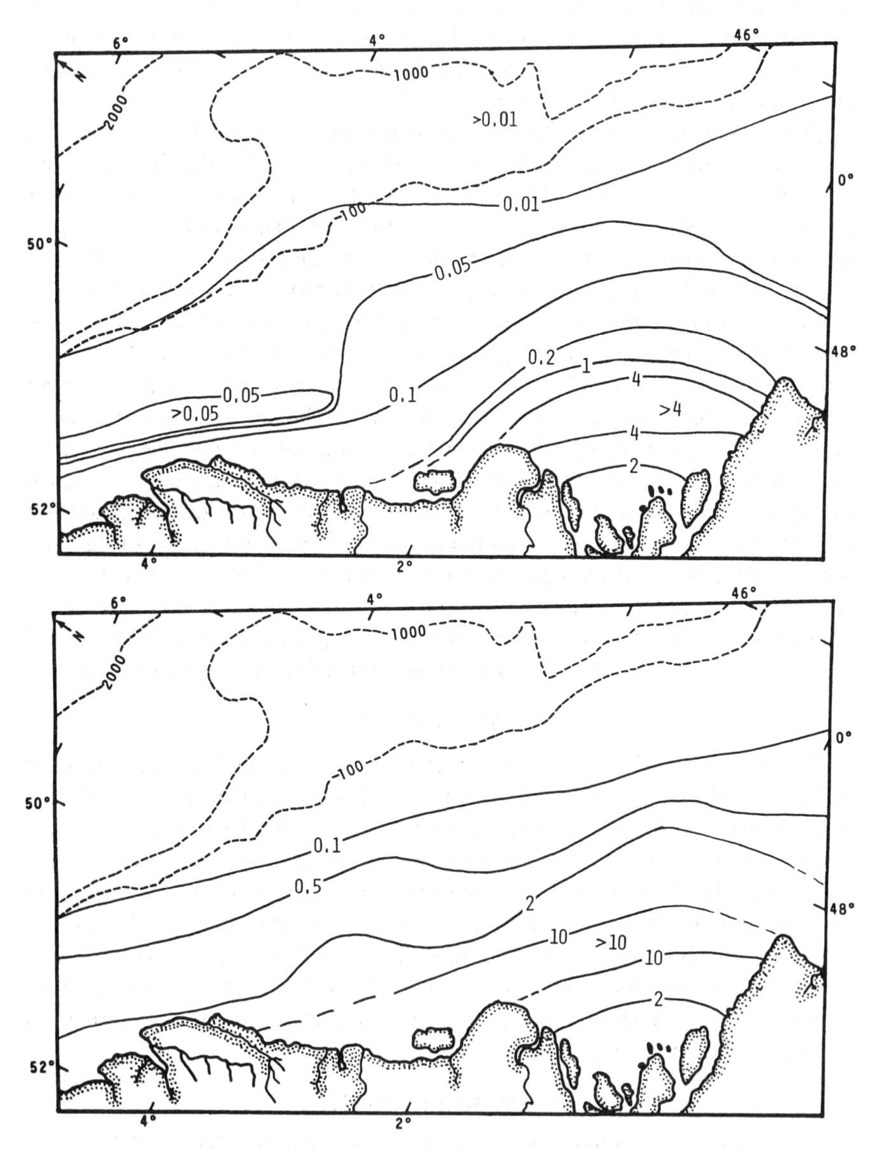

Figure 8. Distribution of carbon (mg/1) of suspended particulate material for the Amazon River/Atlantic Ocean for June-July 1971: top, in the surface waters; lower, in the bottom waters.

REFERENCES

1. Gibbs, R.J. 1970. Circulation in the Amazon River estuary and adjacent Atlantic Ocean. J. Mar. Res., 28: 113-123.
2. ——. 1974. The suspended material of the Amazon Shelf and Tropical Atlantic Ocean, p. 203-210, *In* Suspended Solids in Water, R. J. Gibbs, (ed) Marine Science Series 4, Plenum Press, N.Y.
3. ——. 1976. Amazon River sediment transport in the Atlantic Ocean. Geology, 4:45-48.
4. Meade, R.H. 1972. Transport and deposition of sediments in estuaries, p. 91-210. *In* Environmental framework of coastal plain estuaries, B.W. Nelson (ed). Memoir 133, Geol. Soc. of America.
5. Milliman, J.E., C.P. Summerhayes, and H.T. Barretto. 1975. Oceanography and suspended matter off the Amazon River February-March 1973. J. Sed. Pet. 45: 189-206.

SUSPENDED SEDIMENT BUDGET

FOR CHESAPEAKE BAY[1]

J. R. Schubel and Harry H. Carter
Marine Sciences Research Center
State University of New York
Stony Brook, New York 11794

ABSTRACT: The Susquehanna is the only river that discharges directly into the main body of the Chesapeake Bay. All the other rivers flow into tributary estuaries and most of their sediment loads are entrapped within these estuaries. During periods of high riverflow the Susquehanna dominates the upper 20-30 km of the Bay; the net flow and sediment transport are seaward at all depths, and there is a marked downstream decrease in suspended sediment. With subsiding riverflow, a turbidity maximum is formed in the upper reaches of the Bay.

In the middle and lower reaches of the Bay, shore erosion is not only a major source of inorganic sediment it may be the largest single source.

The distributions of suspended sediment along the axis of the entire Bay are presented for a twelve month period in 1969-1970. These data, and others, were used to formulate a single-segment model of the main body of the Bay that indicates that there is a net movement of sediment into the Bay from the ocean, and that the tributary estuaries are sinks for suspended sediment in the Bay.

INTRODUCTION

The modern Chesapeake Bay estuarine system comprised of the Bay proper and its tributary estuaries was formed by drowning of its ancestral Susquehanna River valley system during the most recent rise in sea level which began 12,000-15,000 years ago. The modern Chesapeake Bay estuary is very young geologically and like other estuaries is an ephemeral feature on a geological time scale. It is being rapidly filled with sediments; sediments from rivers, from shore erosion, from primary productivity, and from the sea. The sediment sources are

[1] Contribution 150 of the Marine Sciences Research Center of the State University of New York, Stony Brook, New York 11794

thus external, marginal, and internal. The particles are organic and inorganic, naturally occurring and anthropogenic. The prevailing mode of sediment transport both into and within the Chesapeake Bay estuarine system is as suspended load.

The primary objectives of this paper are: (1) to briefly review our knowledge of the suspended sediment system of the main body of the Bay; (2) to graphically summarize the most complete set of data available on the distribution of suspended sediment along the axis of the entire Bay, and; (3) to present the results of a simple single-segment model constructed to estimate the net suspended sediment exchange rates between the Bay and its tributary estuaries, and between the Bay and the ocean; and the mean annual Bay-wide sedimentation rate.

HYDROLOGICAL AND SEDIMENTOLOGIC SETTINGS

The Susquehanna, which enters at the head of the Bay at Havre de Grace (Md.), Fig. 1, is the only river that discharges directly into the main body of the Bay. All of the other rivers discharge into estuaries formed by drowning of the lower reaches of these rivers during the most recent rise in sea level. With a long-term average discharge of about 985 m^3/sec the Susquehanna discharges more than 50 percent of the total fresh-water input to the entire Chesapeake Bay system, and more than 85 percent of that introduced north of the mouth of the Potomac. The Potomac has a long-term average discharge of about 320 m^3/sec and is the second largest source of fresh-water accounting for approximately 17 percent of the total input. The James is the next largest river contributing nearly 16 percent of the total fresh-water input. The combined fresh-water input from all the Eastern shore tributaries accounts for less than 7 percent of the total.

The Susquehanna is the major source of fluvial sediment to the main body of the Bay, and the bulk of that material is deposited in the upper 30 km of the estuary. In a similar manner the bulk of the sediment discharged by the other rivers is deposited in the upper reaches of their estuaries and does not reach the Bay proper. The sediment being discharged by the Susquehanna is predominantly clay and silt, with minor amounts of fine sand. All of the coarser sediment is entrapped in the reservoirs along the lower reaches of the River. The only significant active sources of sand to the Bay proper are shore erosion, and the movement of relict sand into the mouth of the Bay from the adjacent shelf. These sources are clearly revealed by the distribution of surficial sediments. Most of the Bay is blanketed with mud. Sand is predominant only: as a near-shore band that fringes the Bay; as a deposit near the mouth of the Bay; and as a large deposit near Smith and Tangier Islands (10). The last deposit is relict (22). The general features of the geological setting of the Chesapeake Bay have been summarized by Ryan (10) and more recently by Wolman (27), Schubel (18) and Folger (5).

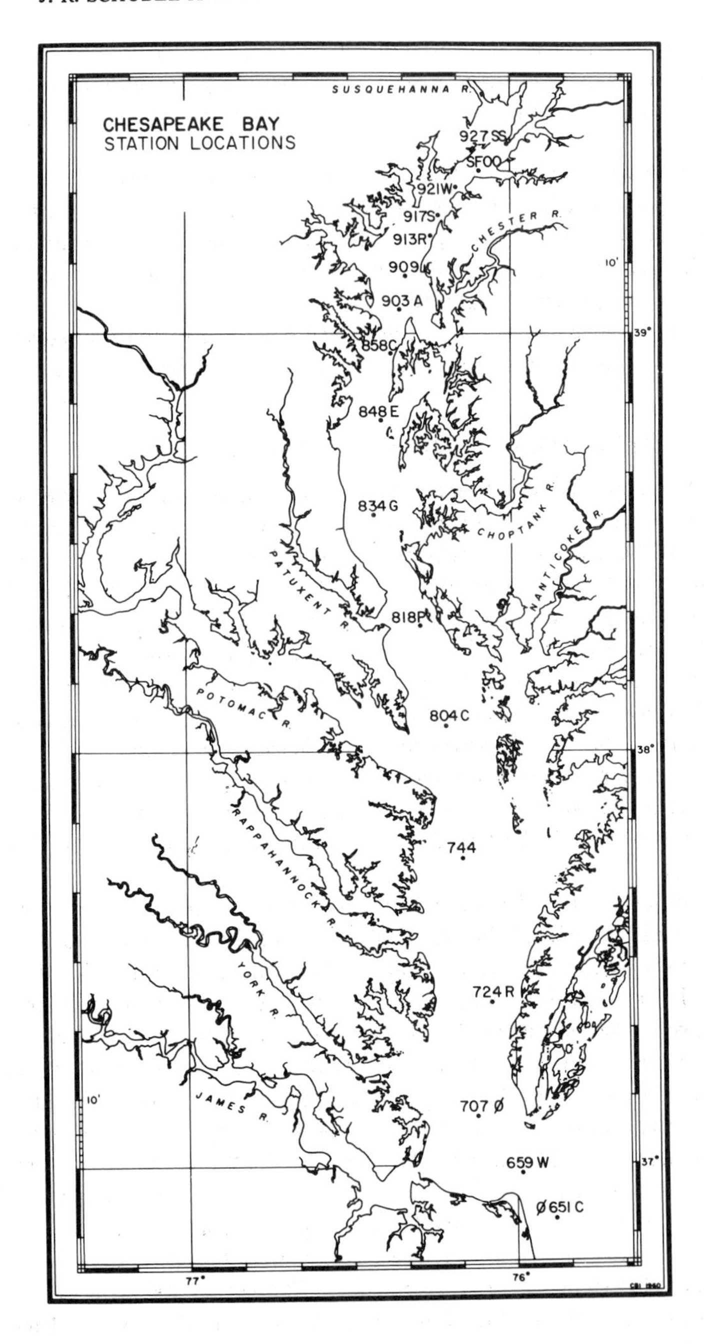

Figure 1. Station location map.

In the upper reaches of the Chesapeake Bay the Susquehanna flow regime and the associated circulation patterns generated within the upper Bay in response to the varying role of the River produce two distinctive distributions of suspended sediment and concomitant patterns of sediment transport (16, 13). The first characterize the spring freshet and other short periods of very high flow. The second, characteristic of periods of low to moderate flow, typify most of the remainder of the year. During periods of high flow the Susquehanna dominates the upper 20-30 km of the Bay; the net flow and sediment transport are seaward at all depths, and there is a marked downstream decrease in the concentration of suspended sediment (16, 13). The bulk of each year's supply of new fluvial sediment is usually introduced during the spring freshet, and about three-fourths of it is deposited in the upper 30 km of the Bay–upstream from Station 913R. With subsiding riverflow a "turbidity maximum" is formed near the head of the estuary. Schubel (14) has shown that the turbidity maximum is generated and maintained by a combination of physical processes; the periodic resuspension of bottom sediments by tidal scour and the sediment trap created in the upper reaches of the net non-tidal estuarine circulation regime.

In the middle and lower reaches of the Bay the coupling of the distribution of suspended sediment to the input of fresh-water and fluvial sediment, and to the estuarine circulation pattern, is much less apparent. Shore erosion and primary productivity are the major sources of suspended sediment to these segments of the Bay (11, 1) except near the bottom, and near the mouth of the Bay where resuspended material commonly dominates.

During periods of extreme flooding such as that following Tropical Storm Agnes (June 1972) the segment of the Bay dominated by the Susquehanna may extend nearly as far seaward as Station 858C, Fig. 1 (20, 19). The effects of Agnes on the suspended sediment system of the Bay have been described in some detail by Schubel (20, 21).

DISTRIBUTION OF SUSPENDED SEDIMENT IN TIME AND SPACE

The distribution of suspended sediment in the upper Bay–north of about 39°10′N–has been described in detail by Schubel (16, 13, 14, 11, 20, 21), and by Schubel and Biggs (23). Biggs (1) studied the suspended sediment as far south as the mouth of the Patuxent estuary, 38°15′N. While there have been extensive reports of the distribution of suspended sediment in the upper Bay, there are relatively few published observations for the middle reaches of the Bay, and a dearth of published data for the lower Bay. Burt (3, 4) made over 25,000 light extinction measurements over the length of the Bay with a Beckman Model DU Quartz Prism Spectrophotometer, but reported only eleven direct determinations of the concentration of suspended sediment, and no direct determinations of either their size distribution or their composition. His optical data are of little value in understanding the suspended sediment system of the Chesapeake Bay. Bond and Meade (2) analyzed eleven *surface* samples of suspended sediment

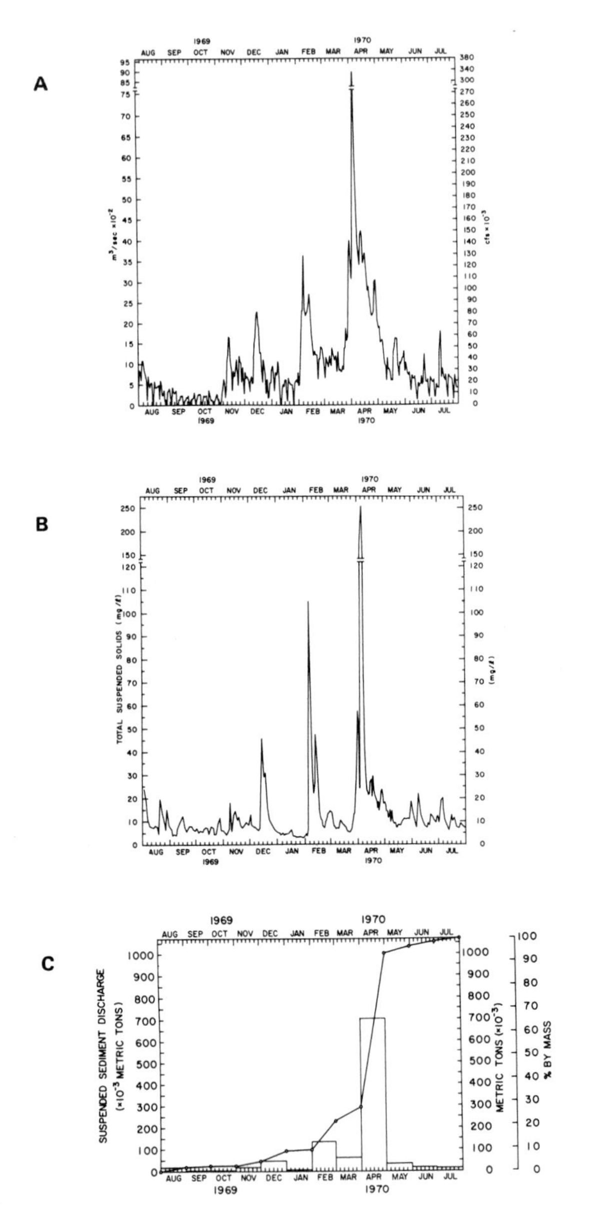

Figure 2. A. Discharge of the Susquehanna River at the Conowingo Hydroelectric Plant, Maryland, between August 1969 and July 1970.
B. Concentration (mg/1) of suspended sediment in the discharge water of the Conowingo Hydroelectric Plant between August 1969 and July 1970.
C. Suspended sediment discharge of the Susquehanna River into the upper Chesapeake Bay during the study period plotted as cumulative mass percent and as the mass of sediment discharged each month.

collected along the length of the axis of the Bay between 6-9 June 1965. Schubel (21) recently presented two longitudinal distributions of suspended sediment over the length of the Bay; one distribution representative of high riverflow, the other typical of low to moderate flow. The last two distributions are the only published longitudinal sections of suspended sediment that extend over the length of the Bay.

From August 1969 through July 1970, Schubel et al. (25) made fairly extensive observations of the distribution and character of suspended sediment along the axis of the Bay. Measurements were made approximately monthly at 17 stations over the length of the Bay (Fig. 1); stations in the upper Bay were occupied more frequently during periods of high riverflow. Determinations were made of the concentration of total suspended solids, and of the temperature and salinity at the surface and at 2m intervals to within about 1m of the bottom. All concentrations were determined gravimetrically using 0.6μm average pore diameter polycarbonate filters (25). For selected samples, estimates were made of the percentage of total suspended solids accounted for by combustible organic matter.

During this twelve month period (August 1969-July 1970) the Susquehanna had a mean flow of about 1020 m^3/sec, and discharged a total of approximately 1.1×10^6 metric tons of suspended sediment into the upper Bay, Fig. 2. The long-term average flow of the Susquehanna is about 985 m^3/sec, and its annual suspended sediment discharge is probably between 1 to 2×10^6 metric tons (12, 17). The flows of the other major rivers, except the James, were also near normal. In August of 1969 Hurricane Camille sent flows of the James to near-record levels. The resulting flood had an estimated recurrence interval of about 100 years. The twelve-month study period was as "typical" as any the authors have experienced in studying the Bay over more than the past decade.

Fig. 3 depicts the distributions of total suspended solids at the surface and at mid-depth along the axis of the Bay over the twelve month study period. Fig. 3 indicates that over most of the year the Bay can be conveniently divided into three segments on the basis of the distribution of total suspended solids: the upper Bay, extending from the head of the Bay at Turkey Point (Station 927SS) to about Station 909; the middle Bay extending from about Station 909 to Station 745; and the lower Bay extending from approximately Station 745 to the mouth of the Bay. The upper Bay is characterized by relatively high concentrations of suspended sediment at all depths throughout the year. As expected, values were generally highest when riverflow was highest (April) and the input of suspended sediment greatest. Nearly 70 percent of the total sediment discharged by the Susquehanna during the twelve-month study period occurred in April. During April a plume of "turbid" water was observed over most of the length of the Bay.

During periods of low to moderate riverflow–August 1969 through January 1970, and June-July 1970–the concentrations of suspended sediment in the upper Bay were somewhat lower than in April, but the marked downstream gradients seaward of about Station 913R persisted. Some evidence of the turbidity maximum that characterizes the upper Bay north of about 913R during periods of low to moderate riverflow (14) can be seen in Fig. 2B. The feature is not readily apparent however, because the sections end at Station 927SS. Comparison of the concentrations of suspended sediment at Stations SF00 and 927SS with those in the mouth of the Susquehanna (25) clearly show that the concentrations of suspended sediment are higher in the upper Bay than farther upstream in the source river–a characteristic feature of turbidity maxima.

Schubel (13) estimated that about 75 percent of the fine sediment introduced into the Bay north of 39°13′N (Station 913R) is deposited in that segment of the Bay. Biggs (1) estimated that about 96 percent of the sediment introduced into the Bay above 39°09′N (Station 909) is deposited within that segment.

Throughout the year, the concentrations of suspended sediment in the middle reaches of the Bay were relatively low. Surface values ranged from about 1 to 6 mg/1, and at mid-depth the approximate range was 1 to 7 mg/1. The principal sources of sediment to the segment of the Bay from about Station 909 to Station 745 are shore erosion and primary productivity (1, 21). For the Bay between 39°09′N and the mouth of the Patuxent estuary (38°18′N) Biggs (1) estimated that shore erosion contributed about 52 percent of the total mass of suspended material, and primary productivity about 40 percent.

Near the mouth of the Bay a secondary maximum in the concentration of suspended sediment was observed between about November 1969 and February 1970. These high values are apparently produced primarily by the resuspension of bottom sediments by tidal scour. The percent organic matter was relatively low (Fig. 3), the salinity data indicate that the water column was well-mixed (25), and there was no other obvious source of fine sediment. The high flow of the James in August 1969 following Hurricane Camille no doubt had some effect on the distribution of suspended sediment in the lower Bay during late August and September, but the high values observed between October and February cannot be ascribed to Camille.

Fig. 4 shows the distribution of combustible organic matter at the surface of the Bay between August 1969 and July 1970. The values plotted are the mass percents of the total suspended solids depicted in Fig. 2A accounted for by combustible organic matter. The samples were combusted at 480°C for 30 minutes (25). Values for other depths have been issued in tabular form (25). Figure 3 shows that the percent of suspended sediment accounted for by organic matter is generally greatest in the middle reaches of the Bay and decreases both farther upstream and downstream. Other data (25) show that at any station the concentration of combustible organic matter normally decreases with depth (25).

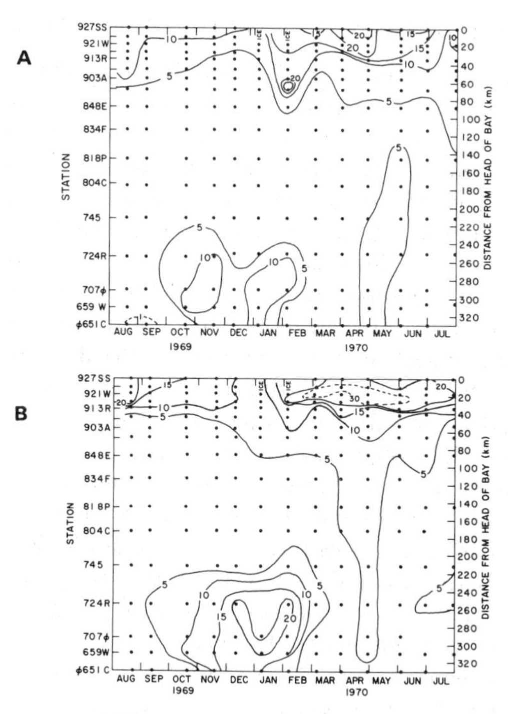

Figure 3. A. Distribution of suspended sediment at the surface along the axis of the Bay from August 1969 to July 1970. All values are in mg/1.
B. Distribution of suspended sediment at mid-depth along the axis of the Bay from August 1969 to July 1970. All values are in mg/1.

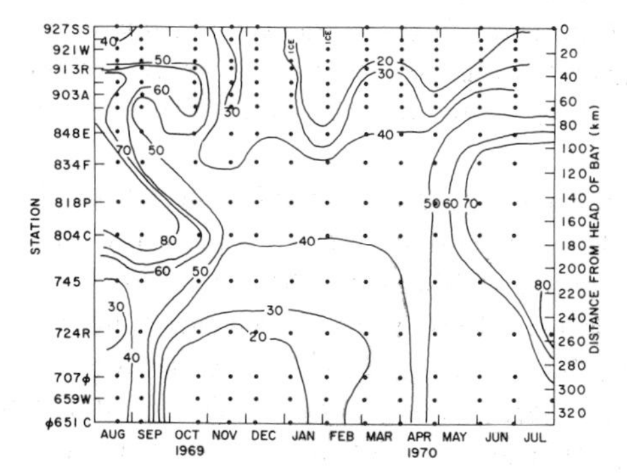

Figure 4. Distribution of combustible organic matter at the surface along the axis of the Bay from August 1969 through July 1970. Values are expressed as the mass percent of total suspended sediment accounted for by combustible organic matter.

MODELING OF SUSPENDED SEDIMENT TRANSPORT

There have been few attempts to mathematically model the suspended sediment of any segment of the Chesapeake Bay. Schubel (13) constructed a simple advective model for a "sediment balance" within the segment of the Bay from its head at Turkey Point to Tolchester (39° 13'N), and Biggs (1) presented the results of a simple advective model, presumably based on the familiar basin equations, for two contiguous boxes extending from the head of the Bay to the mouth of the Patuxent. The formulation of his model was unfortunately not specified. By far the most ambitious attempt to model the suspended sediment of any segment of the Bay is that of Hunter (7). Hunter developed a three-dimensional numerical model for the main body of the Bay from its head to about 39° N. His model is a pseudo-dynamic model designed to predict, from its concentration field, the *source* field of any passive contaminant subject to advection, diffusion, and settling. Hunter used the senior author's suspended sediment data (24, 26) to run the model, and after considerable adjustment he was able to demonstrate that there was an input of sediment from the Susquehanna and a loss of sediment through the southern open boundary near 39°N. As Hunter (7) points out (p. 66), "These results do not look very promising as they do not tell us any more than we know already—that sediment enters the Bay via the Susquehanna and leaves it by the southern open boundary."

In the authors' opinion it is time that we critically assess our panacean view of models and our apparent preoccupation with formulating complicated mathematical models for which there are not sufficient data for construction and independent verification. Scylla and Charybdis seem to have been reincarnated in today's scientific world as "mindless monitoring and factless modeling."

Despite these admonitions, or perhaps because of them, we have constructed another model, a very simple, single-segment model of the entire Bay. The objectives we hoped to attain through the model were estimates of the net exchanges of suspended sediment between the Bay and its major tributary estuaries, and between the Bay and the ocean; and an estimate of the average Bay-wide sedimentation rate.

The model was restricted to the inorganic fraction of the total suspended sediment to avoid the additional problem of estimating the production of suspended matter by primary production. The concentrations used in the model were the combusted values. These residual concentrations do, of course, include some skeletal material that is not destroyed by combustion. In our zeroth order model no corrections were made for this skeletal material since preliminary calculations indicated that the required adjustments would not affect the conclusions of the model.

In our model the Bay consists of a single segment, Fig. 5. Water, salt, and suspended sediment are transported in the upper and lower layers across vertical cross-sections separating the Bay from the ocean and from its major tributaries.

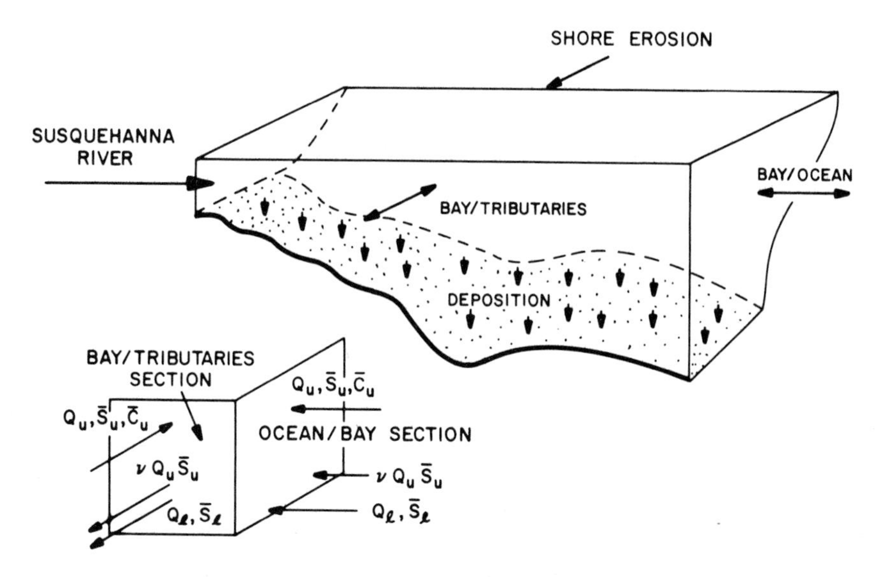

Figure 5. Pictorial representation of the simple, single-segment model used to estimate the exchanges of suspended sediment between the Bay and its tributary estuaries, and between the Bay and the ocean.

Transport of water and suspended sediment are by advection (gravitational convection); and transport of salt is by advection, and by longitudinal diffusion in the direction of decreasing salinity. The advective fluxes of salt through each section were calculated using modified basin equations assuming that over the twelve month period there was no net flux of salt. The modification of the basin equations consisted of exactly balancing a fraction, ν, of the net downstream advective flux of salt in the upper layer by a diffusive flux in the opposite direction. The advective flux of water in the upper layer was then solved for explicitly utilizing these equations and the conservation of mass. This advective flux of water was then used along with the mean concentrations of inorganic suspended sediment in the upper and lower layers to calculate the flux of suspended sediment. In the manipulative process the vertically averaged salinities of the upper and lower layers, $\bar{S}_u$ and $\bar{S}_\ell$, were separated into two parts—an annual average and a deviation term. The advective fluxes of water, Q_u and Q_ℓ, the ratios of the average salinities and the suspended sediment concentrations of the lower to the upper layer, f and f*, were treated in a similar fashion. All non-zero products containing products of deviations were lumped together as a diffusive flux. For salt this flux was assumed to take place in the direction of decreasing salinity. Diffusion of suspended sediment was neglected because the longitudinal gradients of suspended sediment were generally small near the boundaries.

The resulting equations are:

$$<(Q_u)_j> = \frac{<R_j(\bar{S}_u)_j f_j>}{<(\bar{S}_u)_j>} \quad \frac{1}{(<f_j> + <\nu> - 1)} \tag{1}$$

for the water flux in the upper layer and

$$<(Flux_{ss})_j> = <(\bar{C}_u)_j> \frac{<R_j(\bar{S}_u)_j f_j>}{<(\bar{S}_u)_j>} \times \frac{(1 - <f^*>)}{(<f_j> + <\nu> - 1)} + \frac{<R_j(\bar{C}_u)_j f^*_j>}{<(\bar{C}_u)_j)>} \tag{2}$$

for the flux of suspended sediment through section j. In equations (1) and (2) $<\nu>$ is the fraction of the downstream flux of salt $<(Q_u)_j(\bar{S}_u)_j>$ that is balanced by diffusive processes, R_j is the riverflow associated with the jth section, and $(\bar{C}_u)_j$ is the concentration of suspended sediment in the upper layer. In computing the fluxes of suspended sediment from equations (1) and (2) it was assumed that our suite of vertically averaged values of $(\bar{S}_u)_j$, $(\bar{C}_u)_j$, $(\bar{S}_\ell)_j$, and $(\bar{C}_\ell)_j$ are good estimates of the monthly average values. The upper layer was defined as the layer above the maximum vertical salinity gradient; observations in the lower one-third of the upper layer and in the upper one-third of the lower layer were neglected in computing the layer averages. This weighting procedure has been recommended by D.W. Pritchard (unpublished manuscript).

Results of these calculations of suspended sediment fluxes are summarized in Table 1 for a reasonable range of $<\nu>$. We estimate that $<\nu>$ probably lies between 0 and 0.1. It should be noted that our definition of ν differs from that of Hansen and Rattray (6); it is smaller by approximately the factor R/Q_u.

Sediment fluxes through sections separating the Bay and four of its major tributary estuaries are summarized in Table 1. The exchanges with the Eastern Shore tributaries south of 38°15'N have been ignored because of the small flows associated with these rivers. The exchanges with all tributary estuaries north of 38°15'N were estimated as a cumulative flux. The average total fresh-water entering the sides of the Bay between a cross-section just south of the mouth of the Susquehanna and a cross-section at 38°15'N was only about 159 m^3/sec during the study period (28), and we had no way of readily and accurately apportioning this cumulative input among the various tributary rivers. The cumulative flow of 159 m^3/sec is approximately equivalent to the combined flows of the York and the Rappahannock Rivers, and somewhat less than that of the Potomac. In the absence of any other information, we estimated a net flux of 0.03×10^6 metric tons of suspended sediment from the Bay to these tributaries–a value somewhat less than the flux to the Potomac estuary, and somewhat less than the combined flux of suspended sediment from the Bay to the York and Rappahannock estuaries.

Table 1. Suspended sediment fluxes in metric tons per year. Values <0 indicate fluxes from the Bay.

Section	$\nu = 0$	$\nu = 0.1$	$\nu = 0.2$
Ocean	$+0.47 \times 10^6$	$+0.22 \times 10^6$	$+0.11 \times 10^6$
James R.	-0.11×10^6	-0.05×10^6	-0.02×10^6
York R.	-0.03×10^6	-0.02×10^6	$<-0.01 \times 10^6$
Rappahannock R.	-0.05×10^6	-0.02×10^6	-0.01×10^6
Potomac R.	-0.07×10^6	-0.04×10^6	-0.03×10^6
All others	----	-0.02 to 0.04×10^6	----

Table 1 summarizes the mean annual suspended sediment fluxes between the Bay and its tributary estuaries, and between the Bay and the ocean. The model indicates that the Bay is a sink for suspended sediment in the ocean and a source of suspended sediment to each of the major tributaries. To formulate a budget for the inorganic suspended sediment of the Bay proper, the other sources—input from the Susquehanna and from shore erosion—must be known. The input of suspended sediment from the Susquehanna during the twelve-month study period was about 1.07×10^6 metric tons. This is based on nearly daily determinations of the concentration of suspended sediment, and on daily average riverflow at the Conowingo Hydroelectric Plant. The annual average input from shore erosion is less well known.

Schubel (11) estimated that the main body of the Bay between the mouth of the Susquehanna and Station 913R receives about 0.12×10^6 metric tons of silt and clay each year from shore erosion of that segment of the Bay. Biggs (1) made a similar calculation for the Bay from about Station 909 to the mouth of the Potomac and estimated that the input of silt and clay to that segment of the Bay from shore erosion was about 0.28×10^6 tons/year. A very crude estimate—a Mark Twain estimate[1]—for the remainder of the Bay indicates an input of about 0.20×10^6 tons of silt and clay/year. According to these data the main body of the Bay receives about 0.60×10^6 tons of silt and clay each year from shore erosion.

Using the data from Table 1 and the inputs of suspended sediment from shore erosion and from the Susquehanna one can formulate a budget to calculate the mean annual loss to the bottom through deposition, Table 2. The value obtained is 1.73×10^6 tons/year. Using an in-place density of 1.25 tons/m^3 this gives a mean Bay-wide sedimentation rate of fine-grained inorganic sediment of about 0.8 mm/year. The sedimentation rate is, of course, not uniform.

[1] "There is something fascinating about science. One gets such wholesome returns of conjecture out of such a trifling investment of fact." Mark Twain, Life on the Mississippi.

Table 2. Suspended sediment budget.

Sources	Mass/yr
Susquehanna R.	1.07×10^6 tons
Shore Erosion	0.60×10^6
Ocean	0.22×10^6
TOTAL	1.89×10^6 tons
Sinks	**Mass/yr**
Tributaries[1]	0.16×10^6 tons
Deposition	1.73×10^6
TOTAL	1.89×10^6 tons

[1] Assuming $\langle\nu\rangle = 0.1$. See text for discussion.

Schubel (13, 12, 17) estimated that the average sedimentation rate in the segment of the Bay from Station 913R to 927SS is about 3-4 mm/year, and Biggs (1) estimated the sedimentation rate in the same segment of the Bay at 3.7 mm/year. For the segment of the Bay from about station 909 to 818P, Biggs (1) estimated an average sedimentation rate of about 1.1 mm/year. Biggs' estimate is based on a simple advective model that did not consider exchange with the tributaries. The Bay-wide average sedimentation rate of about 0.8 mm/year indicated by our simple model appears reasonable.

The most significant features of the Bay's suspended sediment system revealed by our simple model are that the major tributaries are sinks for suspended sediment in the Bay, and that the ocean is a source of suspended sediment to the Bay. A number of investigators (9, 15) have shown that there is an upstream flow of sediment into the Bay in the lower layer from the ocean, but there has been no previous attempt to demonstrate quantitatively whether the net transport of sediment through the mouth of the Bay is landward or seaward.

Meade (9) pointed out in a discussion of the estuaries of the middle Atlantic coast of the United States that "Although we have information on many aspects of the sedimentary processes and products in this area–more than in any other coastal area of the world–we are unable to answer the question posed at the beginning of this discussion." That question was "On a net basis, does more sediment move out of estuaries onto the continental shelf than moves into the estuaries from the shelf?"

The degree of quality obtained with our simple suspended sediment model is of the same order as that of the salt and water budgets. At this time this is not only acceptable, but it represents a significant improvement in our understanding of Chesapeake Bay sedimentation processes. The next logical step is to produce a similar model with three segments to estimate the partitioning of sediment accumulation within the Bay. Construction of that model is underway.

ACKNOWLEDGMENTS

Preparation of this report was supported by the Marine Sciences Research Center of the State University of New York. The data used in this study were collected under a program sponsored by the former Department of Game and Inland Fish, State of Maryland, and the Bureau of Commercial Fisheries, Department of the Interior through a jointly funded project under Public Law 89-304. We are indebted to Akira Okubo for his advice, and for his absolution for our sin of attributing all our ignorance to diffusion.

LITERATURE CITED

1. Biggs, R.B. 1970. Sources and distribution of suspended sediment in northern Chesapeake Bay. Marine Geol. 9:187-201.
2. Bond, G.C., and R.H. Meade. 1966. Size distributions of mineral grains suspended in Chesapeake Bay and nearby coastal waters. Chesapeake Sci. 7:208-212.
3. Burt, W.V. 1955. Distribution of suspended materials in Chesapeake Bay. Jour. Mar. Res. 14:47-62.
4. Burt, W.V. 1955. Interpretation of spectrophotometer readings on Chesapeake Bay waters. Jour. Mar. Res. 14:33-46.
5. Folger, D.W. 1972. Characteristics of estuarine sediments of the United States. U.S. Geol. Survey Prof. Paper 742. 94 p.
6. Hansen, D.V., and M. Rattray, Jr. 1966. New dimensions in estuary classification. Limnol. Oceanogr. 11(3):319-326.
7. Hunter, J.R. 1975. A three-dimensional kinematic model of suspended sediment transport in the upper Chesapeake Bay. Special Rept. 46, Chesapeake Bay Institute of The Johns Hopkins University, Balto., Md. 72 p.
8. Meade, R.H. 1969. Landward transport of bottom sediments in estuaries of the Atlantic Coastal Plain. Jour. Sed. Petrol. 39:222-234.
9. ———, 1973. Net transport of sediment through the mouths of estuaries: landward or seaward? p. 207-210. *In* International Symposium on Interrelationships of Estuarines and Continental Shelf Sedimentation, Institut de Geologie du Bassin d'Aquitaine, Bordeaux, France.
10. Ryan, J.D. 1953. The sediments of Chesapeake Bay. Maryland Dept. of Geology, Mines and Water Resources Bull. 12. 120 p.
11. Schubel, J.R. 1968. Shore erosion of the northern Chesapeake Bay. Shore and Beach 36(1):22-26.
12. ———. 1968. Suspended sediment discharge of the Susquehanna River at Havre de Grace, Maryland, during the period 1 April 1966 through 31 March 1967. Chesapeake Sci. 9:131-135.
13. ———. 1968. Suspended sediment of the northern Chesapeake Bay. Tech. Rep. 35, Chesapeake Bay Institute of The Johns Hopkins University, Balto., Md. 264 p.
14. ———. 1968. Turbidity maximum of northern Chesapeake Bay. Science 161:1013-1015.
15. ———, (ed.). 1971. The Estuarine Environment, Estuaries and Estuarine Sedimentation. Amer. Geol. Inst., Washington, D.C. Short course lecture notes. 324 p. (not numbered consecutively).
16. ———. 1972. Distribution and transportation of suspended sediment in upper Chesapeake Bay, p. 151-167. *In* B. W. Nelson, ed. Environmental

Framework of Coastal Plain Estuaries. Geol. Soc. Amer. Memoir 133. 619 p.

17. ———. 1972. Suspended sediment discharge of the Susquehanna River at Conowingo, Maryland, during 1969. Chesapeake Sci. 13:53-58.
18. ———. 1972. The physical and chemical conditions of the Chesapeake Bay. Jour. Wash. Acad. Sci. 62:56-87.
19. ———. 1974. Effects of Agnes on the suspended sediment of the Chesapeake Bay and contiguous shelf waters, p. B1-B26. *In* J. Davis, ed. Report on the Effects of Tropical Storm Agnes on the Chesapeake Bay Estuarine System. Publ. No. 34 of the Chesapeake Research Consortium. A report to the U.S. Army Corps of Engineers, Baltimore District.
20. ———. 1974. Effects of Tropical Storm Agnes on the suspended solids of the northern Chesapeake Bay, p. 113-132. *In* R.J. Gibbs, (ed.) Suspended Solids in Water. Plenum Press, N.Y. 320 p.
21. ———. 1975. Suspended sediment in Chesapeake Bay. Civil Engineering in the Oceans III, Proceedings Amer. Soc. of Civil Engineers Specialty Conference held at Newark, Del. on June 9-12, 1975 (in press).
22. ———. (in preparation). The Smith-Tangier Islands sand deposit—a relict feature.
23. ———, and R.B. Biggs. 1969. Distribution of seston in upper Chesapeake Bay. Chesapeake Sci. 10:18-23.
24. ———, and W.B. Cronin. 1975. Suspended sediment in the waters of upper Chesapeake Bay, p. 4-34 to 4-63. *In* T.O. Munson and D.K. Ela, (eds.) A report to the Maryland Department of Natural Resources by Westinghouse Ocean Research Laboratory, Annapolis, Md.
25. ———, C.H. Morrow, W.B. Cronin, and A. Mason. 1970. Suspended sediment data summary, August 1969-July 1970 (Mouth of Bay to Head of Bay). Spec. Rep. 18, Chesapeake Bay Institute of The Johns Hopkins University. 39 p.
26. ———, ———, ———, and ———. 1970. Suspended sediment data summary 24 February 1969-May 1969, upper Chesapeake Bay (Matapeake to Havre de Grace). Special Report 17 of the Chesapeake Bay Institute of The Johns Hopkins University, Balto., Md. 44 p.
27. Wolman, M.G. 1968. The Chesapeake Bay: geology and geography, p. II-7 to II-48. *In* Proceedings of the Governor's Conference on Chesapeake Bay, held at Wye Institute, Maryland, on Sept. 12-13, 1968.
28. United States Geological Survey, August 1970. Estimated stream discharge entering Chesapeake Bay. District Chief, U.S. Geological Survey, 8809 Satyr Hill Road, Parkville, Maryland 21234.

TRANSPORT AND DEPOSITION OF SUSPENDED SEDIMENT IN THE GIRONDE ESTUARY, FRANCE

G.P. Allen
Centre Océanologique
de Bretagne – CNEXO
B.P. 337
29273 BREST CEDEX
France

G. Sauzay
S.A.P.R.A.
Centre d'Etudes
Nucléaires – C.E.A.
B.P. 2
91190 GIF-SUR-YVETTE
France

P. Castaing
J.M. Jouanneau
Institut de Géologie du
Bassin d'Aquitaine
351, Cours de la Libération
33405 TALENCE,
France

ABSTRACT: A highly concentrated (1-10 g/1) turbidity maximum develops at the upstream limit of the salinity intrusion in the Gironde estuary, nourished by the large seasonal influx of alluvial suspended sediment.

The estuary contains two distinct channel systems. In the deeper southern channel, a marked tidal cycle occurs: during neap tides, fluid mud accumulates from settling in the core of the turbidity maximum; during spring tides it is eroded and resuspended. The maximum and fluid mud undergo a seasonal upstream-downstream migration in response to varying river discharge.

These phenomena have been studied using hydrological and radioactive tracer techniques. Very little movement and diffusion occurs within the fluid mud. At each cycle of accumulation and resuspension, a residual lamination is deposited in the channel. During high river flow the resuspended mud is transported by lateral advection to the north channel, where part settles out, and part is evacuated out to sea. This lateral migration and seaward escape appears to be amplified by dredging.

Sediment renewal in the fluid mud-turbidity maximum system appears to be related to its position in the estuary; downstream, lateral losses induce a rapid turnover; upstream, it behaves more like a closed system, with internal recycling.

INTRODUCTION

This paper summarizes the essential hydrological and sedimentological processes acting in the Gironde estuary, as deduced by a number of workers during the past decade (1, 2). These studies were based on more than 50 tidal stations of current velocity and suspended sediment measurement at various points in the estuary and for different tidal amplitudes and conditions of river flow. In addition, two tracer experiments were conducted on the suspended sediments in order to analyze their large scale circulation patterns within the estuary.

The Gironde is located on the south west coast of France (Fig. 1) and is formed by the junction of the Garonne and Dordogne rivers. Areally, it is the largest estuary in France (625 Km^2 at high tide), and forms a regularly flaring indentation in the coast, reaching 100 km inland. The Garonne and Dordogne drain respectively the western Pyrenees and the Massif Central. The total drainage basin has an extent of 71.000 km^2, making it the 3rd largest drainage basin of the country, after the Loire and the Rhone. The geometry of the estuary is extremely smooth and the major morphological features (cross section, width, mean depth, etc) decrease exponentially upstream.

The upper estuary, extending from the Bec d'Ambes to approximately km 50 (distance measured downstream from Bordeaux (Fig. 1)), is characterized by numerous bars, shoals, and islands. The 7 to 10 m deep (below LLW) navigation channel follows the left bank (Fig. 1). The lower estuary (km 50 to 95) exhibits a simpler morphology consisting of two channel systems separated by a succession of bars and a submerged dike. The navigation, or southern channel follows the southern shore. Its depths vary between 6 and 8 m. A break in slope occurs at km 80. Downstream, the depths increase and attain 30 m at the mouth. The north channel is wider and shallower (3-5 m) than the navigation channel, and is also marked by a slope break at km 80 where the depths increase and attain 30 m in the mouth, where the two channels merge.

The pattern of sediment distribution in the Upper and Lower Estuary consists of silt and clay in the channels, medium sand on the shoals and bars, and silty clay on the tidal flats along the bank (Fig. 2).

In the downstream extremity of the Lower Estuary, the channel sediments consist entirely of medium to coarse sand. This sand-mud transition is located between km 75 and 80 (Fig. 2).

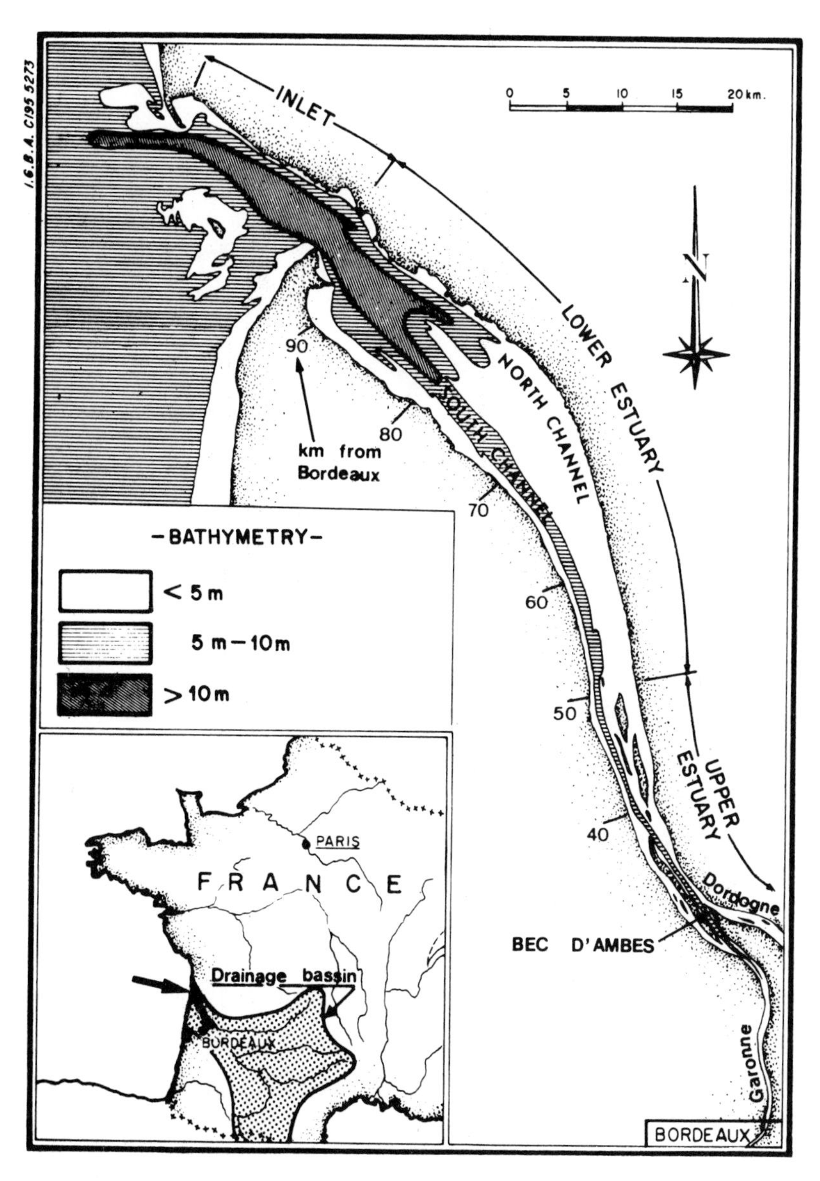

Figure 1. Location and bathymetry of the Gironde estuary.

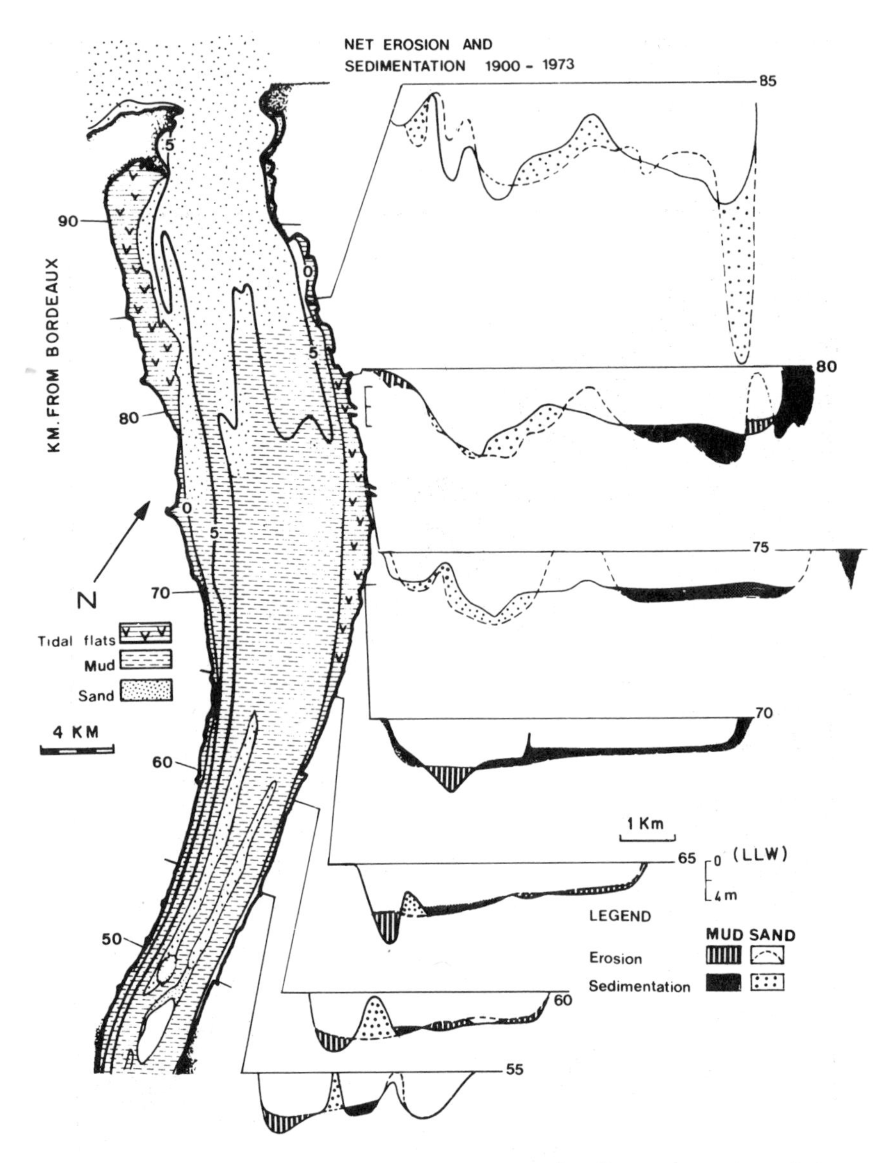

Figure 2. Bottom sediment distribution and erosion and sedimentation patterns between 1900 and 1973.

INPUTS INTO THE ESTUARINE SYSTEM

River Flow

The average yearly combined discharge of the Garonne and Dordogne varies between 500 and 1.000 m^3/sec with a mean of 725 m^3/sec. The highest monthly flow occurs in January (1.450 m^3/sec) and the lowest in August (235 m^3/sec). During river floods, which occur between January and April, the instantaneous discharge can often exceed 3.000 m^3/sec. Generally, two fluvial seasons exist: a period of high discharge in winter and spring, and low discharge during summer and autumn.

Sediment Influx

All of the silt and clay presently accumulating in the estuary are supplied by the Garonne and Dordogne rivers. This influx of suspended sediment has been evaluated by the Port Autonome de Bordeaux at 2.2×10^6 tons/year. Alluvial sediment transport is closely related to discharge, and most of the influx occurs during the period of high river flow, between January and May.

Tides

The tides in the Bay of Biscay are semi-diurnal, with a 12 h 25 min period. The tidal amplitudes in the inlet vary from 1.55 m during neap tides to 5 m during spring tides. The tidal range is amplified in the upper estuary, where it can attain 6 m during equinox springs. Time-height curves are symmetrical in the inlet, but a marked assymetry develops upstream, prolongating the ebb and reducing the flood. The tidal prism, or volume of water introduced at the flood tide varies from 1.1×10^9 m^3 in neap tides to more than 2×10^9 m^3 in spring tides. This volume is of the same order of magnitude as the total subtidal volume of the estuary, which is about 2.2×10^9 m^3.

Tidal currents vary with bottom morphology, river flow and tidal range. Generally the currents are strongest in the channels, attaining 3 m/sec on the surface during spring tides, and 1.5 m/sec at one meter from the bottom. During low river flow, the head of tides occurs at 165 km from the Inlet; during high river discharge, this distance decreases by more than 50 km.

SEDIMENTATION PATTERNS

In order to evaluate the long term evolutive sedimentation trends, and to compare natural and man-influenced processes, bathymetric data in both the north and south channels for the period 1677 to 1973 were studied. This period was seperated into two parts: 1677-1853, and 1853-1973, since the earlier of these two periods is indicative of natural processes, although it includes indirect human influences such as deforestration. The more recent period reflects the results of man's direct interference in the estuarine system, since from 1850 on, the south channel has been artifically deepened and maintained by dredging and improvement works.

Fig. 3 shows the results of this study, expressed as the rate of change of the maximum channel depths as function of position in the estuary. Between 1677 and 1853, there existed a zone of general shoaling in the lower estuary between km 55 and km 75 (downstream from Bordeaux). Although shoaling occurred in both channels, the rate appears to have been much higher in the south channel than in the north channel. This zone corresponds to the position of the turbidity maximum during periods of high river discharge, and the muddy lithology of this zones indicates that shoaling resulted from the accumulation of suspended sediment. Upstream of km 55, the channels were deepening. This erosion can probably be associated with the downstream accretion of sand bars, resulting in channel deepening so as to equilibrate cross sectional areas.

Between 1853 and 1973, the pattern of erosion and sedimentation underwent a complete change. Artificial deepening, and subsequent maintenance of the south channel by dredging, created a strong erosive trend in this entire channel. In the northern channel, the previous high sedimentation zone between km 55 and 70 became a zone of low shoaling, while upstream and downstream the

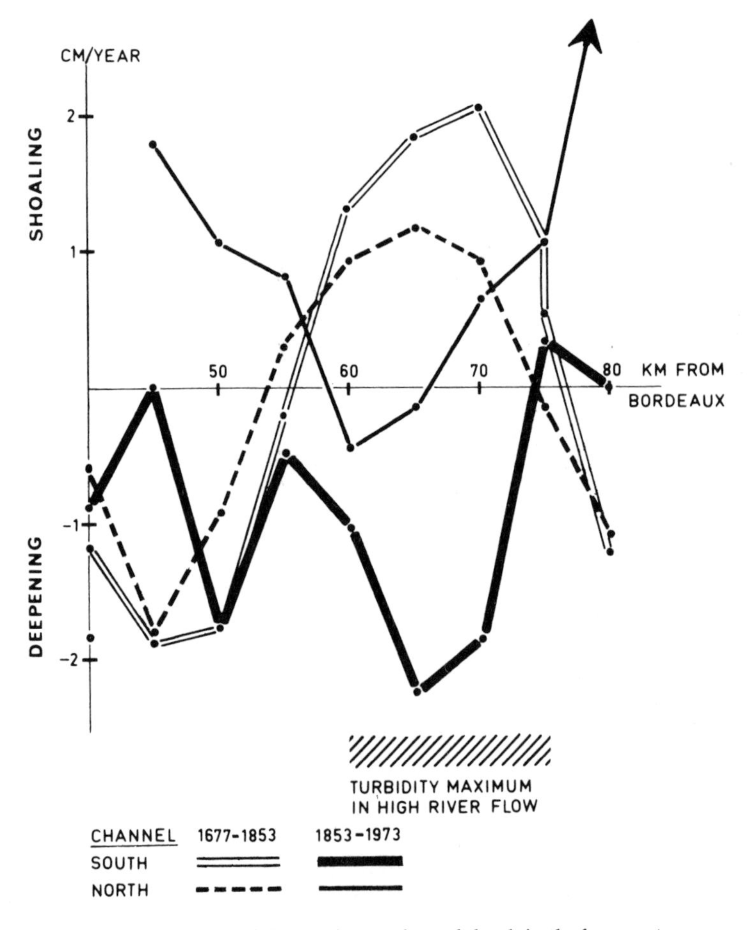

Figure 3. Time rate of change of the maximum channel depth in the lower estuary.

shoaling rates greatly increased, attaining values greater than 2 cm/year in the lower end of the channel.

The present sedimentation patterns in the Gironde are shown in Fig. 2. This pattern consists of continued erosion in the south channel due to dredging; extensive sedimentation of silt and clay in the lower north channel, attaining rates of 5 cm/year, and influx of alluvial sand in the upper estuary, shown by the downstream progradation of the sand bar systems.

A study of the sedimentation and erosion volumes in the estuary between 1900 and 1973, in connection with the superficial sediment patterns and core lithologies, indicates that the mean yearly sedimentation of silt and clay is equivalent to about 1.6×10^6 tons. This estimate, based on scattered borings and fragmentary compaction data, and hence a preliminary evaluation probably subject to some error, indicates that approximately 70% of the yearly river suspended influx is trapped in the estuary, and the remainder (about 6×10^5 tons per year) is evacuated to sea.

ESTUARINE HYDROLOGY

Salinity intrusion

Seasonal variations in river flow cause marked changes in the salinity intrusion and resulting vertical stratification. The estuary varies from type B [salt wedge with tides, Pritchard (11)] during very high river flow to type C (partially mixed) during low river discharge. The upstream limit of the mean position of the salinity intrusion is located about 75 km from the Inlet during low river flow and 40 km from the Inlet during peak fresh water discharges. As in most estuaries of the northern hemisphere, a distinct lateral salinity gradient exists in the lower estuary, with higher salinity values along the south bank.

Water circulation

Schematically, the residual or non-tidal circulation in the salinity intrusion consists, as in most other partially mixed estuaries, of landward flow on the bottom and seaward flow toward the surface. A zone of convergence of the bottom residual flow field exists at the upstream limit of the salinity intrusion.

In the lower estuary, however, the two main channel systems behave differently with respect to bottom density flow. The south channel generally exhibits flow stratification, while in the north channel the bottom residual flow is oriented seaward during high river discharge, and landward during low river discharge. The different flow characteristics of these two channels is probably due to the Coriolis effect (11) and the lateral acceleration of ebb flow due to the lateral curvature of the estuary. These cause the north channel to be the major outlet for river water, and the south channel to be the major inlet for marine saline water. Also, the greater depth further upstream in the south channel (artificially maintained by dredging) induces a more pronounced salinity intrusion in this channel, resulting in an enhanced vertical density stratification.

The bottom residual flow field can be locally extremely complicated due to fragmentation in zones of complex topography, which create successive zones of convergence and divergence.

Variations in the intensity of the residual flow occur with changes in river discharge. An increase of the ratio between the fresh water influx and the tidal prism will increase the vertical density gradients and amplify the residual flow. In the south channel of the Lower estuary, a direct correlation exists between the near bottom upstream oriented residual current and river discharge. Furthermore, changes in density gradients also occur during the fortnightly neap-spring tidal range cycle, approaching in amplitude the changes brought about by varying river flow. An increase in tidal range reduces the ratio of fresh to salt water, amplifies turbulence and mixing, and acts in the same way as a decrease in river discharge, bringing about a less pronounced vertical salinity gradient.

This change in vertical stratification appears to bring about a corresponding change in the bottom residual flow. During decreasing tidal ranges, the bottom landward flow is intensified, strengthening the convergence at the limit of the salt intrusion. This phenomenom is reflected in the assymetry of the bottom velocity-time curves, and during neap tides this assymetry is more pronounced, resulting in a longer flood slack. At times of strong vertical density gradients, such as during high river discharge and neap tides, very marked assymetries can develop, and the flood slack on the bottom can last for several hours, forming the "lentille immobile" (motionless lens effect), discovered by Berthois (3), in the Loire estuary.

SUSPENDED SEDIMENT TRANSPORT AND ACCUMULATION

The turbidity maximum and fluid mud system

The convergence of bottom residual density flows creates a trap for suspended sediments which accumulate in the form of a "turbidity maximum" at the limit of the salinity intrusion. This pehnomenon, which controls the sedimentation of suspended sediment in estuaries, was first studied in the Gironde by Glangeaud (5). Estuaries with small tidal ranges and relatively low sediment influx, such as those of the eastern coast of the US, show relatively short lived turbidity maxima, forming in conjunction with periods of high river flow and sediment influx. After a "rebound cycle", described by Nichols (8) the maximum tends to dissipate by settling of the suspensions. In estuaries such as the Gironde, where tidal action is strong and there is a relatively large influx of suspended sediment, the turbidity maximum is a permanent feature, and its density and size varies with tidal conditions. Sedimentation in such an estuary does not occur in short pulses linked to the yearly "rare event" of river floods, but is a continuous phenomenon, modulated by the superposition of the semidiurnal and fortnightly tidal cycles.

The turbidity maximum of the Gironde consists of a large zone, 20-60 km long, where the concentrations of suspended sediment are considerably higher than in the adjacent fluvial or marine waters. These concentrations vary between 0.2 and 1 g/l on the surface and 1 to 10 g/l on the bottom. In the core of the maximum there is a pronounced downward gradient of increasing sediment concentrations; near the bottom, these can attain extremely high values giving rise to the phenomenon of "fluid mud", similar to that described by Inglis and Allen (6) in the Thames. This fluid mud, which accumulates as distinct pools on the bottom forms only in the south channel and contains sediment concentrations ranging up to 250 g/l (1). The fluid mud is genetically linked to the turbidity maximum, with which it forms a complex sedimentological system, continuously evolving in response to varying tidal conditions.

Tidal variations

During the semi-diurnal tidal cycle, an important oscillation of the estuarine water masses occurs and the tidal excursion can reach 10 to 20 km. This tidal oscillation brings about a similar to and fro movement of the turbidity maximum. In addition to this tidal translation, a semi-diurnal cycle of growth and decay of the maximum occurs as a result of bottom scour and sedimentation. An example of this cycle is shown in Fig. 4, which represents data from 4 synoptic measuring stations in the south channel during a spring tide and high river flow. At each station was measured current velocity, salinity and suspended sediment at 5 different depths.

At the beginning of ebb tide, the core of the turbidity maximum, with concentrations >1 g/l is reduced to a small zone centered on a channel depression between km 50 and km 55. During the ebb, the bottom currents accelerate and attain velocities in excess of 1.5 m/sec at one meter from the bottom. These strong currents erode the bottom mud, causing the turbidity maximum to grow, reaching peak density and extent at low tide. During the low tide slack, part of the suspended mud settles and the turbidity maximum decays. Once the flood currents are strong enough, bottom erosion once again nourishes the maximum which grows in concentration and volume, reaching a maximum an hour before high tide. At the high tide slack, deposition occurs, with a corresponding decay in the turbidity maximum.

If the tidal velocity-time curve were symmetrical, at any point on the bottom an equal amount of sediment would be deposited and eroded during the ebb and flood of each semi-diurnal cycle. However, the assymetry of the velocity-time curves induces a corresponding assymetry in the erosion and deposition cycle (9, 10). Near the bottom, the flood currents dominate over the ebb both in strength and duration, and more erosion and resuspension occurs during flood tide than during the ebb. Furthermore, the longer duration of bottom flood slack favours sedimentation at the end of the flood tide. This appears to be the case, as shown in Fig. 4 where the turbidity maximum appears to be more depleted at flood slack than at ebb slack. The overall effect is to preferentially erode sediments during the flood tide, and deposit them during flood slack, and therefore bring about an upstream movement of sediment.

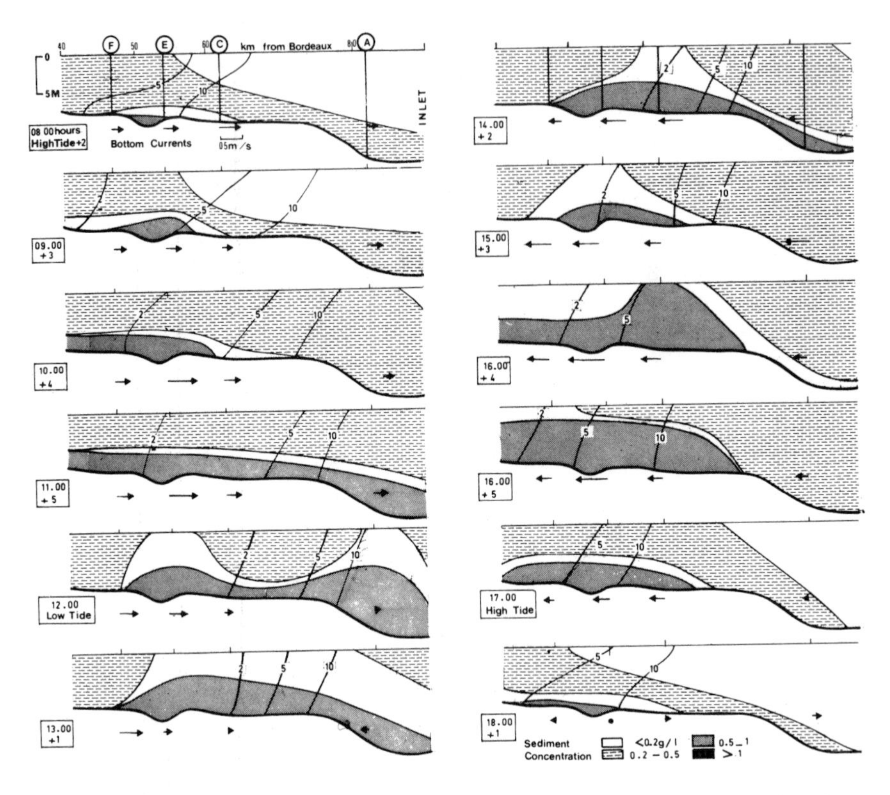

Figure 4. Longitudinal evolution of the turbidity maximum in the south channel during a semi-diurnal tidal cycle; spring tide and high river flow (May 1974).

The amount of sediment eroded at each tide, as well as the ratio at any point between total sedimentation and erosion during the semi-diurnal cycle, varies considerably during the fortnightly cycle of tidal amplitudes. These variations are cyclic and are brought about by the changing current velocities during the fortnightly cycle, as well as the variation in vertical density gradients and accompanying bottom time-velocity assymetry, as previously outlined. With increase in tidal range, current velocities increase, augmenting the erosion and resuspension occurring at each tide. This increases the concentration of the turbidity maximum. During decreasing tidal ranges, the average suspension concentrations decrease, attaining a minimum during neap tides.

Increasing current velocity-time assymetry, brought about by decreasing tidal range, with corresponding longer flood slacks, will augment the sedimentation potential during neap tides. The ratio of the duration of bottom scour to duration of complete settling is directly proportional to tidal amplitude, and a cycle of net erosion and sedimentation occurs during the neap-spring cycle. In a more general sense, during a neap-spring cycle the estuary varies from an accumulative, or entrapment mode, to a dissipative mode, with respect to suspended sediments.

During decreasing tidal ranges, the increasing rate of sedimentation in the core of the turbidity maximum due to increased duration of bottom flood slacks causes the formation of pools of fluid mud. This fluid mud undergoes an accumulation-erosion cycle in response to changes in tidal amplitudes (1, 2). During neap tides, the pools attain their maximum extent, and during spring tides appear to be totally eroded and resuspended into the turbidity maximum (Fig. 5). Fluid mud accumulates only in the southern channel, where the core of

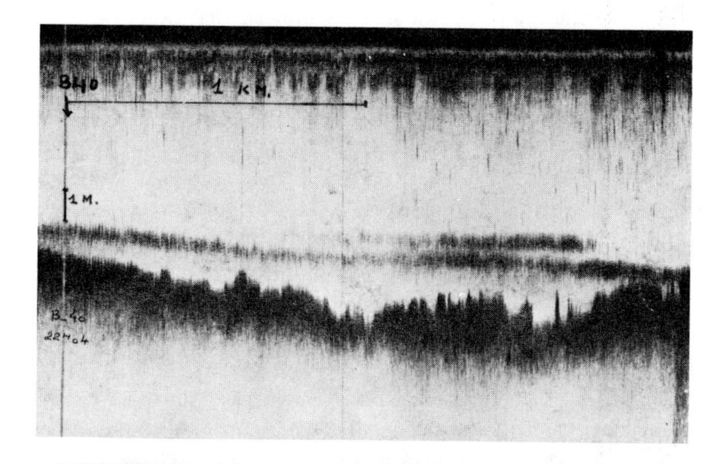

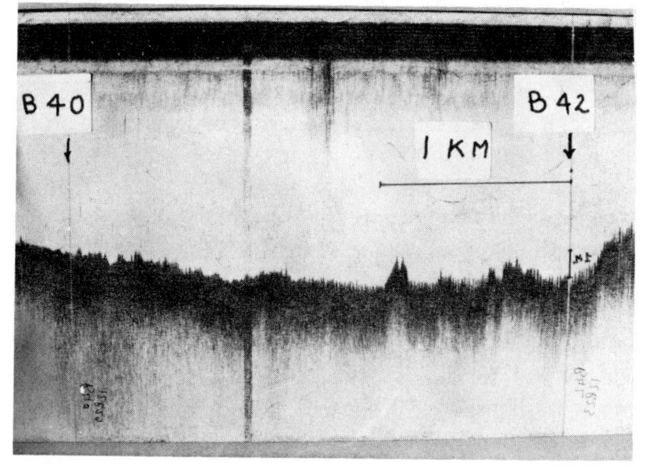

Figure 5. Depression in the south channel of the lower estuary, showing the accumulation of fluid mud during neap tide (top photo) and its resuspension a few days later during spring tides (bottom photo).

the turbidity maximum tends to be most dense, and the bottom tide velocity assymetries and corresponding slack periods the most pronounced. Although no fluid mud occurs in the north channel, sedimentation there follows this same neap-spring cycle.

In order to study the resuspension and dispersion of the fluid mud in the turbidity maximum, a tracer study was made using sediments labelled with radioactive Scandium 46. Five kilos of labelled, naturally occuring fluid mud, were injected into a fluid mud pool in the south channel during a neap tide in May 1974. At this time the river flow was transitional between high and low discharges, and the turbidity maximum was located in the lower estuary. During the few days following the injection, the increasing tidal ranges were insufficient to erode the fluid mud, and in situ surveys showed that practically no internal movement or exchanges with the overlying turbidity maximum occured within the fluid mud. Nine days after the injection, echo-sounder transects showed that the fluid mud had been totally eroded and resuspended. However, bottom surveys showed that approximately 20% of the total injected activity remained on the bottom, indicating that a fraction of the fluid mud had been sufficiently compacted during neap tides so as to resist erosion during the following spring tides. At the same time, several simultaneous sampling stations were made in both channels over several tidal periods to monitor the transport of the tagged sediment which was eroded and suspended into the turbidity maximum. These indicated that 9 days after the injection in the south channel, 50% of the activity which was eroded and suspended had been transported to the north channel. This lateral transfer of sediment occurs by advective transport rather than by diffusion, since the tagged sediment occured as discontinous pulses of suspended sediment, formed during periods of bottom scour. These discrete pulses appeared to retain their identity for at least one tidal cycle, indicating that the rate of dispersion and mixing of suspended sediment in the turbidity maximum is very low ($D_x \sim 0.1 \ m^2/s$). Direct lateral advection through connecting channels in the shoals separating the main channels (Fig. 6) and a possible northward drift of the surface ebb-dominant water over a tidal period, rather than diffusive spreading, appear to be the mechanisms responsible for this lateral transport.

The vertical gradients of the concentration of tagged mud also indicates the lack of rapid vertical mixing of sediment in the turbidity maximum. In the south channel, the concentration of tracer exhibits a downward increasing gradient (Fig. 6), whereas in the north channel the concentrations decrease downward. This is probably due to the bathymetry of the connecting channels which are relatively shallow and form sills between the main channels. The suspended sediments which are transported from the south to the north channel consist of those occuring in the upper part of the water column in the south channel and hence an inverse activity gradient occurs in the north channel. This gradient indicates that the rate of vertical mixing of sediment in the turbidity maximum over a semi-diurnal cycle is low.

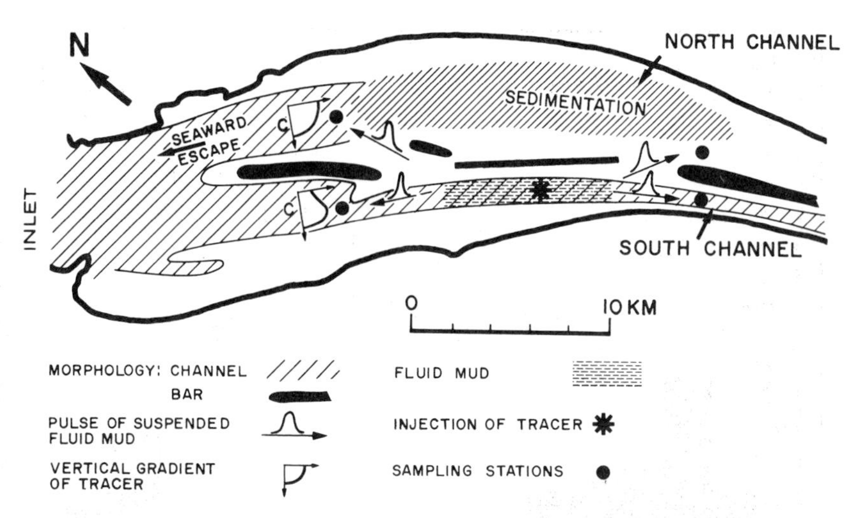

Figure 6. Schematic diagram of resuspension and lateral transfer of tagged mud.

The relative dilution of the tagged sediment was lower in the north channel than in the south. This indicates that resuspended fluid mud, originating in the south channel, forms a proportionally greater fraction of the turbidity maximum in the north channel than in the south channel. This suggests that suspended sediment in the Gironde undergoes a net lateral northward drift from the south channel, where it is eroded from the bottom or brought in by the river influx, to the north channel, where it is either evacuated to sea, or deposited. This lateral flux is probably enhanced by dredging in the south channel, which artificially impedes sedimentation and thereby keeps sediments in suspension, since little dredged sediment is removed from the estuary.

Seasonal Cycles

The turbidity maximum-fluid mud system migrates longitudinally in the estuary in response to changing river flow and salt intrusion (Fig. 7). During periods of low river discharge (May to November) the maximum migrates upstream, between Bordeaux and the Bec d'Ambes, and during high inflow (December to May) is located downstream, between km 60 and 75. The transition periods during which the maximum migrates are relatively short, lasting on the order of several weeks. This seasonal shifting of the maximum is responsible for the seasonal shoaling pattern in the south channel, discovered and studied by Migniot (7), and consisting of sedimentation in the upper estuary in summer and autumn and in the lower estuary in winter and spring.

The tracer experiment permitted the analysis of the upstream movement of the turbidity maximum-fluid mud system during a period of decreasing river

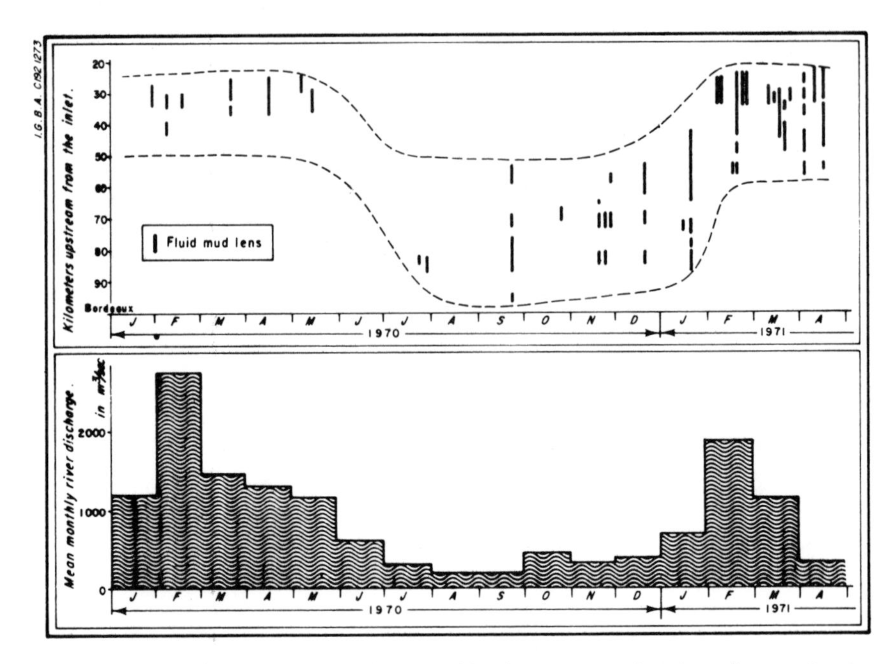

Figure 7. Longitudinal migration of fluid mud in the estuary as function of seasonal variation of river flow.

flow. During the 4 months following the initial injection of tracer, the turbidity maximum and fluid mud pools migrated upstream in response to the reduced river inflow. At each neap tide, when the fluid mud pools had reformed in the core of the turbidity maximum, fluid mud samples were taken and analysed for Scandium concentration. Fig. 8 shows the time variation of the average concentration of tracer in the fluid mud, corrected for radioactive decay. This curve shows the progressive dilution of the tracer, indicating a progressive renewal of the sediment in the fluid mud and turbidity maximum as the system works its way upstream. However, at the beginning of July, when the system reached its upstream position (between Bordeaux and the Bec d'Ambes) an important increase in the mean concentrations occurred. This peak, representing an average of 10 samples, can only be explained by an addition of high tracer concentration sediment to the system due to a rapid initial upstream movement of tagged sediment shortly after injection in the lower estuary, and admixing with the main body of the turbidity maximum which arrived more than a month later. This phenomenom is still unclear, and might represent a rapid upstream movement of undiluted pulses of suspended sediment, perhaps in the north channel and through the complex of islands and bars in the upper estuary.

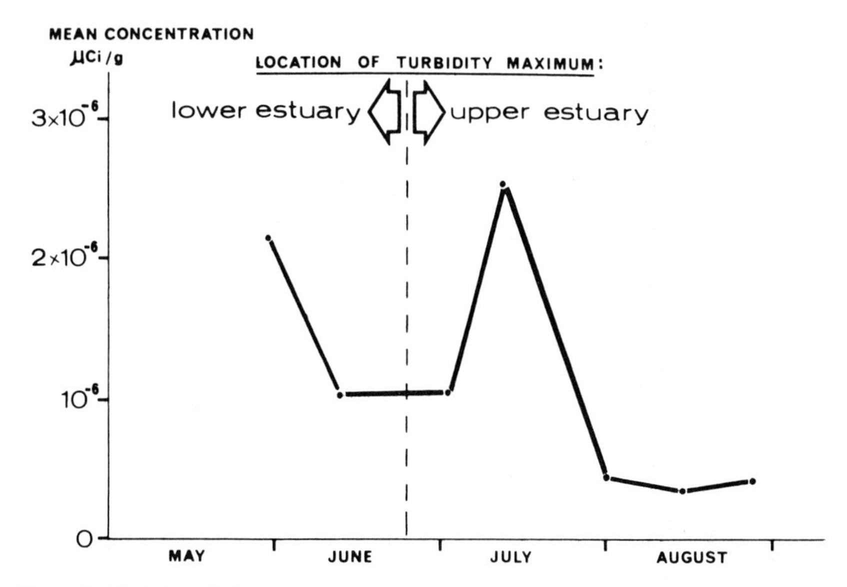

Figure 8. Variation of the mean concentration of radioactive tracer in the fluid mud.

General Sedimentological Model

This model, based on the tracer study, schematises the sediment transfer and renewal rates between the two channel systems. It is presented here in a preliminary form, and assumes a bipolar system, consisting of the upper estuary, where the maximum is located during low river flow, and the lower estuary, when it occurs during high river flow. The flux and transfer rates are established with respect to the south channel and the term "lateral transfer" includes all sediment escaping that channel.

In view of the nature of the data furnished by the tracer study, the total sediment budget in the estuary can be established by the following relation:

$$R + E = Mk(I - \alpha)$$

where:
- R = river influx
- E = net sediment exchange with the bottom in the south channel
- M = total mass of sediment in the fluid mud and turbidity maximum in the south channel
- k = the lateral transfer rate to the north channel
- α = the recycling coefficient in the south channel, i.e., the amount of sediments transported laterally to the north channel, which return to the south channel.

By considering the sediment fluxes occurring during a fortnightly tidal cycle, the equation takes the following form:

$$24\,R + 14\,E_d + 10\,E_u = 14\,M_d\,k_d\,(I - \alpha_d) + 10\,M_u\,k_u\,(I - \alpha_u)$$

The subscripts u and d refer respectively to the upstream and downstream position of the turbidity maximum; furthermore, since there were a total of 24 tidal amplitude cycles in 1974 (14 during the downstream position of the turbidity maximum, and 10 during the upstream position), all the budget parameters are multiplied by the corresponding number of tidal cycles.

After the first cycle of erosion and deposition of fluid mud (May 15 - May 29), the average tracer concentration in the fluid mud was 2.1×10^{-6} μCi/g. Since 20% of the total activity remained sedimented on the bottom, and the remaining 80% was equally distributed between the two channels, the mass of sediment in the maximum and fluid mud of the south channel was:

$$Md = \frac{0.4 \times 10^{7}\ \mu\text{Ci}}{2.1 \times 10^{-6}\ \mu\text{Ci/g}} = 1.9 \times 10^{6}\ \text{tons.}$$

As mentioned previously, the suspended tracer distribution indicated that, in the downstream position, at each tidal amplitude cycle, 50% of the suspended sediment of the south channel transits to the north channel, therefore the corresponding lateral transfer coefficient (k_d) is equal to 0.5. At each tidal cycle, therefore, the tracer is diluted by a factor of two, and one half of the turbidity maximum of the south channel is renewed by a combination of bottom erosion, lateral transfer from the north channel and influx of fluvial sediment. In the upstream position, the lateral transfer coefficient can be deduced from the rate of decrease of tracer concentration in the fluid mud after n cycles. This is given by the relation, $Cn = Co\ (1-k)^n$, where Co is the initial concentration. This relation is valid only for a closed system with no new input of tracer, and can thus only be applied before or after July, since at this time, there was an apparent injection of activity due to the rapid upstream transfer of tarcer as mentioned previously. Application of this relation for the period when the maximum occupied the upper estuary, gives a value of 0.05 for k_u.

At the present time, the depth of the south channel is artificially maintained constant by dredging, both in the upper and lower estuary, therefore, $E_d = E_u =$ O. By making the simplifing assumptions that $M_u = M_d$, the sediment budget equation can be solved for α, which turns out to be slightly less than 8%.

For the downstream position of the turbidity maximum (high river flow) this indicates that at each fortnightly tidal cycle, of the total sediment transferred to the north channel (equivalent to 50% of the total sediment in the south channel), 42% is recycled back to the south channel. Therefore, the net lateral drift

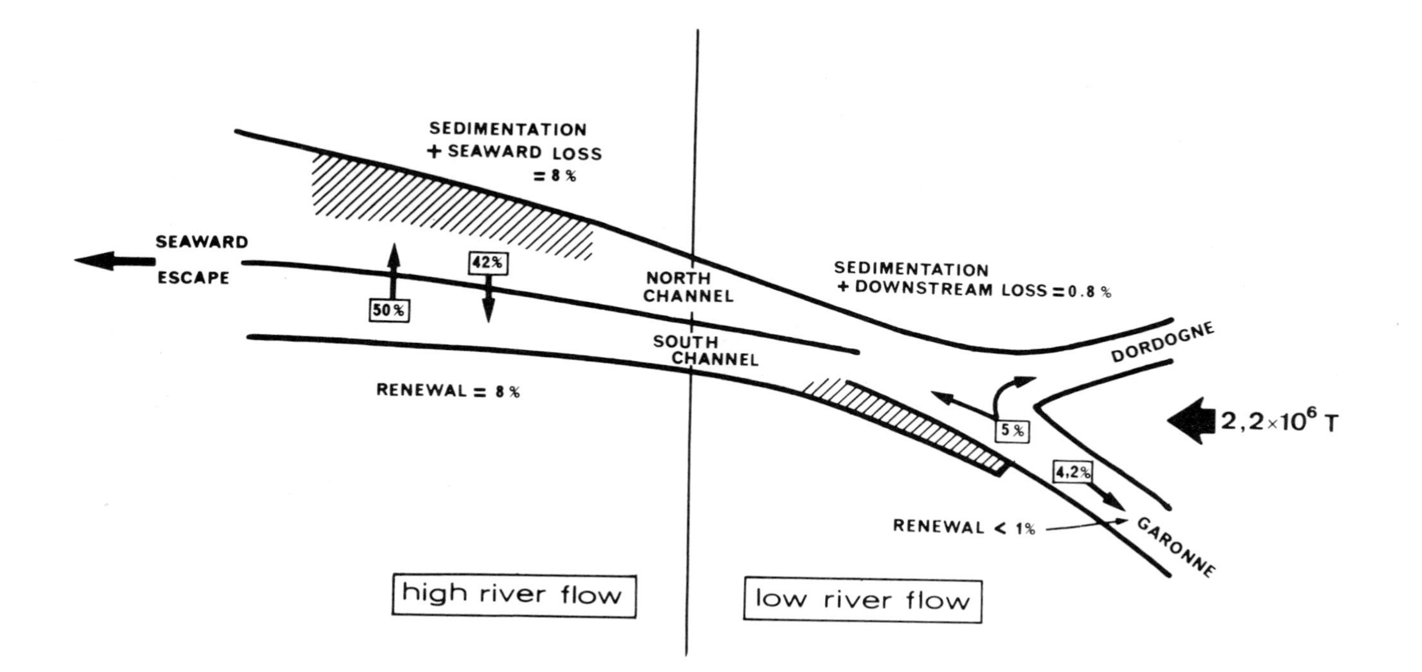

Figure 9. Schematic diagram illustrating the lateral suspended sediment flux between the channels during a fortnightly cycle of tidal ranges. Sediment flux values refer to the percent of the total mass of sediment in suspension in each channel.

of sediment is equivalent to 2.0×10^6 T/year, and represents the total mass sedimented in the north channel, plus that escaping to sea.

When in the upstream position (low river flow), sediment recycling within the turbidity maximum appears to be much lower, as during each fortnightly cycle only 5% of the total amount of sediment in the south channel (in reality the Garonne river, see Fig. 9) is transferred to the north channel. Therefore, less than 1% of the total amount of sediment in the south channel escapes or sediments. This is equivalent to 1.4×10^5 T/year.

Although this model is still preliminary, and based on several optimistic assumptions, it is interesting to note that the values deduced from the tracer study are of the same order of magnitude as those which are obtained by the analysis of sedimentation volumes in the estuary over the last 70 years.

CONCLUSIONS

The numerous studies which have been made by a number of workers during the past decade in the Gironde have shown that the major sedimentological processes governing the transport and deposition of suspended sediment are a continuous but cyclic phenomenon related to the cycles of river discharge and tidal amplitudes. Seasonal variations in river discharge control the position of the turbidity maximum and hence the locus of accumulation of suspended sediment, while the neap-spring tidal cycle controls the actual sedimentation processes.

A large scale lateral transport of suspended sediment occurs within the turbidity maximum, resulting in a general northward drift of sediment in the estuary. This lateral transport occurs by advection, and the rates of lateral sediment diffusion appear to be very low.

ACKNOWLEDGMENTS

The authors would like to thank the Port Autonome de Bordeaux, and in particular the Service d'Etudes and the Service Hydrographique of that organisation, who made available the considerable data and documents on the estuary, and permitted their publication.

The radioactive tracer experiment and associated studies were carried out jointly by the Centre National pour l'Exploitation des Oceans, the Commissariat a l'Energie Atomique, the Institut de Geologie du Bassin d'Aquitaine, and the Port Autonome de Bordeaux.

REFERENCES

1. Allen, G.P., P. Castaing, and A. Klingebiel. 1973. Suspended sediment transport in the Gironde estuary and adjacent shelf. *In* Proc. Intern. Symp. on Interrelationships of Estuarine and Continental Shelf Sedimentation, Bordeaux: 27-37.
2. ______, R. Bonnefille, G. Courtois, and C. Migniot. 1974. Processus de sedimentation des vases dans l'estuaire de la Gironde. Contribution d'un

traceur radioactif pour l'étude du deplacement des vases. La Houille Blanche 1/2:129-136.

3. Berthois, L. 1960. Etude dynamique de la sedimentation dans la Loire. Cah. oceanogr. 9:631-657.
4. Bonnefille, R. 1971. Remarques sur les ecoulements moyens a l'aval de la Gironde. Bull. Inst. Geol. Bassin Aquitaine 11:361-364.
5. Glangeaud, L. 1938. Transport et sedimentation dans l'estuaire de la Gironde (Caracteres petrographiques des formations fluviatiles, saumatres, littorales et neritiques). Bull. Soc. Geol. Fr. 8:599-631.
6. Inglis, C. C., and F.H. Allen. 1957. The regimen of the Thames estuary as affected by currents, salinities and river flow. Proc. Inst. Civ. Engin. 7:827-868.
7. Migniot, C. 1969. L'evolution de la Gironde au cours des temps. Bull. Inst. Geol. Bassin Aquitaine 11:221-281.
8. Nichols, M. 1973. Development of the turbidity maximum in the Rappahanock estuary. *In* Proc. Intern. Symp. on Interrelationships of Estuarine and Continental Shelf Sedimentation. Bordeaux:19-25.
9. ——, and G. Poor. 1967. Sediment transport in a coastal plain estuary. Proc. Am. Soc. Civ. Engin. J. Waterways and Harbors Div., New York, 93:83-95.
10. Postma, H. 1961. Transport and accumulation of suspended matter in the Dutch Wadden sea. Netherlands Jour. Sea Res. 1:148-190.
11. Pritchard D.W. 1955. Estuarine circulation patterns. Proc. Am. Soc. Civ. Engin. New York, 81:1-11.

SUSPENDED-PARTICLE TRANSPORT AND CIRCULATION IN SAN FRANCISCO BAY: AN OVERVIEW

T. J. Conomos and D. H. Peterson
U.S. Geological Survey, 345 Middlefield Road
Menlo Park, California 94025

ABSTRACT: Differences in the relative magnitude and timing of wind stress and river inflow in the northern and southern reaches of San Francisco Bay create different sedimentary conditions. The northern reach is a partially to well mixed estuary receiving most of the total annual fresh-water input (840 m^3 sec^{-1}) and suspended sediment input (4×10^6 metric tons) into the bay; more than 80% of the sediment is received during winter. Density-driven nontidal estuarine circulation (~5 cm sec^{-1}) maintains a turbidity maximum which changes seasonally in particle concentration (40 to 80 mg $litre^{-1}$). Strong tidal currents (⩽225 cm sec^{-1}) and wind-generated waves resuspend sediment from the shallow bay floor: some of the riverborne sediment deposited during winter is resuspended during summer and transported landward to the turbidity maximum. Long-term sediment data (extrapolated from bathymetric charts) indicate that the northern reach is an effective sediment trap. In contrast, long-term sediment data suggest that the southern reach is experiencing net erosion. The southern reach receives little river inflow or riverborne suspended sediment, and the average nontidal circulation is weak (⩽2 cm sec^{-1}). The principal source of suspended sediment (25 mg $litre^{-1}$) in the southern reach is the shallow bay floor (average depth 6 m).

INTRODUCTION

The impact of man's modification of San Francisco Bay has been extreme in its effects on the sedimentological aspects of the estuarine system. Large-scale modification began when placer mining (1848 to 1884) introduced huge quantities of sediment to the bay (11). These sediments caused extensive shoaling with as much as 1 m of sediment deposited in the northern reach. In addition to shoaling, the area of the bay has been reduced substantially by filling in and diking of the margins (22). The resulting volume decrease reduced the volume of the tidal prism which in turn has decreased the tide-related flushing of the bay waters.

The flushing problem is worsened by significant and continuing diversions of the Sacramento-San Joaquin River discharge, the major source of fresh-water inflow to the bay (12). This inflow adds large quantities of suspended sediment that are necessary for present ecological balance (14, 17), and generates an estuarine circulation cell which causes significant nontidal exchange with ocean water and which generates and maintains a turbidity maximum (4, 25).

Our purpose is to describe the suspended-sediment dispersal and the processes controlling this dispersal in order to provide an overview of the sedimentary environment. Particular emphasis is placed on summarizing previous studies and data from our own studies into a conceptual model that conforms with recent scientific advances in estuarine sedimentology (19).

In this paper we 1) describe the bay environment, emphasizing the agents that supply, resuspend, and transport sediment; 2) present a scenario that describes dispersal patterns within the bay and the nearby ocean, comparing and contrasting seasonal differences between the dissimilar northern and southern reaches of the bay; and 3) examine expected future changes in the sedimentary regime.

ENVIRONMENTAL SETTING

The San Francisco Bay system occupies a structural trough formed during the late Cenozoic. During the Pleistocene glaciation, the bay was part of a great drainage basin of the ancestral Sacramento, San Joaquin, and Coyote rivers (Fig. 1) in which sediment accumulated. The most recent sediments were deposited during the Wisconsin transgression which began 14,000 BP (31).

The bathymetry reflects the subaerial stream processes during the Pleistocene. The bay is relatively shallow, having an average depth of 6 m at mean lower low water (Table 1) or 2 m if the large expanses of mudflat are included (Fig. 2). The deepest point is Golden Gate where water depths exceed 100 m. The area has been reduced by 37% in the last 100 years from its natural state to its present 1.24×10^9 m^2 by shoaling caused by the inflow of hydraulic mining debris (11) and by land reclamation (22).

The prevailing wind flow, northwesterly and westerly maritime air, is strongest during summer, reaching average speeds greater than 4 m sec^{-1} (Fig. 3E, F). Although prevailing winter wind speeds are lower, biweekly storms cause southeasterly and southerly winds that can exceed 18 m sec^{-1} (Fig. 3F). Diurnal wind variations are greatest during summer, with typical afternoon speeds (9 m sec^{-1}) three times faster than morning speeds (20).

Prevailing summer winds generate waves with significant wave periods of 2 to 3 sec (29). During winter storms, 5-sec waves can be generated. Offshore, swell with periods 8 to 12 sec are common; during winter, 18-sec waves are moving landward (21). In addition to generating waves, the wind stress creates nontidal water movements with speeds a few to several percent that of the wind speeds (cf. 30).

The tides are mixed and predominantly semidiurnal (6). The diurnal tidal range varies from 1.7 m at Golden Gate to 2.7 m at the south end of the southern reach. This creates a large tidal prism (Table 1) that is about 24% of the bay volume. The tides create currents that are strongest in the channels and that maintain the original Pleistocene stream valley topography. Maximum speeds of 225 cm sec^{-1} are present at Golden Gate and Carquinez strait and 100 cm sec^{-1} near station 32 (Fig. 1). There is tidal mixing between waters of the northern and southern reaches, with typical excursions of 10 to 12 km.

More than 90% of the mean annual river discharge (840 m^3 sec^{-1}) entering the bay is contributed to the northern reach by the combined flows of the Sacramento and San Joaquin rivers (Table 1); the remaining 10% is contributed

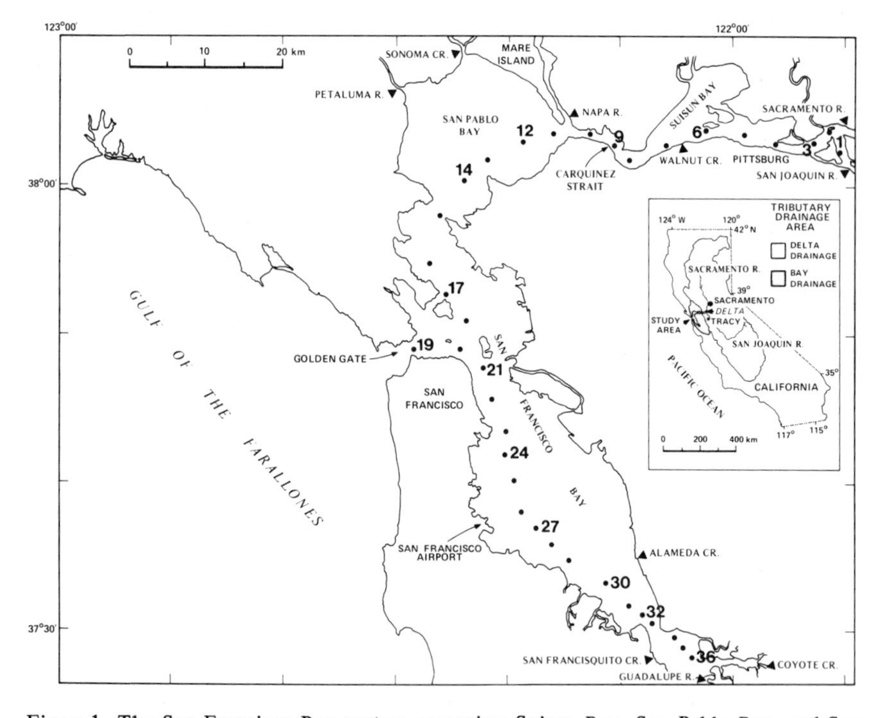

Figure 1. The San Francisco Bay system comprises Suisun Bay, San Pablo Bay, and San Francisco Bay, but is termed San Francisco Bay herein. The northern reach is Suisun Bay, San Pablo Bay and the northern portion of San Francisco Bay (to Golden Gate). The southern reach is San Francisco Bay south of Golden Gate. Station numbers are established hydrographic stations occupied near-monthly from 1969 to 1975. The drainage basins of the Sacramento-San Joaquin River system and of the peripheral streams are in inset.

by small tributary streams and sewage inflow. River runoff is greatest during winter (Fig. 3A).

Dilution of Pacific Ocean water entering Golden Gate by seasonally varying river discharge controls the salinity distribution (Fig. 3B). Water density is controlled by salinity, as bay-wide synoptic water temperature variations rarely exceed 3°C. The geographic distribution of river discharge and vertical salinity field (Fig. 4A, C) shows that the northern reach varies as a partially mixed estuary with vertical salinity differences often 10°/oo during winter and as a well mixed estuary with a vertical salinity difference less than 5°/oo during summer. The southern reach is an embayment with seasonally varying water properties that are largely controlled by water exchanges from the northern reach and the Pacific Ocean. Intrusion of low salinity water into the southern reach is particularly evident during winter periods of wet years (Fig. 4A; 18). Some salinity stratification is present during winter, whereas during summer the water is nearly isohaline with depth because of vertical mixing caused by tidal currents and wind.

Analyses of current meter data (unpublished) indicate dissimilar long-period (several days) fluctuations in water movement in the northern and southern reaches. The waters of the southern reach have little net motions throughout the

Table 1. Geostatistics of San Francisco Bay

Statistic	Value
Area (MLLW)[1]	1.04×10^9 m^2
Including mudflats	1.24×10^9 m^2
Volume[1]	6.66×10^9 m^3
Tidal prism[2]	1.59×10^9 m^3
Average depth[3]	6.1 m
From hypsometric curve[4]	2 m
River discharge (annual)	20.9×10^9 m^3
Delta outflow[5]	19.0×10^9 m^3
All other streams	1.9×10^9 m^3
Suspended sediment inflow (annual)[6]	
Into delta	4.7×10^6 metric tons
From delta into bay	3.3×10^6 metric tons
All other streams	0.9×10^6 metric tons
Total into bay	4.2×10^6 metric tons
Sediment accumulation rate[7]	350 mg cm^{-2} yr^{-1}

[1] Planimetered from Fig. 2; at mean lower low water
[2] From (7).
[3] Volume divided by area; at mean lower low water
[4] Obtained graphically from hypsometric curve and includes mudflats (Fig. 2).
[5] From (9).
[6] From (27); measured from 1957-1959
[7] Assuming uniform deposition throughout bay, no dredging, and no sediment loss to ocean; obtained from annual suspended sediment inflow divided by area of bay (including mudflats).

column (herein termed nontidal), whereas in the northern reach and at Golden Gate, estuarine circulation is clearly defined (25).

Our 3-year study using bimonthly releases of surface and seabed drifters (3, 5) has verified these observations (Fig. 5). The northern reach-ocean section has a permanent estuarine circulation cell maintained by the density difference between Sacramento-San Joaquin River water and seawater. This density difference produces a constant net landward bottom flow of dense seawater in opposition to net seaward flow of less dense river water. These currents have equal and opposite effect on the nontidal flow in the null zone (25). The null zone can be portrayed graphically by the convergence of seabed drifters (Fig. 5B). In contrast, the southern reach, because of the small supply of river inflow and therefore the weaker salinity stratification, does not exhibit two-layer estuarine circulation, but has seasonally reversing near-bottom and surface currents. The strong prevailing winds of summer alter any weak density-induced circulation and control the nontidal drift (Fig. 5, inset). The effect of winter storms causing strong episodic water movements in the southern reach is major but has not been evaluated.

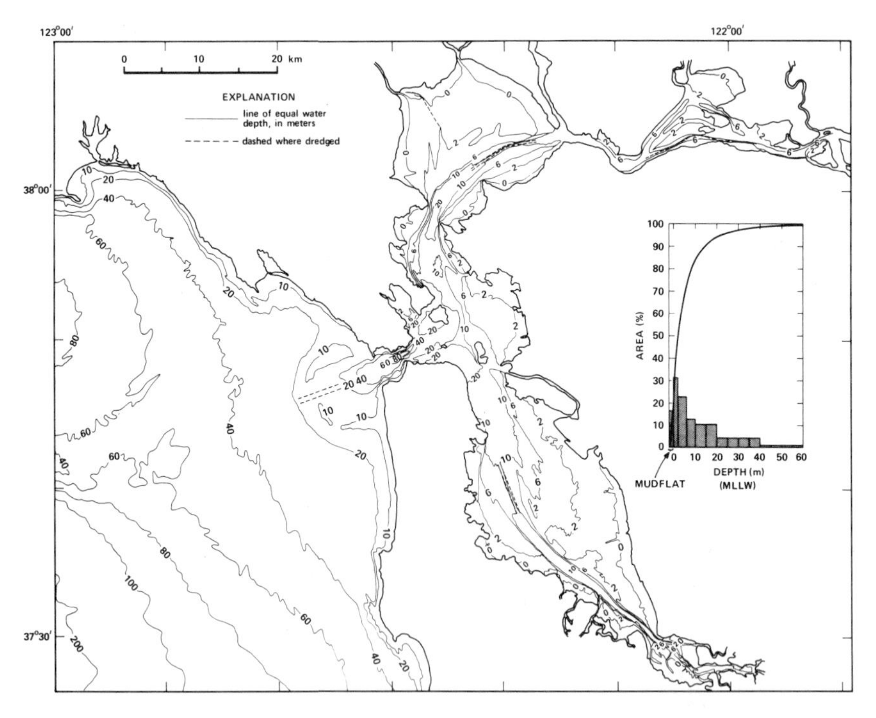

Figure 2. Bathymetric chart of San Francisco Bay. Compiled from CGS charts 5531, 5532, 5533, 5534, and 5072. Datum is mean lower low water. Hypsometric curve (inset) constructed from bathymetric contours, and includes mudflats.

Nontidal current speeds in the northern reach, estimated by drifter data, average 4 and 5 cm sec^{-1} for the near-bottom landward drift and surface seaward drift, respectively. Speeds determined by vector addition of current-meter data are approximately double these values (25, unpublished data). In the southern reach, the sluggish movements, regardless of direction, are between 1 and 2 cm sec^{-1} (3).

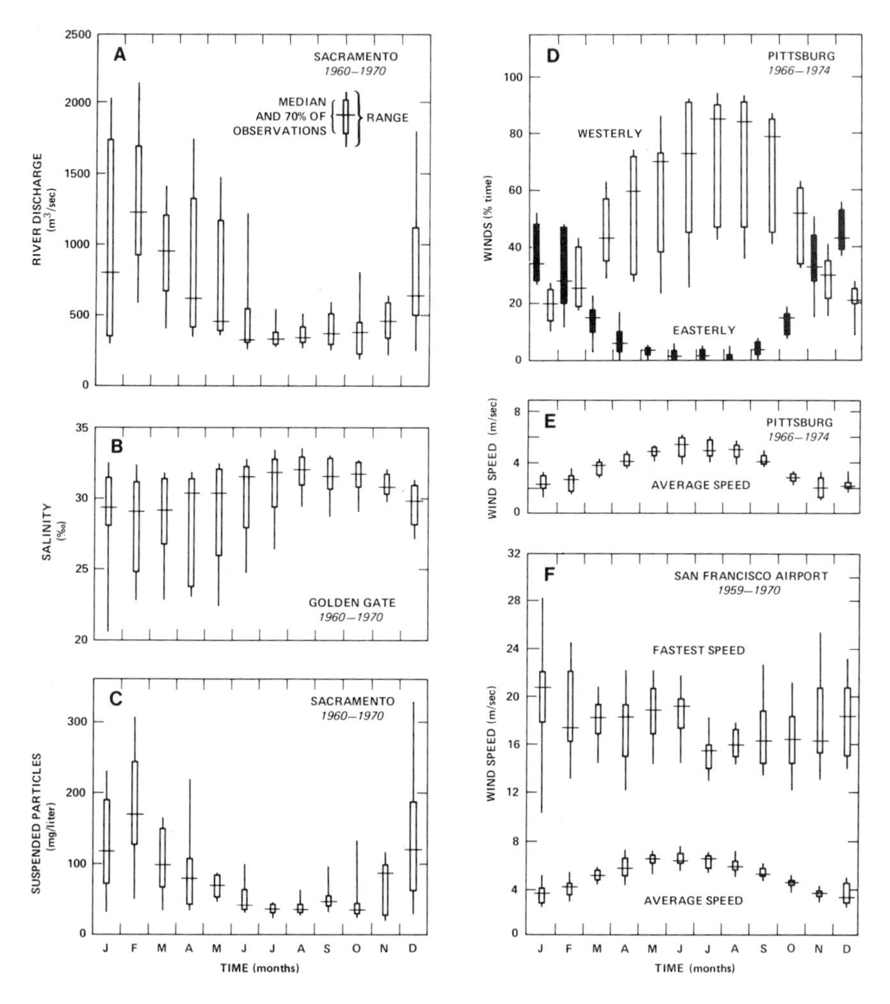

Figure 3. Monthly means of selected environmental variables. (A) Sacramento River discharge at Sacramento. (B) Surface salinity at Golden Gate. (C) Suspended particulate matter at Sacramento. Wind direction (D) and prevailing wind speeds (E) at Dow Chemical Facility, Pittsburg. (F) Prevailing and fastest wind speed at San Francisco International Airport.

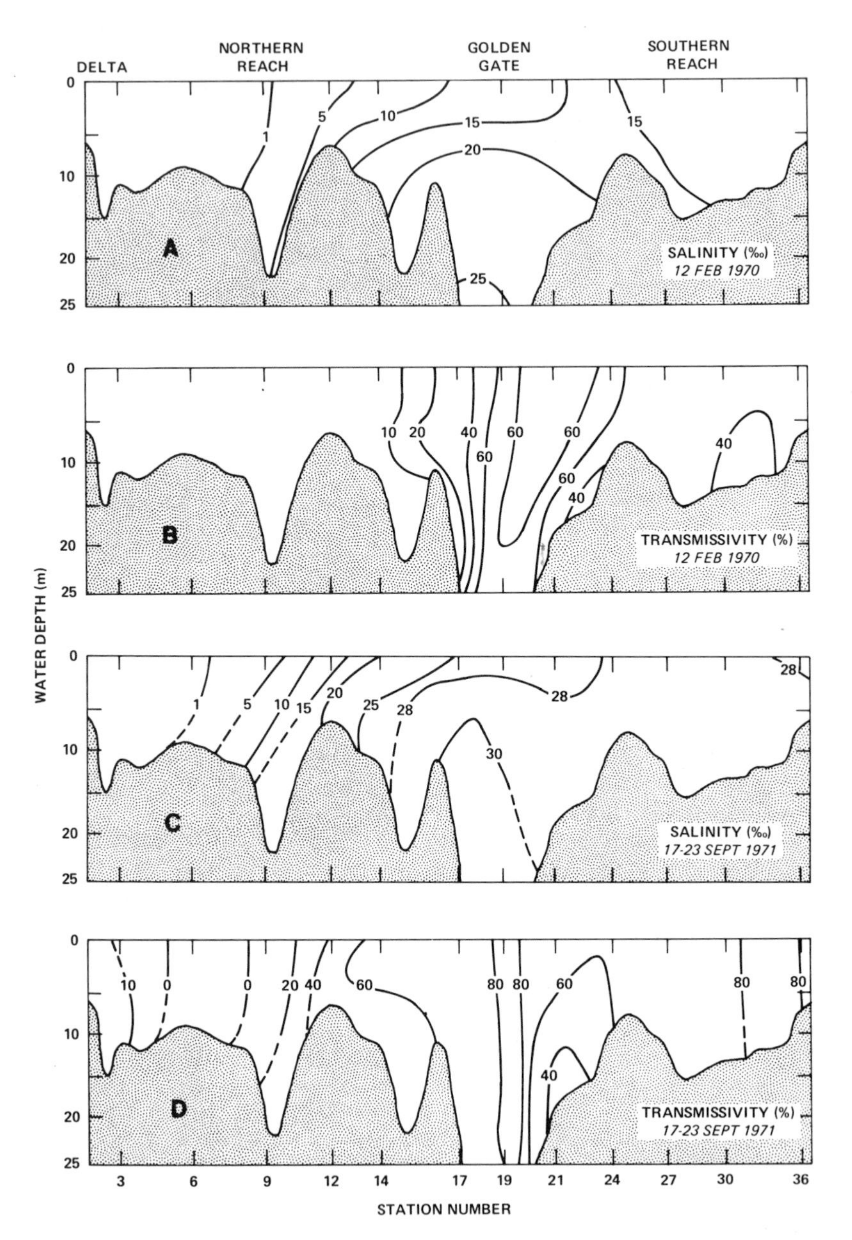

Figure 4. Vertical distribution of salinity (A, C) and transmissivity (B, D), during high and low river discharge periods of Sacramento-San Joaquin (S-SJ) and southern reach streams (SRS). Data obtained at hydrographic stations (Fig. 1) with methods described by Peterson and others (26). February (mean monthly) river discharges: S-SJ = 2170 m^3 sec^{-1} ; SRS = 16 m^3 sec^{-1}. September river discharges: S-SJ = 460 m^3 sec^{-1} ; SRS = 2 m^3 sec^{-1}.

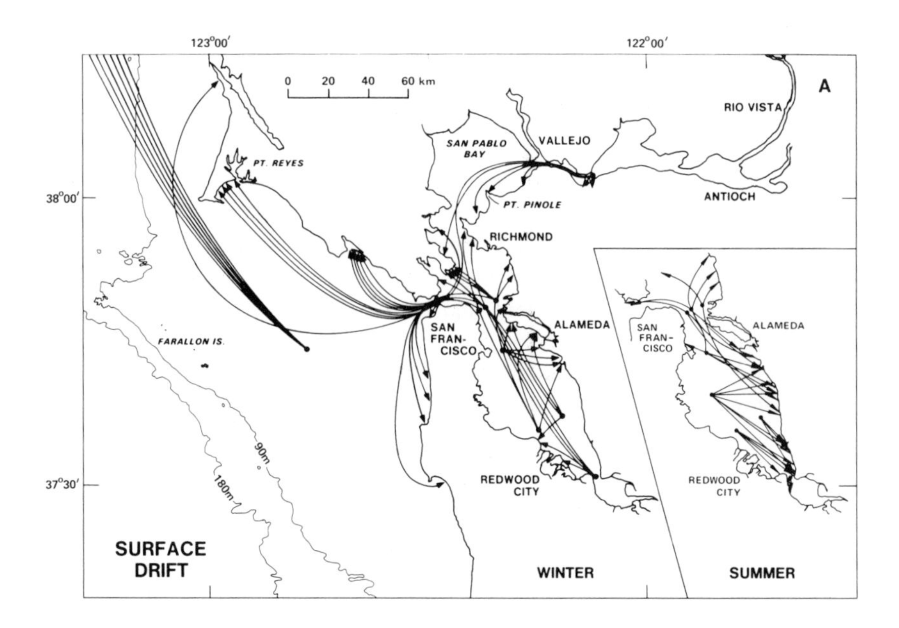

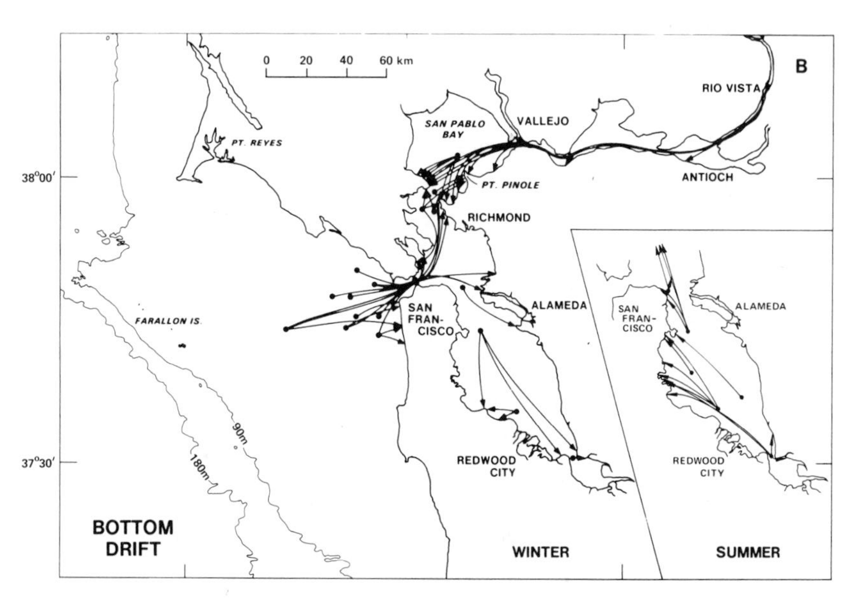

Figure 5. Release and recovery points for surface (A) and seabed (B) drifters in the bay and adjacent ocean. Drifter movements are shown as arrows drawn from release points to recovery locations and portray simplified paths of movement occurring within 2 months of release. Winter release: December 1970 (modified from 3). Summer release (southern reach only shown as inset): September 1971. Data are typical of 18 releases over a 3-year period (1970-1973).

SEDIMENTS

Source

The rivers are the major source of sediments to the bay and delta, contributing 5.6×10^6 metric tons annually (27). Of the 4.2×10^6 metric tons that flow into bay waters, 81% originate from the Sacramento-San Joaquin River drainage (Table 1), while the remainder is contributed by the local streams (Fig. 1). Eighty-five percent of all sediment enters the bay as suspended load (27). This suspended fraction is classified as silty clay (34), typically having a sand-silt-clay ratio of 15:30:55 (27). Sediment input varies greatly during the year, being proportional in concentration to the river discharge (Fig. 3A, C). Over 80% of the suspended riverborne sediment from the Sacramento River is contributed during winter.

Surficial Sediments

Near-equal amounts of silt and clay with various amounts of sand comprise the upper 5 cm of modern sediment (data sources: U.S. Geological Survey and references in 10). Poorly sorted silty clay, clayey silt and sand-silt-clay (Shepard classification; 34) are present in the southern reach and the shallow part of the northern reach, while sand and silty sand cover the deeper areas of the central portion of the bay and of the northern reaches. Gravelly sands are found at Golden Gate, and grade seaward to a well sorted sand that covers most of the continental shelf.

Suspended Sediment

Most (70 to 97%) of the suspended particulate matter in the turbidity maximum is lithogenous sediment. The remaining fraction, which changes seasonally and spatially in concentration and composition, includes both living and detrital biogenous matter (4,25,26).

The turbidity maximum is the dominant feature of the suspended sediment distribution in the northern reach (4,25,26). The turbidity (Fig. 4D) and suspended sediment concentrations (Fig. 6B) are higher in the null zone than in either the upper or lower part of the reach, a situation not unlike other partially mixed estuaries (13,15,19,24,28,32,35). This maximum is a consequence of the typical response of the longitudinal distribution of suspended sediment to estuarine circulation: some riverborne suspended sediment settles by gravity from the seaward-flowing surface layer to the landward-flowing bottom layer where it is entrained, transported to and trapped in the null zone (19).

Particle concentrations of near-surface waters are greatest in the northern reach, having typical concentrations of 15 to 20 mg litre^{-1} in the river ($0°/_{oo}$ salinity) and 90 mg litre^{-1} in the turbidity maximum (Fig. 6B). The lowest concentration, 10 mg litre^{-1}, is at Golden Gate (station 19). The southern reach has water of intermediate (25 mg litre^{-1}) concentrations. The median concentrations are highest during winter at Golden Gate and the southern reach reflecting

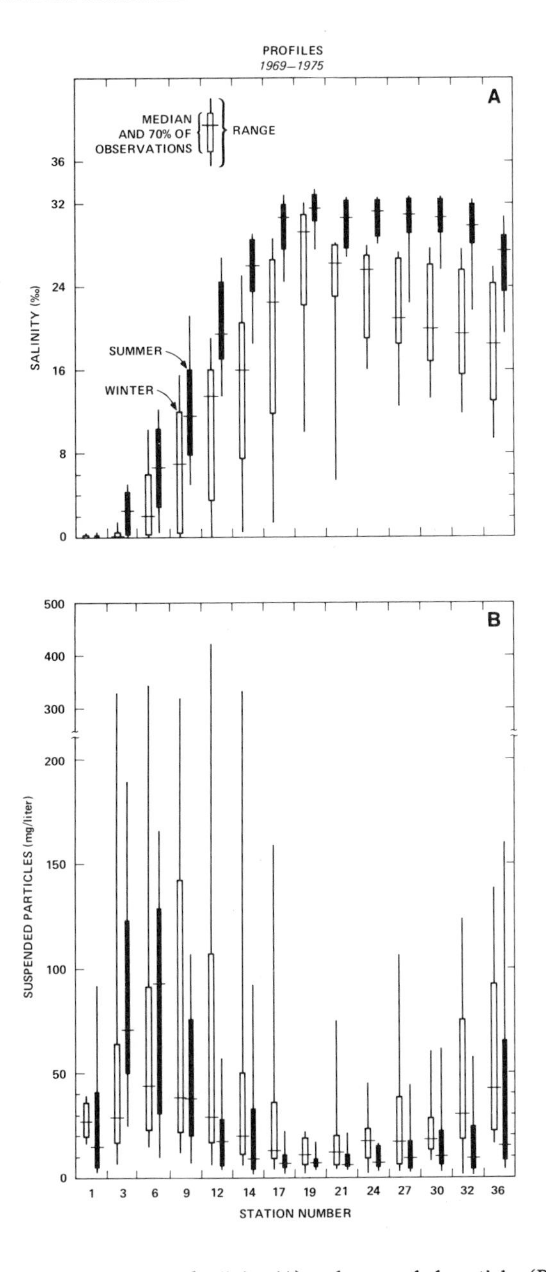

Figure 6. Longitudinal distribution of salinity (A) and suspended particles (B) at 2 m during winter (December through April) and summer (July through October) at near-monthly intervals (1969 to 1975) at hydrographic stations (Fig. 1). Water collection methods and salinity determinations described by Peterson and others (26). Suspended particle concentrations determined from air-dried suspensate retained on a 0.45-μm pore diameter silver filter.

the seasonality of the river input (Fig. 6), and are often highest in shallower water. Although the extreme concentrations are highest in the turbidity maximum during winter, the median concentration there is highest during summer.

Typical particle modal diameters of suspended sediment (preliminary measurements by particle counter) are 4 μm in river water (0°/oo salinity), 8 μm in the turbidity maximum, Golden Gate and the southern reach. These measurements agree with data gathered by the U.S. Corps of Engineers (38) which show median diameters of particles suspended in bay waters ranging between 1 and 6 μm.

SEDIMENT TRANSPORT PATTERNS

The bay is a dynamic system: the large tidal prism causes strong tidal motion, the strong wind field creates large waves and substantial nontidal currents, and the high annual river inflow (three times that of the bay volume) causes estuarine circulation in the northern reach and contributes density-induced advection in the southern reach. This high energy environment, coupled with the large sediment inflow and the shallowness of the bay, causes the suspended and surficial sediment to be quite mobile. This mobility is evidenced by the fact that more sediment is dredged annually from channels than is contributed to the bay by rivers (33).

The dispersal of these sediments through the interaction of transport, deposition and resuspension is on a seasonal cycle, with riverborne sediment supply and deposition dominant during winter and sediment resuspension and redeposition dominant during summer (8). These seasonal differences, combined with the differences in hydrologic, hydrographic and sedimentologic processes and rates between the northern and southern reaches make the bay a complicated system to evaluate. Enough is known from previous data and from our ongoing studies, however, to present a simple conceptual model of the basic sediment transport patterns. As the processes are seasonally modulated, we begin with winter in the northern reach.

Winter Conditions

Sediment enters the northern reach in great quantities (Table 1) during winter (Fig. 3C). The bedload material, the coarser-grained fraction, and some of the aggregated finer-grained fraction of the suspended load deposit soon after entering the estuary (8, 11). Some of the deposited material is periodically resuspended by the tidal currents. Our seabed drifter data indicate that virtually no sediment entrained in the near-bottom river currents is transported seaward of the null zone: the sediment motion is arrested by the landward flowing density current. Of the seabed drifters released landward of the null zone, none of the hundreds recovered have been found seaward of the zone (Fig. 5B; 3,5). The deposited particles cause shoaling in the null zone, which, at this time of the year, is located in San Pablo Bay (25), while the suspended portion is maintained in the null zone and constitutes the turbidity maximum. The suspended particles with lower settling velocities are maintained in the seaward flowing surface layer (Fig. 5A). The concentrations in this layer are determined by relative rates of 1) resuspension caused by tidal currents and by wind waves (8); 2) settling of particles, partly enhanced by particle aggregation (16, 19), and 3) dilution of turbid low salinity water by progressive mixing with less turbid high salinity ocean water (Figs. 4, 6). Some of the deposited portion is later suspended and entrained in the landward flowing density current and transported landward to the turbidity maximum (Fig. 5B). Most of the seaward flowing near-surface sediment is transported through Golden Gate as a lobe-shaped effluent plume and dispersed seaward, while another portion, visible as a turbid water mass, drifts into the southern reach of the bay (2).

The southern reach is also accumulating riverborne sediment from the local streams during winter. Sediment not deposited is transported with the near surface waters through Golden Gate, or is dispersed into the northern reach (Fig. 5). The high winds accompanying periodic winter storms generate waves that resuspend the sediment and allow it to be transported by currents.

Summer Conditions

Summer is marked by a much decreased sediment influx and a concomitant increase in wind speed. This creates a relative increase of wind-wave induced resuspension over deposition (8). Sediment in the northern reach and the northern portion of the southern reach that had been deposited during winter is resuspended by waves and tidal currents (Fig. 8 in 19) and transported to the null zone and the turbidity maximum. As the null zone has migrated landward into the Suisun Bay region because of the diminished river discharge (25), the Mare Island area, which was largely bypassed during high discharge conditions, receives landward moving sediments (8) and shoals dramatically (35).

We do not know much about the disposition of sediments in the southern reach during summer. It appears that the southern reach does not accumulate much new sediment and, in the last several decades, is probably losing sediment to the northern reach. Long-term sediment budgets based on comparisons of bathymetric charts (1856-1957) (37) and our field observations suggest that large expanses of the northern, subtidal part of the southern reach appear to be kept scoured of erodable sediment. Shell debris covers the bottom, and benthic faunal communities consist in large part of species represented by mature, well-established specimens and appear stable with time (F. H. Nichols, oral communication). Net accumulation of fine-grained sediment, however, occurs in the margins and southern portion of the southern reach. This accumulation is apparently controlled by the tidal-current generated particle settling and scour lag effects similar to those described in the Wadden Sea by van Straaten and Kuenen (39) and Postma (28). But there is seasonal erosion in the margin areas as well: at a mudflat at the southern end of the bay, up to 9 cm of sediment has been eroded away within one summer month (23).

Transport to Ocean

Our drifter data, at variance with hydraulic model studies (33), suggest that the bay maintains a pronounced estuarine circulation cell and is an effective sediment trap during normal (i.e., Fig. 3A) river discharge conditions. Of the seabed drifters released at Golden Gate (regardless of tidal stage) and landward, only a few of the thousands recovered were found seaward of Golden Gate. Conversely, of the surface drifters released at Golden Gate (regardless of tidal stage) and seaward, none of the thousands recovered were found landward of Golden Gate. It follows then, assuming that bottom sediment transport directions are similar to those of the seabed drifters, that little if any sediment is transported seaward along the bottom. Virtually all the sediment lost to the ocean is clay or fine-grained silt and is suspended in the seaward-flowing surface current (Fig. 5A), with concentrations similar to those at Golden Gate (Fig. 6B). The concentrations are somewhat proportional to the river discharge levels.

Schultz (33) estimated the annual sediment loss to the ocean, based on a 36-year (1924-1960) average discharge and a suspended sediment concentration at Golden Gate of 50 mg $litre^{-1}$, to be about 30% of the annual riverborne load (Table 1). His estimate is inspired by his hydraulic model studies: sediment retention curves based on the dispersal of gilsonite (simulating surficial sediment), showed that at least 35% of the sediment immediately landward of Golden Gate is tidally dispersed seaward, and the percentage retained increases landward. We suggest, in light of our suspended sediment data (10 mg $litre^{-1}$ at station 19-Golden Gate, Fig. 6B) and our conceptual model that emphasizes the importance of the estuarine circulation cell and null zone, that his 30% loss estimate be revised downward to 6% during normal discharge conditions. This estimate is closer to Gilbert's (11) original estimates and predictions of 4% loss.

Large-scale loss of sediment throughout the water column to the ocean should occur only if river discharge is sufficient to force the null zone seaward through Golden Gate. Such discharge levels would be at least 4 times normal (normal indicated in Fig. 3A); such discharges occur statistically at 5- to 10-year frequencies (40).

Seaward flowing suspended sediments apparently bypass the Gulf of the Farallones and are dispersed at sea or are returned to the bay in the bottom inflowing currents, as no clays or fine silts are found on the continental shelf. Although the sediment on the shelf bottom is demonstrably Sacramento-San Joaquin River debris (41), it apparently represents relict sand stranded during the Holocene transgression or contributed from the bay during exceptionally high but infrequent river discharges. The fine-grained fraction is winnowed away by the strong sea, swell, and currents.

FUTURE OUTLOOK

It is difficult to predict the sediment dispersal patterns that will prevail in future decades because of the difficulty in predicting the course of future water-supply development in the area tributary to the bay (12, 14) and of potential ship channel deepening. Large freshwater diversion projects that would seriously deplete the annual flow of the Sacramento-San Joaquin River system would reduce the riverborne suspended sediment mass (17).

In addition to reducing the suspended sediment supply, river diversion would damp the density-induced estuarine circulation cell in the northern reach (3) and, hence, would affect the position of the turbidity maximum. River-generated two-layered nontidal flow through Golden Gate would diminish and tidal movements would become relatively more dominant. The mode and degree of sediment exchange with the ocean would thus be altered in a yet unknown manner.

Complex and as yet undefined interrelations must also be evaluated when predicting the importance of reduced river flow on the locations and rates of shoaling and on the suspended sediment concentrations and composition. For example, the implications of suspended sediment on availability of incident light (water transparency) and in turn on the phytoplankton growth rates and plant nutrient cycles are not clear (14, 17). Similarly, deepening the ship channel in the northern reach may alter the water circulation patterns and rates by enhancing the density induced circulation (1), and, as in the Savannah Harbor case, may result in an increase in shoaling (36). This deepening may also reduce the near-surface suspended sediment concentration by increasing the water column depth.

REFERENCES

1. Burgh, Ir. P. van der. 1968. Prediction of the extent of saltwater intrusion into estuaries and seas. J. Hydraulic Res. 6(4):267-288.
2. Carlson, P.R., and D.S. McCulloch. 1974. Aerial observations of suspended sediment plumes in San Francisco Bay and the adjacent Pacific Ocean. J. Res. U.S. Geol. Surv. 2(5):519-526.
3. Conomos, T.J. 1975. Movement of spilled oil as predicted by estuarine nontidal drift. Limnol. Oceanogr. 20(2):159-173.
4. Conomos, T.J., and D.H. Peterson. 1974. Biological and chemical aspects of the San Francisco Bay turbidity maximum. Mémoires de l'Institut de Géologie du Bassin d'Aquitaine 7:45-52.
5. Conomos, T.J., D.S. McCulloch, D.H. Peterson, and P.R. Carlson. 1971. Drift of surface and near-bottom waters of the San Francisco Bay system: March 1970 through April 1971. U.S. Geol. Surv. Open-file Map.
6. Disney, L.P., and W.H. Overshiner. 1925. Tides and currents in San Francisco Bay. U.S. Dep. Com., U.S. Coast Geod. Surv., Spec. Pub. 115. 125 p.
7. Edmonston, A.D., and Raymond Matthew. 1931. Variation and control of salinity in Sacramento-San Joaquin delta and upper San Francisco Bay. California Dep. Public Works Bull. 27. 440 p.
8. Einstein, H.A., and R.B. Krone. 1961. Estuarial sediment transport patterns. Amer. Soc. Civil Eng. Proc., J. Hydraulics Div. 87(HY2):51-59.
9. Federal Water Pollution Control Administration. 1967. Effects of the San Joaquin Master Drain on water quality of the San Francisco Bay and Delta. Centr. Pacific Basins Comprehensive Water Pollution Control Proj. Rep. 101 p.

10. Folger, D.W. 1972. Characteristics of estuarine sediments of the United States. U.S. Geol. Surv. Prof. Paper 742. 94 p.
11. Gilbert, G.K. 1917. Hydraulic-mining debris in the Sierra Nevada. U.S. Geol. Surv. Prof. Paper 105. 154 p.
12. Gill, G.S., E.C. Gray, and David Sechler. 1971. The California water plan and its critics: A brief review, p. 3-27. *In* D. Sechler (ed.), California water; A study in resource management. Univ. Calif.
13. Glangeaud, L. 1938. Transport et sédimentation dans l'estuarie et a l'embouchure de la Gironde. Caractères pétrographiques des formations fluviatiles, saumatres, littorales, et néritiques. Bulletin de la Société Géologique de France 8:599-630.
14. Goldman, C.R. 1971. Biological implications of reduced freshwater flows on the San Francisco Bay-Delta system, p. 109-124. *In* D. Sechler (ed.), California water; A study in resource management. Univ. Calif.
15. Inglis, Clyde, and F.H. Allen. 1957. The regimen of the Thames estuary as affected by currents, salinities and river flow. Min. Proc. Inst. Civil Eng. (London) 7:827-878.
16. Krone, R.B. 1962. Flume studies of the transport of sediment in estuarial shoaling processes—final report. Univ. California (Berkeley) Hydraulic Eng. Lab. and Sanitary Eng. Res. Lab. 110 p.
17. Krone, R.B. 1966. Predicted suspended sediment inflows to the San Francisco Bay system. Report to Central Pacific River Basins Comprehensive Water Pollution Control Proj., Federal Water Pollution Control Proj., Federal Water Pollution Control Admin. (Southwest Region). 33 p.
18. McCulloch, D.S., D.H. Peterson, P.R. Carlson, and T.J. Conomos. 1970. Some effects of fresh-water inflow on the flushing of south San Francisco Bay: A preliminary report. U.S. Geol. Surv. Circ. 637A. 27 p.
19. Meade, R.H. 1972. Transport and deposition of sediments in estuaries, p. 91-120. *In* B.W. Nelson (ed.), Environmental framework of coastal plain estuaries. Geol. Soc. Amer. Mem. 133.
20. Miller, Albert. 1967. Smog and weather—The effect of the San Francisco Bay on the bay area climate. San Francisco Bay Conserv. Development Comm. 40 p.
21. National Marine Consultants. 1960. Wave statistics for seven deep water stations along the California coast. Prepared for U.S. Army Corps Eng. District, Los Angeles. 20 p.
22. Nichols, D.R., and N.A. Wright. 1971. Preliminary map of historic margins of marshland, San Francisco Bay, California. Open-file rep. U.S. Geol. Surv. 10 p.
23. Nichols, F.H. In press. Infaunal biomass and production on a San Francisco Bay mudflat. *In* B.C. Coull (ed.), Ecology of Marine Benthos, Belle W. Baruch Library in Marine Sci., v. 6, Univ. South Carolina Press.
24. Nichols, M.M. 1972. Sediments of the James River estuary, Virginia, p. 169-212. *In* B.W. Nelson (ed.), Environmental framework of coastal plain estuaries. Geol. Soc. Amer. Mem. 133.
25. Peterson, D.H., T.J. Conomos, W.W. Broenkow, and P.C. Doherty. 1975a. Location of the nontidal current null zone in northern San Francisco Bay. Estuarine Coastal Marine Sci. 3(1):1-11.
26. Peterson, D.H., T.J. Conomos, W.W. Broenkow, and E.P. Scrivani. 1975b. Processes controlling the dissolved silica distribution in San Francisco Bay, p. 153-187. *In* L.E. Cronin (ed.), Estuarine Research. Chemistry and Biology, v. 1. Academic Press, New York.

27. Porterfield, George, N.L. Hawley, and C.A. Dunnam. 1961. Fluvial sediments transported by streams tributary to the San Francisco Bay area. Open-file rep. U.S. Geol. Surv. 70 p.
28. Postma, H. 1967. Sediment transport and sedimentation in the estuarine environment, p. 158-179. *In* G.H. Lauff (ed.), Estuaries. Amer. Ass. Advance. Sci. Pub. 83.
29. Putnam, J.A. 1947. Estimating storm-wave conditions in San Francisco Bay. Trans. Amer. Geophys. Union 28(2):271-278.
30. Rattray, Maurice, Jr. 1967. Some aspects of the dynamics of circulation in fjords, p. 52-62. *In* G.H. Lauff (ed.), Estuaries. Amer. Ass. Advance. Sci. Pub. 83.
31. Schlocker, Julius. 1974. Geology of the San Francisco North Quadrangle, California. U.S. Geol. Surv. Prof. Paper 782. 109 p.
32. Schubel, J.R. 1968. Turbidity maximum of northern Chesapeake Bay. Sci. 161:1013-1015.
33. Schultz, E.A. 1965. San Francisco Bay dredge spoil disposal. Prepared for presentation to the Committee on Tidal Hydraulics, 53rd meeting San Francisco, May, 1965. 48 p.
34. Shepard, F.P. 1954. Nomenclature based on sand-silt-clay ratios. J. Sedimentary Petrology 24(3):151-158.
35. Simmons, H.B. 1955. Some effects of upland discharge on estuarine hydraulics. Amer. Soc. Civil Eng. Proc. 81 (Separate 792). 20 p.
36. Simmons, H.B. 1965. Channel depth as a factor in estuarine sedimentation, p. 722-730. *In* Proc. Federal Interagency Sedimentation Conf. 1963. U.S. Dep. Agr. Misc. Pub. 970. 933 p.
37. Smith, B.J. 1965. Sedimentation in the San Francisco Bay system, p. 675-708. *In* Proc. Federal Interagency Sedimentation Conf. 1963. U.S. Dep. Agr. Misc. Pub. 970. 933 p.
38. South Pacific Division Laboratory. 1957. Report of soil tests, Bay Model sediment samples. U.S. Army Corps Eng., South Pacific Div. Lab. 17 p.
39. Straaten, L.M.J.U. van, and Ph. H. Kuenen. 1958. Tidal action as a cause of clay accumulation. J. Sedimentary Petrology 28(4):406-413.
40. U.S. Army Corps of Engineers. 1963. Flood control. Append. B. Tech. Rep. on San Francisco Bay barriers. U.S. Army Eng. District, San Francisco. 80 p.
41. Yancey, T.E., and J.W. Lee. 1972. Major heavy mineral assemblages and heavy mineral provinces of the central California coast region. Geol. Soc. Amer. Bull. 83:2099-2104.

MATHEMATICAL MODELING OF SEDIMENT TRANSPORT IN ESTUARIES

Ranjan Ariathurai and Ray B. Krone
Department of Civil Engineering
University of California at Davis
Davis, California 95616

ABSTRACT: The elements to be considered in a mathematical model for estuarial sediment transport are presented. These elements include convection–diffusion terms and source and sink functions based on previous laboratory experiments. The experimental results yielded descriptions of deposition from suspension and erosion of cohesive beds.

Mathematical models that simulate the transport of cohesive sediments are reviewed with recommendations for future work.

INTRODUCTION

Management of water quality and the design and maintenance of navigation facilities in estuaries would be greatly enhanced by detailed predictions of the concentrations of suspended sediments and the rates of sediment deposition that would result from proposed alterations of the fresh water and sediment inflow or the shape of the estuary. Turbidity from suspended sediments often limits the primary productivity of an estuary and is a factor determining the quality of waters for aesthetic enjoyment. The shoaling rate in channels and turning basins is often the critical factor in their design. Furthermore, selection of sites for open water disposal of material dredged during channel maintenance needs to be made with knowledge of the effects of such disposal on the surrounding waters. An accurate predictive model of estuarial sediment transportation would be useful.

Descriptions of estuarial sediment transportation processes have been obtained only during recent years. This paper presents a distillation of these descriptions obtained during a succession of laboratory and field studies. Prediction of suspended sediment concentrations and rates of deposition and scour in estuaries will require a model that explicitly incorporates such descriptions.

Existing models are briefly reviewed including a new two-dimensional model that uses the transport process relations that are presented.

TRANSPORT PROCESSES

Settling. The most important sediment property in the modeling of cohesive sediment transport is the settling velocity. In the absence of continuing aggregation, and at suspended sediment concentrations below a few grams per liter, Stoke's law may be applied to these small particles to obtain the settling velocity V_s as

$$V_s = \frac{gD^2}{18\nu} \frac{(\rho_a - \rho_w)}{\rho_w} \quad (1)$$

where D = diameter of particle, g = acceleration of gravity, ν = kinematic viscosity of the water, ρ_a = density of the sediment particle or aggregate, and ρ_w = density of the water.

Settling velocity at higher concentrations is described by the Richardson–Zaki relation (12),

$$\overline{V}_s = V_s (1 - \phi)^5 \quad (2)$$

where $\overline{V}_s$ = "hindered" settling velocity, and ϕ = volume concentration of the aggregates.

In situ settling velocities in the Thames Estuary were measured by Owen (10). He found that for the same concentration of suspended sediment the mean *in situ* settling velocities were of an order of magnitude greater than those determined in quiescent settling tests in the laboratory. The *in situ* settling velocities were also found to change with the tides. Varying flow conditions and aggregation are the probable causes for the changes in the floc settling velocities.

Settling velocities obtained for a number of flume and standing cylinder tests conducted by Krone are presented in (6, 7). It was found that when aggregation had proceeded to the point that subsequent collisions were infrequent, the settling velocity

$$V_s = K\, C^{4/3} \quad (3)$$

where K = empirical constant.

The same power law was derived from theoretical considerations. Depending on the situation, in situ tests, lab tests, or the power law of Eq. 6 can be used to obtain settling velocities for modeling.

Deposition. When the shear stress on the bed is not sufficient to resuspend particles that contact and bond with the bed, deposition occurs. The shear stress at which there is an incipient net rate of deposition is termed the critical shear stress for deposition. This value may be the same or less than the critical shear stress for erosion, depending on the history of the bed surface. Extensive research has been conducted by Krone (6,7) and Parthenaides (11) on the deposition of cohesive sediments. Krone described the depositional behavior of cohesive sediments in the following manner:

The probability P of particles sticking to the bed increases linearly with a decrease in the bed shear and is given by

$$P = 1 - \tau_b/\tau_{cd} \tag{4}$$

where τ_b = bed shear stress and τ_{cd} = critical shear stress for deposition. In the absence of continuing aggregation the rate of loss from suspension is

$$\left.\frac{dC}{dt}\right|_d = - \frac{PV_sC}{\bar{d}} \tag{5}$$

where C is the suspended sediment concentration, and $\bar{d}$ = average depth through which the particles settle. Integration of Eq. 5 leads to

$$\log \frac{C}{C_o} = - K_o t \tag{6}$$

where $K_o = V_sP/(2.3\ \bar{d})$. This relation was verified in a recirculating flume where the aggregation rate was found to be negligibly slow at concentrations below 300 mg/1 so long as unusual eddy-producing disturbances to the flow were avoided.

At high concentrations, or under flow conditions where collisions of suspended particles are frequent relative to the time of observation, a relation that includes the effect of continuing aggregation was demonstrated that simplifies to

$$\log \frac{C}{C_o} = - K_2 \log t \tag{7}$$

where K_2 = empirical constant, C_o = initial concentration, and t = elapsed time. $K_2 = K_3 V_s P/\bar{d}$, where K_3 includes properties of the aggregating aggregate. For practical purposes $K_3 V_s$ can be combined to give an empirical constant.

Erosion. Most natural cohesive beds are hydraulically smooth in the range of flow conditions of concern. Hence, the hydraulic shear stress at the bed is an accurate measure of the entrainment force. Experiments have shown the existence of a critical shear stress that must be exceeded before erosion of a cohesive bed surface takes place.

The resistance of a cohesive bed to erosion depends mainly on the clay type and structure, and on the chemical composition of the pore and eroding fluids. Extensive studies have been made by Alizadeh (1) and Kandiah (5) on the critical shear stress of saturated cohesive soils. At bed shear stresses just above the critical value, erosion occurs particle by particle and this process is called "surface erosion". At higher levels of stress, however, the bulk shear strength of the bed may be exceeded. The portion of a bed in such a state is susceptible to "mass erosion", i.e., as the bed shear exceeds the critical shear strength of that portion of the bed it fails totally and is instantly suspended. This is most often the case in transient estuarial deposits.

The erosion rate for surface erosion is given by Parthenaides (11) as

$$\left.\frac{dc}{dt}\right|_e = M\left(\frac{\tau_w}{\tau_{ce}} - 1\right) \tag{8}$$

where τ_{ce} = critical shear stress for erosion, and M = erosion rate constant. The erosion rate for mass erosion is theoretically infinite. With the present state of knowledge the critical shear stress for erosion and the erosion rate constant must be determined by laboratory flume tests, or for stronger beds by testing in the rotating cylinder apparatus (13).

GOVERNING EQUATIONS

The mass balance equation for a conservative substance such as sediment in a turbulent flow can be written as

$$\frac{\partial c}{\partial t} + u\frac{\partial c}{\partial x} + v\frac{\partial c}{\partial y} + w\frac{\partial c}{\partial z} = \frac{\partial}{\partial x}e_x\frac{\partial c}{\partial x} + \frac{\partial}{\partial y}e_y\frac{\partial c}{\partial y} + \frac{\partial}{\partial z}e_z\frac{\partial c}{\partial z} + S \tag{9}$$

where

$C = \dfrac{\text{mass of suspended sediment}}{\text{mass of suspension}}$

u, v, w = x, y, z components of the sediment velocity

e_x, e_y, e_z = turbulent diffusion coefficients which include molecular diffusivity

S = rate of change in concentration produced by sources and sinks.

The vertical velocity of the sediment v will in general be different from that of the fluid V, by the settling velocity v_s, i.e., $v = V\text{-}v_s$. The horizontal velocities u and w of the fluid and sediment are assumed to be identical.

If a two-dimensional solution to Eq. 9 is desired it must be integrated along the third dimension. Since the velocity and concentration profiles in the third dimension are not constant in general, dispersion terms arise. If the average value of the concentration along the line on which it is integrated is c (this is already time averaged), then the concentration $\bar{c}$ at any point along the line may be written as

$$\bar{c} = c + c'' \tag{10}$$

where c'' is the deviation from the average.

Similarly,

$$\begin{aligned} \bar{u} &= u + u'' \\ \bar{v} &= v + v'' \\ \bar{w} &= w + w'' \end{aligned} \qquad (11)$$

When averaging over the depth (y-direction),

$$c = \frac{1}{d}\int_0^d \bar{c}\, dy \qquad (12)$$

where d is the depth at the point. Similar integrals give the values of u, v, and w. Also,

$$\int_0^d c''\, dy = \int_0^d u''\, dy = \int_0^d v''\, dy = \int_0^d w''\, dz = 0 \qquad (13)$$

since the sum of the deviations from the mean must be zero.

Substituting these relationships into Eq. 9 and depth averaging (y-direction) yields

$$\frac{\partial c}{\partial t} + u\frac{\partial c}{\partial x} + w\frac{\partial c}{\partial z} = \frac{\partial}{\partial x}(e_x + E_x)\frac{\partial c}{\partial x} + \frac{\partial}{\partial z}(e_z + E_z)\frac{\partial c}{\partial x} + S \qquad (14)$$

Here the overbars have been removed since every term is averaged and again Fick's Law is assumed to apply for dispersion, i.e.,

$$\overline{u''c''} = - E_x\frac{\partial c}{\partial x} \quad \text{etc.} \qquad (15)$$

where E_x, E_z are the dispersion coefficients. If the effective turbulent diffusion coefficients are defined as:

$$\begin{aligned} D_x &= e_x + E_x \\ D_z &= e_z + E_z \end{aligned} \qquad (16)$$

then the final two-dimensional equation is

$$\frac{\partial c}{\partial t} + u\frac{\partial c}{\partial x} + w\frac{\partial c}{\partial z} = \frac{\partial}{\partial x} D_x\frac{\partial c}{\partial x} + \frac{\partial}{\partial z} D_z\frac{\partial c}{\partial z} + S \qquad (17)$$

where

$$S = \frac{dc}{dt}\bigg|_e + \frac{dc}{dt}\bigg|_d \tag{18}$$

The erosion or deposition rate, as the case may be, can be obtained from Eq. 8 or Eq. 5.

REVIEW OF EXISTING MODELS

The authors are aware of three models for cohesive sediment transport, all reported in the last three years. The first for the Thames Estuary by Odd and Owen (9), the second by Ariathurai (2) for general two-dimensional application, and the third by Christodoulou, Leimkuhler and Ippen (3) for dispersion in coastal waters. A brief description of these models follows.

Odd and Owen. The sediment model developed by Odd and Owen for the Thames Estuary was the first mathematical model for cohesive sediment transport. These authors considered the flow to be divided into two unequal horizontal layers: a lower layer of constant thickness much smaller than the upper layer of varying thickness. The Thames Estuary was assumed rectangular in section with the width increasing exponentially along the length. The model was one dimensional and the equations proposed by Parthenaides (11) and Krone (7) for scour and deposition were used. The bulk flow equations for each layer were solved numerically neglecting the convective acceleration term. The sparse leap-frog finite difference method was used to give the water level, water surface slope, and depth averaged velocity, at each time step.

In their words, "Because of its relatively narrow and exponentially varying width, virtually constant depth, good tidal mixing and tides at the mouth close to sine curves, several simplifications could be made without significantly affecting the accuracy of the model." Most estuaries are shallow and extensive horizontally, and a two-dimensional model would provide far more general application.

Ariathurai. The finite element model developed by Ariathurai solved the two-dimensional depth averaged convection-diffusion equation (Eq. 17) with a source-sink term that was evaluated at each time step from the previous concentration. The model required a specified flow field, diffusion coefficients, and sediment characteristics, which were obtained from the laboratory studies by Krone (6, 7). The flow domain was subdivided into a series of triangular elements in which a quadratic approximation was made for the suspended sediment concentration. Galerkin's weighted residual method was used to solve the transient convection-diffusion equation.

The bed was subdivided into a number of layers each 2.5 cm thick for each element. Each layer had a specified bulk density and shear strength as measured by Krone (6). At each time step, depending on whether erosion or deposition took place, the layers were renumbered and new densities and shear strengths assigned. This yielded the bed elevation at each time step and accounted for consolidation of the bed. Since diffusion coefficients and sediment properties could be specified for each element, continuing aggregation could be accounted for by specifying the appropriate settling velocity in each element. However, an intuitive knowledge of the aggregating zone or a Lagrangian solution side by side would be required to predict the orders of aggregation.

This model was tested by predicting the deposition pattern downstream from a permeable barrier obstructing half of the width in a laboratory flume. The model predicted bed contours and suspended sediment concentrations that compared well with the observed values.

This model has since been modified under contract with the Dredge Materials Research Project, U.S. Army Corps of Engineers, using isoparametric quadrilateral elements that permit the use of curved sides. An efficient grid generator and contour plotting routine have also been incorporated. Work is in progress to verify the model with field measurements made in the Savannah Estuary (8). The modified model can be used with the vertical and axial dimensions for rivers, mixing zones, and other regions where breadth averaging is possible, or with the two horizontal dimensions for relatively shallow extensive areas where a logarithmic velocity profile is assumed to compute bed shear.

Christodoulou, Leimkuhler, and Ippen. This model was developed with the main assumptions that: the vertical distribution of sediments is independent of concentration; the velocity field is composed of a net drift and a superimposed sinusoidal tidal velocity; the depth of flow is constant; the dispersion coefficients bear a simple relation to the shear velocity. All three dimensions are considered: the horizontal distribution first solved by Harleman's method (4), and then the vertical distribution was computed from the average concentration. The quasi steady state solution due to the assumed vertical distribution does not yield reliable results for sediment concentrations for times shorter than about two tidal cycles. This model is therefore applicable to long term sedimentation in a velocity field that fluctuates much slower than that which the previous models consider.

Suspended sediments were grouped by size ranges, each with its own settling velocity. Furthermore, the model assumes a single vertical line source and that all the particles reaching the bed stick to it.

Though the source-sink term used has provision for erosion, the concept of critical shear stress is not used and in no application is erosion considered. It seems that this model is suited for application to a problem such as dredge spoil disposal in open waters with the kind of velocity profile assumed. It would be inappropriate for modeling deposition and resuspension in an estuarine environment.

RECOMMENDATIONS AND CONCLUSION

Any effort to simulate sediment motion requires a knowledge of the currents and tides in the estuary. The sediment model can at best be only as good as the flow model. In addition to local velocities, the bed shear is required to determine the rate of erosion or deposition, as the case may be. The internal shear in the fluid affects the rate of aggregation which in turn has a profound effect on the rate of deposition. Local velocities and tides can be obtained by field measurement, from physical models, or from mathematical models. For reasons of cost and convenience the latter method is preferred even though it is still in an early stage of development.

Although estuarine flows are generally three-dimensional in nature, in most cases averaging over the depth or width can provide a great deal of useful information. Until two-dimensional models have reached a greater degree of sophistication, the effort required to develop a three-dimensional model does not seem to be warranted. Satisfactory numerical solutions to the two-dimensional convection-diffusion equation have been developed, especially with the finite element method which provides sufficient stability and versatility.

One of the biggest problems in modeling cohesive sediment transport is to account for aggregation. The time elapsed since the particles became cohesive, the internal shearing in the fluid through which the particles move, the suspended sediment concentration, and the size distribution of the aggregates, determine the aggregation rate. This is a Lagrangian problem since the particle has to be followed. One method for predicting regions in which aggregation would occur is to send out mathematical tracers at each time step to obtain aggregation potential. The regions in an estuary within which aggregation rates are important, however, are easily identified by inspection of data on currents. Those portions of the fresh-salt water mixing region, flows downstream from pilings, and eddies where local velocity gradients are high, for example, are regions in which aggregation rates may be significant.

More theoretical and experimental studies with a view to obtaining new descriptions for the following are indicated:

- scour characteristics of recently deposited beds, i.e., critical shear stress, density, effects of thixotropy;
- settling velocity in turbulent flows, with and without continuing aggregation;
- effective diffusion coefficients in homogeneous and stratified flows.

It is anticipated that useful suspended sediment transport models for general applications to estuarial water quality management and to the design of navigation facilities will be in use in the very near future.

REFERENCES

1. Alizadeh, A. 1974. Amount and type of clay and pore fluid influences on the critical shear stress and swelling of cohesive soils. Ph.D. Thesis, University of California, Davis.

2. Ariathurai, C.R. 1974. A finite element model for sediment transport in estuaries. Ph.D. Thesis, University of California, Davis.
3. Christodoulou, G.C., W.F. Leimkuhler, and A.T. Ippen. 1974. A mathematical model for the dispersion of suspended sediments in coastal waters. Ralph M. Parsons Laboratory, Cambridge, Mass., Report No. 179.
4. Harleman, D.R.F. 1971. One-dimensional models. *In* TRACOR, Inc., Estuarine Modelling: An Assessment. Report to Water Quality Office, EPA.
5. Kandiah, A. 1974. Fundamental aspects of surface erosion of cohesive soils. Ph.D. Thesis, University of California, Davis.
6. Krone, R.B. 1963. A study of rheologic properties of estuarial sediments. Hydraulic Engineering Laboratory, University of California, Berkeley. Report 63-8.
7. ——. 1962. Flume studies of the transport of sediment in estuarial shoaling processes. Hydraulic Engineering Laboratory, University of California, Berkeley. Final Report.
8. ——. 1972. A field study of flocculation as a factor in estuarial shoaling processes. Committee on Tidal Hydraulics, U.S. Army Corps of Engineers. Technical Bulletin No. 19.
9. Odd, N.V.M., and M.W. Owen. 1972. A two-layer model of mud transport in the Thames Estuary. Proc. Institution of Civil Engineers, London. Paper 75175.
10. Owen, M.W. 1969. A detailed study of the settling velocities of an estuary mud. Hydraulics Research Station, Walingford, U.K. Report No. INT 78.
11. Parthenaides, E. 1962. A study of erosion and deposition of cohesive soils in salt water. Ph.D. Thesis. University of California, Berkeley.
12. Peirce, T.J., and D.J. Williams. 1966. Experiments on certain aspects of sedimentation of estuarine muds. Institution of Civil Engineers, London. Paper 6931.
13. Sargunam, A., et al. 1973. Physico-chemical factors in erosion of cohesive soils. Journal of the Hydraulics Division, ASCE, March, 1973.

INTRODUCTION TO SEDIMENTARY PROCESSES II

Convened by:
Maynard M. Nichols
Virginia Institute of Marine Science
Gloucester Point, Virginia 23062

Transport processes along the bed require an understanding of the bed material as well as the flow in the transition or boundary layer above the bed. Both sediment and flow are in a continual state of adjustment through erosion or deposition. In the following papers our knowledge of bed transport is advanced through new field observations and through new instruments for continuous measurements.

The potential for man's intervention in sediment processes is illuminated by case histories for the Veerse Mer and for San Diego Bay. In these estuaries, modifications by man outweigh natural sedimentation and become important geologic processes. These examples, together with the syntheses of lagoons should provide students an insight into the problems just beginning to be recognized.

SHEAR STRESS AND SEDIMENT TRANSPORT

IN

UNSTEADY TURBULENT FLOWS

W.F. Bohlen
Marine Sciences Institute
and
Department of Geology
University of Connecticut
Avery Point
Groton, Connecticut 06340

ABSTRACT: Predictions of sediment transport rates in coastal waters often display large errors. The source of the error is in part the result of insufficient attention to the character of the shear stress field in unsteady-nonuniform flows. A review of available laboratory and field data is used to show that shear stress magnitude and distribution will vary in response to the sense and amplitude of the horizontal pressure gradient. The response appears sufficient to alter both bed and suspended load transport. The consistency of the data indicates that present predictive techniques, based on uniform flow data, should be modified so as to permit inclusion of probable pressure gradient effects.

INTRODUCTION

Quantitative predictions of sediment transport rates in coastal and estuarine waters require an accurate evaluation of the role of the local velocity field in the erosion, transport and deposition of sediment. Typically such evaluations are prepared using the classical transport curves of Shields (15) or Hjulström (10) or the more recent relationships developed by Bagnold (1). The resultant estimates often prove to be highly inaccurate and display limited correlation with subsequent field observations.

The primary cause of these errors rests in the fact that the laboratory conditions used to derive the transport relationship often represent a poor analogue of the coastal environment. In particular the accuracy of laboratory estimates is limited by an inability to adequately scale sediment-flow interactions, by variations induced by biogenic cohesiveness (a factor affecting even the sand sized sediments) and by insufficient attention to transport effects induced by spatial and temporal variations in the velocity field. While a number of investigators have attempted to treat scaling problems and the question of mechanical cohesiveness (6, 17) relatively few studies have discussed the significance of accelerations and decelerations in the sediment transport process. The limited attention to these latter factors is primarily the result of research emphasis being placed first on investigations of transport in steady-uniform flows, the class of flows representative of a major portion of river and open ocean areas. In contrast, the velocity field characteristic of coastal and estuarine areas displays a high degree of spatial and temporal variability, and is more properly considered unsteady and non-uniform. Quantification of sediment transport in such a system requires an understanding of the influence of accelerations on transport mechanics. The results of these analyses will permit determination of the necessity for modification of the classical transport curves and serve to establish their range of applicability.

This paper will discuss some of the prominent features of sediment transport in unsteady-nonuniform flows. Primary emphasis will be placed on the probable influence of pressure gradients and the resultant accelerations and decelerations on the flow field. The results of this discussion will be used to interpret a variety of available field data including a recent set of measurements obtained in eastern Long Island Sound. The work is intended to provide an initial assessment of the significance of these phenomena and to encourage consideration of these factors in future investigations.

FLOW CHARACTERISTICS

Geophysical sediment transport is generally agreed to be a turbulent flow phenomenon. The ability or competence of a given flow to move sediments varies primarily as a function of the magnitude and distribution of shear stress. This parameter governs both the rate of transport and the mode (i.e., bed or suspended load). The character of the shear stress in turbulent flows can be examined using the time averaged momentum equation. For incompressible-viscous fluid in a non-rotating frame this may be written:

$$\underbrace{\rho\left(\underbrace{\frac{\partial U_i}{\partial t}}_{\text{Local Acceleration}} + \underbrace{U_j\frac{\partial U_i}{\partial X_j}}_{\text{Convective Acceleration}}\right)}_{\text{Acceleration}} = \underbrace{-\frac{\partial P}{\partial X_i}}_{\text{Pressure Gradient}} + \underbrace{\frac{\partial}{\partial X_j}\left(\underbrace{\mu\frac{\partial U_i}{\partial X_j}}_{\text{Viscous Stress}} - \underbrace{\rho u_i u_j}_{\text{Turbulent Reynolds Stress}}\right)}_{\text{Shear Stress}} + F_i \qquad (1)$$

where ρ = Fluid density
μ = Molecular viscosity
t = Time
P = Time average pressure
U_i, U_j = Time average velocity
u_i, u_j = Fluctuating velocity components
i = 1,2,3 corresponds respectively to
j = 1,2,3 coordinate axes x,y,z
F_i = Body forces (gravity).

In the case of steady-uniform channel flows Eq. (1) degenerates to a simple balance between gravitational forces and shear stress. The internal pressure field is hydrostatic and displays no horizontal or lateral variability. Shear stress varies linearly from zero at the free surface to a maximum at the bed. Velocity varies only over the vertical and, near the sediment-water interface, is adequately described using logarithmic relationships. These are the conditions under which the majority of the sediment transport relationships have been derived. Field applications of these relations assume similar flow conditions. The assumption is useful in that it permits simple calculation of boundary shear stress (16) and direct application of the laboratory data. Unfortunately geophysical flows generally fail to display simple spatial and temporal uniformity. Examination of oceanographic characteristics suggests that the assumption of uniformity may be overly simplistic.

With the exception of a relatively limited class of river flows the majority of geophysical flows are driven by variations in the internal field of pressure. Horizontal pressure gradients produced by spatial variations in the density field or by tide or wind induced free surface slopes, tend to produce a three dimensional field characterized by some degree of temporal variability. The addition of the pressure forces to the total force field (Eq. (1)) causes an alteration in the distribution and magnitude of both mean velocity and shear stress. Mean velocity profiles no longer display simple logarithmic variability (5, 13, 14). Concurrent stress distributions become nonlinear with stress magnitude varying in response to the sense and magnitude of the pressure gardient. A typical response of the shear stress field is displayed in Fig. 1. This curve, taken from Bradshaw (5), shows normalized shear stress distributions ($\tau/\rho U_1{}^2$) within the flat plate boundary layer for three different longitudinal pressure distributions: Zero longitudinal pressure gradient, a=0; a mild adverse pressure gradient a= −0.15; and a strong adverse pressure gradient a= −0.255. In the absence of a pressure gradient the shear stress distribution is nearly linear with maximum values in the vicinity of the boundary (y/δ_{995} =0). The introduction of an adverse pressure gradient induces an evident nonlinearity and an increase in peak stress levels. The position of maximum stress moves away from the boundary. Boundary stress values are consistently lower than those observed in the zero pressure gradient case.

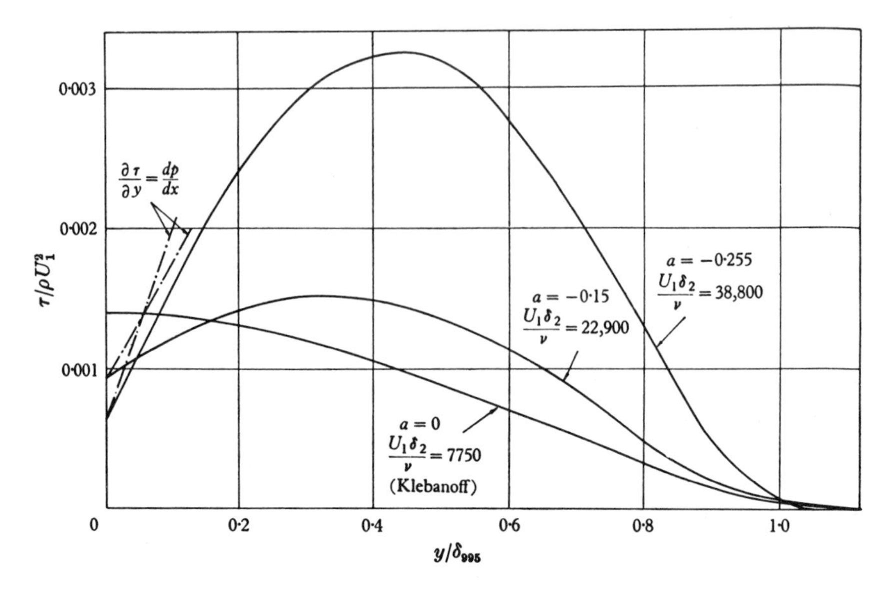

Figure 1. Shear stress distributions within the boundary layer (from Bradshaw (5)).

The effects of favorable or positive pressure gradients will be similar in magnitude to those produced by negative pressure gradients but differ in sense. Mean velocity profiles depart from a logarithmic distribution due to higher near boundary velocities. Boundary shear stress levels will be higher than the zero pressure gradient value and vertical distributions will display a nonlinearity characterized by upward concavity.

The trends indicated by the available data suggest that pressure gradient effects can significantly alter sediment transport mechanisms. The noted variations in boundary shear stress, if of sufficient magnitude, could serve to modify the onset of initial motion and the rates of bedload transport. Stress distributions in the interior of the flow could influence suspended load transport through modifications of the eddy diffusion characteristics. It seems worthwhile, therefore, to investigate the evidence for and the magnitude of pressure gradient effects in geophysical flows.

SHEAR STRESS DISTRIBUTIONS IN GEOPHYSICAL FLOWS

A review of the available literature reveals relatively few field observations of sufficient detail to determine alterations in shear stress magnitude or distribution induced by pressure gradients. Indirect evidence of the significance of this factor is provided by the current measurements described in Wimbush and Munk (18) and the suspended load observations presented by Postma (11). The former study noted that during the tidal cycle current velocities increased "rapidly and smoothly" during the accelerating phase and decreased "slowly and irregularly" during the decelerating phase. These data suggest that the adverse pressure gradient affecting the flow retardation during the declerating phase of the tide is of sufficient magnitude to alter the high frequency structure of the velocity field. Such modifications should result in measurable redistribution of shear stress.

The suspended material concentrations discussed by Postma (11) vary substantially over the tidal cycle. Comparison of flood distributions with those obtained during the ebb reveals an irregular assymetry characterized by a lack of simple dependence on the mean velocity. The data suggest a regular variation in competence. The role of the pressure gradient in this phenomenon however is obscured by a variety of other processes dominated by the advection of "clouds" of fine grained suspended materials past the sampling station.

Direct determinations of the distribution of shear stress in coastal currents have been reported by Bowden, Fairbairn and Hughes (4). A portion of their data, obtained near Anglesey, North Wales, in the Irish Sea, is reproduced in Fig. 2. The shear stress is observed to display a regular variation in magnitude and distribution over a tidal cycle. Near periods of maximum flood and ebb, shear stress displays a nearly linear variation over the vertical. Distributions are similar to those expected in uniform flow conditions.

Decelerating tidal current induces a progressive nonlinearity similar in character to that presented by Bradshaw (5). Maximum distortion occurs near the time of slack water. The initiation of flow acceleration causes a reversal in the shape of the shear stress curves. Distributions become concave upwards. The magnitude of the stress at a given level in the interior of the flow is typically lower than that observed during periods of deceleration. This nonlinearity persists until approximately 1.5 to 2.0 hours after slack water at which time nearly linear conditions are reestablished.

These data indicate that in this location the tidally induced pressure gradients are of sufficient amplitude to produce variations in shear stress distributions similar in character to those observed in laboratory studies of positive and negative pressure gradient effects. The influence is confined to a period around slack water representing approximately 30% of the tidal cycle. It seems reasonable to assume that mean velocity distributions will also be perturbed during this period. This fact was not discussed in the work of Bowden et al. (4) and seems to have been disregarded since the calculations of bed shear stress proceeded on the assumption that the velocity profiles remain logarithmic.

Additional evidence of the significance of pressure gradient effects has recently been presented by Gordon and Dohne (9). Measurements of the high frequency structure of the velocity field in the Choptank River, obtained using a pivoted vane, mechanical current meter revealed a regular variation in turbulence over the tidal cycle. As shown in Fig. 3 (taken from Gordon (8)) kinetic energy levels or turbulence intensities were typically observed to be higher during decelerating phases than during periods of acceleration. The expected response

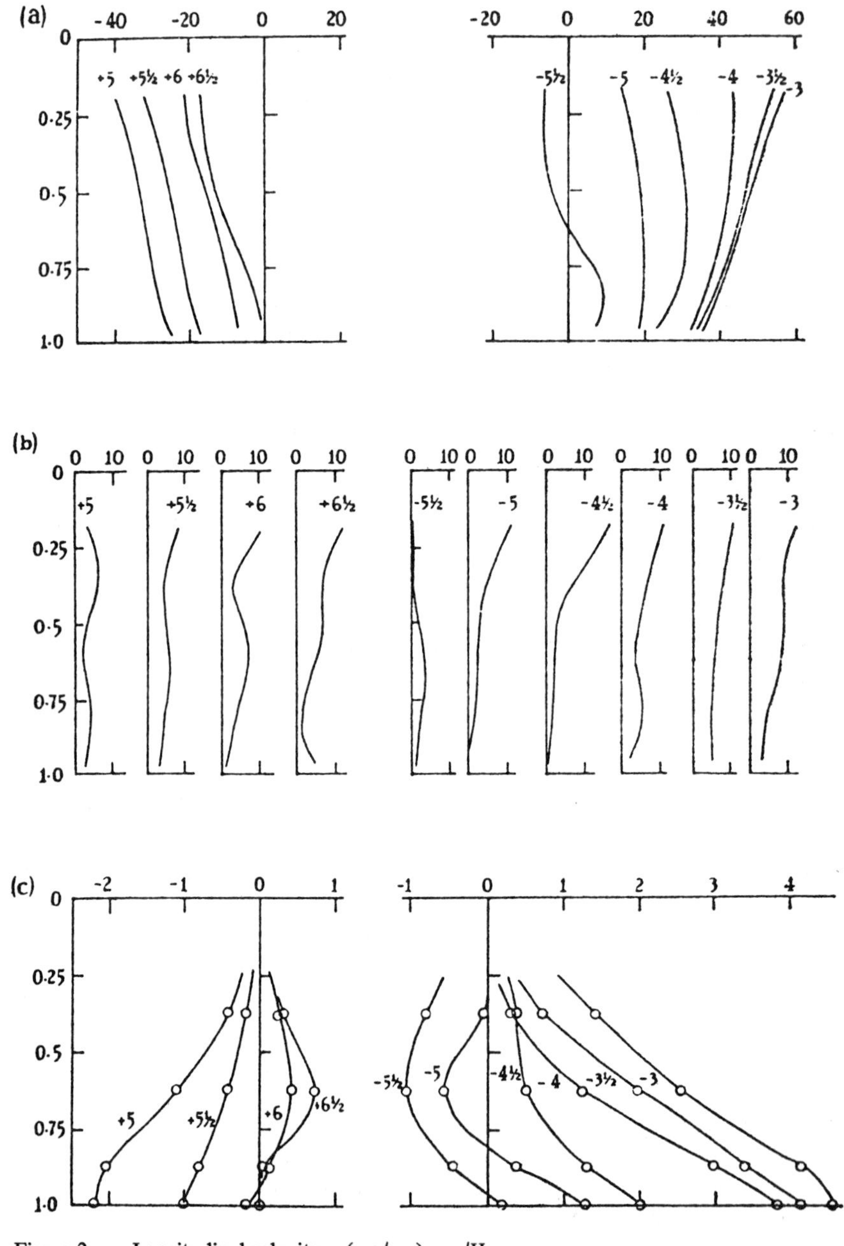

Figure 2. a. Longitudinal velocity u (cm/sec) vs z/H.
b. Cross-stream velocity v (cm/sec) vs z/H.
c. Shear stress distribution F_{ZX} (dynes/cm^2) vs z/H half-hourly intervals.
Station 2 July 19, 1957, 0730-1200 GMT (5 hr after HW to 3 hr before next HW) (From Bowden, Fairbairn and Hughes, (4)).

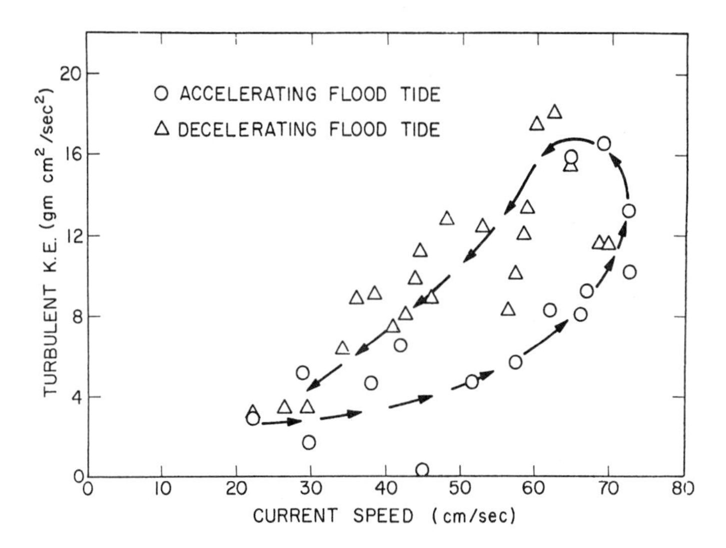

Figure 3. The variation in turbulent kinetic energy over a tidal cycle. (from Gordon, (8)).

of the shear stress can be inferred from these data by recalling that turbulent kinetic energy $= 1/2\ (\overline{u^2} + \overline{v^2} + \overline{w^2})$. The laboratory observations of Bradshaw (5) have shown that the pressure induced increase in shear stress noted above (Fig. 1) is accompanied by a simple increase in the magnitude of the mean square velocity level for each of the fluctuating components u, v and w. Increasing kinetic energy levels therefore suggest concurrent increases in turbulent shear stress.

The observed trends in stress magnitude and distribution are consistent with expected pressure gradient effects. In contrast to the data presented by Bowden et al. (4) however the measurements of Gordon and Dohne indicate that within the Choptank, gradient effects persist over the entire tidal cycle.

The general agreement within the available data suggests that pressure gradient effects should be considered within efforts to describe sediment transport characteristics in coastal waters. An example of an effort to evaluate these effects specifically to complement sediment transport studies is provided by a recently completed set of measurements in eastern Long Island Sound.

LONG ISLAND SOUND MEASUREMENTS

Eastern Long Island Sound (Fig. 4) represents a complex transition zone separating Block Island Sound to the east from central Long Island Sound to the west. The region receives the discharge from two major rivers, the Connecticut and the Thames. These waters are discharged into an area characterized by high energy tidal currents dominated by the semi-diurnal tide (M_2). Peak velocities

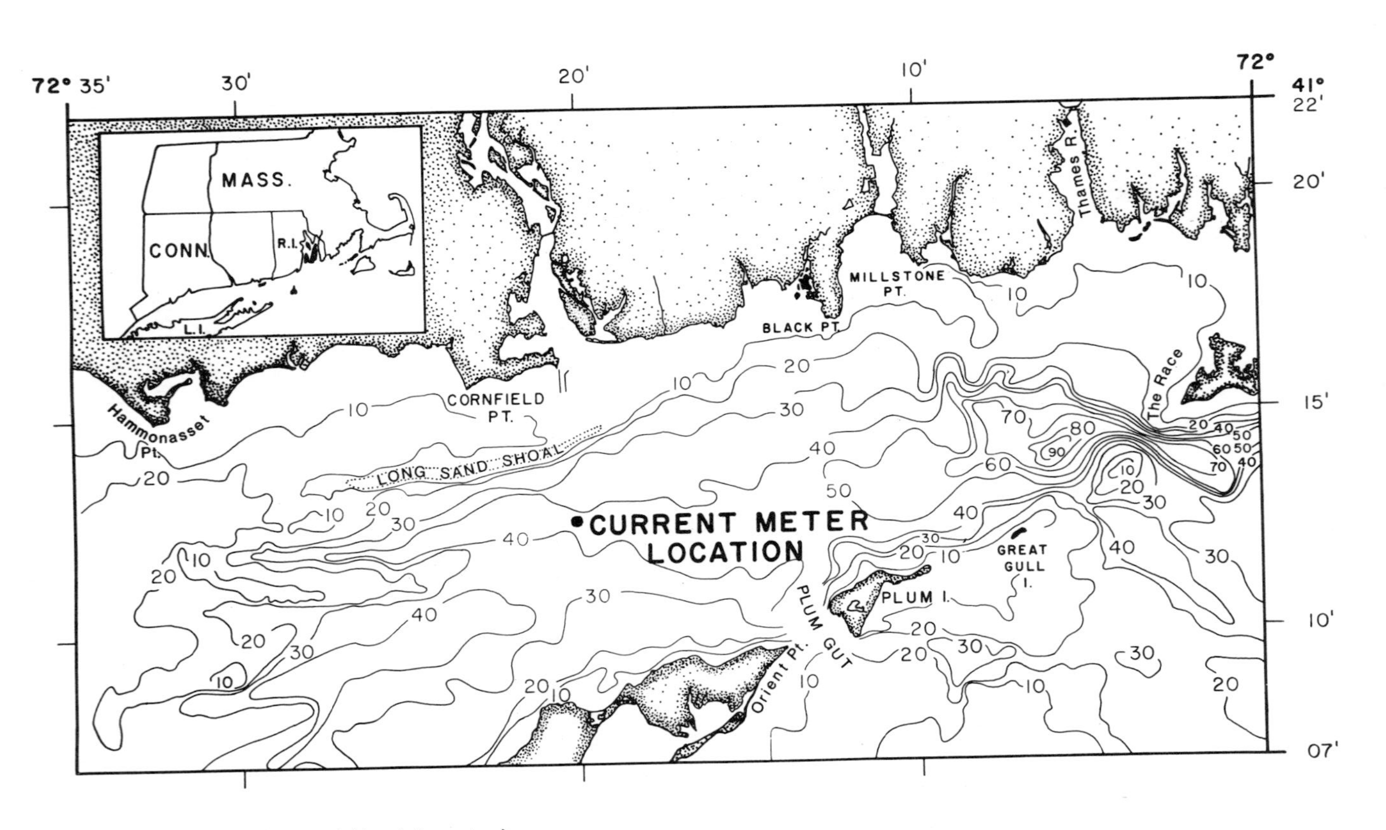

Figure 4. Eastern Long Island Sound (depth in meters).

exceed 200 cm/sec in the Race decreasing to approximately 125 cm/sec in the vicinity of the Connecticut River. The resultant density field generally displays limited vertical stratification except during periods of maximum stream flow (12).

In October 1971 a program designed to investigate the characteristics of suspended material transport in eastern Long Island Sound was initiated. Efforts were made to describe both long term, low frequency variability (3) and high frequency perturbations induced by storm events. This latter feature was to be detailed using a moored instrumentation array located approximately 5km south of the entrance to the Connecticut River in 52m of water (Fig. 4). Site selection followed detailed bathymetric surveys which showed the bottom in the area to be essentially planar with no evidence of significant sand waves.

To complement the variety of data obtained by this array (2) the high frequency structure of the near bottom velocity field was examined using an EGG Model CT3 electromagnetic current meter. The instrument was located 1.0m above the bed in a fixed mount that provided full freedom in rotation (360°) with negligible limitations in roll and pitch. Ballast was employed in the current meter to maintain near neutral buoyancy and accurate alignment in the horizontal plane. In this configuration the instrument provides measurements of the longitudinal component of the velocity field using a vertically oriented magnetic field to sense speed and an internal compass to monitor instrument heading. These data are internally recorded in digital format on magnetic tape.

The measurements to be discussed were obtained during two, three day, observation periods in late February and early March 1975. (February 28-March 3, 1975 designated B01012 and March 5-March 8 designated B01019.) The instrument was programmed to sample the velocity field at a rate of two samples (ea. V-θ pairs) per second for a period of 8.5 minutes. Sampling was repeated each 0.5 hr. Cassette capacity allowed approximately 2.5 days of observation using this sampling scheme.

Following processing by the manufacturer, these data were analyzed using an IBM 360/370 computer. Mean (U) and root mean square $[(\overline{u'^2})^{1/2}]$ velocities were calculated by least squares fitting the data to a second order polynomial. The resultant curve permits definition of a mean value and analysis of the fluctuations around the mean in the presence of time variant average flows (9).

The data (Figs. 5 and 6) display a regular semi-diurnal variability. Rms velocity levels appear to be simply correlated with concurrent mean velocity. As a result, turbulence intensity $(\overline{u'^2})^{1/2}/U$ remains nearly invariant over the entire tidal cycle for both periods of observation. The sharp peaks in the vicinity of slack water are presently considered artifacts produced by the difficulty in defining mean velocity during this period. The evident differences in the amplitude and distribution of velocity observed during the two sampling periods are the result of the semi-monthly inequality. No significant meteorological events occurred during the sampling period.

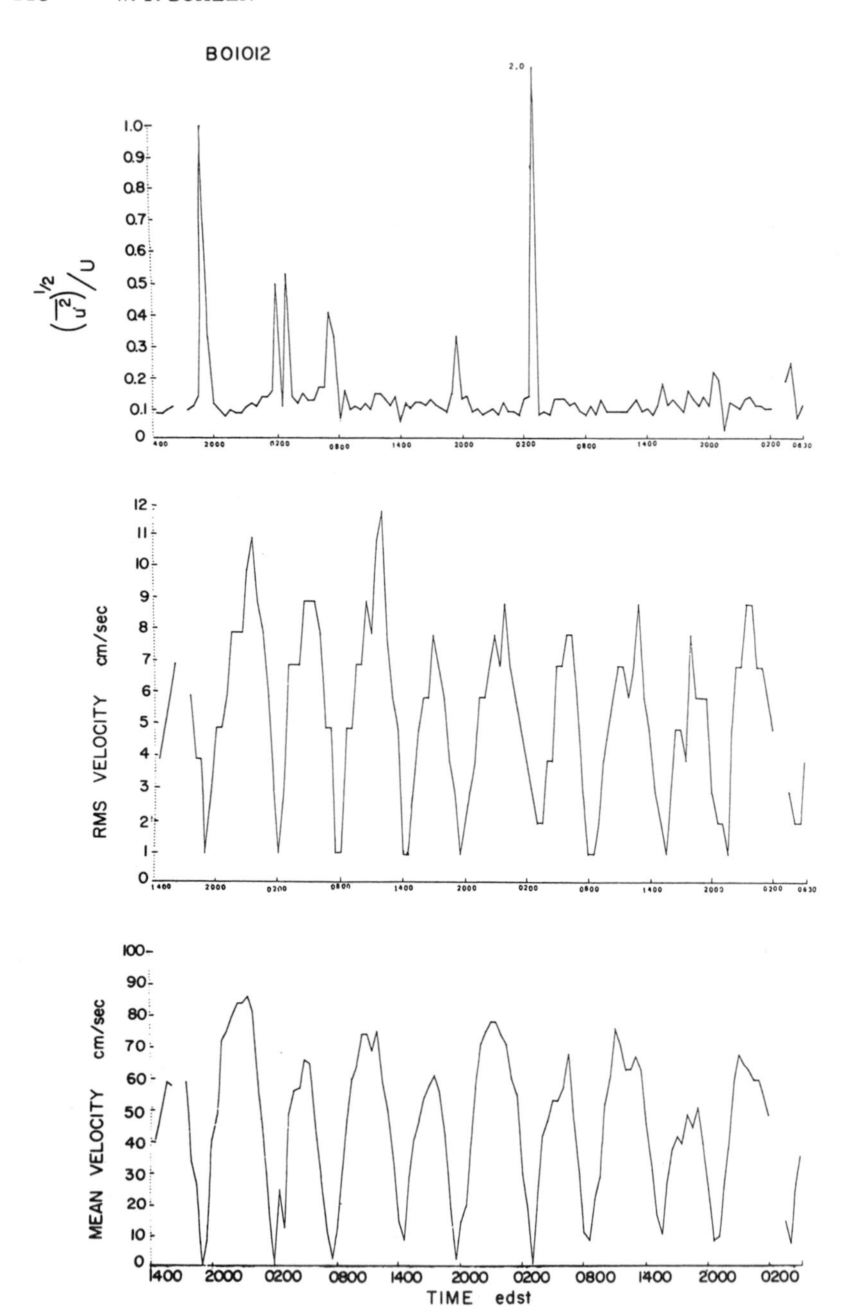

Figure 5. BO 1012 Temporal variations in near bottom turbulent velocity field–eastern Long Island Sound, Feb. 1975.

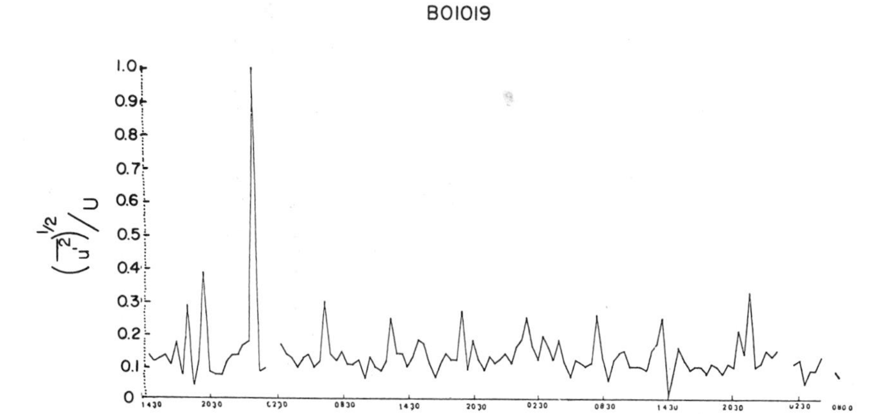

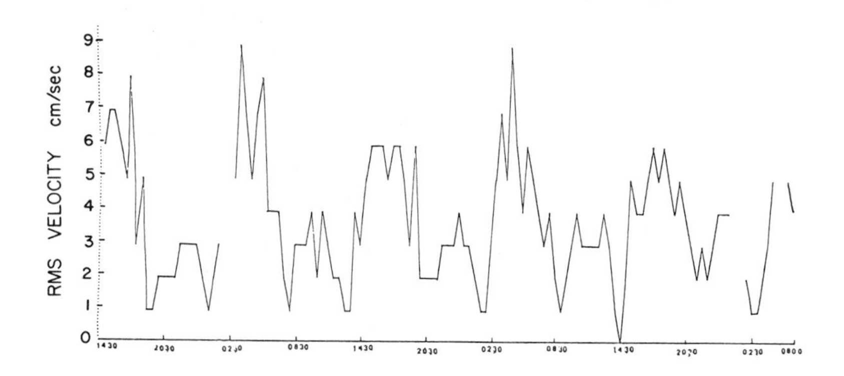

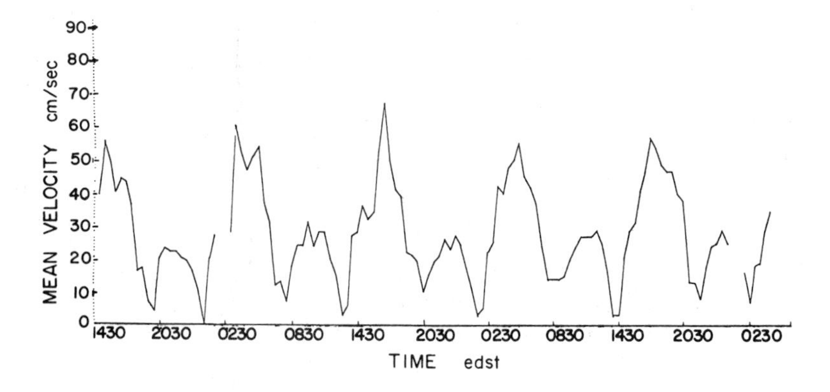

Figure 6. BO 1019 Temporal variations in near bottom turbulent velocity field–eastern Long Island Sound, March, 1975.

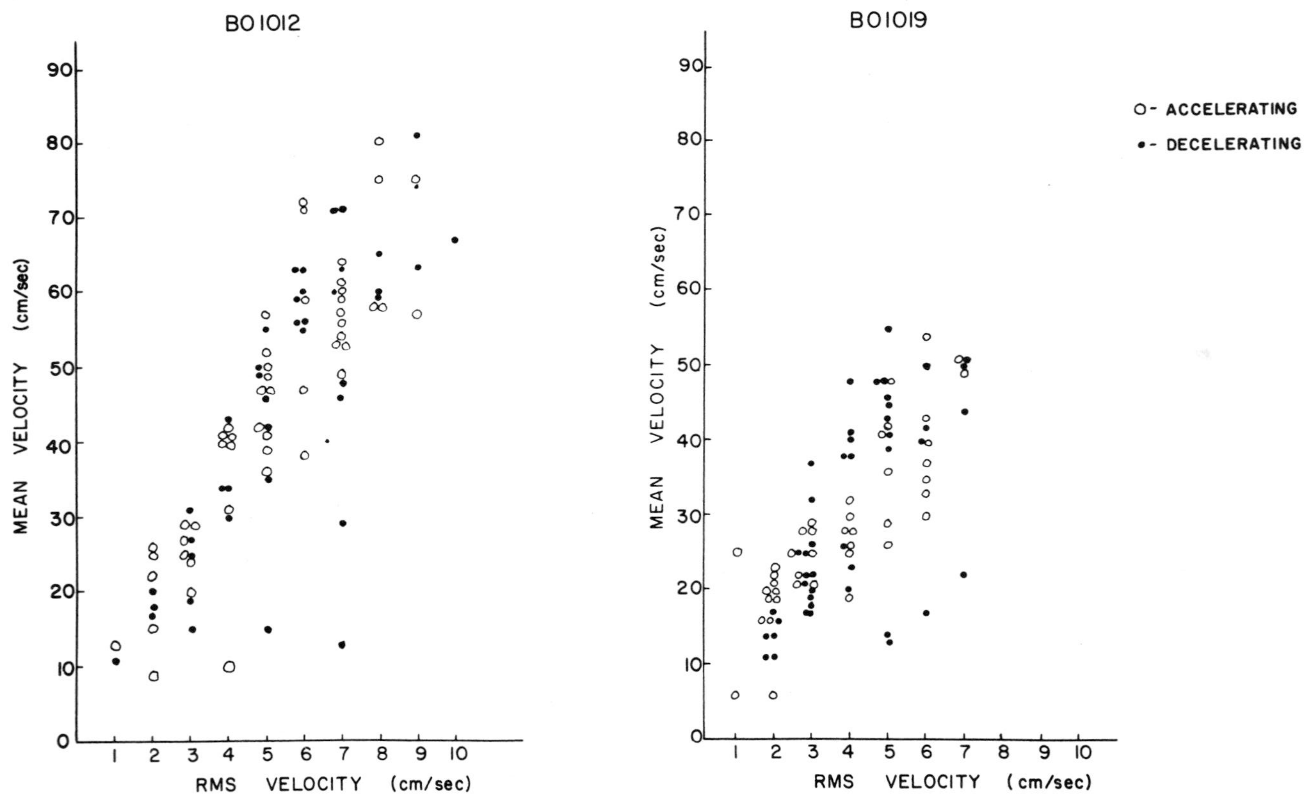

Figure 7. The relationship between mean velocity and rms velocity levels.

An estimate of the influence of pressure gradient effects on shear stress magnitude and distributions can be prepared by examining the relationship between rms velocity and concurrent mean velocity during accelerating and decelerating tidal phases. Although accurate evaluation requires time series observations of each velocity component at several locations on the vertical, previous empirical studies (7) have shown that the ratios between u,v, and w remain essentially invariant despite variations in mean velocity and that therefore shear stress behavior can at least be qualitatively evaluated by simply monitoring the longitudinal velocity components U and u. These data, plotted in Fig. 7, show the rms velocity varying in a nearly linear fashion as a function of mean velocity. Rms levels fail to display any obvious dependence on the character of the acceleration field. Similar levels can be observed during both accelerating and decelerating tidal phase. This behavior is similar to that observed by Bowden et al. (4) at times other than the slack water period and seems characteristic of near uniform flow conditions.

The behavior of the near bottom velocity field around periods of slack water is not shown in Fig. 7. Precise evaluation of data obtained during this interval is difficult because of inaccuracies induced by the directional instability of the current meter at low velocities. Despite this limitation, review of the data set indicates generally higher rms levels during periods of deceleration before slack water. Although the trends suggest variability similar to that observed by Bowden et al. (4) more work is required to establish the extent and significance of this variation.

DISCUSSION

The variety of field observations indicate that shear stress distributions in coastal flows can be expected to display significant nonlinearity over some portion of the tidal cycle. During these intervals both mean velocity profiles and turbulent velocity structure measureably depart from the uniform flow conditions typically studied in laboratory experiments. The quantitative effects of these variations on sediment transport and the accuracy of the predictions based on empirical data remain to be demonstrated. Present indications only suggest that these effects should not be neglected.

Modification of the classical sediment transport formulae to permit application in unsteady flows requires careful evaluation of the factors governing the observed stress distributions. While it is generally agreed that stress alterations result primarily from variations in the horizontal pressure gradient, the differences between the systems observed in each of the study areas suggest that stress behavior is not a simple function of the magnitude of the pressure gradient. Despite the fact that the pressure gradients in each case are of comparable magnitude ($0(10^{-2})$ dynes/cm^3) stress distributions within the Choptank River (13) display a more prominent variability over the tidal cycle than those observed in the Irish Sea (4) and eastern Long Island Sound. These results suggest that the influence of flow non-uniformity on sediment transport may vary in response to a variety of local (i.e. site specific) factors. Additional variability may be introduced by instrumental errors, the factor undoubtedly responsible for portions of the erratic behavior observed in several investigations (9).

An estimate of the probable factors governing local variability can be developed by examining Eq(1). Analysis of the spatial derivatives of this expression indicates that nonlinear stress distributions must be accompanied by finite temporal variability in the shape of the mean velocity profile. Periods characterized by increasing or decreasing magnitude but nearly constant profile will tend to display linear stress distributions. The rate of development of the mean velocity profile is governed by the variety of boundary layer characteristics. These features, in turn, represent the resultant of interactions between local hydrographic conditions, basin configuration and boundary roughness. This combination will determine the intensity and duration of the non-uniform flow conditions developed during each tidal cycle. The resultant stress distributions and sediment transport must, therefore, be expected to display measurable dependence on local characteristics. Future studies of the effects of nonlinear stress distributions on coastal sediment transport must establish the extent and nature of this dependence so as to permit general applicability of the experimental results. Such studies may very well indicate that the present dependence on the magnitude of the horizontal pressure gradient represents an unjustified oversimplification.

ACKNOWLEDGMENTS

The measurements in eastern Long Island Sound were made possible through the loan of the electromagnetic current meter and data processing assistance by the Environmental Equipment Division of EGG in Waltham, Massachusetts. The service was arranged by Dr. Lloyd Lewis with technical assistance provided by Mr. Norm Berry. At the University of Connecticut, field operations, data processing and analysis were supervised by Mr. Fred Everdale. The bottom mount was constructed by Mr. David Good. Assistance of especial value was provided by Captain Jack Blumie and the crew of R/V T-441. The author gratefully acknowledges this assistance.

REFERENCES

1. Bagnold, R.A. 1963. Mechanics of marine sedimentation, p. 507-528. *In* M.N. Hill (ed.), The Sea. Interscience Publishers, New York. Vol. 3.
2. Bohlen, W.F. 1974. Continuous monitoring systems in Long Island Sound: Description and evaluation. Proc. of I.E.E.E. Intl. Conf. on Engineering in the Ocean Environment, Halifax, Nova Scotia, Aug. 1974. Vol. 2:61-69.
3. ——. 1975. An investigation of suspended material concentrations in eastern Long Island Sound. J. of Geophys. Res. 80(36):5089-5100.

4. Bowden, K.F., L.A. Fairbairn and P. Hughes. 1959. The distribution of shearing stresses in a tidal current. Geophys. J. Roy. Astron. Soc. 2:288-305.
5. Bradshaw, P. 1967. The turbulence structure of equilibrium boundary layers. J. of Fluid Mech. 29 (4):625-645.
6. Frankel, L., and D.J. Mead. 1973. Mucilaginous matrix of some estuarine sands in Connecticut. J. of Sed. Pet. 43(4):1090-1095.
7. Gordon, C.M. 1975. A simplified, empirical method for estimating shear stress in the benthic boundary layer. E⊕S Trans. Amer. Geophys. Un. 56(6):377.
8. ———. 1975. Sediment entrainment and suspension in a turbulent tidal flow. Mar. Geol. 18:M57-M64.
9. ———, and C.F. Dohne. 1973. Some observations of turbulent flow in a tidal estuary. J. of Geophys. Res. 78(12):1971-1978.
10. Hjulström, F. 1939. Transportation of detritus by moving water, p. 5-31. *In* P.D. Trask (ed.), Recent Marine Sediments, Amer. Assoc. of Pet. Geol., Tulsa, Oklahoma.
11. Postma, H. 1967. Sediment transport and sedimentation in the estuarine environment p. 158-179. *In* G.H. Lauff (ed.), Estuaries. Amer. Assoc. Adv. of Sci., Washington, D.C.
12. Riley, G.A. 1967. Transport and mixing processes in Long Island Sound. Bull. Bingham Oceanog. Coll. 19:35-71.
13. Schraub, F.A., and S.J. Kline. 1965. A study of the structure of the turbulent boundary layer with and without longitudinal pressure gradients. Rpt. MD-12, Dept. of Mech. Eng., Stanford University, Stanford, California, 157p.
14. Schubauer, G.B., and C.M. Tchen. 1961. Turbulent Flow, Princeton Univ. Press, Princeton, New Jersey, 122 p.
15. Shields, A. 1936. Anwendung der Ähnlichkeitsmechanik und der Turbulenzforschung auf die Geschiebebewegung. Mitteil. Preuss. Versuchsanstalt für Wasserbau u. Schiffbau. Heft 26.
16. Sternberg, R.W. 1972. Predicting initial motion and bedload transport of sediment particles in the shallow marine environment, p. 61-82. *In* D.J.P. Swift, D.B. Duane, and O.H. Pilkey (eds.), Shelf Sediment Transport: Process and Pattern. Dowden, Hutchinson and Ross, Stroudsburg, Pa.
17. Webb, J.E. 1969. Biologically significant properties of submerged marine sands. Proc. of Roy. Soc. of London, Ser B. 174:355-402.
18. Wimbush, M., and W. Munk. 1970. The benthic boundary layer, p. 731-758. *In* A. Maxwell (ed.), The Sea. Interscience Publishers New York. Vol. 4(1).

THE MEASUREMENT OF BED SHEAR STRESSES AND BEDLOAD TRANSPORT RATES

K. R. Dyer
Institute of Oceanographic Sciences, Crossway,
Taunton, UK

ABSTRACT: Recent advances in the understanding of the characteristics of water flow near the sea bed and of the effect of the flow on sediment grains suggests that the bedload movement of sediment needs examination using more sophisticated techniques than those used in the past. Measurements of bedload transport rates have previously been carried out by a variety of sediment traps and samplers. These interfere with the flow, are selective in their retention and do not provide continuous samples. Measurements of the rate of advance of bedforms has been used, but this only gives a minimum transport rate. Techniques are required that will give quantitative estimates of the transport rate on a variety of time and space scales which can be related to the intermittancy in the bed shear stresses and the variations in bed form topography. Recent developments in electronics and instrumentation have suggested a number of techniques which are being applied to these problems. Those selected as being potentially most useful are; a particle impact counter, the measurement of the self-generated noise of the sediment movement, and the development of a transponder simulating gravel particles. These techniques have limitations in their grain size response, but amongst their advantages is that of providing continuous measurements over a considerable period.

INTRODUCTION

One of the biggest hurdles in studies of marine sedimentation is the prediction of sediment transport rates. This requires relating the fluid power to the work rate for the moving solids. The fluid power is normally defined as the bed shear stress times the mean velocity and the sediment transport rate as the mass of grains passing through a unit width of the cross section per unit time. Thus the transport rate involves the number of moving grains, their size and the speed at which they are moving. As we are dealing here specifically with bedload movement in the sea we are considering grains coarser than about 150 μm

moving at the rippled and duned stages just above the threshold. Consequently suspension is negligible.

The relationship between grain speed and fluid shear has not been extensively examined. Notable, however, are the laboratory studies (26, 15). There are many theories available from which the rates of bedload transport in steady unidirectional flows can be calculated. These have been developed from physical arguments (2), empirical measurements (35), probabilistic considerations (14) and dimensional reasoning (1). Many of them have been proved with laboratory measurements with which they are consistent but there are large discrepancies when tested with other laboratory and field measurements. This has been shown by the recent comparison of many theories (53). At high transport rates most theories predict that the total transport of solids is proportional to the second or third power of the bed shear stress, but the differences between the theories are marked at low transport rates and may be due to a variety of causes. There are difficulties in scaling the laboratory results up to natural size, depth variations in particular have .to be considered (54), and cross sectional form and sinuosity may also be important. The form drag of the boundary changes drastically at the lower flow stages so that there are difficulties in scaling the fluid power correctly. However, one of the most crucial factors is the lack of suitable devices for measurement of transport rates in the field.

The application of these theories to the unsteady oscillatory flows in the sea brings further problems. There are times during the tidal cycle when the currents and sediment transport may be quasi steady, but at other times acceleration and deceleration may produce hysteresis effects (18). The interaction of these effects over a tidal cycle produces movements which are unpredictable at the moment. Directions of movement become particularly important. Also the particle sizes may be markedly non uniform and the sea bed may become armoured with coarser, lag material introducing problems of lack of sediment supply. However Sternberg (48) has adapted Bagnold's theory and applied it to the sea by using a variable proportionality coefficient relating the fluid power to the bedload transport rate. The coefficient was empirically a function of excess shear stress above the threshold.

A significant improvement could be made to present predictions if the velocities and stress fields could be computed from knowledge of the mean pressure gradient and the shape of the boundary, and then applied with existing sediment transport theory. For this an understanding of the turbulent shear stresses and a means of separating skin friction from form drag is essential. However experimental verification would eventually be required. Grass (21) in laboratory experiments has pointed to the importance of considering turbulent effects in the interaction between the fluid and the grains. However it is not possible at the moment to predict the turbulent effects from the mean pressure gradient or the boundary shape. Consequently extensive field measurements are necessary. The measurement of sediment transport rates is a coupled problem involving measurement of bed shear stresses and the sediment's response to these driving forces. The characteristics of both must be considered. The purpose of this review is to examine the methods of measuring both the shear stresses and the bedload transport and some of the criteria that would appear to be important in the sea.

BED SHEAR STRESS

There are several ways in which the shear stress at the boundary can be estimated:

1. From the surface water slope. The bed shear stress $\tau = \rho gh \sin \beta$ where β is the slope of the water of depth h and density ρ. This technique is used in flumes where the water slope can be measured accurately with micrometer gauges. However this measure of shear stress is the total channel resistance to flow which can be considered as the sum of the surface frictional drag (ie the drag on the grains of the bed), drag on the walls and the form drag. Standard procedures exist for allowing for wall drag in flumes (28). The form drag is the sum of the horizontal pressure forces on the bed topography. Over a ripple or a dune feature the form drag commonly has values of a half of the friction drag, but because it varies with height and steepness (57), this ratio is not constant as discharge increases and bedforms develop. Raudkivi (41) shows that, because of varying drag, three different mean flow velocities can be obtained for one value of surface water slope. Form and friction drag have been calculated from mathematical models of flow over sand wave features (50). Friction drag varies considerably over sand wave features, reaching a maximum value about equal to that for the same flow over a flat bed (40) and a minimum value of about zero in the troughs, though a crest to trough shear stress ratio of 4:1 appears to hold for some circumstances in the sea (10). Consequently the measurement of water slope is related to the mean shear stress over a large area and over a time of the order of the mean velocity divided by the length of the channel. In the field there are additional problems in areas with longitudinal density gradients and accelerations caused by non-uniform cross-section. However water velocities in a tidal channel have been calculated from observations of water slope (12).

2. Measurement of the profile of mean velocity. Using the assumption of constant shear stress with height, the profile of mean velocity is given by the von Karman-Prandtl formula

$$\frac{u}{u_*} = \frac{1}{\kappa} \ln \frac{Z}{Z_0} \tag{1}$$

where u is the velocity at a height Z, the friction velocity $u_* = \sqrt{\tau/\rho}$ and Z_0 is the bed roughness length. The von Karman constant $\kappa = 0.4$. The logarithmic profile can be measured using miniature rotors (38), and Equation 1 has been commonly observed to hold within 1-2m of the sea bed in unstratified conditions (7,

6). However logarithmic profiles can only be observed by taking an average over a minute or more and even then there are variations in mean current velocity of up to 10% due to large scale longitudinal turbulent effects. However the variations in the slope of the profiles are small and the values calculated for shear stress are consistent to within about ½ dyne cm^{-2}. Additionally the near bed flow is strongly dependant on the small scale topography. Acceleration on the upstream side of the features and deceleration and even separation in their lee provides profiles that are non-logarithmic yet steady when averaged over a minute (9). The effect of ripples and dunes on the velocity profiles has effectively been ironed out by about one wave height above the crest and drastically different values of bed shear stress can be obtained on different sections of the profile. Interpretation in terms of frictional drag is difficult.

The bed roughness length has been related to the grain size diameter D of the sediment. In pipe flow experiments Nikuradse found that $Z_0 \sim 1/30\,D$, but in practice because of the presence of ripples, graded sediment and other non-uniformities of the bed the relationship is not simple. Values for Z_0 are generally in the range 0.01–0.1 cm. Often in the sea, bed shear stresses are calculated from velocities measured at a standard height above the bottom (generally 1 m) via the quadratic stress law and an assumed drag coefficient. This is the same as using a logarithmic profile with a constant Z_0 value, but the value of the roughness length is not too critical. Drag coefficient C_{100} values are generally of the order 10^{-3} (46). However, Ludwick (34) has measured large variations in C_{100} and because the drag coefficient changes continuously advocates using the complete velocity profile.

3. Direct measurement of shear stress on the boundary. These techniques involve the use of load cells or Preston (pitot) tubes and have been reviewed (20). For marine applications the first is difficult to emplace and is affected by moving sediment. The Preston tube measures the dynamic pressure very close to the sediment bed. The relationship between the dynamic pressure and the boundary shear stress is obtained by direct calibration in a flume or pipe, or by some assumption about the velocity profile near the bed and an independent measurement of velocity. This technique is commonly used in laboratory experiments without mobile sediment. Nece and Smith (38) have applied the technique to the sea and found a satisfactory comparison between the Preston tube measurements and calculations from the near bed slope of the velocity profile measured by miniature rotors.

4. Measurement of turbulent shear stresses. The shear stresses in a turbulent flowing liquid are caused by the momentum exchanges involved in fluid interactions in a velocity gradient. The turbulent or Reynolds stresses on a plane parallel to the sea bed are defined as

$$\tau = -\rho\,\overline{u'w'} \tag{2}$$

where u' and w' are the turbulent deviations from the mean longitudinal and vertical velocities respectively, and the bar denotes an average over some suitable time. Laboratory experiments have confirmed the theory that the Reynolds stresses are constant in the wall layer (32), but measurements in the sea do not always show this (4). Nece and Smith (38) found reasonable comparison between Reynolds stresses at about 1.5 m and boundary shear stresses by Preston tube providing averages in excess of 30 min were taken. The widespread observation of logarithmic velocity profiles suggest that the Reynolds stresses should be near constant over a layer near the sea bed in those situations.

Measurement of turbulence can be carried out with ducted rotor flowmeters (43), acoustic doppler current meters (56), or electromagnetic flowmeters (51, 30). The use of hot film probes in the sea have been reviewed (16).

Laboratory measurements have shown that the Reynolds stresses are not produced continuously, but occur intermittently in a "bursting" sequence. The general level of Reynolds stress is low, but there are periodic events creating very high shear stress. Flow visualisation studies have shown the events to be associated with distortions in the velocity profiles (8). The great contributions arise from a locally low horizontal velocity associated with an upward movement (an ejection or burst) and from a high horizontal velocity coupled with a downward movement (a sweep) (52). The bursts are about 1.35 times more intense than the sweeps, the bursts providing 77% of the stress and the sweeps 55% (33) and this proportion does not seem to change much over the outer region of the flow. The intervening periods between events produce a negative contribution to the shear stress. The large bursts give contributions of between 10 and 30 times the mean value of the local Reynolds stress (33, 13). The events travel downstream at about 0.8 of the mean velocity (33) and appear to emanate from the wall and travel outwards and grow larger, though less intense as they are convected downstream. The mechanism appears to be that of a transverse vortex rotating with the flow, which may be of a "hairpin" shape with the ends at the boundary (55) and inclined at 16-20° to the horizontal, which is convected downstream slightly slower than the mean flow (39) and which breaks down presumably by being stretched and because of intense dissipation in the steep velocity gradients with the surrounding fluid. By about 20 times the height above the bed downstream; the burst appears to have lost its distinctiveness on average (33). Even though the bursts emanate from the boundary, their occurrence appears to be determined by the outer-flow conditions (37), and the burst frequency a function of Reynolds number (8).

The same mechanisms appear to occur in the sea (17, 22). The latter author reports burst durations of 5-10 sec with 20-100 sec periods of lesser stress between, with 57% of the stress being produced in only 7% of the time. Similar intermittent turbulent events are often visible at the sea surface on calm days, especially in areas of rough topography, and are often associated with trains of sand waves. Measurements of other turbulent characteristics such as turbulent intensities and spectra are reported by Bowden (3), Bowden and Howe (5), Gordon and Dohne (19), Seitz (42) and others.

The implications of the bursting phenomenon for sediment movement in the sea are great because of the large instantaneous shears which are produced. Because most sediment movement in the sea takes place only a little above the threshold, spatial and temporal variations in burst intensity and bursting rate are likely to have significant effects on the sediment transport rates. The propagation speeds of the bursts appear to be similar to the speeds of grain movement so that individual grains could be moved relatively large distances in one go. The effect of a ring vortex on the grain bed has been described by Sutherland (49) who found that sediment could be set in motion by increasing the vortex frequency. Muller at al (36) have also considered the threshold lift produced by rotating vortices.

BEDLOAD

A great deal of qualitative information has been obtained on bedload movement of sediment by visual observations and photography in the laboratory (49, 21) and by television in the sea. Near the threshold grains move discontinuously both in space and in time, as bursts. At first a few grains are involved, but at higher stresses increasing numbers move. The grains appear to move downstream sometimes at small angles to the mean flow direction. The bursts of sediment movement travel downstream at something less than the mean flow velocity, sometimes also with a slight angle to the mean flow higher up. Individual particles do not travel more than a few centimeters but as one stops another further downstream is set in motion. Thus there is a continual exchange with the bed of particles in motion and there can be cross channel components of movement. As the mean flow velocity increases the bursts of movement occur more frequently, the grain motion is longer and more generally distributed over the bed. Ultimately grain suspension occurs. On a rippled bed the grain motion occurs first in the troughs and in the stagnation region and at shear stresses that would be insufficient to cause movement on a flat bed (44). This effect is also controlled by the height of the ripple or dune and appears to be related to high intensities of turbulence which more than compensate for the lower mean shear stresses in these areas. Hsu and Kennedy (23) have shown that turbulence intensities are greater in the troughs than near the crests. In the sea Sternberg (45) has shown strong unsteady currents along the troughs between the ripples to be important in moving sediment in an across-channel direction.

Obviously it is difficult to define the threshold of movement and a lot of the scatter in the flume results is probably the result of this. However visual observations in the sea have shown reasonable correspondence with flume results (47).

BEDLOAD MEASUREMENT

In flumes the bedload is normally measured by collecting the total sediment throughput in a stilling tank at the end of the flume. In rivers a variety of trap and pit devices are used. These are reviewed by Hubbel (24) and Graf (20). Traps have a number of limitations: they disturb the layer in which the transport takes place, their alignment is critical, they are selective in the sizes they trap and are consequently difficult to calibrate. Efficiencies are quoted ranging from 40-70%. In addition, they provide a time average transport over a small area on a 'once-off' basis.

The movement of bedforms has also been used. However this gives a minimum value of transport averaged over a considerable time as movement of the features is slow and many grains travel faster than the shape of the bedform. Kachel and Sternberg (29) have used stereophotography for examining ripple migration and related the bedload transport to shear stresses obtained from velocity profiles. Echo-sounding can be used on larger features in the sea. However, the bedforms move in a complex way, forward movement and backward movement occurring on the same crest (45, 31) and in the sea their movement is related to long term residual transport.

Radioactive tracers are useful for examining movement over periods in excess of a few days and show direction of movement particularly well, but there are difficulties in quantifying the transport rates mainly because it is also necessary to know the depth of mobile sediment, and have the tracer uniformly distributed through it. The use of tracers is reviewed in the International Atomic Energy Agency Technical Report 145 (25).

For use in the sea we require instruments that can measure bedload transport rates continuously and remote recording would be an advantage as most movement is likely to take place when direct recording is impossible. In use they would need to measure over spatial scales that are compatible with those involved in the shear stress measurements. The Institute of Oceanographic Sciences (IOS) have selected a number of techniques which appear to meet some of these requirements. They are all acoustic techniques and are undergoing development at the moment.

Particle impact counter. This consists of a piezoelectric ceramic disc transducer about 1 cm in diameter with a natural resonant frequency of about 2 mHz. A particle hitting the transducer will cause it to ring at its resonant frequency and mechanical damping restricts the vibration to about 10 cycles. Each pulse exceeding a certain threshold level, which is limited by noise, triggers a 2 msec monostable circuit giving a saturation count rate of about 500 counts per second. The small size should allow deployment close to the boundary within the zone of saltating grains and omnidirectionality could be achieved by having a vertically mounted cylindrical shell transducer. However not all grains will hit the transducer because of the divergence of the flow, the smaller, slower grains may not register and more than one grain may hit at one time. Consequently there will be calibration difficulties and absolute measurements may not be possible. However it should be particularly useful for indicating whether sediment is moving or not and for investigation of the threshold of movement. It should be usable in a continuous self-recording mode, but would most likely be used in a burst-sampling mode when deployed for long periods. Because of its small size and fast response time use in conjunction with electromagnetic flowmeters is anticipated and in principle this would allow investigation of the eddy diffusion of sediment. Counts averaged over longer periods would allow use with velocity profile measurements.

Self-generated noise of sediment movement. Collisions between one particle and another make each particle resonate at a frequency which should be dependent on the grain size and mineralogy. The frequency should rise with decreasing size and the total output of noise should be related to the number of grain impacts. Johnson and Muir (27) have investigated the noise emitted by a uniform 5 mm gravel in a flume over a frequency range up to 150 kHz. In similar work, ambient noise in the flume from the pump has been found to be a large problem. However measurements at IOS in a rotating drum in still water with a variety of sizes of natural and artificial grains and over a wider frequency range, have shown that the noise is generated over a broad spectrum, but the peak frequencies are grain-size dependent. Experiments in the field have experienced problems with other ambient, electronic and ship noise, but have not affected the feasibility of the technique. This technique would have the advantage of not interfering with the flow at the point where the movement is taking place and directional transducers could provide a definable and variable field of view. Measurement could be continuous, but calibration would be required for calculation of transport rates.

Acoustic pebble. For gravel size particles it is possible to build an acoustic transponder with about the size and density of a pebble and to monitor its movement using a side scan sonar. This technique has been described by Dyer and Dorey (11) and will measure the speed of grain movement, though giving no information on how many natural grains are moving.

There are a number of other techniques under consideration. These include measuring the attenuation of an acoustic signal caused by the mass of grains in the water. Attenuation will be increased with increasing sediment concentration and could be measured on paths of varying lengths parallel to and close to the sea bed. Also use of a gated doppler flowmeter would give information on the speed of moving grains near the sea bed, though again it would not measure how many were moving above a certain limiting value. However, at the moment, it seems that our most useful tool is still the underwater television.

DISCUSSION

Because of the time and space scales inherent in the ways of measuring the shear stresses and the sediment transport rates it is necessary to pair the techniques so that the same characteristics are measured. Consequently use of a particle impact counter recording continuously would need to be coupled with measurements of flow using an electromagnetic flowmeter, for instance. By averaging, however, one could use velocity profiles averaged over the same period, but different characteristics of the flow and the sediment's response would be measured. The velocity profile contains components of flow caused by disturbances some way upstream, whereas the grains respond to a much more localized situation. Similarly there is not much to be gained by measuring turbulent bursts at one height and the sediment movement at another if the burst does not reach the bed within the area of interest. In areas of ripples or dunes the measurements need to be averaged over a multiple of wavelengths. Point measurements on a dune form require additional information on the local bed slope for analysis in physical terms.

Streamwise changes of depth or mean flow velocity, and the presence of internal boundary layers formed by topographic or sea bed roughness changes will create spatial variations in the characteristics of the outer flow. Because the occurrence of "bursts" appears to be controlled by the outer flow characteristics, it is possible that the intermittent bursts in shear stress may not be randomly distributed on the sea bed. This could create areas of potentially greater mobility on the sea bed and the consequent areal variation in erosion and deposition could cause the formation of dunes, providing there is sufficient sediment available. Thus it is necessary that the distribution of shear stresses and transport rates be examined on scales up to several hundreds of metres in extent. Additionally much of the sea bed has possible achieved some equilibrium with the naturally occurring forces and significant transport may only occur under extreme conditions. Consequently measurements require to be made continuously for long periods. There are also problems concerning the effect of wave motion additional to the tidally-induced shear stresses. This introduces a whole new range of variables which, taken with those discussed here, provides a wide field of difficult research problems.

ACKNOWLEDGEMENTS

I wish to thank many of my colleagues for stimulating discussions on the topics reviewed here, particulary Dr. W. R. Parker, R. L. Soulsby and M. J. Tucker. The development of the instruments for measuring bedload transport is being carried out by A. P. Salkield, G. Le Good and N. Millard.

REFERENCES

1. Ackers, P., and W.R. White. 1973. Sediment transport: new approach and analysis. J. Hyd. Div. Amer. Soc. Civ. Eng. 99 (HY11):2041-2060.
2. Bagnold, R. A. 1966. An approach to the sediment transport problem from general physics. U.S. Geol. Surv. Prof. Paper 422-1.
3. Bowden, K.F. 1962. Measurements of turbulence near the sea bed in a tidal current. Jour. Geop. Res. 67:3181-3186.
4. ——, and L.A. Fairbairn. 1956. Measurements of turbulent fluctuations and Reynolds stresses in a tidal current. Proc. Roy. Soc. Lond. A237: 422-438.
5. ——, and M.R. Howe. 1963. Observations of turbulence in a tidal channel. Jour. Fluid Mech. 17:271-284.

6. Channon, R.D., and D. Hamilton 1971. Sea bottom velocity profiles on the continental shelf southwest of England. Nature 231:383-385.
7., Charnock, H. 1959. Tidal friction from currents near the sea bed. Geophys. J. R. astr. Soc. 2:215-221.
8. Corino, E.R., and R.S. Brodkey. 1969. A visual investigation of the wall region in turbulent flow. Jour. Fluid Mech. 37:1-30.
9. Dyer, K.R. 1970. Current velocity profiles in a tidal channel. Geophys. J.R. astr. Soc. 22:153-61.
10. _____. 1972. Bed shear stresses and the sedimentation of sandy gravels. Mar. Geol. 13:M31-M36.
11. _____, and A.P. Dorey. 1974. Simulation of bedload transport using an acoustic pebble. Mem. Inst. Geol. Bassin Aquitaine No. 7:377-380.
12. _____, and H. Lasta King. 1975. The residual water flow through the Solent, S. England. Geophys. J. R. astr. Soc. 42:97-106.
13. Eckelmann, H. 1974. The structure of the viscous sub-layer and the adjacent wall region in a turbulent channel flow. Jour. Fluid Mech. 65:439-459.
14. Einstein, H.A. 1950. The bed-load function for sediment transportation in open channel flows. U.S. Dept. Agric. Soil Cons. Serv. Tech. Bull. 1026.
15. Francis, J.R.D. 1973. Experiments on the motion of solitary grains along the bed of a water-stream. Proc. Roy. Soc. Lond. A332:443-471.
16. Frey, H.R., and G.J. McNally. 1973. Limitations of conical hot platinum film probes as oceanographic flow sensors. Jour. Geop. Res. 78:1449-1461.
17. Gordon, C.M. 1974. Intermittant momentum transport in a geophysical boundary layer. Nature 248:392-394.
18. _____. 1975. Sediment entrainment and suspension in a turbulent tidal flow. Mar. Geol. 18:M57-M64.
19. _____, and C.F. Dohne. 1973. Some observations of turbulent flow in a tidal estuary. Jour. Geop. Res. 78:1971-1978.
20. Graf, W.H. 1971. Hydraulics of sediment transport. McGraw-Hill, New York 514 p.
21. Grass, A.J. 1971. Initial instability of fine bed sand. J. Hydr. Div. Amer. Soc. Civ. Eng. 96 (HY 3):619-632.
22. Heathershaw, A.D. 1974. "Bursting" phenomena in the sea. Nature 248:394-395.
23. Hsu, S-T., and J.F. Kennedy. 1971. Turbulent flow in wavy pipes. Jour. Fluid Mech. 47:481-502.
24. Hubbel, D.W. 1964. Apparatus and techniques for measuring bed-load. U.S. Geol. Surv. Water Supply Paper 1748.
25. International Atomic Energy Agency. 1973. Tracer techniques in sediment Transport. Tech. Reports Series No. 145. IAEA, Vienna 1973.
26. Ippen, A.T., and R.P. Verma. 1953. The motion of discrete particles along the bed of a turbulent stream. Proc. 3rd Minn. Int. Hyd. Cong.
27. Johnson, P., and T.C. Muir. 1969. Acoustic detection of sediment movement. Jour. Hyd. Res. 7:519-540.
28. Johnson, J.W. 1942. The importance of side-wall friction in bed-load investigations. Civil Engineering, Amer. Soc. Civ. Eng. 12:329-331.
29. Kachel, Nancy B., and R.W. Sternberg. 1971. Transport of bedload as ripples during an ebb current. Mar. Geol. 10:229-244.
30. Kanwisher, J., and K. Lawson. 1975. Electromagnetic flow sensors. Limnol. Oceanogr. 20:174-182.

31. Langhorne, D.N. 1973. A sandwave field in the Outer Thames Estuary, Great Britain. Mar. Geol. 14:129-143.
32. Laufer, J. 1954. The structure of turbulence in fully developed pipe flow. Nat. Advisory Comm. Aeronautics Tech. Report 1174.
33. Lu, S.S., and W.W. Willmarth. 1973. Measurements of the structure of the Reynolds stress in a turbulent boundary layer. Jour. Fluid Mech. 60:481-511.
34. Ludwick, J.C. 1975. Variations in the boundary-drag coefficient in the tidal entrance to Chesapeake Bay, Virginia. Mar. Geol. 19:19-28.
35. Meyer-Peter, E., and R. Müller, 1948. Formulae for bed load transport. Int. Assoc. Hydro. Res. 2nd meeting, Stockholm.
36. Müller, A., A. Gyr, and T. Dracos. 1971. Interaction of rotating elements of the boundary layer with grains of a bed; a contribution to the problem of the threshold of sediment transportation. Jour. Hyd. Res. 9:373-411.
37. Narahari Rao, K., R. Narasimha, and M.A. Badri Narayanan. 1971. The "bursting" phenomenon in a turbulent boundary layer. Jour. Fluid. Mech. 48:339-352.
38. Nece, R.E., and J.D. Smith. 1970. Boundary shear stress in rivers and estuaries. Jour. Wat. Harb. Div., Amer. Soc. Civ. Eng. 96 (WW2):335-358.
39. Nychas, S.G., H.C. Hershey, and R.S. Brodkey. 1973. A visual study of turbulent shear flow. Jour. Fluid Mech. 61:513-540.
40. Raudkivi, A.J. 1963. Study of sediment ripple formation. Jour. Hyd. Div. Amer. Soc. Civ. Eng. 89 (HY 6):15-33.
41. ——. 1967. Loose boundary hydraulics. Pergamon Press, Oxford. 331 p
42. Seitz, R.C. 1971. Results of a field study using the 3-axis Doppler shift current meter. Tech Report 72. Chesapeake Bay Inst., Johns Hopkins Univ.
43. Smith, J.D. 1974. Turbulent structure of the surface boundary layer in an ice-covered ocean. Rapp. P. -v. Réun., Cons. Int. Explor. Mer 167:53-65.
44. Southard, J.B., and J.R. Dingler. 1971. Flume study of ripple propagation behind mounds on flat sand beds. Sedimentology 16:251-263.
45. Sternberg, R.W. 1967. Measurements of sediment movement and ripple migration in a shallow marine environment. Mar. Geol. 5:195-205.
46. ——. 1968. Friction factors in tidal channels with differing bed roughness. Mar. Geol. 6:243-261.
47. ——. 1971. Measurements of incipient motion of sediment particles in the marine environment. Mar. Geol. 10:113-119.
48. ——. 1972. Predicting initial motion and bedload transport of sediment particles in the shallow marine environment. *In* D.J. P. Swift, D.B. Duane, and O.H. Pilkey (eds). Shelf sediment transport process and pattern. Dowden, Hutchison & Ross. Penn. 656 p.
49. Sutherland, A.J. 1967. Proposed mechanism for sediment entrainment by turbulent flows. J. Geop. Res. 72:6183-6194.
50. Taylor, P.A., P.R. Gent, and J.M. Keen. 1976. Some numerical solutions for turbulent boundary-layer flow above fixed, rough, wavy surfaces. Geophys. J.R. astr. Soc. 44:177-201.
51. Tucker, M.J. 1972. Electromagnetic current meters. Proc. Soc. Underwater Tech. 2:53-58.
52. Wallace, J.M., H. Eckelmann, and R.S. Brodkey. 1972. The wall region in turbulent shear flow. J. Fluid Mech. 54:39-48.

53. White, W.R., H. Milli, and A.D. Crabbe. 1975. Sediment transport theories: a review. Proc. Instn. Civ. Engrgs. 59:265-292.
54. Williams, G.P. 1970. Flume width and water depth effects in sediment transport experiments. U.S. Geol. Surv. Prof. Paper 562-H.
55. Willmarth, W.W., and B.J. Tu. 1967. Structure of turbulence in the boundary layer near the wall. Physics Fluids. 10:S134-S137.
56. Wiseman, W.J. 1969. On the structure of high-frequency turbulence in an estuary. Tech. Report 59. Chesapeake Bay Inst. Johns Hopkins Univ.
57. Vanoni, V.A., and L-S. Hwang. 1967. Relation between bed forms and friction in streams. J. Hyd. Div. Amer. Soc. Civ. Eng. 93(HY 3):121-144.

SEDIMENTOLOGY AND CHANNEL SLOPE MORPHOLOGY

OF AN

ANOXIC BASIN IN SOUTHERN NETHERLANDS

Richard W. Faas
Lafayette College
Easton, Pennsylvania
and
Stanislas I. Wartel
Royal Belgian Institute for Natural Sciences
Brussels, Belgium

ABSTRACT: The Veerse Meer is a brackish, non-tidal lake in southern Netherlands. It was separated from the tidal OosterSchelde in 1961 by dams at each end. Environmental and sedimentological changes are occurring that are different from the parent water body.

Bottom conditions range from well-oxidized to completely anoxic. Physical properties of the bottom sediments vary, depending upon the bottom water conditions. Bioturbation in aerobic areas has mixed the pre- and post-1961 sediments. Areas which became anoxic shortly after enclosure show a lack of mixing, with post-1961 clayey sediments lying directly over pre-1961 sandy sediments. Extensive levelling of the pre-1961 bottom is occurring with fine-grained sediments accumulating in scour pits and depressions.

Channel slope measurements show three distinct morphologies, each reflecting changes in the dynamic history of the estuary. "Cliffed" slopes are relict features from the pre-1961 tidal system. "Steep" slopes are found adjacent to channel sides in anoxic environments and contain fine-grained sediments. "Gentle" slopes found adjacent to mid-channel sand islands are composed of coarser-grained sediments.

Nature of post-1961 sedimentation appears directly influenced by 1) biologic activity, 2) wind wave and boat wake winnowing, 3) proximity to polder drainage channels and tidal flats, and 4) lack of tidal activity.

INTRODUCTION

The Delta region of southern Netherlands was studied because of its many varied estuarine environments. This work examines the sediments and sedimentary processes in three adjacent, but very different, water bodies: 1) the WesterSchelde, 2) the Veerse Meer, and 3) the OosterSchelde (Fig. 1). Sampling of each was done in 1973, with a Shipek bottom sampler. Several sites in each environment were selected for detailed studies involving *in situ* bottom measurements and extensive utilization of SCUBA techniques.

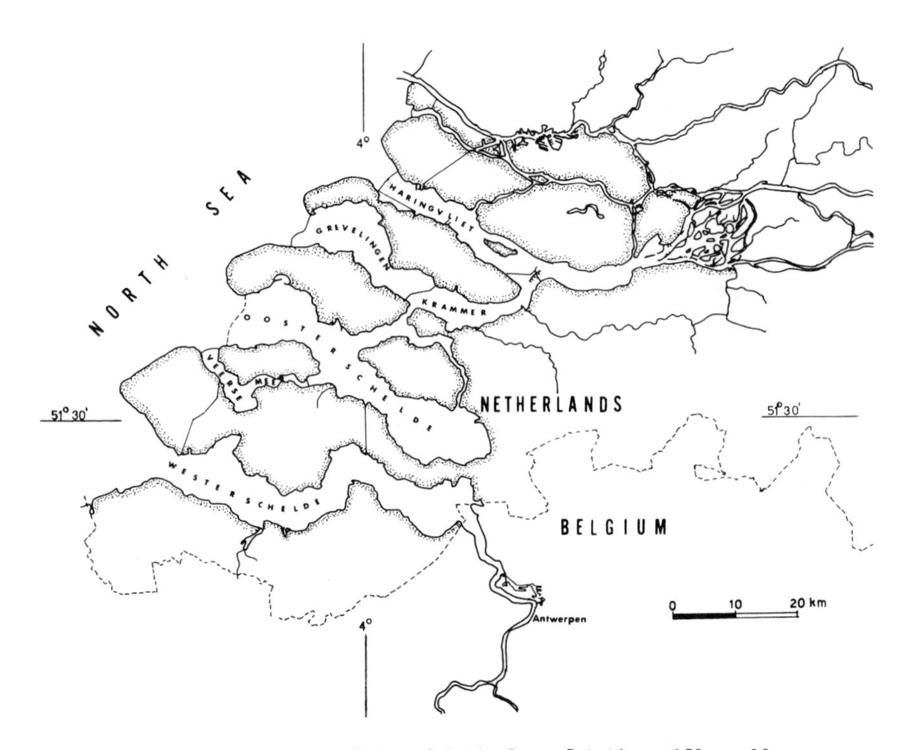

Figure 1. Map showing location of WesterSchelde, OosterSchelde, and Veerse Meer.

THE PROBLEM

The Veerse Meer, originally a segment of the tidal OosterSchelde, a sea arm in the southern Netherlands, was created in 1961 by the installation of the Veerse Gat dam. This completed the transformation of the segment to a non-tidal environment begun in 1960 by the Zandkreek-dam at its western edge. Since 1961 the Veerse Meer has become a stagnant, brackish-water lake in which severe salinity and oxygen fluctuations occur (1). Drainage of agricultural surface water from the surrounding polderland, and raw or partially-purified sewage from small towns brings high contents of dissolved phosphate and other nutrients into the water. Pronounced spring phytoplankton blooms and oxidation of the dead organisms depletes dissolved oxygen of the deeper waters. By midsummer large sections of the lake become anoxic and, in the deep former tidal channels, salt-stratification occurs. Anoxic environments probably persist throughout the entire year in these deep, cold waters.

The biological consequences of drastic interruption of the tidal regime have been studied extensively by the Delta Institute for Hydrobiological Research at Yerseke, Netherlands in connection with the implications for biological change to the adjacent OosterSchelde, scheduled to be dammed completely by 1978.

Because of the proximity of the Veerse Meer to two dynamic water bodies, the Wester- and OosterSchelde, the authors began a comparative sedimentological study in the summer of 1973. A surficial sediment sampling program revealed differences in sediment types that seemed related to tidal processes. The basic question was concerned with the sedimentological effects resulting from the conversion of a water body with a high tidal range to a tideless basin. Questions pertaining to the main problem were:

1. How had the texture and composition of the sediments been changed by the lack of tidal flow?

2. How were sediments presently being distributed in the non-tidal environment?

3. What were the physical characteristics of the sediments being deposited in the new environment?

REGIONAL CHARACTERISTICS OF THE WATER BODIES

The WesterSchelde is the present mouth of the river Scheldt and extends about 80 km. up the river to Antwerpen, Belgium, where the boundary of the brackish water area is usually found (Fig. 1). It varies in width between 1 to 7.8 km. and has a maximum depth of about 57 meters. Extensive salt marshes occur along the southeastern margin. At low tide, about 40% of the area normally covered at high tide emerges as shoals (13). The average discharge is about 90 m^3/sec., ranging between less than 10 m^3/sec., to 500-600 m^3/sec. Average mud content (particles less than 50 micron diameter) is roughly 300 mg./1, on a line from Breskens to Vlissingen at the mouth of the WesterSchelde. This represents 80% of the suspended sediments, the remainder being sand-sized material (7). Numerous large sand islands lie within the main channel and exhibit a variable surface morphology of large and small-scale ripple marks, mega-ripples, and gullies, formed by run-off at ebb tide.

The OosterSchelde, formerly the mouth of the Scheldt, is a sea arm about 50 km. long and varies in width between 3 and 9.2 km. (Fig. 1). Its maximum depth is 49 meters. During low tide, 43% of the area normally covered at high tide emerges as shoals. A few areas of salt marsh can be found along the margin, particularly in the eastern end.

The yearly fluctuation of water temperature is 16° C, ranging from 3.0°C in January to 19.1°C in July (12). The sea arm is classed as a vertically homogeneous estuary, as is the WesterSchelde (5). Salinity varies from normal marine at the mouth to 16.5°/₀₀ at the entrance to the Keeten.

Mud content of the suspended sediments is much less than that in the WesterSchelde. More is found on the northern shore near Zierikzee than the southern, near Yerseke, with greatest quantities being deposited in the eastern end (7).

Sand islands are abundant and cause the channel to bifurcate. These islands exhibit a less variable surface morphology than those in the WesterSchelde and may indicate less active sand movement.

The Veerse Meer, a tideless lake located between the OosterSchelde and WesterSchelde, is 24 km. long and varies from 0.25 to 1.5 km. in width (Fig. 1). Depths range from 6 meters in the channel to 30 meters in the old tidal gullies. The volume of the Veerse Meer fluctuates between 93 million m^3 (summer) and 83 million m^3 (winter) when the water level is lowered by 0.7 meters below N.A.P. (Nieuw Amsterdam Peil - Dutch Ordnance Level).

Prior to 1961, salt marshes and sand and mud flats bordered the Veerse Meer. Several large sand islands also existed within the main channel. Since closure, these islands have become stabilized with various forms of vegetation. Changes in the sand islands, tidal flats, and marshes resulted in increased erosion and sedimentation and included: 1) a lowering of the ground water table, 2) dessication and subsidence of the bar surface with horizontal and vertical cracking, 3) development of a surficial crust caused by hypersalinity and a blue algal mat, 4) winnowing of silt and clay particles, leaving a sand lag mantle; the finer particles being transported offshore (2).

Since closure, the chlorinity of the Veerse Meer has steadily decreased, ranging between 6.5 and 12.5 °/₀₀ Cl. Eutrophication, associated with an increase in organic substances and particulate and soluble phosphates has developed, encouraged by the continuous supply of run-off waters from the surrounding polders. (9)

SEDIMENT CHARACTERISTICS

WesterSchelde

During 1973, forty-five bottom samples were taken with the Shipek sampler in the WesterSchelde. Only ten contained clay in appreciable amounts (often clay balls). The sediments are predominately sand, with considerable variation in median diameters (Table 1). The finer-grained material is found adjacent to the shorelines, whereas the coarser-grained sediments are found in the deeper channels. A definite areal gradation in grain size is seen; coarser sizes are located in the westermost reach, with finer sizes occurring in the central portion near

Table 1. Size distribution characteristics, carbonate and organic carbon content of all surface samples.

WesterSchelde

Sample #		Sand %	Md. (μ)	So.	Sk.	$CaCO_3$(%)
73B	30	96.6	235	0.72	0.00	3.27
	31	85.4	183	0.29	+0.34	7.95
	32	94.8	370	0.71	+0.45	5.14
	33	96.9	205	0.40	+0.23	5.14
	35	69.7	128	2.88	+0.79	3.50
	36	99.0	345	0.42	+0.14	8.65
	37	99.0	280	0.35	−0.29	5.61
	38	97.9	270	0.42	+0.09	6.54
	39	95.5	390	0.41	0.00	12.86
	40	97.4	340	0.42	+0.19	8.41
	41	99.9	280	0.46	+0.17	4.68
	42	98.5	187	0.31	+0.16	3.28
	43	86.7	180	0.74	+0.92	16.60
	44	78.4	150	1.38	+0.85	7.24
	45	97.9	282	0.24	+4.41	1.40
	46	97.7	345	0.47	+0.27	1.64
	47	97.8	250	0.65	−0.34	5.14
	48	96.4	248	0.32	−0.19	1.64
	49	97.0	250	0.41	+2.34	5.14
	50	97.5	235	0.23	+0.17	2.10
	51	96.3	390	0.30	+0.23	9.82
	52	95.7	285	0.50	+0.02	5.14
	53	95.4	310	0.52	+0.23	4.44
	54	82.4	230	3.07	+0.85	11.22
	55	96.8	380	0.60	+0.12	9.32
	56	98.2	195	0.24	+0.04	2.80
	57	97.1	345	0.56	−0.05	9.55
	58	98.1	290	0.30	+0.27	1.30
	59	97.7	320	0.42	+0.10	3.26
	60	98.2	220	0.32	+0.28	1.63
	61	97.8	210	0.27	+0.33	2.80
	62	88.8	132	0.38	+0.26	7.46
	63	68.7	107	2.25	+0.85	–
	64	98.3	290	0.48	+0.17	3.28
	67	99.2	338	0.22	+0.14	3.03
	68	97.2	144	0.30	+0.40	1.86
	69	97.7	185	0.36	+0.25	3.03
	70	98.6	185	0.33	+0.06	1.16
	71	97.8	230	0.52	+0.08	1.86
	72	98.2	190	0.32	+0.13	0.93
	73	97.7	185	0.42	+0.10	1.86
	74	97.7	237	0.16	+0.25	0.93
	75	98.3	216	0.84	+0.06	1.30

OosterSchelde

Sample #		Sand %	Md. (μ)	So.	Sk.	$CaCO_3$(%)
73D	01	97.2	135	0.29	+0.14	1.88
	02	98.9	190	0.58	+0.02	1.88
	03	92.1	185	0.90	−0.21	1.41
	04	99.3	252	0.54	+0.17	1.88
	05	99.2	275	0.55	+0.20	7.16

06	99.4	240	0.30	+0.03	0.94
07	84.4	355	1.06	+0.54	4.57
08	99.5	335	0.26	+0.15	0.35
09	98.7	320	0.28	−0.07	3.05
10	99.5	330	0.33	−0.09	4.22
11	99.5	300	0.32	+0.01	0.24
12	99.3	251	0.40	0.00	0.59
13	98.7	185	0.33	+0.14	2.11
14	85.3	112	0.57	+0.38	7.87
15	98.7	288	0.25	+0.16	1.04
16	98.7	170	0.37	−0.16	2.20
17	97.8	188	0.28	+0.14	1.97
18	99.4	230	0.40	+0.17	1.16
19	98.8	180	0.31	+0.10	1.62
20	99.6	285	0.33	−0.06	0.93
21	99.3	232	0.28	+0.04	0.81
22	99.5	330	0.33	+0.39	0.81
23	98.5	228	0.31	+0.13	1.27
24	99.3	252	0.27	+0.04	0.70
25	99.3	240	0.27	−0.04	1.04
26	99.4	217	0.31	+0.16	0.58
27	99.6	315	0.25	0.00	0.81
28	99.6	300	0.34	+0.03	1.85
29	98.7	163	0.34	+0.24	2.55
30	94.2	248	0.30	+0.27	3.94
31	99.4	260	0.26	+0.09	0.46
32	98.8	214	0.41	+0.10	2.79
33	99.2	255	0.32	+0.13	0.58
34	99.1	315	0.28	0.00	1.27

Veerse Meer

Sample #	Sand %	Md. (μ)	So.	Sk.	$CaCO_3$(%)	Org. C(%)
73C 01	96.5	212	0.54	+0.27	2.78	3.12
02	5.6	2	–	–	13.47	4.85
03	7.3	2.2	–	–	16.94	4.95
05	85.1	220	0.95	+0.27	3.02	1.98
06	81.1	218	1.78	+0.10	4.64	0.69
07	58.8	80	1.44	+0.45	6.97	2.82
08	95.4	148	0.32	+0.00	3.02	–
09	38.3	35	–	–	10.91	3.21
10	30.9	20	–	–	11.60	2.05
11	14.0	3.8	–	–	8.82	4.98
12	92.6	203	0.43	+0.04	2.75	0.37
13	84.2	117	0.61	+0.52	6.65	–
14	83.8	106	0.58	+1.24	6.93	0.63
15	68.7	99	2.12	+0.71	7.81	1.38
16	93.5	108	0.29	+1.55	6.06	0.60
17	30.8	25	–	–	9.40	2.84

Sand % – percent of sample between 2 mm and 62 μ.
Md. (μ) – particle diameter corresponding to 50%.
So. – sorting: $\frac{\phi 84 - \phi 16}{2}$
Sk. – skewness $\frac{M - Md.}{So.}$, where M = mean diameter $\frac{\phi 84 + \phi 16}{2}$
$CaCO_3$ – total carbonate as analyzed by Dietrich-Scheibler calcimeter.
Org. C – total organic carbon as analyzed by Walkley and Black method of wet oxidation.

Hansweert and upstream toward the Belgian border. With few exceptions, sediments are well-sorted and positively skewed (8).

Twenty-four samples contained less than 5% carbonate (Table 1). A definite carbonate gradient exists with highest values being found in the western half of the estuary, suggesting either that the shelly material was derived from the adjacent North Sea, or, because of the marine nature of the western portion, that shelled fauna existed within that part of the estuary. Because of the nature of the substrate, i.e., coarse clastics and extreme mobility, the former is thought more likely as the decrease in salinity inland, increased turbidity, and substrate instability may inhibit population development (13).

OosterSchelde

Thirty-four bottom samples were taken from the OosterSchelde. Fine-grained deposits were distinctly lacking, yet on the whole, the sands appeared finer-grained than the WesterSchelde. However, the percentage of sand-sized particles was higher. The lack of finer material is probably because freshwater inflow is limited to polder run-off, whereas the WesterSchelde has a large freshwater drainage basin. The particles are well-sorted and positively skewed (Table 1). Fine to medium sands were found in the deeper channels, at depths to 36 meters.

Bedded clay deposits consisting of a series of small escarpments which outcropped at -15 meters were observed on the north side of a sand island (Roggenplaat) in the western end of the OosterSchelde. They appeared to underlie large portions of the island. They were very cohesive and extensively bored by clams and crabs, causing blocks to drop out and move downslope. Several large masses, several meters across, were observed at 20 meters depth. This material gradually became eroded into smaller, rounded pieces (buttons) by the tidal currents.

Quantities of mud occur in the southeastern end of the sea arm, adjacent to the salt marshes. This material forms distinct layers between sandy units, and also as discrete "clay buttons" mixed with the sand as individual clasts. It appears to be recently deposited sediment as opposed to the bedded clays of the Roggenplaat which appear to be older, exhumed material.

Many samples contained considerable shell material; however, the general trend seemed to be toward less carbonate than in the WesterSchelde. Only two samples possessed carbonate contents greater than 5% (Table 1).

Veerse Meer

Seventeen Shipek samples were taken from the Veerse Meer. In general, the sediments ranged from clay to clayey sand (6). Calcium carbonate and organic carbon analyses were performed on surface samples that were taken from areas of fine-grained sediment (Table 1).

The Veerse Meer exhibits extreme sedimentary diversity and is dominated by fine-grained sediments.

Distribution Patterns – Post 1961

The most notable change in the Veerse Meer since 1961 has been the sudden accumulation of fine-grained sediments where previously only coarse clastics had been deposited. Changes in the average value of physical properties, organic matter, and carbonate content between pre- and post-1961 bottoms are shown in Table 2 (Data from Cores C-4 and C-11).

Present day sedimentation in the Veerse Meer is still influenced by pre-1961 conditions. Sedimentation is non-uniform, being more variable in areas adjacent to eroding underwater slopes and sand islands. Ponding of extremely fine-grained sediments occurs within depressions and old tidal channels within the main channel.

Fine-grained sediments dominate most areas, particularly those associated with the steeper channel slopes. Some of the fine-grained material has been eroded from the inter-laminated peat-clay deposits of the salt marshes. However, run-off from the surrounding polderland may be responsible for a large percentage of fine-grained material. Some muds may also be introduced as suspended sediment in OosterSchelde water which enters from the Katseveer sluice each time it is opened for boat traffic.

Sediment cannot be introduced from the land other than from specific drainage channels. The entire Veerse Meer is enclosed by dikes, which, from the shoreline to the base are covered with a pavement of basalt and limestone blocks. This provides an impenetrable surface that effectively limits shoreline erosion.

In some areas, bioturbation seems to be mixing pre- and post-1961 sediments; in others, post-1961 sediments form a sharp discontinuity above the pre-1961

Table 2. Textural and geochemical characteristics associated with pre- and post-1961 sediments.

	Sand	Silt	Clay	w%	$CaCO_3$	Org. C
Post-1961						
$\overline{X}$	24.3	49.3	26.3	182.9	13.14	2.63
s	17.9	17.6	11.9	78.0	3.04	0.74
Pre-1961						
$\overline{X}$	91.6	5.2	8.0	34.6	6.14	0.38
s	5.6	1.1	–	2.8	0.49	0.15

Where $\overline{X}$ = mean value
s = standard deviation

Sand – percent of sample between 2 mm and 62 μ.
Silt – percent of sample between 62 μ and 2 μ.
Clay – percent of sample finer than 2 μ.
w% – percent moisture content, on dry wt. basis.
$CaCO_3$ – total carbonate as determined by Dietrich-Scheibler calcimeter.
Org. C – total organic carbon as determined by Walkely-Black wet oxidation procedure.

sediments. Mixed deposits invariably contain juvenile *Mya* and *Cardium* clams. Worm burrows exist abundantly in the bottom sediments and fecal material was observed surrounding the openings of the burrows. Non-mixing occurs only in anoxic areas, primarily the deep pits along the southern shoreline. However, the channel adjacent to the town of Veere contained only black anoxic mud as did two sample sites (C-2 and C-3) in the extreme western end of the lake.

Slope Stability

A detailed study of selected underwater slopes shows a relationship of slope morphology to sediment type and geochemical environment (4). Slopes, defined as *regular steep*, are found in anoxic portions of the Veerse Meer. An example of a regular steep slope is found at Gebroken Dijk (Fig. 2). Here the slope becomes concave below -8 meters. Sediment moves down the steep upper slope and accumulates in the lower concave region. Mud thickness increases below -10 meters to 1.5 meters at -20 meters. Bulk density decreases down slope, with unusually low values occurring below -15 meters, associated with high water contents and low shear strengths. Organic carbon content averages 1.88%. Extreme anoxic conditions prevail, beginning about -5 meters and extending to the base at -20 meters. Redox potential decreases from -273 millivolts at -5 meters to -475 millivolts at -20 meters. The sediments become more acidic with depth (Table 3). A bubbly texture persists throughout the length (70 cm.) of each core which is believed due to gas generation. Gas could easily be released by jarring the bottom which responded as a gel.

At several sites a peculiar form of down-slope movement was observed. Sheets of sediment, 1.5 m. across and several cm. thick, were outlined by zig-zag cracks

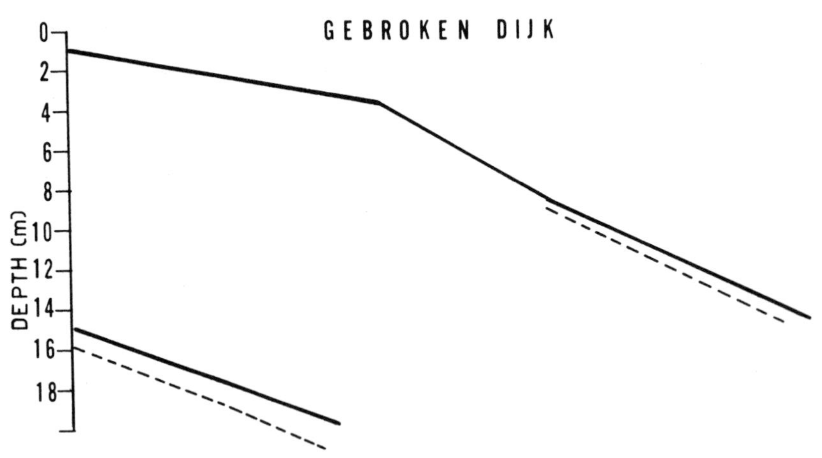

Figure 2. Profile of slope at Gebroken Dijk, (Gd). Dashed line shows mud thickness downslope.

Table 3. Redox potential and pH values for slope sediments.

Middelplaat, T = 14°C.				Wohlfahrtsdijk, T = 18°C.				Gebrokendijk, T = 14°C.			
D(m)		pH	Eh(mv)	D(m)		pH	Eh(mv)	D(m)		pH	Eh(mv)
0	W	8.5	+155	0	W	8.3	+110	-5	W	8.3	-150
					M	7.8	-171		M	8.0	-273
-5	W	8.5	+106	-3	W	8.4	+100	-8.5	W	8.1	-140
	M	7.8	-376		M	8.2	-199		M	7.9	-365
-10	W	8.0	-118	-8	W	8.2	-150	-20	W	7.7	-300
	M	7.9	-315		M	7.8	-298		M	7.2	-475
-15	W	7.8	-182	-10	W	8.1	-106				
	M	7.2	-482		M	7.8	-402				
-20	W	7.8	-181								
	M	7.1	-453								

W = Measurement made in water above sediment.
M = Measurement made in sediment.

which occurred as the mass moved down slope. Other evidence of down-slope movement was shown by slumping which left concave pits 0.5 meter wide. Several long linear high areas or crests separated by troughs 6-8 cm. deep trend perpendicular to the slope and appear to be accumulations of material resulting from down-slope movements, perhaps related to storms or boat wakes.

Instability of the steep slopes results from: 1) the kind of sediment being deposited, and 2) the geochemical environment. The dominance of clay-sized particles imparts an internal cohesion (shear strength) dependent upon weak chemical bonding rather than frictional effects. Gas generation creates void spaces and inhibits gravitational consolidation.

Some evidence of post-1961 sediment distribution patterns still can be found on the slopes of Veerse Meer. One should expect to find a vertical size gradation, with coarser deposits found in the channel bottom and finer-grained deposits grading up the slope to the tidal flats (10). The sediment distribution on most slopes does not follow this pattern. Generally, sandy sediments are found near the top of the slope (upper 5 meters) with silt and clay increasing toward the base. However, some places still exhibit the pre-1961 sediment distribution (e.g., Middelplaat, Mp). This is a stabilized intra-channel sand island and no fine-grained sediments are being contributed to its slopes (2). Hence, nothing is available for reworking and the pre-1961 size down-slope distribution persists.

Clay Minerals

X-ray analysis of surface samples indicates that illite dominates, followed by kaolinite and smectite in about equal abundances, and chlorite, the latter occurring in most of the samples. Analysis of cores from anoxic regions showed that these minerals persisted to a depth of 45 cm in the sediment.

Clay balls from the OosterSchelde reveal only trace quantities of chlorite and none from the WesterSchelde. Although some OosterSchelde water does enter the Veerse Meer periodically, it is unlikely that sufficient chlorite enters to form a significant portion of the clay minerals. It is suggested that authigenic chlorite may be forming in the deeper, anoxic environments. Chlorite might be derived from smectite by removal of Mg^{++} from sea water (3), producing a decrease in alkalinity. The reaction is pH sensitive and increases as pH rises above 8.0. Chlorite is also believed to form from smectite but is inhibited by organic acids which block interlayer sites (12). It appears that special conditions, i.e., high pH, presence of smectite, and low dissolved organic matter contents are necessary for authigenic chlorite to form.

pH measurements on cores from several localities (Wd, Mp, and Gd) during June 1975 revealed generally high values. Interface samples showed high pH values in the water overlying the muds, with 8.5 recorded at −5 meters at Middelplaat (Mp)(Table 3).

In view of the consistently high pH values, it is suggested that at least two of the parameters necessary for authigenic chlorite development (i.e., high pH and smectite) are present in the Veerse Meer. No data is available on dissolved organic matter content.

DISCUSSION

The general effect on the nature, distribution, and physical properties of sediments being deposited in a dynamic environment when that environment is suddenly changed into a non-tidal environment is to increase the amount of fine-grained material which is incorporated into the bottom sediments as permanent deposits. Simultaneous changes in the water column which lead to the establishment of anoxic conditions in the bottom waters further modify the character of the sediments. Sediments which formerly would be carried out to sea accumulate rapidly and in anoxic regions, form a blanket of soft, saturated black muds in which generation of gases takes place through the activities of anaerobic bacteria. Wind wave and boat wake erosion of underwater channel slopes provides sediment which is transported basin-ward, rather than being resuspended and transported laterally, to be either redeposited in tidal flats or carried out to sea. This material, high in organic matter, accumulates either toward the base of the slope or further out in the main channel, filling scour pits and depressions formed under the previous tidal regime.

Generation of gases accelerates transportation across steep slopes by retarding the normal effect of gravitational consolidation. Sediments possess high water content and low shear strength and are subject to mass down-slope movement. In the basin, continued gas generation maintains the sediment in an unconsolidated state, perhaps causing it to attain a gel-like consistency.

Mixing of pre- and post-1961 sediments takes place in aerobic environments due to the action of infaunal organisms, and wind wave and boat wake energies.

Samples from such areas are clayey sands, sandy clays, and sand-silt-clay (6). Similar areas in the Wester- and OosterSchelde are sands and gravels. Due to the increase in finer particle concentrations the substrate becomes changed. In anoxic areas, silt and clay deposits overlie the previously deposited coarse sands, with sharp discontinuity. In anaerobic areas, the fine-grained materials are mixed downward and no visual evidence of change in sedimentary processes can be seen.

The fine-grained sediments in the Veerse Meer appear to be derived from two primary sources. The erosion of the interlaminated tidal flat clays, silts, and peats from storm waves and boat wakes is obvious. At Wulpenburg (Wb), undercutting and slumping of these older deposits is seen, with large and small blocks moving as discrete masses down the slope. Not so obvious, but of significant importance is the fact the clay-sized sediments must be entering the Veerse Meer through the drainage channels from the surrounding polderlands and tidal marshes (Fig. 3). Due to the lack of tides, sediments enter the basin from the channels and accumulate locally. Some redistribution must occur because of the dominant western storm winds which shift the mud gradually to the eastern part of the lake.

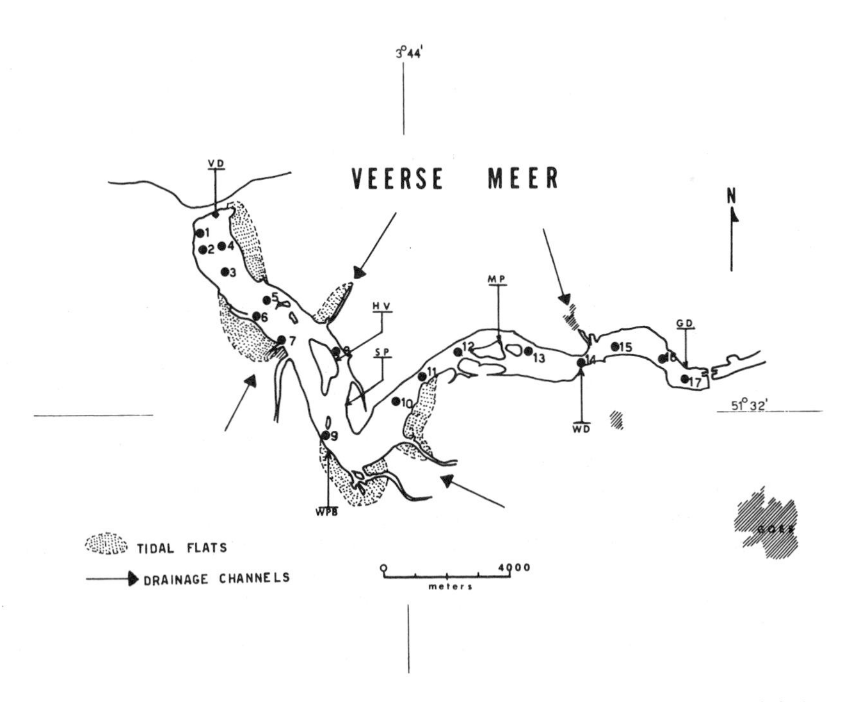

Figure 3. Map of Veerse Meer, showing location of grab samples, tidal flats, and drainage channels.

CONCLUSIONS

Changes in sedimentary processes of the type described in this paper must be expected to occur in any tidal environment which is so modified as to become non-tidal. The degree to which fine-grained sediments become dominant will depend upon the possible sources of sediment and the morphology of the channel slopes. Steep slopes provide unprotected reaches within which wind wave and boat wake erosion can remove sediment. Salt marsh deposits, primarily peat and interlaminated fine-grained material, provide a source of sediment. Land drainage channels contribute both sediments and nutrients which gives rise to anoxic conditions in the bottom waters. Consideration should be given to these factors in planning such modifications in order to predict sedimentologic changes.

ACKNOWLEDGEMENTS

The majority of this work was performed while the senior author was on sabbatical leave from Lafayette College at the Catholic University of Louvain, Leuven, Belgium. Financial support was provided by a Lafayette College Summer Faculty Research Fellowship and a grant from the National Geographic Society. Dr. K. F. Vaas, Director, Institute for Hydrobiological Research, Yerseke, Netherlands provided us with the research vessel, "H. G. de Man." Sediment analyses and geochemical analyses were performed by the Royal Belgian Institute for Natural Sciences, Brussels, and the Department of Geology and Geography, Catholic University of Louvain, Leuven, Belgium. We are grateful to Dr. Vaas, Dr. Capart, and Dr. F. Gullentops for their support of this work.

REFERENCES

1. Bakker, C. 1972. Milieu en plankton van het Veerse Meer, een tien jaar oud brakwatermeer in Zuidwest-Nederland. Mededelingen van de Hydrobiologische Vereniging 6: 15-38.
2. Beeftinck, W.G., M.C. Daane, and W. de Munck. 1971. Tien jaar botanisch-oecologische verkenningen langs het Veerse Meer. Natuur en Landschap 25 e (2): 50-65.
3. Berner, R.A. 1971. Principles of Chemical Sedimentology, McGraw-Hill Book Company, New York. 240 p.
4. Faas, R.W., and S.I. Wartel. 1975. Sedimentation and stability in an artificially enclosed estuarine segment. In prep.
5. Pritchard, D.W. 1967. Observations of circulation in coastal plain estuaries, p. 37-44. *In* G.H. Lauff (ed), Estuaries. AAAS Pub. No. 83.
6. Shepard, F.P. 1954. Nomenclature based on sand-silt-clay ratios. Jour. Sed. Pet. 24:151-158.
7. Terwindt, J.H.J. 1967. Mud transport in the Dutch Delta area and along the adjacent coastline. Neth. Jour. of Sea Res. 3(4):505-531.
8. Trask, P.D. 1932. Origin and Environment of Source Sediments of Petroleum. Gulf Publishing Co., Houston, Texas. 323 p.

9. Vaas, K.F. 1970. Studies on the fish fauna of the newly created lake near Veere, with special emphasis on the plaice (*Pleuronectes platessa*). Neth. Jour. of Sea Res. 5(1):50-95.
10. Van Straaten, L.M.J.U., and P.H. Kuenen. 1957. Accumulation of fine-grained sediments in the Dutch Wadden Sea. Geol. en Mijnbouw, N.S., 19:329-354.
11. Walkley, A., and I.A. Black. 1934. An examination of the Degthareff method for determining soil organic matter and a proposed modification of the chromic acid titration method. Soil Science 27:29-38.
12. Whitehouse, U.G., and R.S. McCarter. 1958. Diagenetic modification of clay mineral types in artificial sea water. Clays and Clay Minerals, Natl. Acad. Sci. Publ. 566:81-119.
13. Wolff, W.J. 1973. The estuary as a habitat. An analysis of data on the soft-bottom macrofauna of the estuarine area of the rivers Rhine, Meuse, and Scheldt. Zoologische Verhandelingen No. 126, 242 p.

DREDGING AND SPOIL DISPOSAL-

MAJOR GEOLOGIC PROCESSES IN SAN DIEGO BAY, CALIFORNIA

David D. Smith
David D. Smith and Associates
Environmental Consultants
P.O. Box 929-E
San Diego, California 92109

ABSTRACT: San Diego Bay is a crescent-shaped, well-mixed estuary 22.5 km long, and initially about 55 km^2 in area, with depths generally less than 4.5 m except for a 7.5-20 m deep channel. The present bay volume is roughly 230 $\times$ 10^6 m^3.

Since the early 1900's, dredging and use of spoil disposal as fill have reworked and shifted 100 to 140 $\times$ 10^6 m^3 of sediment, with a resulting 27% reduction in the bay's water area and an approximate doubling in depth of 55% of the original water area. Only 17 to 18% of the original area remains undisturbed by dredging or fill.

Since the bay reached its approximate present configuration in Holocene time, the only significant sediment source has been river/stream deposition which delivered an estimated 0.8 to 1.1 $\times$ 10^6 m^3 annually, until diversion and damming of principal tributaries between 1875 and 1919 reduced sedimentation by more than 80%. For the 30 year period of maximum dredging (1940-1970), the average dredging rate was 3 to 6 times the original sedimentation rate, and roughly 17 to 34 times the sharply reduced present sedimentation rate.

Thus, dredging and spoil disposal as geologic processes are substantially more important than all other erosional and depositional processes presently operating in San Diego Bay.

INTRODUCTION

As an outgrowth of engineering and environmental geological investigations associated with dredging projects in several estuaries, the writer has been investigating the importance of dredging and spoil disposal as estuarine geological processes. As will be shown in this paper, in estuaries with limited sediment input, dredging and spoil disposal can be particularly important, and in some cases are probably the dominant geologic processes operating.

Since 1950, more than 35 papers have considered one or more aspects of dredging in estuaries; selected examples of the more comprehensive previous studies include Cronin's (4) landmark review of human influences, Bassi and Basco (1), Boyd et al. (2), Brehmer (3), Hargis (10), Hutton et al. (13), Ingle (14), Lee and Plumb (17), Mackin (19), May (20), O'Neal and Sceva (25), and Saila et al. (31). The majority of the studies cited deal with the effects of dredging on estuarine processes and water quality, biologic communities, or sedimentation patterns. To the writer's knowledge, none has addressed dredging specifically as a geologic process.

The twofold purpose of this paper is to call attention to the importance of dredging and spoil disposal as estuarine geologic processes, and to assess their importance in San Diego Bay, a moderate sized, relatively simple California estuary, which has been modified extensively by these processes. This assessment focuses on the broader, large scale results of dredging and spoil disposal, i.e. effects on the basin geometry and sediment distribution pattern. The assessment is expressed in terms of a) the extent of the areas dredged and filled compared to the original size of the bay, b) the volume of sediment moved compared to the natural sediment influx to the bay, and c) the type of sediment removed and type of sediment exposed by dredging. An assessment of the importance of extensive dredging and spoil disposal in Corpus Christi Bay, a larger, more complex Texas estuary, is currently in progress (35, 36).

Many of the numeric data used in this paper are clearly "guesstimates" based on the best information available. As a result, this paper constitutes a first approximation assessment of the importance of dredging and spoil disposal as geologic processes rather than a quantitatively precise mass balance-type analysis. More precise data pertinent to any aspect of the assessment would be welcomed by the writer.

Before discussing dredging and spoil disposal in San Diego Bay, by way of background, it is appropriate to review dredging procedures, and to examine briefly the concept of dredging and spoil disposal as geologic processes.

DREDGING PROCEDURES

For purposes of this paper, dredging can be categorized as either hydraulic or mechanical. Most typically, hydraulic dredging involves the use of a large centrifugal pump to lift a sediment/water slurry from the bottom through a pipe to the surface where it is then pumped via pipeline for discharge at a disposal site. Alternatively, hopper-type hydraulic dredges discharge material into their own hold and then transport it to a disposal site. Various aspects of the principles and procedures of hydraulic dredging are discussed in Bassi and Basco (1), Boyd et al. (2), Herbich (11), and Huston (12), and in the many papers cited by these authors. Because the amount of mechanical dredging in San Diego Bay has been relatively minor, mechanical dredging is not discussed here.

Disposal of the dredged material can be categorized as either confined or open water. Confined disposal involves discharge of the sediment/water slurry behind levees or dikes. The suspended sediment settles, the supernatant water is decanted, and the diked enclosure is eventually filled with spoil. In open water disposal, the sediment/water slurry is discharged without any type of confinement, typically forming underwater mounds (1) which, as disposal continues, may or may not become emergent. About two-thirds of all dredge spoil in the United States is disposed of in open water, the remainder being placed in diked or other disposal areas on shore (2). San Diego Bay is atypical in this regard, in that relatively little open water disposal has occurred; most disposal has been for land fill, and thus has been confined or semi-confined.

HYDRAULIC DREDGING AND SPOIL DISPOSAL AS GEOLOGIC PROCESSES

Conceptually, hydraulic dredging is analogous to erosive excavation, fluid transport, and standing water or fluvial deposition of sediment. In a broader sense, however, dredging and spoil disposal can be considered to be legitimate geologic processes which pick up and relocate substantial volumes of estuarine sediment. Thus, dredging is an estuarine reworking-reshaping mechanism; estuarine sediments are reworked, and the geometry of the estuary floor, channel, and margins is reshaped.

As a reworking mechanism, hydraulic dredging and spoil disposal affect natural sediment distribution patterns in such ways as: 1) exposing deeper sediment horizons in dredged areas, 2) burying surface sediment with spoil (which is a variable blend of the original sediment layers), 3) converting bedded clays to clay ball conglomerates, 4) loading of soft estuary floor muds with resulting settlement and local lateral displacement in the form of mudwaves, and 5) potential release of pollutants from the dredged sediments as a result of mixing with water during the dredging process.

As pointed out by Shideler (34), modern sedimentation patterns within an estuary ". . . . represent composite responses to a complex combination of physical, chemical, and biological processes that are regulated by the bay's hydraulic regime and geologic framework. . . . " It is evident that extensive reshaping of the geometry of an estuary by dredging and spoil disposal would affect the hydraulic regime, which would in turn affect the sedimentation processes operating in the estuary.

More specifically, as a reshaping mechanism, dredging and spoil disposal permanently modify the bottom and shoreline geometry of the estuarine basin, with associated changes in the hydraulic regime and resulting changes in physical conditions and processes including: 1) circulation patterns, 2) flushing, 3) salinity, dissolved oxygen, and temperature distribution, 4) sediment transport, and

5) areas of sediment erosion and deposition within the estuary. The biological effects associated with such changes are documented in many of the papers cited in the Introduction.

As geologic processes, dredging and spoil disposal, with associated land fill, have modified the hydraulic regime and influenced sedimentation processes in a number of estuaries (see, among others, the examples cited in 3, 4, 10, 26, 34, 35). In some estuaries, these modifications have been so extensive that, as Meade (22) and Shideler (34) point out, differentiating between natural and man-induced effects is a major obstacle impeding the understanding of estuarine sedimentary processes. This paper focuses on assessing the magnitude of man-induced effects in San Diego Bay, a moderate size, relatively simple estuary.

SAN DIEGO BAY

Setting

Located in the extreme southwest corner of the United States, just a few miles north of the U.S.-Mexico border, San Diego Bay (Fig. 1) is one of the finest natural harbors in western North America. The bay is a crescent-shaped body about 22.5 km long, ranging from 0.4 to 4 km in width. Mean lower low water depths (except in the 7.5 to 20 m deep channel) range from generally deeper than 9 m in the north bay, from 3.0 to 4.5 m in the central bay, and from 0 to 2.4 or 3 m in the south bay. At half tide, the bay now has an area of about 43 km^2 and, according to the Corps of Engineers (38), a volume of more than 230 $\times 10^6$ m^3 of water; the tidal prism is about 73 $\times 10^6$ m^3 (7).

San Diego Bay is unique among equivalent or larger size bays in the United States in that it presently receives minimal fresh water runoff and has a high evaporation rate (5, 39). The area is characterized by a semi-arid, subtropic Mediterranean-type climate with seasonal rainfall averaging 25 to 28 cm annually. The bay is a well-mixed estuary with no significant density currents (38).

Tides in San Diego Bay are mixed diurnal (38); the extreme range of tides within the bay is 3.2 m (6) and the mean tidal range is 1.3 m. Ford (5) reports that tidal action, together with the geometry of the bay, produces a natural flushing action.

The narrow entrance to the bay completely blocks the action of ocean seas and swells upon the enclosed bay waters. Local winds generate steep, short-period waves or chop within the bay, which rarely exceed 0.6 or 0.9 m in height. Prevailing winds are from the northwest through southwest quadrants, and average monthly wind velocities range from 7.5 to 11 km/hr; strong winds are very rare (38).

Geologically, San Diego Bay occupies a structurally controlled depression (23, 24) in which are located the seaward portions of the ancestral valleys of the San Diego, Sweetwater, and Otay rivers that were trenched during Pleistocene low stands of the sea. The bay portion of these valleys flooded when sea level rose concurrent with the melting of the last Pleistocene continental ice sheet (7).

Development of the Silver Strand barrier (24) or sand spit (38) which extends from Imperial Beach north to Coronado helped separate the bay from the sea. More or less concurrently, the sediment load of the three rivers caused progressive shoaling and partial filling of the bay depression.

Prior to dredging activities (38), sands were most common near the mouth and along the westward margins of the bay; muds characterized the eastern

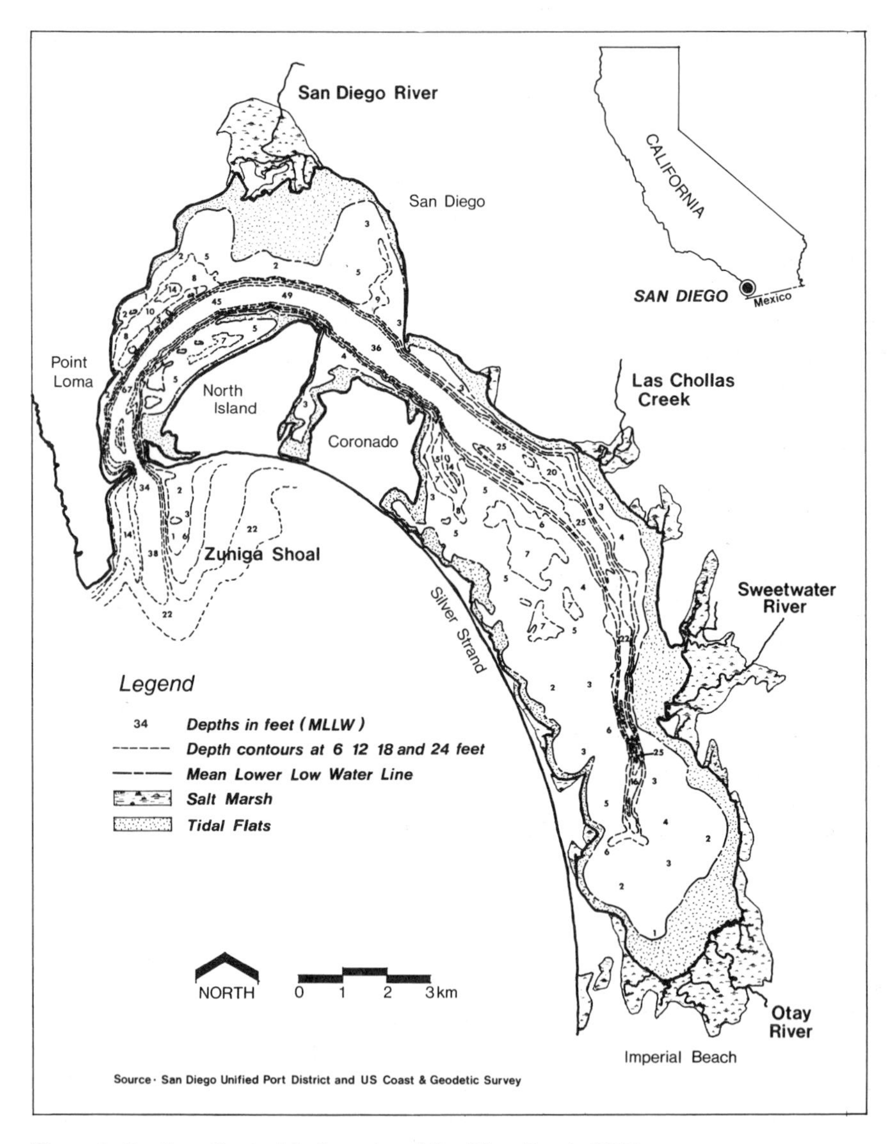

Figure 1. Configuration and bathymetry of San Diego Bay in 1857.

margins and southern end of the bay. Subbottom profiling work in the relatively undisturbed southern portion of the bay (18) indicates that the bay floor muds average about 3 to 3.6 m in thickness, and that these muds rest on a 12 to 18 m section of unconsolidated sands and silty sands, which in turn rest on somewhat older semi-consolidated sediments [according to Moore and Kennedy (24), probably the early Pleistocene Linda Vista formation].

Extent of Fill

Prior to major filling activities, San Diego Bay had an area of 54 to 57 km^2, as defined by the mean high tide line of 1918 (7). Filling activities, primarily using dredged material, began in 1888 (27) and intensified markedly shortly before and during World War II. As illustrated in Fig. 2, approximately 15.5 km^2

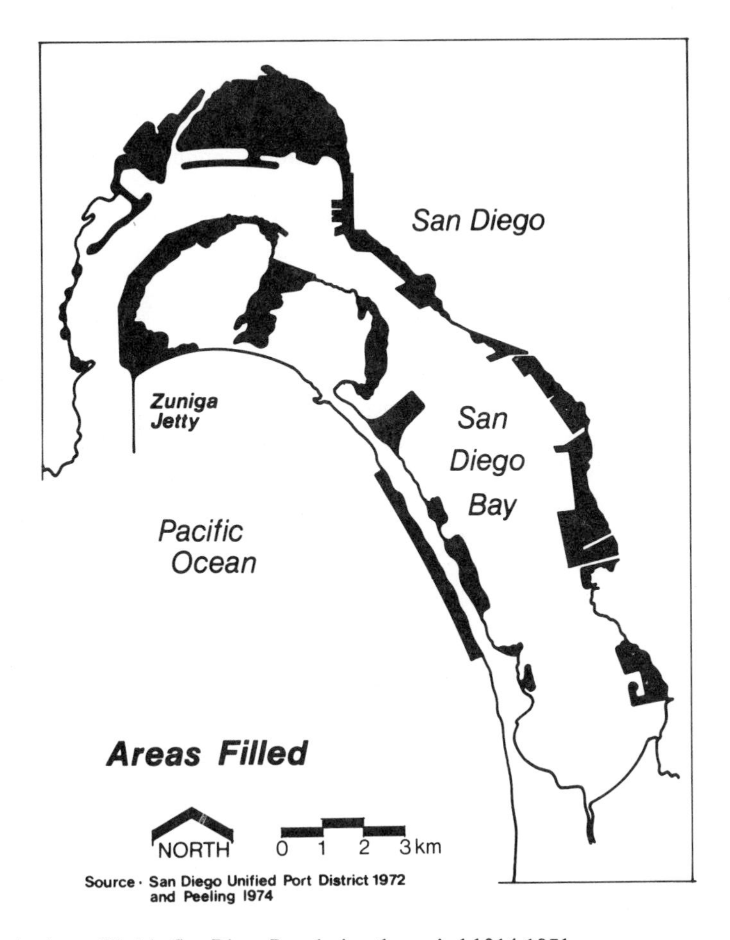

Figure 2. Areas filled in San Diego Bay during the period 1914-1971.

of the bay, amounting to about 27%, has been filled. Using a different datum (i.e. half tide), the Corps of Engineers (38) estimates " . . . that about 21 percent of the original high water bay area has been filled; most of the filled areas originally consisted of biologically rich salt marsh tidelands . . . ."

Extent of Dredging

The areal extent of dredging in the bay is illustrated in Fig. 3. According to (38):

> Starting near the beginning of this century the channels in the north and central parts of the bay were straightened, widened, and deepened to facilitate navigation; and practically all of the marshlands in the north bay were filled to create lands for airfields, highways, docks, shipyards, parks, tourist and recreational facilities, and other uses.
>
> . . . Extensive dredging since 1940 has . . . altered the character of the bay floor. . . . In areas that have been dredged, the bay floor mud layer has been removed exposing the underlying sandy strata.

Approximately 31 km^2 of the 1918 bay floor, amounting to 55%, has been dredged. As evident in Fig. 3, except for an area of about 10 km^2 in the extreme south bay area, most of the bay floor has been dredged. Further, this dredging has virtually doubled the depth of most of the shallower portions of the north and central bay.

If dredging and filling are both considered, only about 17 to 18% of the bay area represented by the 1918 mean high tide line remains undisturbed.

As to the volumes moved, it is estimated that dredging has shifted about 100 to 140 $\times 10^6$ m^3 of bay floor sediment approximately as follows: a) about 2 $\times 10^6$ m^3 dredged mechanically and transported to deepwater ocean disposal sites[1], b) about 22 $\times 10^6$ m^3 pumped to Silver Strand as beach replenishment (27), and c) roughly 75 to 115 $\times 10^6$ m^3 pumped to bay marginal areas as fill[1], as illustrated in Fig. 2. By comparison, the San Diego River, the former chief tributary to the bay before diversion in 1875-77, is estimated to have delivered about 3.8 to 5.3 $\times 10^5$ m^3 to the bay annually.

It is evident that the marked reduction in area of the bay has reduced the volume of the tidal prism (say, by roughly one-fourth), and it is probable that increasing the depth of the bay, together with the reduction in the tidal prism, has reduced the flushing rate.[2] In this regard, the effects of forthcoming dredging and fill projects on the tidal prism and on flushing in San Diego Bay are discussed in (37, 38, 40).

[1] Estimates based on review of available data with Mr. Donald R. Forrest, San Diego Unified Port District, whose extensive knowledge of San Diego Bay dredging projects has greatly facilitated preparation of this paper.

[2] Because of the imprecision of the data presently available, it is not possible to develop a meaningful estimate of the magnitude of the reduction in flushing rate.

Further, although adequate data are not available to test the supposition, it seems likely that the tidal circulation pattern in the bay may have been modified as a result of the change in bay geometry discussed here.

NATURAL SEDIMENT INPUT

As discussed by several workers (9, 21, 22, 29, 33, 34), an estuary may receive sediment contributed from various external, marginal, and internal sources. In the case of San Diego Bay, the principal *potential* sources are: external—fluvial and coastal sedimentation; marginal—shoreline erosion; and internal—substrate erosion. The term potential is stressed because, as will be shown, present contribution of sediment to San Diego Bay from all sources is minimal.

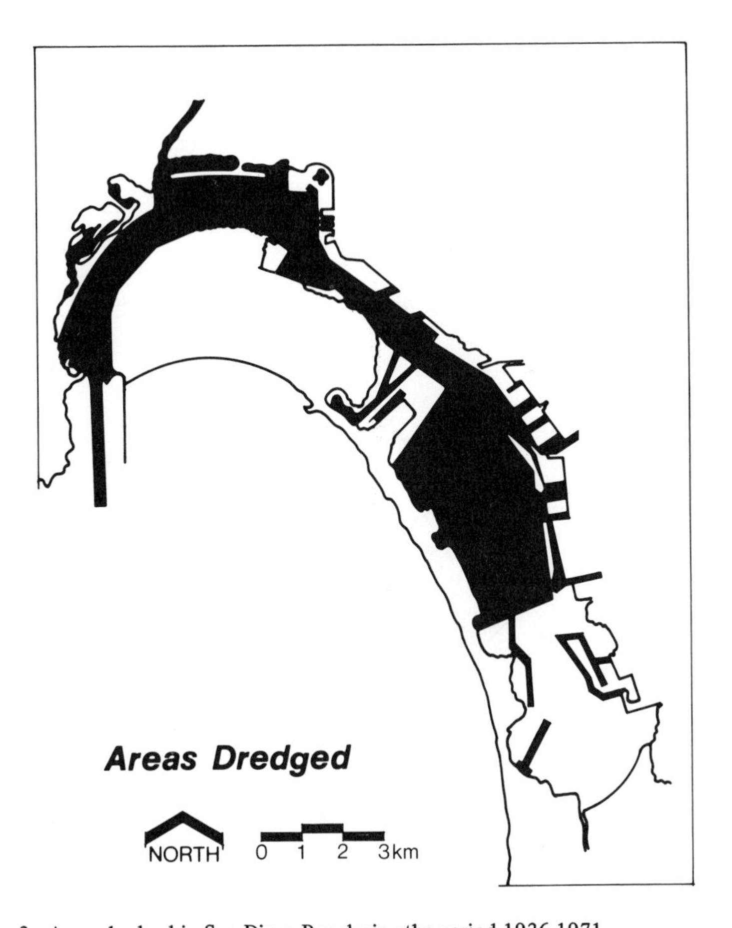

Figure 3. Areas dredged in San Diego Bay during the period 1936-1971.

An assessment of these various potential sources suggests that prior to man's intervention (which began in the 1870's), fluvial deposition was the predominant source of sediment delivered to the bay. Earlier in Holocene times, littoral drift played a key role in building the narrow, elongate Silver Strand barrier or spit which separates most of the bay from the Pacific Ocean. A brief review of sediment influx from various sources follows.

River/Stream Sedimentation

The principal tributaries to San Diego Bay (Fig. 4) were the San Diego River, the Sweetwater River, and the Otay River, and a number of smaller streams; the drainage basins tributary to the bay totalled about 2330 km^2. As evidenced by the prominent delta of the San Diego River in northern San Diego Bay (see Fig. 1), as well as the deltas of the Sweetwater and Otay rivers, as late as the 1850's natural sedimentation (primarily fluvial deposition) was gradually filling the bay. Beginning in the 1870's, however, fluvial sediment influx was almost eliminated as a result of the 1875-77 diversion of the flow of the San Diego River away from San Diego Bay and into nearby Mission Bay (28, 38), and construction of water storage reservoirs on the Sweetwater River in 1888, and the Otay River in 1919 (see Fig. 4). The diversion terminated the sediment contribution of the San

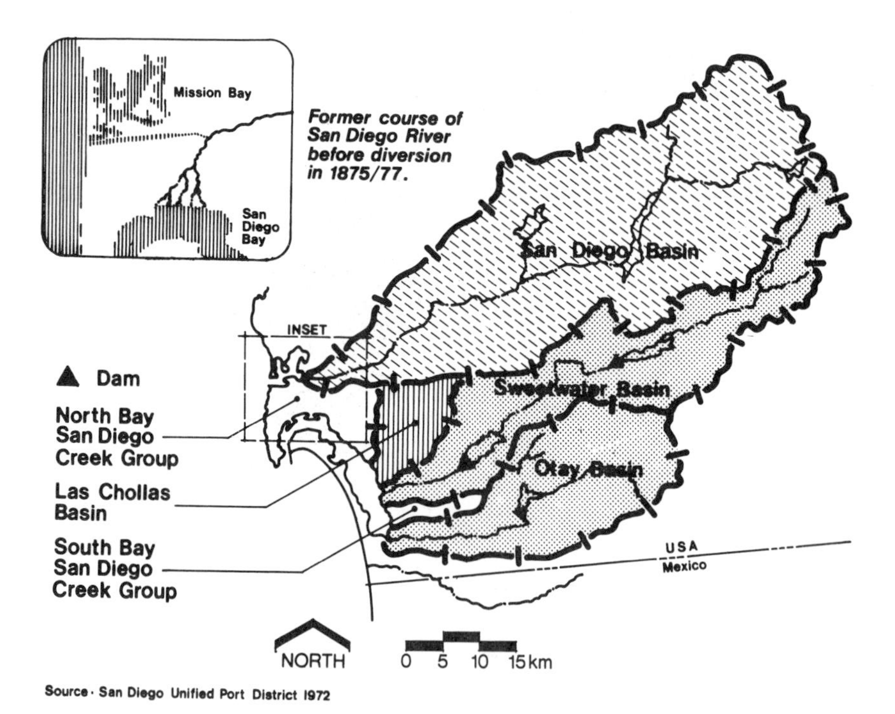

Figure 4. Tributaries and drainage basins contributing sediment to San Diego Bay.

Diego River, and, as will be shown, the reservoirs reduced the input from the Sweetwater and Otay rivers by about 75%.

Major Tributaries

Prior to diversion, the San Diego River, with a drainage area of about 1125 km^2, delivered an estimated 3.8 to 5.35 $\times$ 10^5 m^3 of sediment (28) each year to San Diego Bay.

Although no specific sediment load data could be obtained for the Sweetwater and Otay river basins, their former and present sediment contribution to San Diego Bay can be approximated. According to data from San Diego County (32), the combined area of the two basins (950 km^2) equals about 85% of the San Diego River basin. Assuming the sediment load was roughly proportional to area, then these two basins, prior to dam construction, together delivered about 3.25 to 4.5 $\times$ 10^5 m^3 to the bay. Since dam construction, about 76% of the Sweetwater-Otay drainage area has been controlled by reservoir storage. The current sediment contribution to San Diego Bay from the undammed, downstream 24% of the Sweetwater-Otay basins is estimated at 0.75 to 1.0 $\times$ 10^5 m^3 per year.

Other Tributaries

The basin of Las Chollas Creek and neighboring streams (Fig. 4) has an area of 120 km^2, which is about 11% of the area of the San Diego River basin; again assuming sediment load to be proportional to area (which is a highly conservative assumption because of the almost completely urbanized character of these drainage basins), the Las Chollas basin probably contributes less than 0.4 to 0.6 $\times$ 10^5 m^3 of sediment annually. According to USACE (38), the runoff (and thus sediment) from the remaining small watersheds draining into the bay is negligible; accordingly, the sediment input from these remaining watersheds is assumed conservatively to be not more than 0.25 $\times$ 10^5 m^3 annually.

Total Fluvial Contribution

Based on the above, the total fluvial sediment delivered to San Diego Bay annually prior to man's intervention was probably on the order of 0.8 to 1.1 $\times$ 10^6 m^3. Following diversion and dam construction, the annual fluvial sediment contribution dropped to 1.4 to 1.9 $\times$ 10^5 m^3; roughly this rate apparently has obtained for the last 55 years. In short, according to USACE (38), fluvial sediment contribution to the bay is minimal because: 1) few major drainages enter the bay, 2) many of these drainages are partially controlled by dams, and 3) stream velocities downstream from the dams are generally non-scouring.

Coastal Sedimentation

Three types of coastal sedimentation mechanisms potentially apply to San Diego Bay: influx via littoral transport and passage through the bay entrance, storm related washovers across the Silver Strand, and wind transport across the Silver Strand.

Littoral Transport

Because of the protection afforded by the Point Loma headland and the natural narrowness of the entrance to the bay (see Fig. 1), it is unlikely that littoral transport has been a significant source of sediment subsequent to accretion of the Holocene sand cover over the Pleistocene Bay Point and Linda Vista formations at Coronado and North Island along the northwest margin of the bay. Once the bay entrance attained the approximate configuration shown on the 1857 chart (Fig. 1), the hydraulic geometry of the entrance as suggested by the shape and depth of channel probably did not favor influx of a significant volume of sediment. In 1898, a training wall (the Zuniga jetty) was built along the south margin of the channel (Fig. 2) to help keep the bay entrance scoured to project depth. This training wall also serves to minimize the introduction of sand to the harbor entrance by littoral transport.

At the present time, the littoral drift along Silver Strand is northwest (15, 16) toward the bay entrance, where some sand accumulates on Zuniga Shoal, and some enters the bay's entrance channel. The sand which enters the channel is then transported offshore by ebb tidal currents to an area of accretion on the shallow shelf.

It is the conclusion of Corps of Engineers (38) that, because of the conditions summarized above, San Diego Bay is not subject to littoral shoaling, i.e. delivery of sediment via the entrance.

Washover Fans and Deltas

Because of the generally benign climate and almost total absence of severe oceanic storms along the Southern California coast, severe wave action is very rare and storm surge buildup is virtually unknown. As a result, development of washover fans and deltas, which are common external sources of estuarine sediment in many Atlantic and Gulf coast estuaries, does not occur in San Diego Bay.

Wind Transport

Onshore winds transport some sand from the Silver Strand beaches across the narrow peninsula and into the western edge of the bay. Prevailing winds are from the northwest through southwest quadrants (38); average monthly wind velocities range from 7.5 to 11 km/hr, but strong winds are very rare. Because of the very gentle character of the winds, the virtual absence of a backshore, and the sand-deficient character of the Silver Strand beaches (16, 38), the amount of sand transported into the bay by wind action at the present time is judged to be small.

Shoreline Erosion

It would be expected that the principal natural marginal source of sediment in the bay would be wave erosion of the bay shoreline. As illustrated in Fig. 5, however, roughly 65% of the shoreline has been protected from erosion by piers and docks, bulkheads, revetments and riprap. Because of the minimal available fetch in the narrow northern end of the bay, the unprotected areas there are not subject to severe erosion. As evident in the figure, in the central and southern parts of the bay, roughly two-thirds of the unprotected shoreline areas are

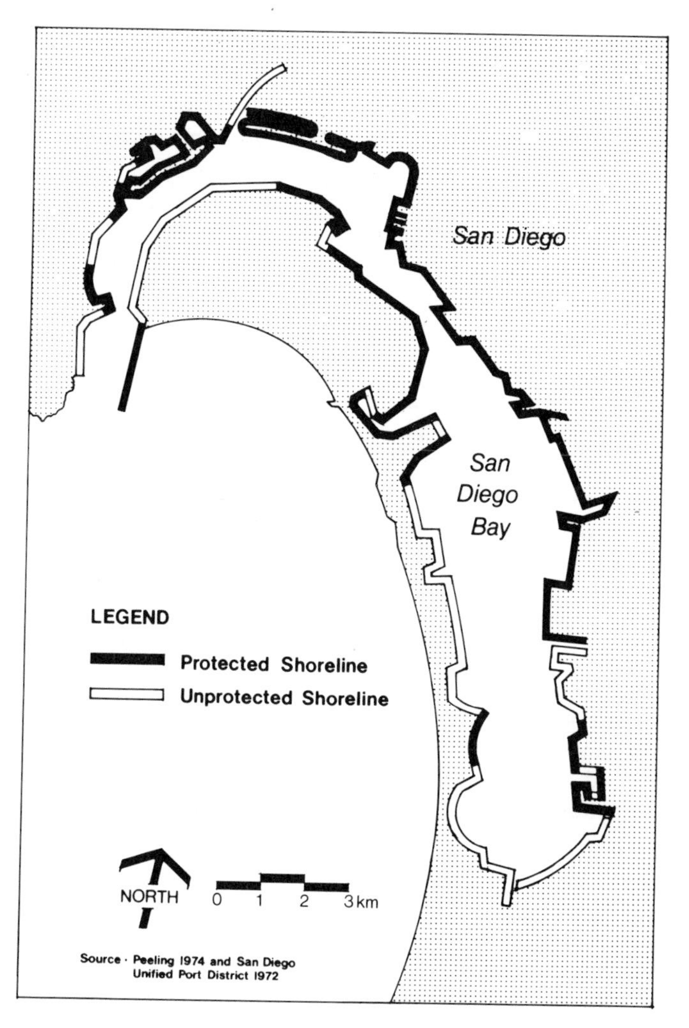

Figure 5. San Diego Bay shoreline characteristics.

located on lee shores with reference to the prevailing northwest through southwest quadrant winds. As a result, only about 18 to 20% of the unprotected shoreline and about 7% of the overall bay shoreline appears subject to significant erosion. Thus, at present, shoreline erosion is judged to provide only a minimal contribution of sediment to the bay.

Substrate Erosion

There is no evidence that erosion of substrate serves as an internal sediment source in San Diego Bay. Conversely, the approximate doubling in depth of virtually all of the central and north bay by dredging argues against substrate erosion functioning as a source of any significant volume of sediment.

On balance, in terms of the 100 to 140 $\times$ 10^6 m^3 volume of sediment dredged from San Diego Bay, natural sediment input to the bay from all sources is minimal and has been for more than 55 years. This conclusion is strongly supported by the fact that dredging in the bay to maintain interior channel depths has been extremely limited–only 3.4 $\times$ 10^5 m^3 in the 29 year period since 1946 (38). Additional evidence supporting this conclusion is the persistence of a dredged channel in the extreme south bay that has not been maintained for more than 30 years.

DISCUSSION

As presented earlier, the total annual fluvial sediment influx to San Diego Bay before man's intervention was probably about 0.8 to 1.1 $\times$ 10^6 m^3. For purposes of this discussion, it is assumed that the other principal external sediment source, i.e. coastal sedimentation, contributed an insignificant volume of sediment to the bay.

Assuming that the 0.8 to 1.1 $\times$ 10^6 volume figure is the correct order of magnitude, then the 100 to 140 $\times$ 10^6 m^3 dredged from San Diego Bay (primarily in the 35 years since 1940) is equivalent to roughly 90 to 175 years of natural sediment input.

Further, because the mouth of the San Diego River periodically shifted between San Diego Bay and nearby Mission Bay (28), it is probable that the 0.8 to 1.1 $\times$ 10^6 m^3 value should be reduced by about 50% for roughly half of the past several thousand years. In effect, this would reduce the average annual rate of sedimentation to about 75% of the 0.8 to 1.1 $\times$ 10^6 figure (i.e. 0.6 to 0.8 $\times$ 10^6 m^3 per year) which in turn would mean that the 100 to 140 $\times$ 10^6 m^3 dredged by man is equivalent to about 125 to 235 years of natural sedimentation in the bay.

But, as demonstrated earlier, man's intervention has reduced natural sedimentation from streams and other sources to probably less than 1.4 to 1.9 $\times$ 10^5 m^3 per year, which is about 18% of the natural influx before man intervened. If one calculates an average annual dredging rate for the 30 year period 1940-1970[3], the resulting rate of 3.3 to 4.7 $\times$ 10^6 m^3 is roughly 3 to 6 times the 0.8 to 1.1 $\times$ 10^6 m^3 annual rate of natural sediment input to the bay prior to man's intervention. With natural sediment input reduced by man to roughly 18% of its previous level, this 30 year dredging rate becomes roughly 17 to 34 times the present annual sedimentation rate.

Although some of the quantity figures used in the above approximation are "guesstimates," because of the major difference in magnitudes between sediment input and the volumes dredged, it seems certain that the overall relationships would not be appreciably different even if the areas and volumes could be determined with more precision. In short, the dominance of man's dredging over natural sedimentation in San Diego Bay is clear-cut. Further, it is evident that in this bay, man has reversed temporarily the natural geologic sequence of estuarine filling by sedimentation described in (8, 29, 30) and others.

CONCLUSIONS

In San Diego Bay, man's dredging and spoil disposal activities have functioned as sediment reworking and estuarine basin reshaping mechanisms, temporarily reversing the natural geologic sequence of estuarine filling by sedimentation. As geologic processes, dredging and spoil disposal have been substantially more important than all other erosional and depositional processes currently operating in the bay.

Since the early 1900's, dredging and spoil disposal as fill in San Diego Bay have reworked and shifted 100 to 140 $\times$ 10^6 m^3 of sediment, with a resulting 27% reduction in the water area of the bay, and an approximate doubling in depth of 55% of the original water area. Only 17 to 18% of the original area remains undisturbed by dredging or fill. These changes have reduced the volume of the tidal prism and probably the flushing rate, possibly modified the circulation pattern, and removed surficial bay muds exposing underlying sands.

Prior to man's diversion of the largest tributary river (1875-77), and damming of the two next largest (1888 and 1919), fluvial transport contributed about 0.8 to 1.1 $\times$ 10^6 m^3 of sediment to the bay annually. Coastal sediment does not reach the bay in significant quantities, and because of protective structures or unfavorable geometry, erosion of the bay shoreline contributes minimal sediment to the bay floor.

For the 30 year period 1940 to 1970 (during which most of this dredging took place), the annual dredging rate averaged 3.3 to 4.7 $\times$ 10^6 m^3, which is 3 to 6 times the former yearly fluvial sediment input to the bay. As a result of

[3] Subsequent to enactment of the California Environmental Quality Act in 1970, very little dredging has been carried out in San Diego or other California bays.

diversion and damming of the major tributaries, and extensive urbanization of the drainage basins of the minor tributaries, the annual volume of stream delivered sediment has dropped by 80% or more. Accordingly, the 30 year average annual dredging rate is roughly 17 to 34 times the current yearly sediment input to the bay.

Effectiveness of dredging as a geologic process in San Diego Bay is evidenced further by the unusually low volume of maintenance dredging in interior channel areas (about 3.4×10^5 m^3 in 29 years), and by the persistence of an unmaintained channel dredged more than 30 years ago.

Even though the changes described here are short-lived in terms of geologic time, they are not short-lived in terms of man's time scale, his use of estuaries and bays, and his environmental impact thereon.

It seems probable that dredging and spoil disposal are important geologic processes in other estuaries, particularly in those a) with limited sediment input such as in arid and semi-arid climates (e.g. along the south Texas coast), or b) where the scale of harbor development is large in proportion to the size of estuary affected.

ACKNOWLEDGMENTS

The writer gratefully acknowledges the following individuals whose varied assistance greatly facilitated this study: Messers. J. E. Liebmann, D. R. Forrest, T. E. Firle, F. H. Trull, R. G. Gautier, and K. L. Scheevel of the San Diego Unified Port District; Drs. M. M. Nichols, Virginia Institute of Marine Science, and C. C. Mathewson, Texas A & M University; Messers. F. A. Kingery, WESTEC Services, Inc., M. P. Kennedy, California Division of Mines, I. C. Macfarlane, E. J. McHuron, D. L. Woodford, Dames & Moore, N. T. Sheahan, Brown & Caldwell, and P. L. Horror, Intersea Research, Inc. Mr. T. J. Peeling, Naval Undersea Center, made available figures from his reports as a basis for figures in this paper. Mss. L. Smiraglia, K. M. Smith and K. F. Graham assisted with documentation and research; Ms. Graham edited the manuscript.

REFERENCES

1. Bassi, D. E., and D. R. Basco. 1974. Field study of an unconfined spoil disposal area of the Gulf Intracoastal Waterway in Galveston Bay, Texas. Texas A & M Univ. TAMU-SG-74-208. Jan. 74 p.
2. Boyd, M. B., R. T. Saucier, J. W. Keeley, R. L. Montgomery, R. D. Brown, D. B. Mathis, and C. J. Guice. 1972. Disposal of dredge spoil–problem identification and assessment and research program development. U. S. Army Engineer Waterways Experiment Station Tech. Rept. H-72-8. Nov. 121 p.
3. Brehmer, M. L. 1967. A study of the effects of dredging and dredge spoil disposal on the marine environment. VIMS Spec. Rept. Marine Sci. and Ocean Eng. 8 to U. S. Army Corps of Engineers. 24 p.
4. Cronin, L. E. 1967. The role of man in estuarine processes, p. 667-689. *In* G. H. Lauff (ed.), Estuaries. Am. Assoc. Adv. Sci. Publ. 83.

5. Ford, R. F. 1968. Marine organisms of south San Diego Bay and the ecological effects of power station cooling water discharge. Environmental Engineering Lab. Tech. Rept. C-188, for San Diego Gas & Electric Co. 278 p.
6. Forrest, D. R. 1975. Personal communication regarding maximum tidal ranges in San Diego Bay. San Diego Unified Port District. Sept. 1975.
7. Gautier, R. G. 1972. Natural physical factors of the San Diego Bay tidelands. Part of Master Plan Revision Program. San Diego Unified Port District Planning Dept. Jan. 88 p.
8. Gorsline, D. S. 1967. Contrasts in coastal bay sediments on the Gulf and Pacific coasts, p. 219-225. *In* G. H. Lauff (ed.), Estuaries. Am. Assoc. Adv. Sci. Publ. 83.
9. Guilcher, A. 1967. Origin of sediments in estuaries, p. 149-157. *In* G. H. Lauff (ed.), Estuaries. Am. Assoc. Adv. Sci. Publ. 83.
10. Hargis, W. J., Jr. 1972. Engineering works and the tidal Chesapeake, p. 105-123. *In* Remote Sensing of Chesapeake Bay. NASA SP-294.
11. Herbich, J. R. 1975. Coastal and Deep Ocean Dredging. Gulf Pub. Co. Houston, TX. 622 p.
12. Huston, J. 1970. Hydraulic Dredging. Cornell Maritime Press, Cambridge, MD. 318 p.
13. Hutton, R. F., B. Eldred, K. D. Woodburn, and R. M. Ingle. 1956. The ecology of Boca Ciega Bay with special reference to dredging and filling operations. Florida Maritime Lab. Tech. Ser. 17:1-86.
14. Ingle, R. M. 1952. Studies on the effect of dredging operations upon fish and shellfish. Tech. Series No. 5. Florida State Bd. Conservation, St. Petersburg.
15. Inman, D. L. 1973. The Silver Strand littoral cell and erosion at Imperial Beach. Presented during California Representative Lionel van Deerlin's Imperial Beach Conference, 1 Feb. 1973.
16. ____, C. E. Nordstrom, S. S. Pawka, D. G. Aubrey, and L. C. Holmes. 1974. Nearshore processes along the Silver Strand littoral cell. As cited *in* Corps of Engineers. 1975. Final Environmental Impact Statement, San Diego Harbor, San Diego County, California.
17. Lee, G. F., and R. H. Plumb. 1974. Literature review on research study for the development of dredged material disposal criteria. For U. S. Army Engineer Waterways Experiment Sta., Vicksburg, MS by Inst. for Environmental Studies, Univ. Texas, Dallas.
18. Lockheed Oceanics Div. 1967. Geological survey of south San Diego Bay. Rept. No. 20867. Aug. For San Diego Unified Port District. 58 p.
19. Mackin, J. G. 1967. Canal dredging and silting in Louisiana bays. Texas Inst. Marine Sci. Publ. 7:262-314.
20. May, E. B. 1973. Environmental effects of hydraulic dredging in estuaries. Alabama Marine Resources Bull. No. 9 April. p. 1-85.
21. Meade, R. H. 1969. Landward transport of bottom sediments in estuaries of the Atlantic coastal plain. J. Sed. Petrology 39: 222-234.
22. ____. 1972. Transport and deposition of sediments in estuaries, p. 91-120. *In* B. W. Nelson (ed.), Environmental framework of coastal plain estuaries. Geol. Soc. Amer. Mem. 133.
23. Moore, G. W., and M. P. Kennedy. 1970. Coastal geology of the California Baja California border area. *In* Pacific slope geology of northern Baja California and adjacent Alta California. Amer. Assoc. Petroleum Geologists (Pacific Section) Fall field trip guidebook.

24. ——, and ——. 1975. Quaternary faults at San Diego Bay, California. J. Research U. S. Geol. Survey. 3(5):589-595.
25. O'Neal, G., and J. Sceva. 1971. Effects of dredging on water quality in the Northwest. Environmental Protection Agency, Office of Water Programs, Region X, Seattle, WA. July.
26. Oppenheimer, C. H. 1973. The development of a multi-purpose deepdraft inshore port on Harbor Island, Texas, to accommodate VLCC vessels, an environmental impact statement. In cooperation with Nueces County Navigation District No. 1, Corpus Christi, TX. Feb. 54 p.
27. Peeling, T. J. 1974. A proximate biological survey of San Diego Bay, California. Naval Undersea Ctr. Rept. No. NUC TP 389. June. 83 p.
28. Rambo, C. E., and W. C. Speidel. 1969. A case study of estuarine sedimentation in Mission Bay–San Diego Bay, California. For Dept. Interior FWPCA, Contract No. 14-12-425 by Marine Advisers, La Jolla, CA. Feb. 58 p.
29. Rusnak, G. A. 1967. Rates of sediment accumulation in modern estuaries, p. 180-184. *In* G. H. Lauff (ed.), Estuaries. Am. Assoc. Adv. Sci. Publ. 83.
30. Russell, R. J. 1967. Origins of estuaries, p. 93-99. *In* G. H. Lauff (ed.), Estuaries. Am. Assoc. Adv. Sci. Publ. 83.
31. Saila, S. B., S. D. Pratt, and T. T. Polgar. 1972. Dredge spoil disposal in Rhode Island Sound. Marine Tech. Rept. No. 2. Univ. Rhode Island, Marine Experiment Sta., Kingston.
32. San Diego County. 1973. Hydrology report, 1973 season. Dept. Sanitation and Flood Control. 180 p.
33. Schubel, J. R. 1971. Sources of sediments to estuaries, p. V1-19. *In* J. R. Schubel (conv.). The estuarine environment–estuaries and estuarine sedimentation. Amer. Geol. Inst. Short Course Lect. notes.
34. Shideler, G. L. 1975. Physical parameter distribution patterns in bottom sediments of the lower Chesapeake Bay estuary, Virginia. J. Sed. Petrology. 45(3):728-737.
35. Smith, D. D., and E. J. McHuron. 1974. Dredge spoil evaluation studies in Corpus Christi Bay, Texas (abst.). Geol. Soc. Amer. Ann. Mtg. Miami, FL. p. 618.
36. ——. In preparation. Dredging and spoil disposal as geologic processes in Corpus Christi Bay, Texas.
37. ——, and M. Wright. 1975. Environmental impact report for the Chula Vista boat basin/wildlife reserve. For San Diego Unified Port District. Nov. 117 p.
38. U. S. Army Corps of Engineers, Los Angeles District. 1975. Final environmental statement, San Diego Harbor, San Diego County, California. Feb.
39. U. S. Navy Volunteer Research Reserve Unit 11-5. 1950. A survey of San Diego Bay. Progress rept. 1. Dec. 31. 21 p.
40. U. S. Navy. 1975. Environmental impact statement, proposed new ammunition facility at the Naval Air Station, North Island Project P-800, San Diego, California. July.

SEDIMENTARY PATTERNS AND PROCESSES IN WEST COAST LAGOONS

John E. Warme, Luis A. Sanchez-Barreda, and Kevin T. Biddle
Department of Geology
Rice University
Houston, Texas 77001

ABSTRACT: We have studied two tidal lagoons, Mugu Lagoon in temperate California and Laguna Potosi in tropical Mexico. Critical for the sedimentology, ecology, and history of the lagoons is their inlet position and condition (open or closed). Sedimentation is tidally controlled when the inlets are open and negligible when they are closed.

Sand distribution is controlled by present and past inlet positions, and in both lagoons sand is the dominant sediment and the foundation material. Open inlets allow tidal currents to move sand from the beach into channels and tidal flats proximal to the inlet and promote transport of mud into distal reaches. Lagoon waters are then marine and support abundant life.

When the inlets are closed, lagoon waters are quiet and sediment input is negligible except for washover and eolian sand from beaches and dust from adjoining land. Lagoon waters may become stagnant and/or hyposaline.

Significant for sedimentation and evolution of the lagoons are differences in climate, stream runoff, orientation with the inlet, and vegetation. Mugu Lagoon is irregularly estuarine, is elongated perpendicular to its inlet, thus minimizing fresh water flushing, and is dominated by an upper tidal zone salt marsh and thus has extensive intertidal flats. In contrast, Laguna Potosi is annually estuarine, is elongated between its inlet and a fresh water effluent, thus being completely flushed, and has vegetation dominated by lower tidal zone mangroves that have fixed its perimeter.

Presently the lagoon inlets are artificially kept open to assure normal marine circulation, thus circumventing the temporal vagaries of the natural inlet cycle, hastening sand deposition and shortening the life of the lagoons.

INTRODUCTION

To better understand the sedimentary processes operating in tidally-controlled lagoons we have studied two west coast examples: Mugu Lagoon, in temperate California, and Laguna Potosi, in tropical Guerrero State, Mexico. Our purpose here is to summarize the sedimentology of both lagoons and compare them, because they exhibit similar hydrography but exist in different climates.

MUGU LAGOON

Warme (11, 12) and Biddle (2) discussed the geologic setting, sedimentary environments, fauna, flora and other ecological and paleoecological aspects of Mugu Lagoon. It is located near Point Mugu, southern California, at the junction of the western most end of the Santa Monica Mountains and the eastern edge of the Oxnard Plain, a subaereal and submarine fan-delta complex (Fig. 1). Approximately 400 m of primarily unconsolidated Pleistocene sediments underlie the western arm of the lagoon; the eastern arm apparently lies on a wave-cut terrace of Miocene rock, which is exposed in the adjoining mountains.

The lagoon averages about 400 m in width and extends 7 km parallel to the coastline behind a narrow barrier beach. It is less than 3 m deep at low water except where artificially dredged, and the average depth within the low-water perimeter is less than 1 m. Calleguas Creek, a small intermittant stream, empties into the central part of the lagoon (Fig. 1). Mugu Lagoon has been divided into several physiographic units each with its own topographic expression, sediments, plants and animals (Fig. 1), which are detailed elsewhere (12).

The eastern arm of the lagoon was most extensively studied because of its natural state, being located on U. S. Navy property where public access is normally prohibited.

Climate

Coastal southern California has a Mediterranean-type climate, with a summer-dry season. Precipitation is usually between 25 and 50 cm per year, and is mostly the result of a few intense storms (12). Prevailing winds are from the north and west near Point Mugu.

Wave Approach and Longshore Drift

Vigorous waves approach the coastline near Mugu Lagoon mainly from the northwest, although this part of the coast is subject to wave attack from both the Northern and Southern hemispheres (4). The predominant northwesterly wave pattern results in a net longshore transport of sediments to the southwest, and provides an apparent continuous supply of sand at the inlet. Some of this sand is transported in and out of the lagoon, and a portion of it is permanently left there.

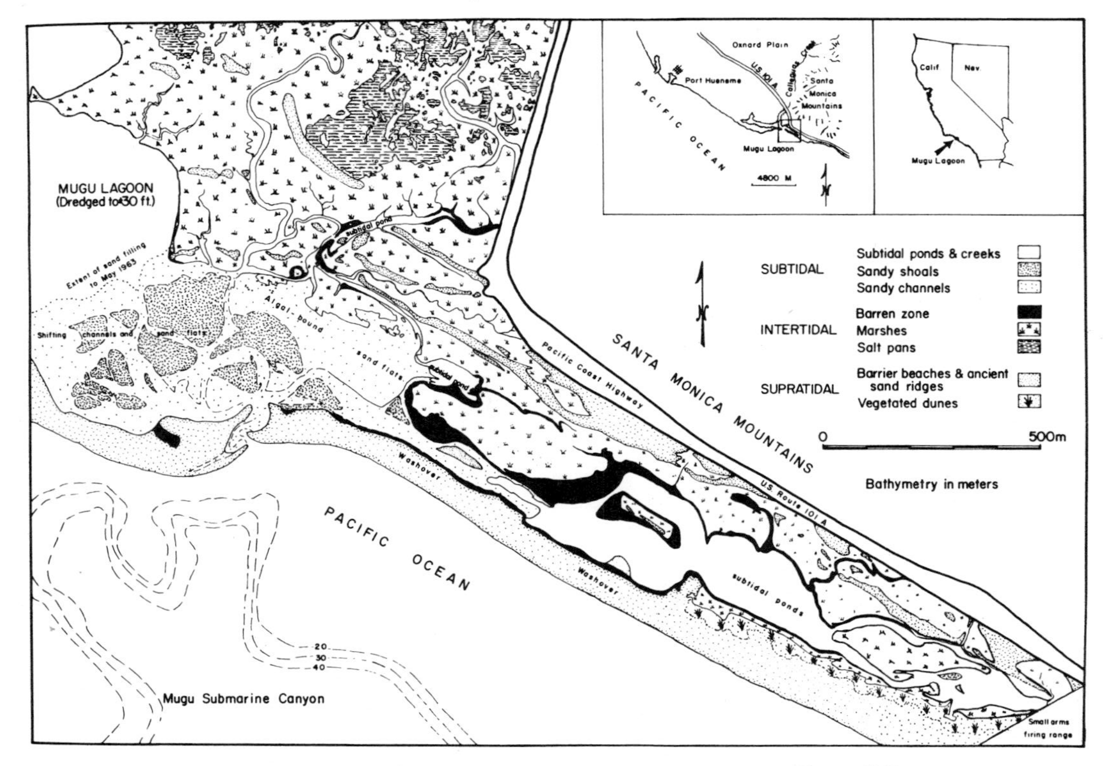

Figure 1. Location of Mugu Lagoon and depositional habitats in the eastern arm. Modified from Warme (12).

Tides and Tidal Currents

Pacific coast tides are dominated by semidiurnal constituents, giving rise to two high and two low tides each day, usually of unequal range. Spring tidal range is about 2 m. However, because the lagoon inlet and extensive inner tidal delta is maintained at about mean sea level (see below), only the upper half of the tidal spectrum affects the lagoon interior (11, 12), with tidal levels below or at mean sea level resulting in lagoon drainage through the inlet, possibly for several days during neap tidal periods. Maximum water exchange occurs during spring tides, when one or two orders of magnitude more water volume is exchanged than during neaps, owing to the greatly expanded tidal prism at mean high water and above. The irregular tides give rise to water levels, currents, and durations of slack water inside the lagoon that are extremely variable during any given month or year. Warme (12) has discussed the effect of tidal heights on the sediments and organisms of the lagoon.

Maximum current velocities (in excess of 2 m per sec.) are attained during flood tides at and near the inlet. Current speeds are maintained in the main channels, but elsewhere decrease with distance from the inlet, becoming sluggish in the farther and upper reaches. Velocities are damped immediately as the water overtops channel banks and spreads over tidal flats and marshes. Ebb currents are slower and flow longer, to account for equal volumes of ebb and flow waters.

LAGUNA POTOSI

Sediments and sedimentation in Laguna Potosi were studied by Sanchez-Barreda (8). Laguna Potosi is located in the state of Guerrero, on the south-western coast of Mexico (Fig. 2). It is 180 km north of Acapulco, and 10 km south of the port town of Zihuatanejo. The lagoon is situated on a narrow coastal plain of marginal marine deposits over shallow bedrock, judging from crystalline rock outcropping near the coast and from the sea stacks ("morros") offshore. The coastal plain borders the rugged Sierra Madre del Sur, which rises to altitudes of 1000 m and more only a few kilometers from the coastline.

Laguna Potosi lies behind two wide (200-250 m) beach ridges, one from the north and one from the east, that converge on a sea stack as tombolo spits (Fig. 2). The lagoon extends for 2.5 km parallel to the southern spit and is about 300 m in width. It has widespread subtidal flats cut by tidal channels, well-developed bordering mangrove and marsh vegetation, and peripheral salt pans. Maximum channel depth is about 3 m, but most of the area enclosed by the low-water perimeter is less than 1 m deep. The Rio Petatlan flows continuously into the eastern end of the lagoon, with greatest discharge during the summer, through a vegetated channel or "moat" (1) leading from the mouth of the river 4 km to the southeast (Fig. 2). It has flowed vigorously at times, as evidenced by a 2 m-deep scour occupying the narrow eastern end of the lagoon where the channel emerges from thick mangroves.

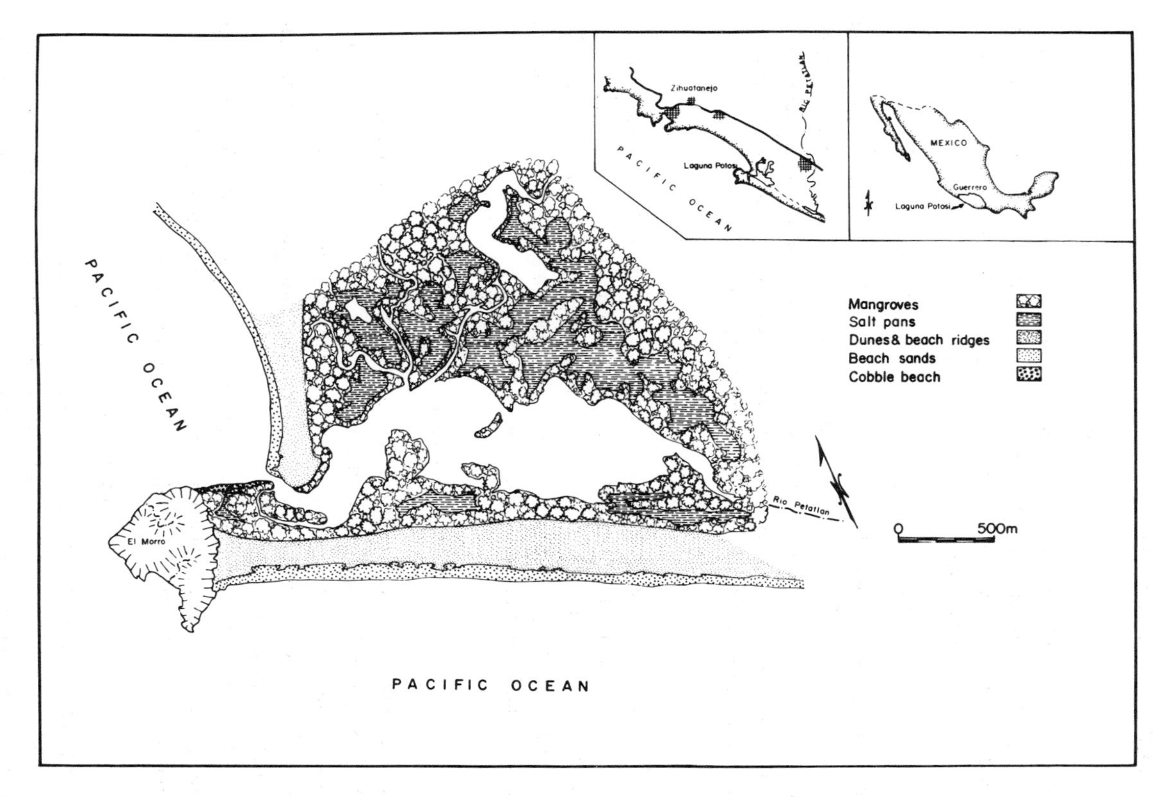

Figure 2. Location of Laguna Potosi and depositional habitats in and adjacent to the lagoon.

Laguna Potosi is largely undeveloped and still in its natural state. The salt pans, however, have been used for years as evaporation ponds, and reworking of sediments from this activity has been extensive. A small fishing village is located on the north side of the lagoon inlet.

Climate

Climate on the Guerrero coast is tropical-humid with a rainy summer and hot dry winter (7). Yearly precipitation averages 210 cm in the mountains and 110 cm along the coast. Prevailing winds are southeasterly from April to October, and shift to the northeast from November to April (6).

Wave Approach and Longshore Drift

The coastline near Laguna Potosi is subject to Pacific Ocean swells (generated in high-latitude storm belts of both hemispheres during their respective winter months), storm waves, and local wind waves. Concommitant shifts of longshore transport direction in response to Pacific Ocean swells are possibly responsible for formation of the tombolo spits that enclose the lagoon.

Tides and Tidal Currents

The Pacific Ocean tides along the Guerrero coast are similar in character to those at Mugu Lagoon but attain only about half the range. Tidal current velocities within the lagoon are maximum, exceeding 1 m per second, in the narrow channel near the inlet, and in the channels between ponds, and decrease across the sand flats and into the peripheral vegetated areas.

SEDIMENTARY PATTERNS

The sediments of both Mugu Lagoon and Laguna Potosi were examined for source areas, grain-sizes, and sedimentary processes, products, rates and history. Methods of study included surface mapping of environments, grain-size analyses, estimates of current velocities, and coring and x-ray radiography for examination of textures and structures.

Grain sizes were determined using the procedures of Folk (3). The sand fraction was dry-seived and the mud fraction analyzed by pipette (Mugu) and hydrophotometric (Potosi; (6)) methods. Results are presented in Fig. 3.

Mugu Lagoon

Grain Sizes. Results of grain-size analyses (Fig. 3) show a linear trend for the Mugu samples, indicating a mixing, in varying proportions, of two basic types of sediment: 1) clean sand, and 2) mud comprised of a mixture of one part silt and two parts clay (12). The only samples that do not follow this trend are the silt-rich sediments of the salt pans and surrounding high marshes, probably derived directly from weathered siltstones of the adjoining mountains.

There is a lack of fine sand and coarse silt within Mugu Lagoon, although abundant grains of these sizes are present just offshore (5). Sorting action of waves apparently moves these size classes offshore and concentrates coarser sand onshore (11).

The framework of the lagoon is sand, derived by traction processes. All cores, even from the muddiest (on the surface) parts of the lagoon, become sandy within 1 m of the surface, usually 95% or more sand (10). The direct source of this sand is the beach, where it is supplied by longshore drift, and derived mainly from rivers northwest of the lagoon (4). Mud, in contrast, is transported by suspension from turbid offshore waters during and after storms, or introduced directly into the lagoon by Calleguas Creek. The mud is then pumped into the distal reaches by tides. The processes that lock mud into positions landward of sand in protected environments were summarized by Warme (12).

Sediment distribution. There are five sediment distribution processes identified in Mugu Lagoon. Listed in order of decreasing importance, they are: 1) tidal currents, 2) barrier washover, 3) wind, 4) runoff from Calleguas Creek, and 5) runoff from the Santa Monica Mountains. The relative importance of the first three processes and the patterns of sediment movement are shown in Fig. 4. The last two processes operate but have not been assessed critically, and will not be further discussed.

Simplifying for convenience Folk's (3) diagram (Figs. 3, 5), the surface sediments of Mugu Lagoon can be labelled as sand (more than 90% sand-sized

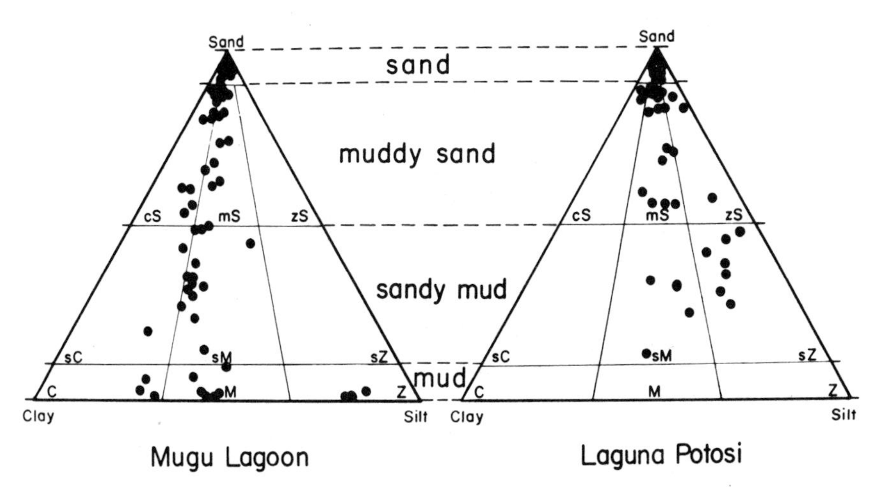

Figure 3. Triangular diagrams [after Folk (3)] showing sand-silt-clay contents of 79 surface sediment samples from Mugu Lagoon and 52 surface samples from Laguna Potosi: S=sand, Z=silt, C=clay, M=mud, cS=clayey sand, mS=muddy sand, zS=silty sand, sC=sandy clay, sM=sandy mud, sZ=sandy silt; Mugu Lagoon data replotted from Warme (12). See text for discussion.

grains), muddy sand (50-90% sand), sandy mud (10-50% sand), and mud (0-10% sand). In Mugu Lagoon the salt-pan silts are distinctive enough to be mapped separately on Fig. 5.

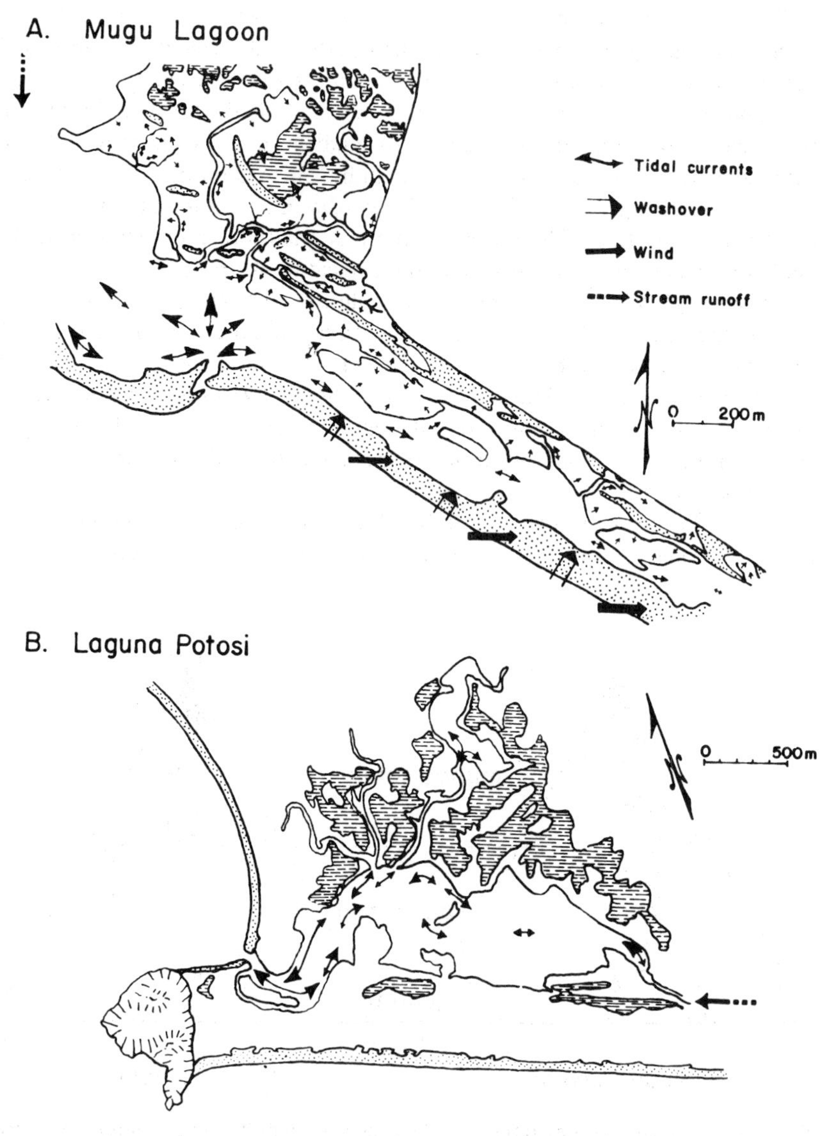

Figure 4. Sediment pathways in (A) Mugu Lagoon and (B) Laguna Potosi. Arrow azimuths represent directions of transport, lengths represent relative current velocities, and sizes or arrowheads represent relative amount of sediment transported in each direction.

Clean sand exists along the present beach, on the ancient sand ridges and on or near the tidal delta and its channels. All of these areas are being or have been subjected to strong tidal currents, ocean waves, and, for the supratidal and intertidal segments, to the wind. Muddy sand exists in the farther reaches of the lagoon where tidal currents are less intense, and organic mixing is very vigorous.

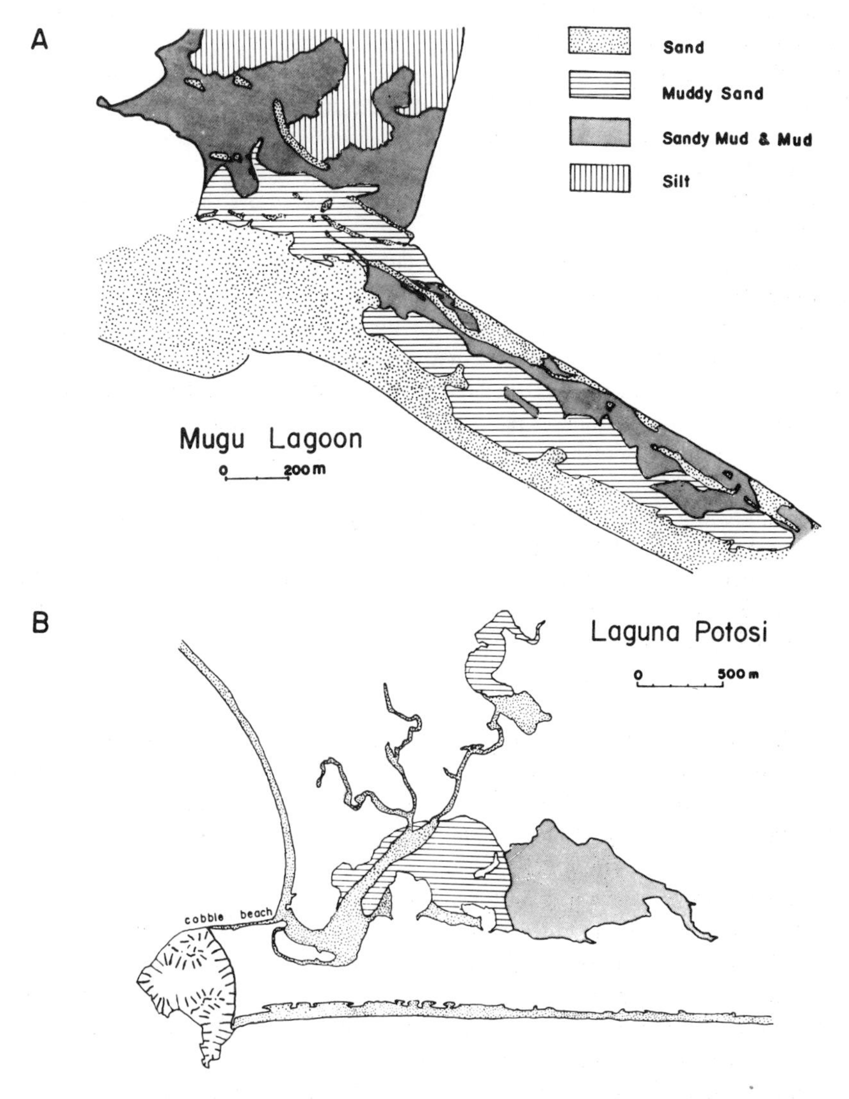

Figure 5. Diagrammatic representation of (A) surface sediments from all environments in Mugu Lagoon, and (B) of surface sediments from the subtidal portions of Laguna Potosi.

Sandy mud occurs only in the higher areas of the lagoon, such as the marsh, or in areas far removed from strong current action, such as the distal ends of tidal creeks. Here the transporting currents have lost most of their competency and can carry sand only rarely. The distribution of silt was mentioned above.

Relative importance of transporting agents. Tidal currents introduce sand into the lagoon through the inlet. Once inside, most of the sand is deposited on the tidal delta or in the artificially-deepened portion of the lagoon, which has been dredged several times in the last decade, yielding thousands of cubic meters of sand. Small amounts of sand are also moved into the more distal parts of the lagoon by tidal currents, much of it by floatation of dried grains on rising tides.

Tidal curerents are also responsible for transportation and deposition of most of the mud in the lagoon. Warme (12) gives several lines of evidence showing that the majority of mud in Mugu Lagoon is deposited by tidal currents and not by runoff.

Barrier washover deposits sand in the lagoon through a combination of waves and high tides, which commonly will result in wave uprush cresting the barrier, washing down the back side across the barrier and into the lagoon, bringing beach sand with it. More than one meter of sand may be deposited in a narrow ribbon in a few hours, and the barrier is widening on the back side by this process (12). Washover takes place anywhere along the beach, any time of the year, and results in sand being deposited on the seaward side of the lagoon well away from the major source of sand at the tidal inlet.

Strong winds usually blow eastward with a small northern component, deflating the barrier and blowing sand into the lagoon. A lag deposit that resembles desert pavement of gravel-sized grains and shells prevents further deflation. The next episode of washover destroys the lag sheet, creating gravel laminae on the gentle back side of the barrier.

Laguna Potosi

Sediments of Laguna Potosi are composed of varying proportions of pure sand and a mud that is primarily silt. In Fig. 3 the samples from Potosi show greater scatter than those from Mugu, suggesting more heterogeneity in provenance and perhaps a more complicated hydrography—the lagoon is subjected to both tidal and river currents.

Sediment distribution. As with Mugu Lagoon, five sedimentary processes can be recognized (see above); however, the relative importance of river runoff at Potosi is much greater. Ranked by decreasing importance these processes are: 1) tidal currents, 2) river runoff, 3) local runoff, 4) wind, and 5) washover. Sand in the lagoon occurs near the inlet and in the scoured tidal channels. Muddy sand exists in shallow flats where current activity is diminished. Sediment in the far eastern portion of the lagoon consists of sandy mud, the finer components of which are brought by the Rio Petatlan. Grain-size analyses appear elsewhere (8).

Normally, runoff from the river enters the ocean at Barra Valentin, 4 km east of Laguna Potosi. Longshore drift closes this outlet during the dry season. The river water then backs up behind the bar forming an elongate moat (1) parallel to the coastline, shunting part of the river water into the eastern end of the lagoon. Because of the long passage through the mangroves only fine-grained suspended sediment reaches the lagoon via this route. From the 2 m deep scour at the narrow eastern end of the lagoon, it is apparent that fresh water enters in volumes and velocities significant enough to flush fine-grained detritus from the distal end on the lagoon.

Runoff from surrounding land during the rainy season is an important sediment contributor. However, much of this material is sand-sized, and deposited in the salt pans before it reaches the lagoon proper. Fig. 4 shows the relative importance of these processes and the paths of sediment movement. Wind and washover were probably more significant earlier in the lagoon's history before dense bordering vegetation was established and before beach ridges were as wide as at present.

IMPORTANCE OF THE TIDAL INLETS

Because tidal currents are the most important distributors of sediment within both lagoons, the position and character of the tidal inlets play an extremely important role in supplying, distributing and retaining sediment. Inlet position controls current patterns near the inlets, and thus controls sediment transport pathways. When the inlet is closed the major source of sediment, shoreline sand, is isolated from the lagoon, arresting sedimentation except for fine-grained material already present settling from suspension.

The inlet at Mugu Lagoon has a natural open-closed cycle. It is normally maintained by tidal flow and periodically closed by sand migrating along the coast in response to the longshore current (4). Combinations of neap tides, berm-building wave regimes, and high rate of sand supply by littoral drift act to close the inlet.

The well-developed inner tidal delta at Mugu (Fig. 1) is a product of inlet migration pumping sand into the lagoon as it moves. The incoming tides, aided by waves, the short but vigorous pulse of the flood tides versus the more sluggish ebbs, and other factors listed by Warme (12), combine to move incoming sand topographically upward and inward to be deposited as bars whose aggregate is the fan-shaped tidal delta opposite the inlet. This slightly concave-upward, tabular body acts as a dam or threshold to retard outflowing water and to trap sediment. Ebb tides, in contrast, are more slowly funnelled through ebb channels that hug the seaward side of the barrier, bringing water laterally from the two arms of the lagoon.

The inlet at Potosi also closes periodically by longshore drift and lack of runoff during the dry season. The inlet is presently fixed and does not migrate extensively, owing to dense vegetation on both sides and to cobbles from the sea stack on the south side that form a shingle beach. There is no well-developed inner tidal delta at Potosi, the lagoon mouth being narrow, the tidal channel fixed by vegetation, and the orientation of the lagoon parallel to the inlet channel (see comparison below).

Once closed these tidal inlets may remain sealed for months or years. Natural reopening results from storm waves cutting through the barrier (we have not observed this process), or breaching of the beach from behind during periods of high runoff and raised lagoon waters. As soon as the barrier is breached the out-flow rapidly erodes a channel and tidal circulation is regained. Presently the inlet at Mugu seals naturally about every six months and is dredged open by the Navy opposite the head of the submarine canyon. The inlet at Potosi is frequently dug open by local fishermen to assure marine circulation and fish migration.

COMPARISON OF MUGU LAGOON WITH LAGUNA POTOSI

The two lagoons have similar settings, being situated on narrow coastal plains bounded by steep onshore and offshore topography. They are developed on a framework of sand over shallow bedrock. The dominant sediments are sands distributed from the inlet or deposited as ancient beaches; mud occurs mainly as a thin and sometimes ephermeral veneer over the sand. Present sedimentation is dominated by tidal currents when the inlets are open, except for the peripheral salt pans that receive runoff from the adjoining land. Both lagoons are shallow, and now largely filled with tidal sediments, reflecting continued deposition at the present stand of sea level.

The regions differ in important and unimportant ways. Washover is not important at Laguna Potosi because the beach is very wide and vegetated. The fact that the mud fraction at Mugu is dominated by clay and at Potosi by silt is probably a trivial matter of provenance. The relative importance of the larger tidal range at Mugu has not been evaluated. Clearly of significance for the evolution and condition of these lagoons, however, are differences in climate, vegetation, stream influence, and orientation and elongation with respect to inlet position.

Temporal vagaries of runoff from Calleguas Creek would leave the Mugu inlet, once closed, inoperable for months or years until the next heavy rainfall filled the lagoon and caused the barrier to be breached from behind. The more predictable annual rains at Potosi assure that it will be opened yearly, unless the barrier is first deeply breached at the mouth of the Rio Petatlan. Climate at Potosi also promotes the flourishing mangrove fringe that has now firmly fixed the lagoon margins, even near the inlet, and extended into the subtidal zone. In contrast, the marsh plants at Mugu do not become established until sedimentation has raised the sediment-water interface to above the low-water datum in the lagoon, leaving intervening widespread intertidal sand- and mud-flats; these are subject to reworking until raised to a height suitable for marsh colonization (10, 12).

Because of the orientation of Mugu Lagoon, parallel to the coast with Calleguas Creek and the tidal inlet bisecting it, any freshwater flushing affects only a narrow central region. A very high proportion of sediments moved laterally into the lagoon by tidal currents remain where they are deposited without removal by land drainage running the length of the lagoon. Laguna Potosi, with the Petatlan effluent at one end and tidal inlet at the other, is capable of more thorough flushing, especially of finer sediments, leaving a broad very sandy central region in the lagoon.

THE DEPOSITIONAL RECORD

In addition to examining the distributions, textures, and processes affecting surface sediments, as discussed above, we have investigated sedimentary structures in one lagoon (Mugu) as indicators of its potential geologic record (2). Cores of various sizes and shapes were collected in all of the sedimentary units of the lagoon (Fig. 1). Two basic associations of textures and primary structures were discovered. One is clean sand with varying kinds and scales of cross-stratification; the other is muddy sand and sandy mud with wavy parallel laminations and very thin beds. As predicted from the discussion of the processes working in Mugu Lagoon, above, oriented cores revealed a dominance of cross-stratification produced by flood tides, inclined landward and away from the tidal inlet (Fig. 6a). However, the burrowing activities of the abundant plants and animals in the lagoon modify or obliterate primary sedimentary structures, even in such "high energy" habitats as the tidal delta and main sand channels. Burrows and bioturbated sediments are clearly perhaps the most characteristic features of marine tidally-controlled lagoons (Fig. 6b).

By incorporating different types and degrees of biological reworking with the two basic associations of sediments and primary structures, the lagoon may be divided into several facies. The facies coincide almost exactly with obvious habitats in the lagoon, such as sand channels, muddy creeks, tidal flats, salt marsh and salt pans. These results imply strongly that if these facies were preserved in the rock record each could be discriminated and would have paleoenvironmental significance.

ACKNOWLEDGMENTS

We thank the U. S. Navy and Commander Robert Baker for continued access to Mugu Lagoon and enthusiasm for our work, the Consejo Nacional de Ciencia y Tecnologia and the Comision del Rio Balsas for supporting work in Laguna Potosi, the Henry L. and Grace Doherty Charitable Foundation for support of the Rice University Marine Geology Program, R. R. Lankford and H. C. Clark for numerous courtesies and assistance, and many colleagues for assistance in the field.

Figure 6. X-ray radiographs (positive prints) of vertical sections of sediment from Mugu Lagoon showing (A) sand and isolated burrow (on left), from the zone of intense physical reworking on the tidal delta (see Fig. 1), and (B) laminated muddy sand from a tidal creek, with shells and almost complete mixing by burrows of bivalves, shrimp, crabs and polychaete worms.

REFERENCES

1. Bascom, W. 1954. The control of stream outlets by wave refraction. Jour. Geology 62: 600-605.
2. Biddle, K. T. 1976. Physical and biogenic sedimentary structures of a Recent coastal lagoon. Masters Thesis, Dept. of Geology, Rice University, 115 p.
3. Folk, R. L. 1961. Petrology of Sedimentary Rocks. Hemphill's, Austin, Texas, 154 p.
4. Inman, D. L. 1950. Report on beach study in the vicinity of Mugu Lagoon, California. Beach Erosion Board Technical Memorandum 14: 1-47.
5. ____. 1950. Submarine topography and sedimentation in the vicinity of Mugu Submarine Canyon, California. Ibid 19:1-45.
6. Jordan, C. F., G. E. Fryer, and E. H. Hemmen. 1971. Size analysis of silt and clay by hydrophotometer. Jour. Sedimentary Petrology 41: 489-496.
7. Lankford, R. R. 1975. Descripcion general de la zona costera Guerrero y Michoacan, Mexico. Centro de Ciencias del Mar y Limnologia, (unpub.).
8. Sanchez-Barreda, L. A. 1976. Sedimentology of Laguna Potosi and environs, State of Guerrero, Mexico. Masters Thesis, Dept. of Geology, Rice University, 82 p. + appendices.
9. Tamayo, J. 1962. Geographia General de Mexico. Inst. Mexicano Invest. Economicas, Mexico.
10. Warme, J. E. 1967. Graded bedding in the recent sediments of Mugu Lagoon, California. Jour. Sedimentary Petrology 37 (2): 540-547.
11. ____. 1969. Mugu Lagoon, coastal southern California: origin, sediments and productivity, p. 137-154. *In* A. Ayala-Castañaras and F. B Phleger (eds.), Lagunas Costaras, un Symposio. UNAM-UNESCO, Mexico, D. F., 686 p.
12. ____. 1971. Paleoecological aspects of a modern coastal lagoon. Univ. Calif. Pub. Geological Sci. 87:1-131.

COASTAL LAGOONS OF MEXICO

THEIR ORIGIN AND CLASSIFICATION

Robert R. Lankford, UNESCO Marine Geologist
Centro de Ciencias del Mar y Limnologia, UNAM
Apartado Postal 70-305
Mexico 20, D. F., Mexico

ABSTRACT: The coastal zone of Mexico extends 10,000 kilometers along the borders of the Pacific, Gulf of California, Gulf of Mexico and Caribbean and contains approximately 125 coastal lagoons defined here as: "a coastal zone depression below MHHW, having permanent or ephemeral communication with the sea, but protected from the sea by some type of barrier." Mexican coastal lagoons vary widely in their physical and environmental characteristics and in their degree of man's use and modification. They are geologically classified according to origin of the depression and barrier characteristics as: I. *Differential Erosion* (usually drowned valleys but includes solution depressions); II. *Differential Terrigenous Sedimentation* (typically associated with fluvial/deltaic systems); III. *Barred Inner Shelf* (offshore barrier on inner continental shelf); IV. *Organic* (usually coralgal but includes also mangrove and other organisms); V. *Tectonic-Volcanic* (directly linked with faulting, folding, or volcanism). Many coastal lagoons are compounds of two or more basic classes. Interrelating the original geomorphic characteristics with geologic development history, coastal oceanography and regional climatology, one may predict major types of existing lagoon environmental systems. Brief case studies illustrate the main mexican coastal lagoon types.

INTRODUCTION

One of the more conspicuous features of the highly varied physiography of Mexico is its 10,000 km-long coastal zone bordering four major water bodies, the Pacific Ocean, the Gulf of California, the Gulf of Mexico and the Caribbean Sea. At least 123 marginal marine depressions occur within the coastal zone, locally designated as: *bahía* (bay), *sonda* (sound), *boca* (mouth, inlet), *estero* (estuary), *estuario* (estuary), *caleta* (rocky cove), *lago* (lake), *laguna* (lagoon), *lagunilla* (small lagoon) and *laguna costera* (coastal lagoon). These variously named features, increasingly referred to collectively in Mexico as coastal lagoons, have assumed growing importance in the marine science community due to their actual or potential economic value and the growing need to improve coastal zone management. This paper is basically a snythesis of the present state of knowledge regarding coastal lagoon origins, patterns of development and general environmental characteristics. An evaluation of the synthesis has led me to classify mexican coastal lagoons mainly on the basis of geological origins of the inundated depressions and of processes which form protective barriers. The term, coastal lagoon, will be used here to refer to all of these multiple named marginal marine depressions, and is defined as:

> *a coastal zone depression below MHHW, having permanent or ephermeral communication with the sea, but protected from the sea by some type of barrier.*

There is no intent that this broad definition be adopted or that it should supplant other existing definitions; it is given solely as a working definition. The 123 coastal lagoons in Mexico are given in Fig. 1 and Appendix A. Mention in the text to specific lagoons is followed by coastal region identification and lagoon number (eg: D-13 signifies Coastal Lagoon 13 located in Region D).

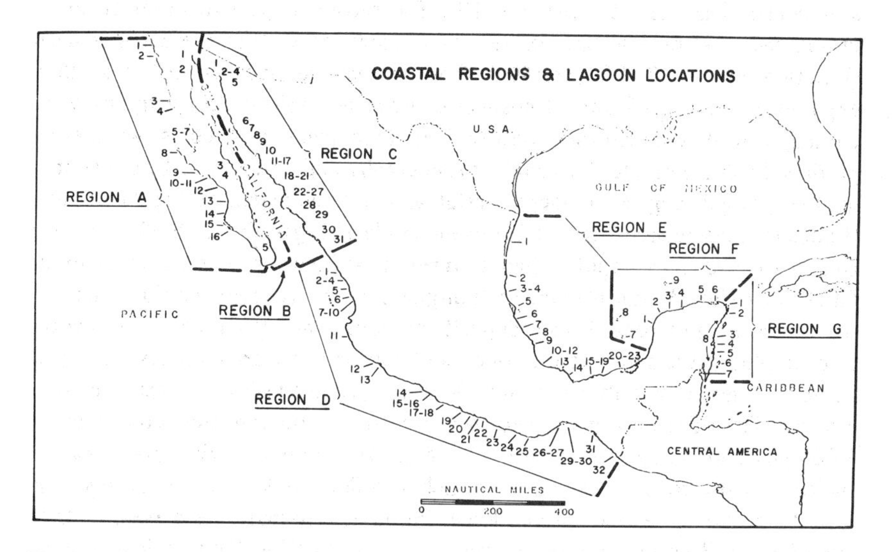

Figure 1. Location map showing the seven coastal regions (A-G) and the location of coastal lagoons within regions. Refer to Table 1 for description of regions and to Appendix A for identification, classification and investigation citations for coastal lagoons.

REGIONAL SETTING

Coastal lagoon environments derive from complex interactions between various marine, terrestrial and atmospheric factors, many of which are poorly understood. Within the purpose of this paper—to understand the origin and development of mexican coastal lagoons—an evaluation of the physiographic and geologic controls, climatic conditions, and coastal oceanography furnishes the basic regional setting whereby coastal lagoon behaviour may better be identified, compared and contrasted. In addition, a summary of Quaternary sea level history provides additional bases for interpretation of origins and development. Analysis of the geographic distribution of these major factors has lead to the division of the coastal zone of Mexico into seven large regions within which coastal lagoons having similar origins have predictably similar environments (Fig. 1; Table 1).

Sea Level History

Numerous studies of late Quaternary sea level fluctuations and shoreline formation have been summarized (6, 18, 55, 88, and others). Three events have particular bearing on modern coastal lagoons. The first was the stabilization of the Pleistocene shoreline during the Sangamon Interglacial about 80,000 years BP between 5 and 8 m above present sea level (98). Stabilization led to the construction of a raised topography of deltaic-lagoon-beach deposits first recognized by Price (72). Remnants of this ridge-like system are preserved on the modern mexican coastal plain margins and are commonly associated with present coastal lagoons (69) (see Fig. 10). The second important event occurred during the -130 m low-stand of sea level associated with maximum Wisconsin glaciation about 18,000 years BP. The present continental shelf was then exposed to terrestrial and atmospheric processes: valleys and canyons were eroded, fluvial sedimentation occurred in flood plains and deltas, weathering produced soil zones, etc. This ancient topography and potential sediment supply set the geologic stage for modern coastal lagoon development during the ensuing Holocene transgression. Rise of Holocene sea level began about 18,000 years BP and proceeded more rapidly than terrigenous sedimentation rates until about 5,000 years BP. A thin blanket of transgressive sand covered the shelf, derived from reworking of coastal plain deposits by advancing littoral zone turbulence. Topographic depressions were flooded and exposed to marine energy and open ocean beaches commonly were built along the mainland shore of present coastal lagoons. The third important event began when the transgression slowed about 5,000 years BP and at a level of −3 to −4 m. Barrier-building processes began to enclose narrow portions of the inner shelf and flooded depressions. Terrigenous as well as marine sedimentation slowly began to prograde the shoreline, thus initiating the Holocene regression. Major rivers soon filled their ancient valleys, and actively prograding deltas now replace earlier marine embayments. This latest event has been continuing to the present, although Curray et al. (19) have noted that the regression has been slowing due to changing climates and an accumulating cover of terrigenous mud over the transgressive shelf sands thus depriving many shorelines of primary sediment sources.

Geology and Physiography

The physiography of Mexico, perhaps as varied as any comparable region of the world, has been described and figured (61, 76, 95, 96). The equally diverse geology, investigated by many authors, is summarized by De Czerna (24) and in the geologic map of the Republic (15). Major aspects of the geology of coastal regions and adjacent marine areas are summarized for the Pacific-Gulf of California margins (1, 9, 29, 50) and for the Gulf of Mexico-Caribbean border (33, 46, 57). A general summary is shown in Fig. 2; the major drainage divides and run-off areas are given in Fig. 3.

The regional geology and high-relief physiography result largely from plate tectonism, the lateral movement of major plates of the earth's crust and upper mantle through continued renewal and destruction. Coastal margins of all continents are affected and have obtained their major physical characteristics recently classified by Inman and Nordstrom (43) and in Mexico by Carranze-Edwards et al. (10). Fig. 2 indicates that coastal margins of Mexico vary widely in their physiography. The Pacific coast typically has high relief with cliffed shorelines or narrow, steeply inclined coastal plains, bordered by mountain ranges with elevations up to 3,300 m; relatively subdued relief occurs along parts of the mainland coast of the Gulf of California, the Isthmus of Tehuantepec area and in

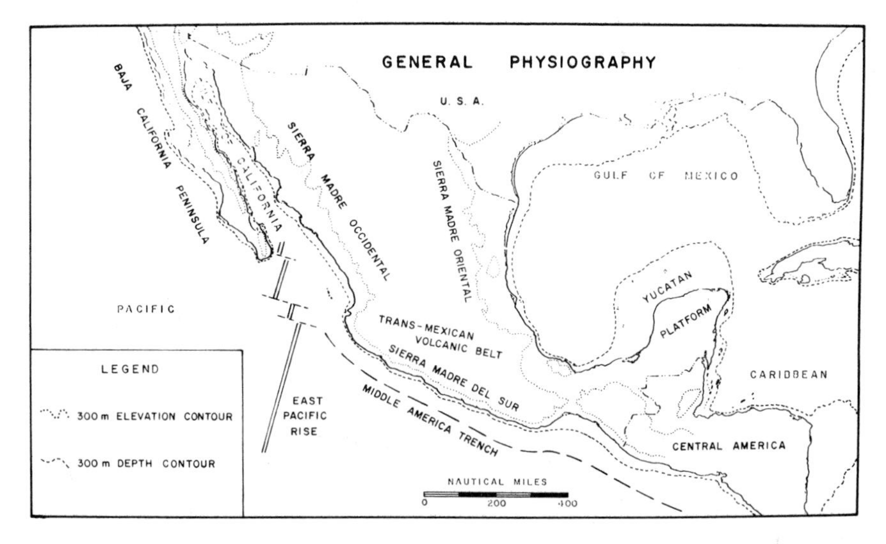

Figure 2. Generalized physiography and major geological features of Mexico. The 300 m elevation contour separates coastal plains and foot hills from mountains; the 300 m depth contour limits the continental shelf-upper continental slope.

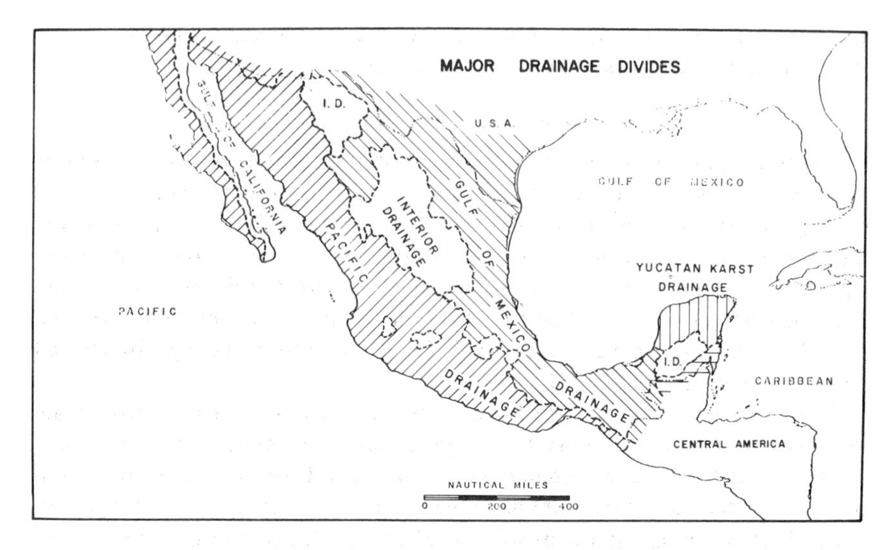

Figure 3. Major drainage divides and water sheds of Mexico. Note that closed dashed lines represent interior drainage basins. Information from Tamayo (95).

the central and southern Pacific coast of Baja California. The Pacific continental shelf typically is narrow or even non-existent and often has rocky, sediment-free substrates; rocky shelf islands and coastal plain prominances often occur. Inman and Nordstrom (43) classify the Pacific margin of Mexico as a "collision coast" resulting from the opposed movement of the American and Cocos plates, physiographically expressed by the Middle American Trench which marks the line of collision and the uplifted Sierra Madre Occidental, Sierra Madre del Sur and their southern Mexico- Central America continuation. The Gulf of California continues to be opened by lateral spreading along the northern extension of the East Pacific Rise. Transform faults associated with the Rise extend ESE under the continent probably to form the Trans-Mexican Volcanic Belt, an active volcanic and seismic zone which extends from the Pacific coast to the margin of the Gulf of Mexico. The volcano, Pico de Orizaba (5,747 m), is less than 100 km from the Gulf of Mexico and modern vulcanism occurs in the Tuxtlas Volcanics directly bordering the Gulf.

The Gulf of Mexico margin, in contrast to the Pacific, decends gently from the Sierra Madre Oriental foothills as a typically broad, low-relief coastal plain classified (43) as a "coast of marginal seas". High-relief coasts occur in two areas, the Jalapa Uplift and the Tuxtlas Volcanics of Murray (57), and constitute the eastern terminus of the Trans-Mexican Volcanic Belt. The Gulf of Mexico continental shelf decreases in width southward and is only 8 to 10 km wide off the Tuxtlas Volcanics but widens again toward the southeast. The shelf surface generally has subdued relief and is blanketed by varying amounts of unconsolidated terrigenous sands and muds; living and dead coral reefs occur off cuspate forelands.

The carbonate Yucatán Platform of the Gulf and Caribbean contrasts sharply in geology with the western and southwestern margin of the Gulf of Mexico. The land surface is almost without relief and river run-off is non-existent due to well-developed karst topography. The northern continental shelf of the Platform (Yucatán Shelf or Campeche Bank of many authors) is up to 200 km wide; most of the shelf is extremely flat and shallow and is covered by various calcareous sediments. Vertical relief occurs near active reefs along the shelf margin. The Caribbean shelf of the Yucatán Platform is relatively narrow and typically irregular due to reef accumulation, faulting and karst solution depressions.

Climate

Mexico's principal climate patterns are shown in Fig. 4. The major zonation, arid-temperate in the north to humid-tropical in the south, results from latitude differences in solar heating and global air mass circulation. Departures in the regional zonation are in response to topographic controls (compare Figs. 2 and 4). More detailed climatic data are in Tamayo (95, 96).

Rainfall and run-off in the southern three-fourths of Mexico are seasonal, typically occurring between May-June and September-October. Larger rivers flow throughout the year with seasonal differences in discharge while smaller

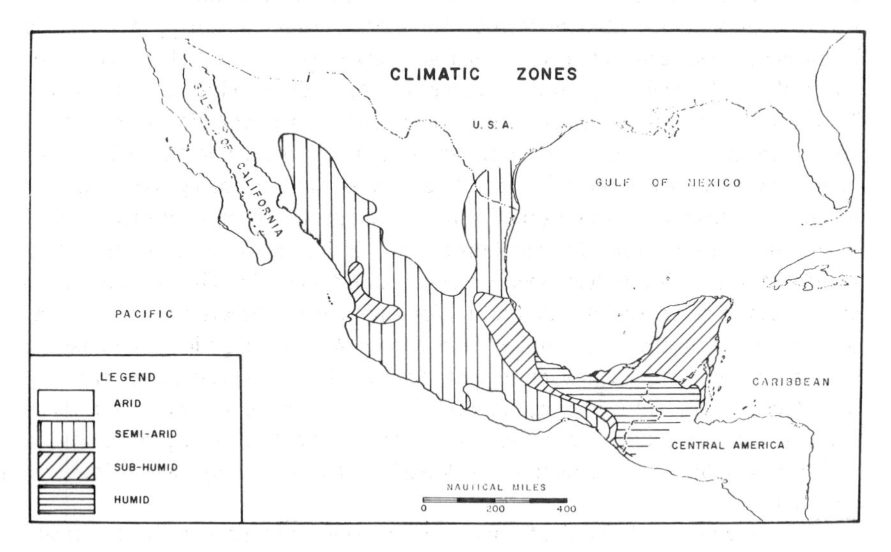

Figure 4. Climatic zones of Mexico. Zones are defined on the bases of precipitation, temperature and humidity. Modified from Garcia (33) and Tamayo (95).

rivers may be ephemeral; high water and flooding often occur during late summer and early fall. Torrential rains associated with summer tropical cyclones frequently inundate the Gulf- Caribbean region as well as the southern half of the Pacific coast. Precipitation in the arid north is more evenly distributed during the year with the exception of the lower half of Baja California where infrequent rain is mainly in summer.

García (33) has summarized average annual precipitation data which clearly indicate topographic and latitude controls. With some minor high-altitude exceptions, there are less than 500 mm of precipitation in the Gulf of California region, along the U.S. border, and in the Central Plateau region. Thus run-off from the arid region is negligible. Precipitation in the Sierra Madre Occidental, the Sierra Madre Oriental and other coast ranges (Fig. 2) increases to the south and southeast, attaining values of more than 4,000 mm/year. The majority of run-off occurs along the southwest Gulf of Mexico due to geologic control of drainage (Fig. 3). With the exception of the Rio Balsas with a very large drainage basin, total volumes of Pacific coast rivers normally are small and become increasingly seasonal and even irregularly intermittant to the north. Note that the Yucatán Platform is sub-humid to arid (Fig. 3). The karst topography and interior drainage basin capture precipitation and rivers do not exist.

Waves

Wind waves which predictably affect the mexican coasts are of five principal types: open ocean swell of the Pacific; local wind waves of the Gulf of California, Gulf of Mexico and Caribbean basins, storm waves associated with tropical hurricanes, waves associated with polar air mass movement called "northers" or "nortes", and local wind waves within coastal lagoons (Fig. 5). Munk and Traylor (56) describe the Pacific swell generated in high-latitude storm belts of both hemispheres during their respective winter months. Exposed Pacific coast sectors of Mexico are typically subjected to seasonally changing wave regimes: northern hemisphere swell from October through March and southern hemisphere swell from April through September; during spring and fall transition months, both swell sets have seen observed simultaneously. The exposed shores are subjected to seasonally alternating longshore drift directions although the southern drift component appears to dominate. Waves are typically long-period (12 to 15 sec) with deep-water heights up to 3-4 m. Exposed coasts typically have high energy. Lower energy sectors exist where protected by headlands or where coast orientation relative to wave direction results in strong refraction. In such topographically controlled areas, longshore transport may have a single net direction.

The protected Gulf of California coast is principally affected by seasonally reversing local wind waves which are short-period and relatively low. Strongest wave action occurs in the extreme northern and southern parts of the Gulf and along deltaic prominances of the mainland; the high-relief and very irregular Baja California coast appears to receive the lowest wave energy.

The Gulf of Mexico and Caribbean, like the Gulf of California, are oceanographic mediterraneans. Waves are produced with limited fetch within the basins by dominant winds. Leipper (45) has reviewed the marine meteorology within the Gulf- Caribbean area and showed that dominant wind directions vary within the northeast-southeast quadrant. Resultant wind waves typically have short periods (5-7 sec) and deep water heights vary up to 1.5 m. Net longshore drift along the Caribbean and central portion of the Gulf coasts is to the north. It is southerly along the shore of Laguna Madre de Tamaulipas (E-1), variable but weak in the southwestern Gulf, and westward along the northern shore of the Yucatán Platform. The entire area may be considered as having intermediate wave energy although relatively higher energy conditions exist along the Caribbean coast and northwestern half of the Gulf coast.

Tropical depressions and hurricanes occur during summer and early fall on both the Pacific and Atlantic sides of Mexico with about equal frequency. Gulf-Caribbean storms are normally of greater intensity. Summary track charts (85) show that the Yucatán Platform and northwestern Gulf coasts are more frequently affected by hurricanes than the southwestern Gulf. On the Pacific side, tropical cyclones originate in the Gulf of Tehuantepec and either move directly westward or follow a northwestern course parallel to the mainland coast. Occasional storms following the latter course extend into the Baja California

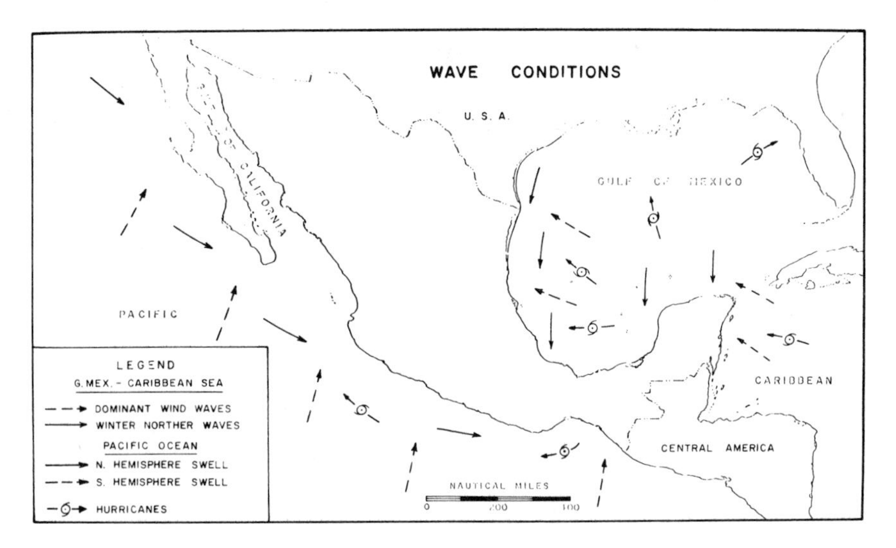

Figure 5. Expected wave conditions affecting the coasts of Mexico. Data from Leipper (45), Munk and Traylor (56), and the Secretaría de la Presidencia (85).

area. More severe storms result in coastal erosion, meteorological tide surges, barrier breaching, etc. Additionally, tropical depressions and hurricanes may produce torrential rains and coastal flooding which frequently have more drastic effects on coastal lagoons than do the storm winds and waves.

The fourth wave category is that produced by "northers", the southward flow of polar air masses from October through March. Each year from 15 to 20 "northers", each lasting from one to five days, leave the U.S. mainland, gain speed as they cross the Gulf of Mexico, and strike the lower Gulf coast with velocities which frequently exceed 165 km/hr (31). The gale- to hurricane-force winds and their associated high, short-period waves raise sea level, erode beaches, build storm berms and build north-south longitudinal dunes which may prograde into coastal lagoons (69). Cromwell (16) has shown that Gulf of Mexico "northers" frequently cross the Isthmus of Tehuantepec with sufficient force to become a major environmental factor in Laguna Superior (D-28) on the Pacific coast; compare Figs. 2 and 5.

The last category are those waves which are generated within coastal lagoons by local winds, including dominant winds, storms and day-time convective winds (sea breezes). Lagoon waves typically are short-period (2-3 sec) and low, although heights of 0.5-1.0 m can occur in larger lagoons. Observations indicate that lagoon wave base (the depth of effective wave orbital velocity at the bottom) is usually 1 m or less. Within wave base depth, however, waves may effectively modify lagoon shorelines, suspend bottom sediments and initiate water mixing processes. Prolonged wind stress on lagoon surfaces also affects water level both positively and negatively.

Tides

There have been relatively few tide studies in Mexico other than published tide tables (25, 26) and general summaries for the Gulf of Mexico (48) and for the Gulf of California (78, 79). However, several generalizations may be made from available information. Fig. 6 shows three-day spring tide records for twelve stations on the four mexican coasts. It is immediately apparent that there are large amplitude variations ranging from about 7 m near the head of the Gulf of California to less than 30 cm at Isla Cozumel in the Caribbean. Pacific tides are predominantly semidiurnal while Gulf of Mexico tides show a stronger diurnal character.

Tidal currents are an important source of coastal lagoon energy, particularly in inlets and channels, for erosion and transportation of sediment and in mixing processes. According to Groen (37), current velocities in lagoons will vary according to tide amplitude, the area of the lagoon, river effects, and the hydraulics of the inlet channel (cross-sectional area, depth and friction). Postma (70, 71) and Phleger and Ayala-Castañares (65) show that maximum currents occur in channels at or shortly after mid-tide and that flow frequently may

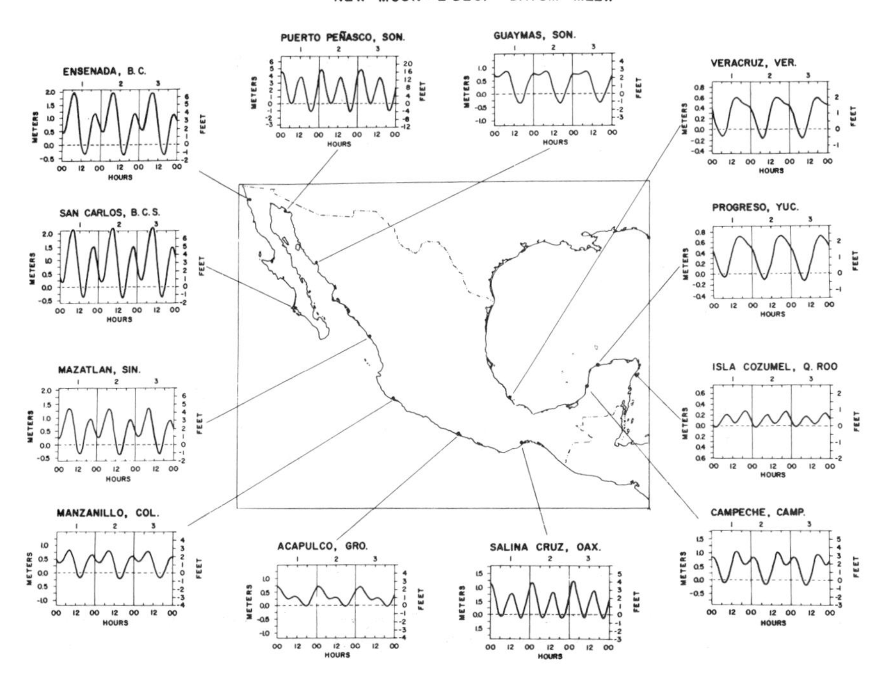

Figure 6. Three-day tide curves for 12 Mexican stations. Note differences in vertical scales; datum is mean lower low water (MLLW). Data from tide tables of Departamento de Oceanografía (25, 26).

exceed threshold velocities of sediment erosion; maximum sediment suspension lags maximum velocity by about one hour.

Most of the semidiurnal tides shown in Fig. 6 are typical in that "lower low water" follows "higher high water" (maximum vertical change). This condition usually produces maximum velocities during ebb tide and, if run-off exists, it will further increase ebb flow (8, 107) (Fig. 7). Differentially higher ebb velocities should have the effect of flushing suspended sediment from coastal lagoons to be removed by littoral drift. This is considered to be one of the major mechanisms which inhibits or slows the sedimentary infilling process typical of most coastal lagoons. If the tide curve is reversed and "higher high water" follows "lower low water", as is the case in the northern Gulf of California and to a lesser degree in the Gulf of Mexico (Fig. 6), the maximum expected velocities occur during flood tide. Net sediment transport will be from the turbulent surf zone into the lagoon; this tendency would be opposed by river run-off. In

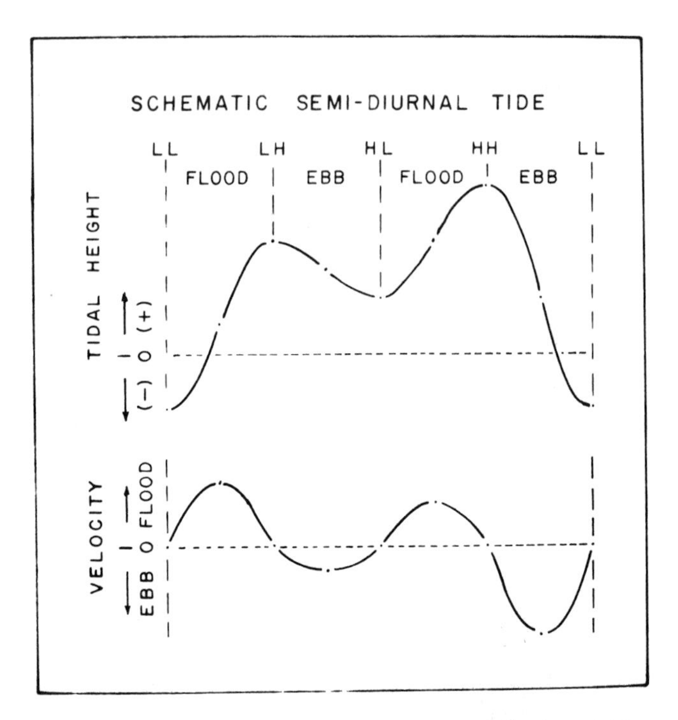

Figure 7. Schematic semi-diurnal tide; LL = lower low water, LH = lower high water, HL = higher low water, HH = higher high water; no vertical scale values are intended.

the arid northern Gulf of California area, high tides with differentially greater flood velocities apparently have almost filled former lagoons such as the shallow tidal flats with choked channels of the "Estero el Moreno" (B-2). The extensive inner tidal deltas of Laguna de Términos (E-22) also illustrate this process.

In addition to astronomic tides, wind stress produces significant positive or negative changes in water level. These storm tides are most pronounced in the Gulf of Mexico due mainly to hurricanes and "northers". The sudden onset of strong winds may result in flooding of lagoon margins or in lowering the water level to the extent that large areas of the bottom are exposed. Whether the storm tide is positive or negative, strong currents occur in inlets and in otherwise low-energy areas of lagoons and the net effect of these events frequently is environmentally catastrophic.

MEXICAN COASTAL LAGOONS

The diverse geographic distribution of varying oceanographic, atmospheric and terrestrial factors described above has resulted in a highly varied assortment of coastal lagoons in Mexico, each having unique environmental behaviour.

Table 1. Mexican coastal regions (see Fig. 1).

Region A. Pacific Coast; Baja California Peninsula to Cabo San Lucas (16 coastal lagoons): intermediate- to high-relief flank of coast range; narrow water shed, many dry valleys with typically small drainage basins; coast and mountain climate arid, precipitation in winter in north, in summer in south, increasing with elevation, very rare run-off; continental shelf very narrow, usually less than 20 km, widens to 50-70 km in central and southern embayments; wave energy very high on open, exposed coasts; tide energy very high with higher ebb velocity.

Region B. Gulf of California Coast; Baja California Peninsula to Cabo San Lucas (5 coastal lagoons): usually very high-relief mountain front, narrow, steep coastal plains in northern and southern extremes; water shed extremely narrow, many steep canyons; coast and mountain climate very arid, precipitation in winter in north, in summer in south, increasing with elevation, very rare run-off except for Río Colorado; continental shelf extremely narrow, usually non-existent, widens to 40-80 km in north; wave energy low to very low; tide energy from high in south with higher ebb velocity to extremely high in north with higher flood velocity.

Region C. Gulf of California Coast; Colorado River to Mazatlán (31 coastal lagoons): intermediate- to high-relief flank of coast ranges with wide to narrow coastal plains; water shed is narrow, many small rivers with small drainage basins; coast and mountain climate arid in north to semi-arid in south; precipitation mostly in summer, increasing with elevation and to the south, most rivers with very small volumes, frequently dry, larger rivers with intermediate volumes and seasonal flow; continental shelf usually narrow and irregular from 5-25 km, widens to 70 km in north; wave energy low, highest along delta fronts and increases to south near Gulf opening; tide energy from intermediate in south with higher ebb velocity to extremely high in north with higher flood velocity.

Region D. Pacific Coast; Mazatlán to Central America (32 coastal lagoons): high-relief flank of coast ranges; water shed narrow, many rivers with small drainage basins; coast climate semi-arid to sub-humid becoming very humid to southeast, precipitation in summer increasing with elevation and to southeast, rivers with small volumes and marked seasonal flow, may become dry in winter; continental shelf very narrow, usually 5-10 km, somewhat wider to northeast and to southeast; wave energy very high on open, exposed coasts; tide energy high with higher ebb velocity.

Region E. Gulf of Mexico Coast; United States to Yucatán Platform (23 coastal lagoons): typically low-relief coastal plain 150-30 km wide, low sea cliffs may occur; wide water shed, major rivers with large drainage basins; coast climate semi-arid in north to sub-humid in south, locally very humid, precipitation in summer, increasing markedly with elevation, rivers with large volumes; continental shelf typically narrow but varies from 150 km to 10 km; wave energy intermediate to low except for summer hurricanes and winter "northers;" tide energy low except for storm surges.

Region F. Gulf of Mexico Coast; Yucatán Platform to Cabo Catoche (9 coastal lagoons): extremely low-relief 250 km by 450 km carbonate platform, low coast scarps occur; water shed interior and internal due to karst topography, no rivers; coast and interior climate arid to sub-humid, summer rain but no run-off; continental shelf very broad averaging 125 km, and shallow averaging less than 10 m; wave energy very low except on shelf margin reefs and during summer hurricanes and winter "northers"; tide energy low except for storm surges.

Region G. Caribbean Coast; Yucatán Platform, Cabo Catoche to Central America (8 coastal lagoons): typically low-relief 250 km by 450 km carbonate platform, low coast scarps occur; water shed mainly interior and internal due to karst topography, few small rivers near British Honduras; coast climate arid to sub-humid, precipitation in summer increasing to south, some submarine fresh water springs, run-off in south; continental shelf irregular, typically 10 km wide to essentially non-existent; wave energy intermediate to low, high during summer hurricanes; tide energy low except for storm surges.

Establishment of seven regions for the mexican coast (Table 1) is a first approximation effort to arrange factor variation into a workable pattern in which coastal lagoons will have generally predictable characteristics. A survey of existing studies in Mexico indicates that this first approximation does function although more detailed analyses of regional control factors will be necessary for better prediction.

Surveys of available literature, maps, charts and atlases, satellite images, and air photos as well as my personal observations show that there are a minimum of 123 coastal lagoons in Mexico (Fig. 1, Appendix A). The inventory would be larger had I included the very small lagoons and the numerous locally named subdivisions of larger coastal lagoons. Lagoon names in quotations indicate that a provisional identification is used in the absence of a formal geographic name. Appendix A also classifies each coastal lagoon and lists significant investigations which are given in References. Appendix A citations generally do not include published taxonomic studies, biological inventories, or water quality investigations.

CLASSIFICATION

Considering the superabundance of existing classification schemes in whatever area of interest, it is not without a sense of apology that I present still another, justifying it only in that it further defines my broad use of the term, coastal lagoon. My classification contains no new ideas, all of which have been presented before in one form or another. I have drawn primarily from ideas and information in the literature (10, 11, 28, 43, 54, 73, 74, 81, 87, 89, 92, 110) as well as from many discussions with colleagues. The classification is basically geological in that consideration is given to origins and subsequent patterns of development of coastal lagoon depressions and to processes or conditions which form or otherwise result in some type of protective barrier. It is designed for coastal lagoons of Mexico but probably applies also to other low-latitude regions of the world; the basic scheme could be modified to include glacial and other ice effects (fjords, moraine dams, pack ice action, etc.) of higher latitudes.

No functional classification of natural phenomena can be precise to the extent that no gradations or mixtures exist between end-members. This is the case here. Many coastal lagoons in Mexico have multiple origins, or a specific lagoon may have more than one type of barrier. The inventory given in Appendix A not uncommonly gives more than one classification for a specific lagoon; if known, the primary classification is given first and the secondary category follows in parenthesis. Principal coastal lagoon types and subtypes and their expected characteristics are given in Table 2 and are schematically illustrated in Fig. 8; type and subtype designations (eg: Type III-B) are the same in Table 2 and Fig. 8.

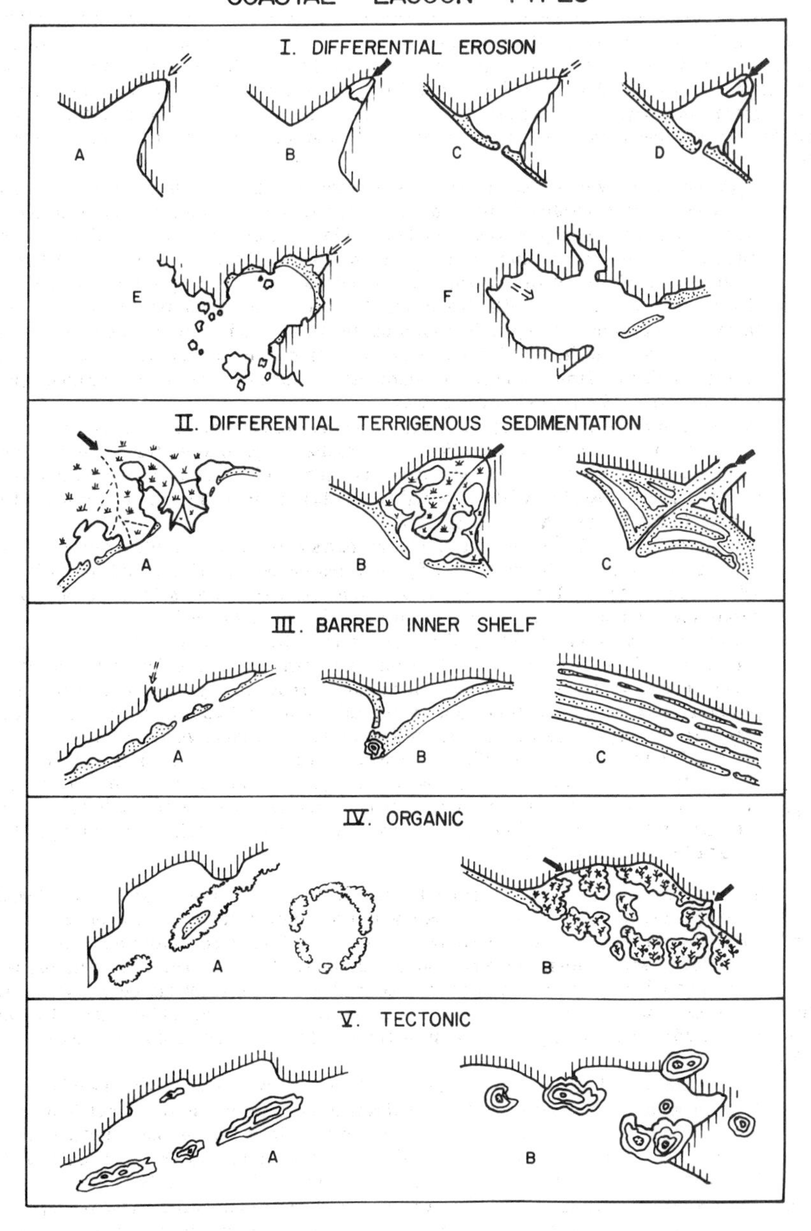

Figure 8. Coastal lagoon types and subtypes in Mexico; numerals and letters refer to classification in Table 2.

Table 2. Classification of mexican coastal lagoons.

Type I. Differential Erosion. Depressions formed by non-marine processes during lowered sea level. Inundated by Holocene transgression. Strongly to little modified since stable sea level stage during last 5,000 years. Forms and bathymetry are variable; typical drowned river valley geomorphology, occurring mainly along broad, low-relief coastal plains; steep, rocky canyon forms along high relief coasts; oval to irregular karst depressions along Caribbean coast.

A. Open Drowned Valley. No physical barrier; run-off absent to infrequent; form and bathymetry little modified by modern processes; energy conditions reduced due to wave refraction; salinity normal but hypersaline gradients may exist. (This subtype usually is small and is known to occur only as satellites of larger coastal lagoon complexes, typically of the I-D subtype; no named examples are given in Table 1).
B. Open Drowned River Mouth. No physical barrier; run-off continuous or seasonal; form and bathymetry typically modified by fluvial and marine processes; energy conditions normally reduced due to wave refraction, may be high due to tidal action; salinity variable, from hyposaline gradients to normal, rarely hypersaline. (Ex: Colorado River Mouth, B-1; see Fig. 9A).
C. Barred Drowned Valley. Physical barrier present; runoff absent or infrequent; form and bathymetry variously modified by littoral-zone processes (tides, wind action, waves); energy mainly due to tidal currents, high in channels and inlet(s), low over shoals; salinity usually with hypersaline gradients, may be nearly normal. (Ex: Laguna Ojo de Liebre, A-1, see Fig. 9B).
D. Barred Drowned River Mouth. Physical barrier present; run-off continuous or seasonal; form and bathymetry usually modified by lagoon delta(s) and formation of sub-lagoons; energy due to both tidal action and river flow; salinity usually with hyposaline gradients. (Ex: Laguna de Alvarado, E-12, see Fig. 9F).
E. Drowned Rocky Canyon. Physical barrier usually absent, rocky islands or spits may occur; run-off absent or seasonal; form and bathymetry usually little modified, pocket beaches/small coarse-grained deltas may occur; energy variable depending on geometry and wave refraction; salinity typically normal, hypo- or hypersaline gradients may occur. (Ex: Bahía Tortuga, A-8, see Fig. 9C; Laguna Verde, E-8).
F. Drowned Karst Depression. Physical barrier usually present as sand shoals, coral growth; no surface run-off, fresh water springs may occur in bottom; form and bathymetry modified near opening by marine processes; energy normally low except for tidal and hurricane effects; salinity usually normal, hyposaline near springs. (Ex: Laguna Chumyaxchal, G-3).

Type II. Differential Terrigenous Sedimentation. Coastal lagoons associated with fluvial/deltaic systems produced by irregular sedimentation and/or surface subsidence due to compaction/loading effects. Have been both formed and variously modified during last 5,000 years; many are geologically very young (hundreds of years). Typically sand barriers are rapidly formed, enclosing very shallow marginal and/or intradeltaic depressions; low sediment supply deltas may have shallow and frequently ephemeral, elongate lagoons between prograding beach ridges. Occur most frequently along deltaic plains in Regions C and E.

A. Intradeltaic and Marginal Depressions. Sand barriers typically present; run-off may be direct or river water may enter lagoons through inlets; form and bathymetric modifications occur rapidly; energy usually low except in channels and inlets; salinity typically low but may show seasonal or short-term variations. (Ex: Laguna de Términos, E-22, see Fig. 9E; Bahía de Topolobampo, C-24, see Fig. 9D).
B. Delta Barred Depressions. Barriers of various types: mud, sand, mangroves, etc.; run-off usually direct from river or distributaries; form and bathymetric modifications occur slowly to very rapidly; energy typically very low, except in channels; salinity usually very low, may vary with river discharge. (Ex: Laguna Tlalixcoyan, E-11 and associated lagoons, see Fig. 9F; Laguna de Pueblo Viejo, E-4 and associated lagoons, see Fig. 9H).

C. Delta Beach Swales. Multiple sand beach-barriers present; run-off across river levee, river or sea water may enter through tidal channels; form and bathymetry rapidly modified, lagoons may become seasonally dry; energy very low except in tidal channels; salinity highly variable, from fresh to hypersaline. (Ex: Laguna Chijol, E-3).

Type III. Barred Inner Shelf. Depressions are the inundated inner margin of the continental shelf bordered by the land surface on their inner margins and protected from the sea by various wave/current-produced sand barriers. Barrier formation dates from stable modern sea level within the last 5,000 years. Major orientation axes parallel to the coastal trend. Bathymetry typically very shallow except for erosial channels, modifications due principally to littoral zone processes including wind and hurricane activity and to localized terrigenous sedimentation. Typical "coastal lagoon" of many authors, occurring along low-relief coastal plains with intermediate to high wave energy.

A. Gilbert-de Beaumont Barrier Lagoon. Sand barriers extensive, occasionally multiple; run-off absent or very localized; form and bathymetry modified by tidal action, storm surges, wind-blown sand, and locally by streams which tend to segment elongate lagoons; energy relatively low except in channels and during storm conditions; salinity variable depending on climatic zones. (Ex: Laguna Madre de Tamaulipas, E-1; Laguna Superior, D-28, see Fig. 9G).

B. Cuspate Lagoon. Sand barriers in triangular orientation, with apex related to offshore wave refraction (islands, reefs, shoals) or joining rocky promontories; run-off absent or very localized; form and bathymetry modified as in III-A; energy typically low except in tidal channels and during storm conditions; salinity variable depending on climatic zones. (Ex: Laguna de Potosí, D-14; Laguna de Tamiahua, E-5, see Fig. 9H).

C. Strand Plain Depressions. Multiple sand barriers separating multiple linear troughs; run-off absent or seasonal or local; form and bathymetry slightly modified by tidal action and nonmarine processes; salinity highly variable, from nearly fresh to hypersaline, may become seasonally dry. (Ex: Laguna de Agua Brava, D-5, see Fig. 9I).

Type IV. Organic. Depressions produced by growth of organic barriers on inner continental shelf since stable sea level stand during last 5,000 years. Forms are variable, from parallel to coastal trend, to oval, to highly irregular; bathymetry shallow and irregular. Typically includes coralgal systems along Yucatán coasts, and mangrove communities in protected, subtropical areas.

A. Coralgal Barrier Lagoon. Rigid carbonate barrier enclosing irregular lagoon; run-off absent or very localized; form and bathymetry modified by organic growth, tides and hurricane effects; energy conditions normally low or variable in channels; salinity normal (Ex: Arrecife Alacrán, F-9; Laguna Nichupté, G-2, see Fig. 9L).

B. Mangrove Barrier Lagoon. Barrier of dense mangroves, normally with trapped clastic sediment; run-off typically present; form and bathymetry rapidly modified by organic growth and tides; energy conditions usually very low or variable in channels; salinity normal to hyposaline for at least part of year. (Ex: Laguna Nichupté, G-2, see Fig. 9L; Laguna Atasta, E-21).

Type V. Tectonic. Depressions and/or barriers produced by faulting, folding or vulcanism in coastal area in the geologic past, independent of sea level history. Forms are variable, from elongate, to oval, to very irregular; bathymetry varies from shallow to deep, often irregular. Occur along high relief coasts.

A. Structural Lagoon. Barrier typically of uplifted rock, may be irregular, discontinuous or continuous; run-off may be present or absent; form and bathymetry usually little modified except by localized run-off, lagoon frequently very deep, nearshore processes may unite rocky barriers by shoals or beaches; salinity usually near normal. (Ex: Bahía Magdalena, A-15, see Fig. 9J).

B. Volcanic Lagoon. Depression and/or barrier formed by volcanos/lava flows independent of sea level history, may be quite young, run-off may be present or absent; form and bathymetry are highly variable or shallow, modified by run-off, tidal action or subsequent vulcanism; salinity variable depending on climatic conditions. (Ex: Bahía de San Quintín, A-4; Laguna de Santecompan, E-13, see Fig. 9K).

DISCUSSION

It has been estimated (63) that approximately one-third of the 10,000 km coast of Mexico is lagoon coast. However, coastal lagoons are not uniformly distrubuted; see Fig. 1. Regions B and C facing the Gulf of California have the same coast lengths but very different numbers of coastal lagoons, five on the precipitous Baja California side, two of which are associated with the Rio Colorado, and 31 along the subdued coastal and deltaic plain topography of the mainland Gulf shore. High-relief coast topography similarly reduces the occurrence of coastal lagoons in other areas such as the northern and southern parts of Region A and the cliffed, mountainous coast of the north-central sector of Region D. Coastal lagoons along the low-relief Gulf of Mexico-Caribbean coast are not only more numerous but also are more uniformly distrubuted. It should be pointed out, however, that distribution frequency and the linear extent of lagoon coast are not related. For example, the numerous small coastal lagoons in Region D generally comprise a limited percent of the total coast length. Conversely, other coasts may contain a nearly continuous system of coastal lagoons such as the 600 km-long Type II complex of southern Region C or the 300 km of continuous reefs of the tropical Caribbean sector (G-8) which continue almost without interruption into Central America.

It is interesting also to note the regional distribution of lagoon types classified in Appendix A which are summarized in Table 3 according to primary types within the coastal regions described in Table 1.

Table 3. Regional Distribution of Primary Lagoon Types

REGION \ TYPE	I	II	III	IV	V	TOTAL
A Pacific Baja Calif	5		8		3	16
B Gulf Cal Baja Calif	1		3		1	5
C Gulf Cal Mainland	6	16	9			31
D Pacific Mainland	4		27			31
E Gulf Mex Mainland	2	14	6		1	23
F Gulf Mex Yucatán			5	4		9
G Caribbean Yucatán	4		2	2		8
TOTAL	22	30	60	6	5	123

Approximately one half of the total are Type III coastal lagoons formed within the last 5,000 years by wave-constructed sand barriers on the inner continental shelf margin. This type is common to all mexican coastal regions but is relatively more abundant in Region D, a high-energy coast. This apparent abundance, discussed in more detail below, reflects a locally high incidence of subdivision by internal river delta growth of originally longer lagoons. It is interesting also to note in Table 3 that Type II coastal lagoons, formed during the Holocene regression by differential terrigenous sedimentation, occur exclusively in Regions C and E, each region having relatively low wave energy. Type II coastal lagoons are conspicuously absent in arid Regions A and B of Baja California, in Region D typified by high energy, and in Regions F and G which border the carbonate Yucatán Platform where run-off and terrigenous sedimentation do not exist. Since Type II coastal lagoons obviously derive directly from irregular terrigenous sediment supply, the regional distribution of these coastal lagoons reflects a dynamic balance between the quantity and grain-size of the terrigenous sediment on one hand and the energy of nearshore processes on the other. Abundant Type II coastal lagoons in Region C are associated with prominant deltas formed by the Rio Yaqui, Rio Mayo and Rio Fuerte in the low-wave regime of the Gulf of California (see Fig. 9D). Phleger and Ayala-Castañares (65) and others show that the quantity of sediment supplied by the rivers is small due mainly to low run-off. Even the limited sediment supply has been sufficient, however, to over-balance the low coastal energy which tends to oppose delta formation. In Region D, Curray et al. (20) in their studies of the Costa de Nayarit show that the Rio Grande de Santiago and other rivers deliver relatively larger sediment supplies than the Region C rivers, but typical deltas are not constructed. High-energy nearshore processes easily sort sediment grain-sizes, removing the fines by suspension and concentrating the coarser sand at the shore to form a very broad strand plain with shallow Type III-C coastal lagoons (D-5 – D-10) (see Fig. 9I). Even in the Gulf of Mexico, this dynamic balance is demonstrated by Psuty (75) and Tanner and Stapor (97) in their studies of the extensive strand plain fronting the multiple delta complex of the Rio Grijalva, Rio Usumacinta and the Rio San Pedro y San Pablo. Here the dynamics between sedimentation and nearshore processes are nearly balanced and eight Type II coastal lagoons occur, particularly along deltaic margins (E-15 – E-22) (see Fig. 9E). Type III coastal lagoons also reflect relatively high wave energy. One of the more striking examples of a Type III-A lagoon system is represented by the 200 km-long complex of Laguna Superior (D-28), Laguna Inferior (D-29), Mar Muerto (D-30) and Laguna La Joya (D-31) on the Pacific Coast (see Fig. 9G). Type III-B coastal lagoons are associated with cuspate forelands, tombolos or other types of sand barriers usually related to offshore reefs, islands or rocky coastal promontories. Laguna de Tamiahua (E-5) is barred by a large cuspate foreland formed by alternating longshore transport directions ("northers" and

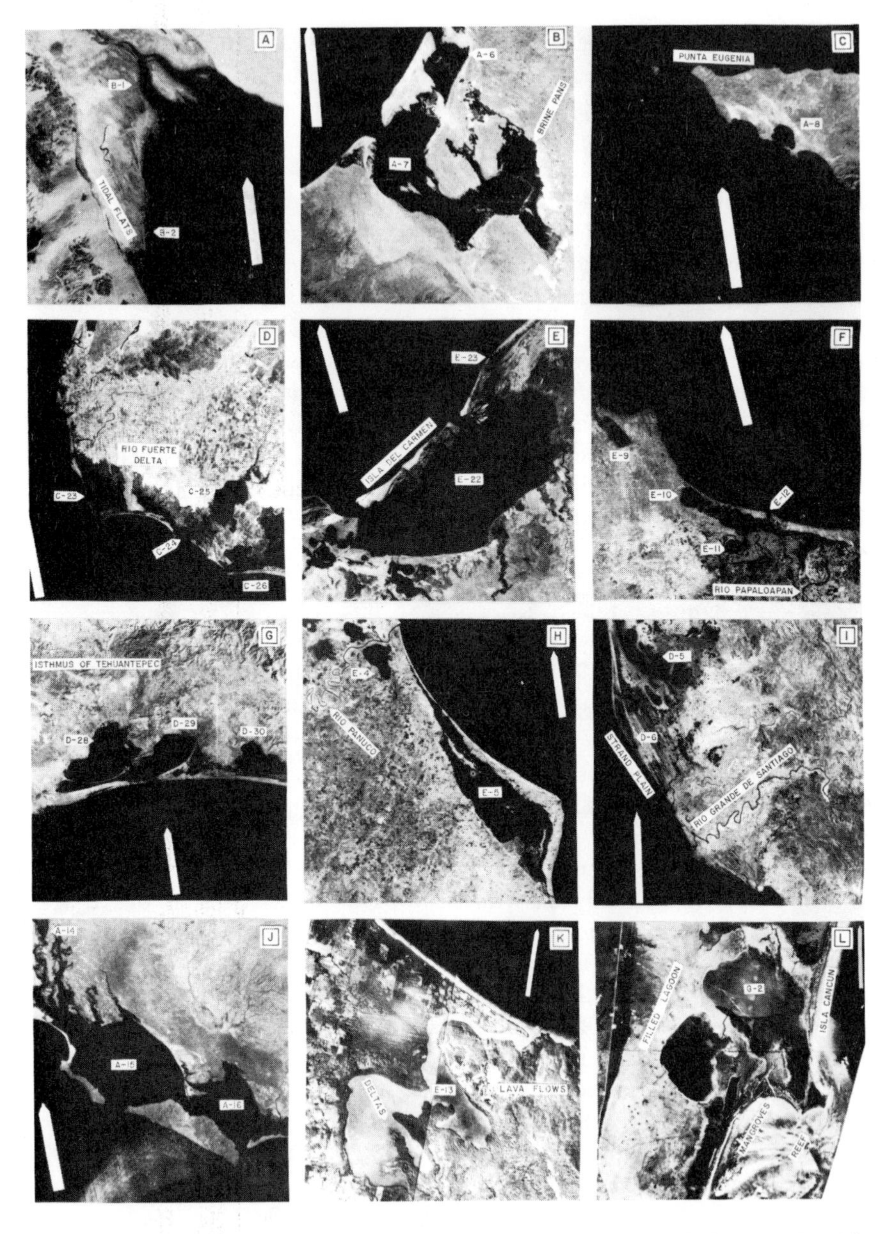

Figure 9. Selected Mexican coastal lagoons. A through J are NASA ERTS images, Band 7; north arrows have a length of 25 km. K and L are standard air photo mosaics; north arrows have a length of 2 km. See Fig. 1 and Appendix A for location and classification.

A. Estuario del Rio Colorado (B-1) and "Estero el Moreno" (B-2); northern Gulf of California.

dominant wind waves from the southeast, Fig. 5) and by wave refraction induced by a cluster of living and dead offshore reefs (see Fig. 9H).

The other types of coastal lagoons are well-known, particularly the Type IV-A coral or coralgal barrier lagoons which occur along the Caribbean coast (Region G), and more discontinuously along the seaward margin of the broad continental shelf of the Yucatán Platform (Region F). A combination of organic barriers is associated with Laguna Nichupté (G-2) (see Fig. 9L). Here a small coralgal barrier has formed at the south end of Isla Cancún thus forming a reduced energy environment in which a secondary mangrove barrier has grown to enclose the southern end of the lagoon. Type IV-B mangrove barrier-lagoon systems occur commonly throughout Mexico but usually occur at a secondary level; an open, exposed coast mangrove barrier such as that near Cape Romano, southwest Florida, does not exist in Mexico. The Type V tectonic coastal lagoons are rare today but probably were more abundant in the geologic past before they were extensively modified into other types. An illustration of the Bahía Magdalena (A-15)-Bahía Almejas (A-16) tectonic system is given in Fig. 9J. The major activity has been NW-SE faulting; the upthrown side of the fault zone forms a series of high-relief barrier islands. Vulcanism such as lava flows and/or formation of small volcanos creates depressions and barriers such as Bahía de San Quintín (A-4) and Laguna de Santecomapan (E-13). The latter is located in the very humid (4,000 mm of rain/year) volcanic coast of the Gulf of Mexico and has been extensively modified by run-off despite the very young age of vulcanism; the last eruption was in the late 18th Century (see Fig. 9K).

Type I coastal lagoons include those with depressions created by terrestrial erosion or by limestone solution during the last glacially lowered sea level and are third in classified abundance. Except for the karst solution depressions (Type

Figure 9. (continued)

B. Laguna Ojo de Liebre (A-7) and Laguna Guerrero Negro (A-6); Pacific coast of central Baja California Peninsula.
C. Bahía Tortuga (A-8); Pacific coast of central Baja California Peninsula.
D. Bahía San Esteban (C-23), Bahía de Topolobampo (C-24), Bahía Ohura (C-25), and Bahía de San Ignacio (C-26); mainland coast of Gulf of California.
E. Laguna de Términos (E-22) and Laguna Sabancuy (E-23); southwestern coast of Gulf of Mexico.
F. Laguna de Alvarado (E-12), Laguna Camaronero (E-10), Laguna Tlalixocoyan (E-11), and Laguna Mandinga (E-9); western coast of Gulf of Mexico.
G. Laguna Superior (D-28), Laguna Inferior (D-29), and Mar Muerto (D-30); Pacific coast of southern Mexico (Gulf of Tehuantepec).
H. Laguna de Tamiahua (E-5) and Laguna de Pueblo Viejo (E-4), western coast of Gulf of Mexico.
I. Laguna de Agua Brava (D-5) and Laguna Mexcaltitlán (D-6); Pacific coast of southwestern Mexico.
J. Bahía Magdalena (A-15), Bahía Almeja (A-16) and Bahía Santo Domingo (A-14); Pacific coast of southern Baja California Peninsula.
K. Laguna de Santecomapan (E-13); western coast of Gulf of Mexico.
L. Laguna Nichupté (G-2); northeastern coast of Yucatan Platform.

I-F), these are the classic drowned valleys with or without run-off and with or without physical barriers; hydrodynamic barriers, however, do exist. Type I-E, the drowned rocky canyon, is more abundant along the more precipitous coasts than I have indicated in Fig. 1 and Appendix A. Bahía Tortuga (A-8) is an example shown in Fig. 9 C. It has rocky shores and an irregular rocky bottom; pre-modern run-off and sedimentation modified the northern lagoon shore. The Estuario del Rio Colorado (B-1) is the only coastal lagoon in Mexico which corresponds to the much-discussed hydrodynamic model of an estuary (see Fig. 9A). Thompson (101), however, points out significant differences imposed by the extremely high tides of the northern Gulf of California (Fig. 6). I also have noted the effect of differentially higher flood tide velocities which have modified the nearby "Estero el Moreno" (B-2) and have formed extensive tidal flats in the lower deltaic plain of the Rio Colorado. Laguna Ojo de Liebre (A-7) is an example of a barred drowned valley without run-off (Type I-C) in the arid Pacific coast of Baja California. Phleger and Ewing (68) and Phleger (62) have described its hydrodynamic and sedimentological characteristics and those of adjacent Laguna Guerrero Negro (A-6, Type III-A). One of the more interesting aspects of the area is the precipitation of salt by monthly spring tide innoculation of brine pans which border the upper lagoon margin (64) (Fig. 9B).

Phleger (63) has pointed out that coastal lagoons continually are being modified geometrically and consequently, environmentally. He mentions migration of barrier dunes into lagoons as a major factor causing segmentation. My observations indicate that aeolian segmentation is limited either to the arid Pacific coast of Baja California, to the northern Gulf of California, and to a few areas of the Gulf of Mexico. In the latter instance, the very high-velocity "northers" erode beaches to form N-S longitudinal dunes which have been shown to prograde into coastal lagoons (69); Laguna Camaronero (E-10) has been separated from Laguna de Alvarado (E-12) by "norther" action (see Fig. 9F). Elsewhere aeolian effects are strongly reduced due to growth of dense coastal dune vegetation which traps transported beach sand and promotes the upward growth and stabilization of coastal dune ridges (Fig. 10). Phleger (63) also mentioned lagoon segmentation by small river deltas. This is one of the more important modifications of Type III-A coastal lagoons. For example, this type of very long coastal lagoon originally characterized much of Region D which has a narrow drainage divide and many small rivers with seasonal discharge (average = 0.6×10^6 m/year), see Table 1. Small delta growth has fragmented the few formerly long lagoons into many small ones (D-16 – D-21). It is interesting to speculate that this process has reduced the area and depth of the original Type III-A lagoons and thereby has reduced the effective tidal prisms applicable to inlet dynamics of the present coastal lagoons. Tidal action was also cited by Phleger (63) as a modification agency, giving as an example Laguna Santo Domingo (A-14, Type III-A) where opposed tidal currents meet between multiple inlets, lose velocity, and deposit tidal flat sediments. Storm surges or meteorologic tides have more drastic effects in coastal lagoons in most of the Gulf of Mexico- Caribbean region. Laguna Madre de Tamaulipas (E-1, Type III-A) is approximately 200 km in length with a single permanent inlet at its southern end. Past storm surges have repeatedly cut the sand barrier forming temporary inlets with internal tidal deltas and wide washover fans. Organic activity, particularly mangrove growth and floral successions (98), is a major modification agency within most mexican coastal lagoons. Additionally, oyster and coral communities and wide-spread *Thalassia* and green alga not only contribute organic and calcareous sediment to

Figure 10.Laguna de Nuxco (D-16) air photograph; located on southern mainland coast of the Pacific. Note that the inlet mouth is closed and that discharge from Río Nuxco is blocked by modern beach ridges.

coastal lagoons but also effectively trap suspended sediment. Lagoon circulation and water chemistry may be completely altered by organisms.

One of the more interesting observations of coastal lagoon evolution is hypothetically summarized in Fig. 11. Stage I represents conditions about 5,000 years BP when sea level attained its present stand and flooded a Type I coastal depression. If the depression was located in an arid, low-energy coast, it would remain as a Type I-A system. If, as in the case of Fig. 11, run-off occurs and coastal process energy is relatively high, Stage II is initiated with lagoon delta growth and the formation of a partial sand barrier. Laguna San Ignacio (A-11) is a "fossil" Stage II lagoon whose development was arrested probably due to late Holocene climate changes to the present aridity of Baja California. Stage III represents the closing of the lagoon by barrier growth, formation of tidal inlet channels and bars, and the infilling of the ancient valley mainly by delta progradation and fringing mangrove growth. This stage is represented today by the numerous inter-connected remnants of former Laguna de Alvarado (E-12) where

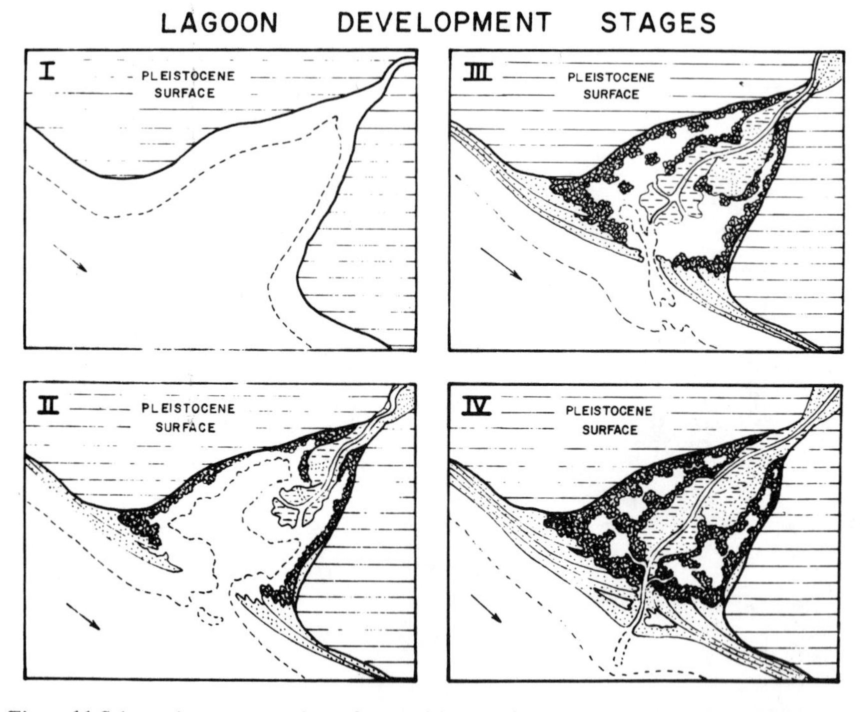

Figure 11. Schematic representation of coastal lagoon development stages. Dashed line is a reference depth contour; the arrow indicates dominant longshore transport direction; the stippled pattern is beach and barrier sand bodies; the dashed pattern is delta-fluvial deposits; the dark irregular pattern is mangrove (or other intertidal vegetation).

the modern deep channel of the Rio Papaloapan essentially by-passes most of the lagoon complex (see Fig. 9 F). If terrigenous sedimentation rates are high, Stage III will evolve into Stage IV. Water and sediment are mainly discharged directly into the sea possibly forming a marine delta. Mangroves and over-bank sediments slowly fill and reduce the area of the lagoons. Laguna de Pueblo Viejo (E-4) represents one of the many shallow remnant coastal lagoons on the mature delta surface of the Rio Panuco; (Fig. 9 H). Depending on terrigenous sediment supply and nearshore processes, a typical drowned river valley may evolve through a continuous series of stages into a completely different classification system, all of which presently exist in Mexico.

Temporal environmental variation, frequently seasonal, is a well-known characteristic of all coastal lagoons but differs in degree and in kind with geographic location. Several small Pacific coastal lagoons (D-14–D-22) now being studied show a rather extreme variation tendency; they have a seasonally shifting hydrologic balance between opened and closed inlet conditions. Laguna de Nuxco (D-16) illustrates this tendency (Fig. 10). It is a small Type III-A coastal lagoon, 2 km by 4.5 km in area, located in rugged granitic terrain. Nearby is the Rio Nuxco which has a drainage area of less than 100 km^2 and small seasonal flow mainly between July and November. The river mouth is blocked by a high, 800 m-wide beach that diverts discharge into Laguna de Nuxco via a linear depression behind the beach. The sand barrier is composed of four vegetated beach ridges up to 20 m in elevation and totaling 900 m in width; the narrow beach which usually blocks the inlet has a berm crest 4 m above sea level. A Pleistocene (probably late Sangamon) barrier parallels the modern barrier and forms part of Nuxco's shore. The remaining shoreline is fringed by mangroves and by granitic outcrops. The normally flat muddy lagoon floor averages about 2.5 m below sea level but contains very numerous vertical outcrops of rock. When the inlet is closed, lagoon water level is variable with rises up to 1.75 m reported locally. The closed lagoon water budget is a positive function of local precipitation plus run-off, and a negative function of evaporation plus seepage through the porous sand barrier; run-off has the greatest short-term effect. I observed in June, 1975, a 1.2 m rise in two days during heavy rains associated with a Pacific tropical depression. High internal water level not only increases the lagoon surface area and floods shore margins but also increases hydrostatic pressure and barrier seepage. If the variable run-off temporarily decreases, lagoon level will subside by seepage and evaporation (about 1,800 mm/year) in a few weeks. However, should run-off continue and hydrostatic pressure increase, the lateral seepage through the narrow inlet beach also increases; this effect may be further augmented by heavy local rain which forms a perched water table within the beach. The lateral seepage over-saturates the otherwise compact sands of the beach foreshore, and normal surf action begins thixotropically to erode the beach which blocks the inlet. Continued run-off and/or rain will maintain lateral seepage so that inlet beach erosion can continue to the point of opening the inlet from the ocean side. Should this occur, Laguna de Nuxco quickly drains and a variable period of open communication with the ocean is initiated. During this time, tidal action introduces saline water and various marine species enter the lagoon. Duration of open inlet conditions usually varies between a few days to a few weeks depending on run-off or tides which tend to keep the inlet open, and the intensity of nearshore processes which tend to close the inlet. Should the latter occur, the cycle is renewed as a closed system. Subsequent precipitation, run-off and water loss through seepage slowly reduce salinity although evaporation tends to oppose salinity reduction. Similar cyclic trends are typical of many of the coastal lagoons of the mexican Pacific coast. Whether or not a particular lagoon proceeds through a complete open-closed cycle depends on local variables which affect the internal water budget, as well as the size and depth of the lagoon, the tidal prism and the nearshore process energy. As a generalization, large coastal lagoons receiving run-off normally have an annual open-closed cycle while either small lagoons with run-off or large lagoons without run-off only sporadically proceed through a complete cycle.

ACKNOWLEDGMENTS

I am indebted to those many persons whose ideas, information and discussions have entered directly or subconsciously in my presentation. I particularly wish to acknowledge the long and stimulating influence of Fred B Phleger who, more than anyone, has encouraged my interest in coastal lagoons. Dr. Agustín Ayala-Castañares, Dr. Alfredo Laguarda-Figueras and colleagues of the Centro de Ciencias del Mar y Limnología of the Universidad Nacional Autónoma de México have variously supported this study. Numerous air flights have been provided by the Comisión del Río Balsas and by the Comisión del Río Papalopan. The Office of Marine Science of UNESCO granted permission to publish the investigation. Miguel Hernández-Pulido assisted with the illustrations and Marta Díaz de Castro-Sampedro very kindly typed the manuscript.

REFERENCES

1. Allison, E. C. 1964. Geology of areas bordering Gulf of California, p. 3-29. *In* Tj. H. van Andel and G. G. Shor, Jr. (eds.), Marine Geology of the Gulf of California. Amer. Assoc. Petrol. Geol., Mem. 3, Tulsa, Okla.
2. Ayala-Castañares, A. 1963. Sistemática y distribucíon de los foraminíferos Recientes de la Laguna de Términos, Campechê, México. Univ. Nal. Autón. México, Bol. Inst. Geol. 67:1-130.
3. ____, A. Garcia-Cubas, Jr., R. Cruz, and L. R. Segura. 1969. Síntesis de los conocimientos de la geología marina de la Laguna de Tamiahua, Veracruz, México, p. 39-48. *In* A. Ayala-Castañares and Fred B. Phleger (eds.), Lagunas Costeras, Un Simposio: Coastal Lagoons, A Symposium. Mem. Simp. Intern. Lagunas Costeras. UNAM-UNESCO, Nov. 28-30, 1967. México D. F.
4. ____, M. Gutiérrez-Estrada and V. M. Malpica. 1976. Geología Marina de la region de Yavaros, Sonora. Univ. Nal. Autón. México, Bol. Inst. Geol. (in prep.).

5. ———, and L. R. Segura. 1968. Ecología y distribución de los foraminiferos Recientes de la Laguna Madre, Tamaulipas, México. Univ. Nal. Autón. México, Bol. Inst. Geol. 87:1-89.
6. Bloom, A. L. 1971. Glacial-eustatic and isostatic controls of sea level since the last glaciation, p. 355-379. *In* K. K. Turekian (ed.), The Late Cenozoic Glacial Ages. Yale Univ. Press, New Haven and London.
7. Brady, M. J. 1972. Sedimentology and diagenesis of carbonate muds in coastal lagoons. Ph.D. Diss. Rice Univ., Houston, Tex. 288 p.
8. Bruun, P. 1973. Port Engineering. Gulf Publishing Co., Houston, Tex. 436 p.
9. Bullard, E. 1971. The origins of the oceans, Art. 21, p. 196-205. *In* Oceanography, Readings from Scientific American. Freeman and Co., San Francisco.
10. Carranza-Edwards, A., M. Gutiérrez-Estrada, and R. Rodriques-Torres. 1975. Unidades costeras de la República Mexicana. Univ. Nal. Autón. México, An. Centro Cienc. Mar. Limnol. (in press).
11. Caspers, H. 1967. Estuaries: analysis of definitions and biological considerations, p. 6-8. *In* G. S. Lauff (ed.), Estuaries. Amer. Assoc. Adv. Sci., Publ. 3, Washington, D.C.
12. Castro-Supúlveda, C. H. 1969. Estudio teórico-físico experimental del Puerto de Yacaltepén en la Ciénaga de Progreso, Yucatán, México, p. 377-396. *In* A. Ayala-Castañares and Fred B Phleger (eds.), Lagunas Costeras, Un Simposio: Coastal Lagoons, A Symposium. Mem. Simp. Intern. Lagunas Costeras. UNAM-UNESCO, Nov. 28-30, 1967. México, D. F.
13. Cervantes-Castro, D. 1969. Estabilidad del acceso a la Laguna del Mar Muerto, Chiapas, México, p. 367-376. *In* A. Ayala-Castañares and F. B Phleger (eds.), Lagunas Costeras, Un Simposio: Coastal Lagoons, A Symposium. Mem. Simp. Inter. Lagunas Costeras. UNAM-UNESCO, Nov. 28-30, 1967. México, D. F.
14. Coll de Hurtado, A. 1969. Estudio geomorfológico preliminar de la costa veracruzana comprendida entre Alvarado y Punta Puntilla, Ver. Univ. Nal. Autón. México, Bol. Inst. Geogr. 1:65-78.
15. Comité de la Carta Geológica de Mexico. 1968. Carta Geológica de la República Mexicana, Escala 1:2,000,000. Mexico, D. F.
16. Cromwell, J. E. 1975. Processes, sediments, and history of Laguna Superior, Oaxaca, Mexico. Ph.D. Diss., Univ. California, San Diego. 143 p.
17. Cruz-Orozco, R. 1968. Geología marina de la Laguna de Tamiahua, Veracruz, Mexico. Univ. Nal. Autón. México, Bol. Inst. Geol. 88:1-47.
18. Curray, J. R. 1960. Sediments and history of Holocene transgression continental shelf, northwest Gulf of Mexico, p. 221-266. *In* F. P. Shepard, F. B Phleger and Tj. H. van Andel (eds.), Recent Sediments, Northwest Gulf of Mexico. Amer. Assoc. Petrol. Geol., Tulsa, Okla.
19. ———, F. J. Emmel and P. J. S. Crampton. 1969. Holocene history of a strand plain lagoonal coast, Nayarit, Mexico, p. 63-100. *In* A. Ayala-Castañares and F. B Phleger (eds.), Lagunas Costeras, Un Simposio: Coastal Lagoons, A Symposium. Mem. Simp. Intern. Lagunas Costeras. UNAM-UNESCO, Nov. 28-30, 1967. México, D. F.
20. ———, and D. G. Moore. 1963. Sedimentos e historia de la Costa de Nayarit. Bol. Soc. Geol. Mexicana 26(2):107-116.
21. ———, and ———. 1964. Holocene regressive sands, Costa de Nayarit, p. 76-82. *In* L. M. J. U. van Stratten (ed.), Developments in Sedimentology,

v. 1, Deltaic and Shallow Marine Deposits, Proc. 6th Internat. Sed. Congr. Netherlands and Belgium, 1963.
22. ——, and ——. 1964. Pleistocene deltaic progradation of continental terrace, Costa de Nayarit, Mexico, p. 193-215. *In* Tj. H. van Andel and G. G. Shor, Jr. (eds.), Marine Geology of the Gulf of California. Amer. Assoc. Petrol. Geol. Mem. 3, Tulsa, Okla.
23. Davis, R. A. 1964. Foraminiferal assemblages of Alacran Reef, Campeche Bank, Mexico. J. Paleontology 38:417-421.
24. De Czerna, Z. 1961. Tectonic Map of Mexico. Geol. Soc. Amer. 1:2,500,000.
25. Departamento de Oceanografía. 1974. Tablas de 1975, predicción de mareas: Puertos del Golfo de México y Mar Caribe. Univ. Nal. Autón. México, An. Inst. Geofís. 20, Apend 1, Pt. A.
26. ——. 1974. Tablas de 1975, predicción de mareas: Puertos del Oceáno Pacífico. Univ. Nal. Autón. México, An. Inst. Geofís. 20, Apend. 1, Pt. B.
27. Ekdale, A.A. 1974. Marine molluscs from shallow-water environments (0 to 60 meters) off the northeast Yucatan coast. Bull. Mar. Sci. 24:638-668.
28. Emery, K.O., and R.E. Stevenson. 1957. Estuaries and lagoons. I. Physical and chemical characteristics, p. 673-693. *In* J.W. Hedgpeth (ed.), Treatise on Marine Ecology and Paleoecology, I. Geol. Soc. Amer., Mem. 67.
29. Fisher, R.L. 1961. Middle America Trench: topography and structure. Bull. Geol. Soc. Amer. 72:703-720.
30. Folk, R.L., and R. Robles. 1964. Carbonate sands of Isla Perez, Alacran Reef complex, Yucatan, J. Geology 72:255-292.
31. Freeland, G.L. 1971. Carbonate sediments in a terrigenous province: the reefs of Veracruz, Mexico. Ph.D. Diss. Rice Univ., Houston, Tex. 236 p.
32. ——, and R.S. Dietz. 1971. Plate tectonic evolution of the Caribbean-Gulf of Mexico region. Nature 232:20-23.
33. García, E. 1969. Distribución de la precipitación en la República de México. Univ. Nal. Autón. México, Bol. Inst. Geogr. 1:2-30.
34. García-Cubas, A., Jr. 1963. Sistemática y distribución de los micro-moluscos Recientes de la Laguna de Términos Campeche, México. Univ. Nal. Autón. México, Bol. Inst. Geol. 67:1-55.
35. Gorsline, D.S., and R.A. Stewart. 1962. Benthic marine exploration of Bahia de San Quintin, Baja California, marine and Quaternary geology. Pacific Nat. 3:282-319.
36. Grady, J.R. 1965. Oceanography of Bahia de San Quintin, Baja California. Pacific Nat. 6:1-35.
37. Groen, P. 1969. Physical hydrology of coastal lagoons, p. 275-280. *In* A. Ayala-Castañares and F. B Phleger (eds.), Lagunas Costeras, Un Simposio: Coastal Lagoons, A Symposium. Mem. Simp. Intern. Lagunas Costeras. UNAM-UNESCO, Nov. 28-30, 1967. México, D.F.
38. Gutiérrez-Estrada, M., and V.M. Malpica. 1975. Fisiografía y sedimentos Recientes de las Lagunas Pom y Atasta, Campeche, México. Univ. Nal. Autón. México, Bol. Inst. Geol. (in press).
39. ——, ——, and A. Ayala-Castañares. 1976. Geología marina de la región de Huizache-Caimanero, Sinaloa, México (in press).
40. ——, J. Martinez and V.M. Malpica. 1975. Sedimentos Recientes carbonatados de la Laguna Sabancuy, Campeche, México. Univ. Nal. Autón. México, Bol. Inst. Geol. (in press).
41. Hoskin, C.M. 1963. Recent carbonate sedimentation on Alacran Reef,

Yucatan, Mexico. Nat. Acad. Sci. – Nat. Res. Council Publ. 1089. Washington, D.C. 160 p.
42. Inman, D.L., G.C. Ewing, and J.B. Corliss. 1964. Coastal sand dunes of Guerrero Negro, Baja California, p. 99-100. *In* Prog. 1964 Ann. Mtg., Geol. Soc. Amer.
43. ____, and E.C. Nordstrom. 1971. On the tectonic and morphologic classification of coasts. J. Geology 79:1-21.
44. Kornicker, L.S., and D.W. Boyd. 1962. Shallow-water geology and environments of Alacran Reef complex, Campeche Bank, Mexico. Bull. Amer. Assoc. Petrol. Geol. 46:640-673.
45. Leipper, D.F. 1955. Marine meteorology of the Gulf of Mexico: a brief review, p. 89-98. *In* Gulf of Mexico, its origin, waters and marine life. U.S. Fish Wildlife Serv., Fishery Bull. 89, Washington, D.C.
46. Logan B.W., J.L. Harding, W.M. Ahr, J.D. Williams, and R.G. Snead. 1969. Sediments and reefs, Yucatan shelf, Mexico. *In* Amer. Assoc. Petrol. Geol. Mem. 11, Tulsa, Okla. p. 1-198.
47. Malpica, V.M., A. Ayala-Castañares, and M. Gutiérrez-Estrada. 1976. Fisiografía y sedimentos Recientes de la región de Agiabampo, Sonora, México. (in prep.).
48. Marmer, H.A. 1954. Tides and sea level in the Gulf of Mexico, p. 101-118. *In* Gulf of Mexico, its origin, waters, and marine life. U.S. Fish Wildlife Serv., Fishery Bull. 89, Washington, D.C.
49. Menard, H.W. 1971. The deep-ocean floor, Art. 17, p. 161-170. *In* Oceanography, Readings from Scientific American. Freeman and Co., San Francisco.
50. Morales, G.A. 1966. Ecology, distribution and taxonomy of Recent Ostracoda of the Laguna de Terminos, Campeche, Mexico, Univ. Nal. Autón. México, Bol. Inst. Geol. 81:1-103.
51. Moore, D.G., and J.R. Curray. 1964. Sedimentary framework of the drowned Pleistocene delta of the Rio Grande de Santiago, Nayarit, Mexico, p. 275-281. *In* L.M.J.U. van Stratten (ed.), Developments in Sedimentology, v.1, Deltaic and Shallow Marine Deposits, Proc. 6th Internat. Sed. Congr. Netherlands and Belguim, 1963.
52. Moore, D.R. 1958. Notes of Blanquilla Reef, the most northerly coral formation in the western Gulf of Mexico. Publ. Univ. Texas Inst. Mar. Sci. 5:151-155.
53. Moorelock, J., and K.J. Koenig. 1967. Terrigenous sedimentation in a shallow water coral reef environment. J. Sed. Petrology 37:1001-1005.
54. Morgan, J.P. 1967. Ephemeral estuaries of the deltaic environment, p. 115-120. *In* G.S. Lauff (ed.), Estuaries. Amer. Assoc. Adv. Sci., Publ. 3, Washington, D.C.
55. Mörner, N.A. 1971. The position of the ocean level during the interstadial at about 30,000 B.P. Canadian J. Earth Sci. 9:132-143.
56. Munk, W.H., and M.A. Traylor. 1947. Refraction of ocean waves: a process linking underwater topography to beach erosion. J. Geology 55:1-26.
57. Murray, G.E. 1961. Geology of the Atlantic and Gulf Coastal Province of North America. Harper and Bros., New York. 696 p.
58. Nichols, M.M. 1962. Tidal flat sediments of the Sonaran Coast, p. 636-637. *In* Proc. 1st. Nat. Shallow Water Conf., Publ. NSF-ONR, Washington, D.C.
59. ____. 1964. Sedimentology of Sonoran coastal lagoons. Ph.D. Diss. Univ. California, Los Angeles.

60. Olivares-Beltrán, G. 1969. Acceso a la Bahía de Topolobampo, Sinaloa, México, p. 407-419. *In* A. Ayala-Castañares and F. B Phleger (eds.), Lagunas Costeras, Un Simposio: Coastal Lagoons, A Symposium. Mem. Simp. Intern. Lagunas Costeras. UNAM-UNESCO, Nov. 28-30, 1967. México, D.F.
61. Ordoñez, E. 1936. Principal physiographic provinces of Mexico. Bull. Am. Assoc. Petrol. Geol. 20:1277-1307.
62. Phleger, F.B. 1965. Sedimentology of Guerrero Negro Lagoon, Baja California, Mexico, p. 205-237. *In* Geology and Geophysics. Colston Res. Soc. 17th Sympos. Butterworth (Colston Papers), v. 17, London.
63. ____. 1969. Some general features of coastal lagoons, p. 1-26. *In* A. Ayala-Castañares and F.B Phleger (eds.), Lagunas Costeras, Un Simposio: Coastal Lagoons, A Symposium. Mem. Simp. Intern. Lagunas Costeras. UNAM-UNESCO, Nov. 28-30, 1967. México, D.F.
64. ____. 1969. A modern evaporite deposit in Mexico. Bull. Am. Assoc. Petrol. Geol. 53:824-829.
65. ____, and A. Ayala-Castañares. 1969. Marine geology of Topolobampo lagoons, p. 101-136. *In* A. Ayala-Castañares and F.B Phleger (eds.), Lagunas Costeras, Un Simposio: Coastal Lagoons, A Symposium. Mem. Simp. Intern. Lagunas Costeras. UNAM-UNESCO, Nov. 28-30, 1967. México, D.F.
66. ____, and ____. 1971. Processes and history of Terminos Lagoon, Mexico. Bull. Am. Assoc. Petrol. Geol. 55:2130-2140.
67. ____, and ____. 1972. Ecology and development of two coastal lagoons in northwest Mexico. Univ. Nal. Autón. México, An. Inst. Biol., 43, Ser. Cienc. Mar Limnol. 1:1-20.
68. ____, and G.C. Ewing. 1962. Sedimentology and oceanography of coastal lagoons in Baja California, Mexico. Bull. Geol. Soc. Amer. 73:145-182.
69. ____, and R.R. Lankford. 1974. Sedimentos y foraminíferos de la Laguna de Alvarado, Veracruz. Abs. v. 5°. Congr. Nal. Oceanografía, Guaymas, Son., 22-25 Oct., 1974.
70. Postma, H. 1965. Water circulation and suspended matter in Baja California Lagoons. Netherlands J. Sea Res. 2:566-604.
71. ____. 1967. Sediment transport and sedimentation in the estuarine environment, p. 158-179. *In* G. S. Lauff (ed.), Estuaries. Amer. Assoc. Adv. Sci., Publ 3, Washington, D.C.
72. Price, W. A. 1933. Role of diastrophism in topography of Corpus Christi area, south Texas. Bull. Am. Assoc. Petrol. Geol. 17:907-962.
73. ____. 1955. Shorelines and coast of the Gulf of Mexico, Its Origin, Waters and Marine Life. U.S. Fish Wildlife Serv., Fishery Bull. 89, Washington, D.C.
74. Pritchard, D. W. 1967. What is an estuary: physical viewpoint, p. 3-5. *In* G. S. Lauff (ed.), Estuaries. Amer. Assoc. Adv. Sci., Publ. 3, Washington, D. C.
75. Psuty, N. P. 1966. The geomorphology of beach ridges in Tabasco, Mexico. Louisiana State Univ. Coastal Studies Inst. Tech. Rept. 30, 51 p.
76. Raisz, E. 1959. Landforms of Mexico (map). Cambridge, Mass.
77. Rice, W. H., and L. S. Kornicker. 1962. Mollusks of Alacran Reef, Campeche Bank, Mexico. Publ. Univ. Tex. Inst. Mar. Sci. 8:366-403.
78. Roden, G. I. 1964. Oceanographic aspects of Gulf of California, p. 30-58. *In* Tj. H. van Andel and G. G. Shor (eds.), Marine Geology of the Gulf of California. Amer. Assoc. Petrol. Geol. Mem. 3, Tulsa, Okla.

79. ____, and G. W. Groves. 1959. Recent oceanographic investigations in the Gulf of California. J. Mar. Res. 18:10-35.
80. Rigby, J. K., and W. G. McIntire. 1966. The Isla de Lobos and associated reefs, Veracruz, Mexico. Brigham Young Univ. Geol. Studies 13:3-46.
81. Russell, R. J. 1967. Origins of estuaries, p. 93-99. *In* G. S. Lauff (ed.), Estuaries. Amer. Assoc. Adv. Sci., Publ. 3, Washington, D. C.
82. Sánchez-Barreda, L. 1972. Transporte de sedimentos a lo largo de la parte interna de una barra, en el complejo de las Lagunas Superior e Inferior de Oaxaca. Prof. Thesis, Univ. Autón. Baja, California, Ensenada, B. Calif.
83. Schlaepfer, C. 1968. Estudios de los minerales pesados en los sedimentos de la Laguna Madre, Tamaulipas, México. Univ. Nal. Autón. México, Bol. Inst. Geol. 84:45-66.
84. Scruton, P.C. 1961. Rocky Mountain Cretaceous stratigraphy and regressive sandstones, p. 242-249. *In* Wyoming Geol. Assoc. Symp. 16th Ann. Field Conf., Green River, Wyo.
85. Secretaría de la Presidencia. 1974. Huracanes en el Océano Pacífico y el Océano Atlántico (Atlas). CETENAL. México, D. F.
86. Segura, L. R., and A. Ayala-Castañares. 1976. Ecología de los foraminíferos Recientes de la Laguna de Tamiahua, Veracruz, México. An. Centro Cienc. Mar Limnol. (in prep.).
87. Shepard, F. P. 1973. Submarine Geology, 3rd Ed. Harper and Row, New York, 516 p.
88. ____, and J. R. Curray. 1967. Carbon-14 determination of sea level changes in stable areas, p. 283-291. *In* M. Sears (ed.), Progress in Oceanography, v. 4. Pergamon, New York.
89. ____, and H. P. Wanless. 1971. Our Changing Coastlines. McGraw-Hill, New York, 579 p.
90. Silva-Bárcenas, A. 1963. Sistemática y distribución de los géneros de diatomeas de la Laguna de Terminos, Campeche, México. Univ. Nal. Autón. México, Bol. Inst. Geol. 67:1-31.
91. Stapor, F. W. 1971. Origin of Cabo Rojo beach-ridge plain, Veracruz, Mexico. Trans. Gulf Coast Assoc. Geol. Socs. 21:223-230.
92. Steers, J. A. 1967. Geomorphology and coastal processes, p. 100-107. *In* G. S. Lauff (ed.), Estuaries. Amer. Assoc. Adv. Sci., Publ. 3, Washington, D. C.
93. Sterns, C. E., and D. L. Thurber. 1967. $Th^{230} - U^{234}$ dates of Late Pleistocene marine fossils from the Mediterranean and Moroccan littorals, p. 293-305. *In* M. Sears (ed.), Progress in Oceanography, v. 4. Pergamon, New York.
94. Stewart, H. B., Jr. 1958. Sedimentary reflections of depositional environments in San Miguel Lagoon, Baja California, Mexico. Bull. Am Assoc. Petrol. Geol. 42:2567-2618.
95. Tamayo, J. L. 1949. Atlas Geográfico General de México. México, D. F.
96. ____. 1962. Geografía General de México. Inst. Mex. Invest. Econ., 2nd Ed., México, D. F., 562 p.
97. Tanner, W. F., and F. W. Stapor. 1971. Tabasco beach-ridge plain: an eroding coast. Trans. Gulf Coast Assoc. Geol. Socs. 21:231-232.
98. Thom, B. G. 1967. Mangrove ecology and deltaic geomorphology: Tabasco, Mexico. J. Ecol. 55:301-343.
99. ____. 1969. Problems of the development of Isla del Carmen, Campeche, Mexico. Zeitschr. Geomorph., N. Folge, Bd. 3, Heft 4:406-413.

100. Thompson, R. W. 1962. Coastal plain of the northwestern Gulf of California, p. 559-562. *In* Proc. 1st. Nat. Shallow Water Conf., Publ. NSF- ONR, Washington, D.C.
101. ——. 1968. Tidal flat sedimentation on the Colorado River Delta, northwestern Gulf of California. Geol. Soc. Amer., Mem. 107. 133 p.
102. Walton, W. R. 1952. Ecology of living Foraminifera, Todos Santos Bay, Baja, California. J. Paleontology 29:952-1018.
103. Wantland, K. F. 1969. Foraminiferal assemblages of the coastal lagoons of British Honduras, p. 621-641. *In* A. Ayala-Castañares and F. B Phleger (eds.), Lagunas Costeras, Un Simposio: Coastal Lagoons, A Symposium. Mem. Simp. Intern. Lagunas Costeras. UNAM-UNESCO, Nov. 28-30, 1967. México, D. F.
104. ——, and W. C. Pusey, III. 1971. A guidebook for the field trip to the southern shelf of British Honduras. 21st Ann. Mtg. Gulf Coast Assoc. Geol. Socs., Oct. 10-13, 1967, New Orleans, La., 87 p.
105. Ward, W. C. 1967. Shelf islands of northeast Quintana Roo, Mexico, p. 124-136. *In* Guidebook Field Trip 7, 1967 Ann. Mtg. New Orleans Geol. Soc., New Orleans, La.
106. ——. 1970. Diagenesis of Quaternary eolianites of NE Quintana Roo, Mexico. Ph.D. Diss. Rice Univ., Houston, Tex. 207 p.
107. Wiegel, R. L. 1964. Oceanographical Engineering. Prentice-Hall, Englewood, N. J. 532 p.
108. Yañez-Correa, A. 1963. Batimetría, salinidad, temperatura y distribución de los sedimentos Recientes de la Laguna de Términos, Campeche, México. Univ. Nal. Autón. México, Bol. Inst. Geol. 67:1-47.
109. ——, and C. Schlaepfer. 1968. Composición y distribución de los sedimentos Recientes de la Laguna Madre, Tamaulipas, México. Univ. Nal. Autón. México, Bol. Inst. Geol. 84:5-44.
110. Zenkovich, V. P. 1962. Processes of Coastal Development. J. A. Steers (ed.), English Trans., 1967. Interscience, New York. 738 p.

Appendix A. Mexican Coastal Lagoon Inventory

Number	Lagoon Name	Classification	Published References
		REGION A	
1	Estero San Miguel	I-C	(94)
2	Estero de Punta Banda	III-B, (V-A)	(102)
3	"Laguna Vicente Guerrero"	III-A	
4	Bahía de San Quintín	V-B	(35,36)
5	Laguna Manuela	III-A, (III-B)	(42,62,68)
6	Laguna Guerrero Negro	III-A	(42,62,63,64,68,70)
7	Laguna Ojo de Liebre	I-C, (III-A)	(42,62,63,64,68,70)
8	Bahía Tortuga	I-E	
9	"Laguna Abreojos"	III-A	
10	"Estero Ballenas"	I-C	
11	Laguna San Ignacio	I-C, (III-A)	
12	"Estero San Benito"	III-A	
13	Laguna San Gregorio	III-A	
14	Laguna Santo Domingo	III-A	(63,68)
15	Bahía Magdalena	V-A, (III-B)	
16	Bahía Almejas	V-A, (III-B)	
		REGION B	
1	Estuario del Río Colorado	I-B	(100,101)
2	"Estero el Moreno"	III-A	(100,101)
3	"Estero de San Lucas"	III-B	
4	Bahía de Concepción	V-A	
5	Ensenada de la Paz	III-A, (III-B)	
		REGION C	
1	Bahía de Aduar	I-C, (III-A)	
2	Bahía Cholla	III-B	
3	"Estero Peñasco"	III-B	
4	Laguna Salada	I-C, (III-A)	
5	"Estero de San Jorge"	III-A	
6	"Estero del Sargento"	III-A	
7	Laguna de la Cruz	I-C, (III-A)	
8	Estero Tastiota	I-C	(58,59)
9	Bahía San Carlos	I-E	
10	Laguna de Guaymas	I-E, (III-B)	
11	"Laguna Vicicori"	II-A	
12	Estero Tortuga	II-A	
13	Estero de Tecolote	II-A	
14	Estero de Algodones	II-A	
15	Estero de la Luna	II-A	
16	Estero de Lobos	II-A	
17	Estero Corga	II-A	
18	Estero de Huivulay	II-A	

(continued)

Number	Lagoon Name	Classification	Published References
Appendix A, REGION C (continued)			
19	Estero Ciaris	II-A	
20	Estero de Santa Lugarda	II-A	
21	Bahía de Yavaros	II-A	(4)
22	Estero de Agiabampo	II-A	(47)
23	Bahía San Esteban	II-A	(60,65)
24	Bahía de Topolobampo	II-A, (I-C)	(60,65)
25	Bahía Ohuira	II-A, (I-C)	(60,65)
26	Bahía de San Ignacio	II-A	
27	Bahía de Navachiste	III-A	
28	Bahía de Playa Colorada	III-A	
29	Bahía de Santa María	III-A, (III-C)	
30	Ensenada del Pabellón	III-A, (I-D)	
31	"Ensenada de Quevedo"	III-A	
	REGION D		
1	Estero de Urias	III-B, (III-A)	
2	Laguna de Huizache	III-A	(39,67)
3	Laguna de Caimanero	III-A	(39,67)
4	Laguna de Escuinapa	III-A	
5	Laguna de Agua Brava	III-C	(19,20,21,22,51)
6	Laguna Mexcaltitlán	III-C	(19,20,21,22,51)
7	Boca Cegada	III-C, (I-C)	(19,20,21,22,51)
8	Estero del Pozo	III-C, (I-C)	(19,20,21,22,51)
9	Estero del Rey	III-C, (I-C)	(19,20,21,22,51)
10	Estero de San Cristobal	III-C,(I-D)	
11	Laguna Agua Dulce	I-C	
12	"Estero de Navidad"	III-A, (III-B)	
13	Laguna Cuyutlán	III-A, (III-B)	(84)
14	Laguna de Potosí	III-B, (III-A)	
15	Las Salinas de Cuajo	I-C	
16	Laguna de Nuxco	III-A	
17	Laguna Mitla	III-A	
18	Laguna Coyuca	III-A	
19	Laguna Tres Palos	III-A	
20	Laguna Tecomate	III-A	
21	Laguna Chautengo	III-A	
22	Laguna Apozahualco	III-B, (III-A)	
23	Laguna de Alotengo	III-A, (I-C)	
24	Laguna de Chacahua	III-A, (III-B)	
25	Laguna de Pastoria	III-A	
26	"Estero de Punta Conejo"	I-C	
27	Estero del Río Tehuantepec	I-D	
28	Laguna Superior	III-A	(16,82)
29	Laguna Inferior	III-A	(16,82)
30	Mar Muerto	III-A	(13)
31	Laguna La Joya	III-A	
32	Laguna del Viejo	III-A	

Number	Lagoon Name	Classification	Published References
	REGION E		
1	Laguna Madre de Tamaulipas	III-A, (II-A)	(5,83,109)
2	Laguna de San Andres	III-A	
3	Laguna Chijol	II-C	
4	Laguna de Pueblo Viejo	II-B	
5	Laguna de Tamiahua	III-B	(3,17,52,80,86,91)
6	Laguna Tampamuchoco	II-B	
7	Laguna Grande	III-A	
8	Laguna Verde	I-E	
9	Laguna Mandinga	III-B	(31,53)
10	Laguna Camaronero	II-B	(69)
11	Laguna Tlalixcoyan	II-B	(69)
12	Laguna de Alvarado	II-B, (I-D)	(14,69)
13	Laguna de Santecomapan	V-B	
14	Laguna de Ostión	I-D	
15	Laguna de Carmen	II-A	(75,97,98)
16	Laguna Machona	II-A	(75,97,98)
17	Laguna Tupilco	II-A	(75,97,98)
18	Laguna Mecoapan	II-A	(75,97,98)
19	Estero de Chiltepec	II-A	(75,97,98)
20	Laguna Pom	II-A,(IV-B)	(38,75,97,98)
21	Laguna Atasta	II-A, (IV-B)	(38,75,97,98)
22	Laguna de Términos	II-A	(2,34,50,66,75,90, 97,98,99,108)
23	Laguna Sabancuy	III-A	(40)
	REGION F		
1	Laguna de Celestum	III-A	
2	"Estero de Progreso"	III-A	(12)
3	"Estero de Telchac"	III-A	
4	"Estero de Punta Arenas"	III-A	
5	Laguna Lagartos	III-A	
6	Laguna de Yalahua	III-A, (IV-B)	
7	Cayo Arcas	IV-A	(46)
8	Arrecifes Triángulos	IV-A	(46)
9	Arrecife Alacrán	IV-A	(23,30,41,44,46,77)
	REGION G		
1	Bahía Contoy	III-A	(27,105)
2	Laguna Nichupté	III-A, (IV-A,B)	(7,28,105,106)
3	Laguna Chumyaxchac	I-F, (IV-A)	
4	Bahía de la Ascención	I-F,(IV-A)	
5	Bahía del Espíritu Santo	I-F, (IV-A)	
6	Bahía Chetumal	I-F, (IV-A)	(103,104)
7	Banco Chinchorro	IV-A	(104)
8	Unamed reef lagoons	IV-A	(103,104)

INTERACTIONS BETWEEN TIDAL WETLANDS AND COASTAL WATERS

Convened by:
Scott W. Nixon
Graduate School of Oceanography
University of Rhode Island
Kingston, Rhode Island 02881
and
William E. Odum
Department of Environmental Sciences
University of Virginia
Charlottesville, Virginia 22903

The functional coupling of coastal wetlands and offshore waters is almost axiomatic in marine ecology. It is often suggested that this coupling may take the form of physical energy inputs, flows of dissolved nutrients and organic matter, net fluxes of living or dead particulate organic material and suspended sediments, or more subtle interconnections such as habitat diversity and refuge. Moreover, there is a general feeling that such interactions represent major processes in the ecology of coastal waters. Unfortunately, however, there is little evidence in the literature to argue for or against such a relationship. For example, while there have been dozens of measurements documenting the high rates of primary production in coastal salt marshes and their vast potential for detrital export, there are only two or three reasonably reliable direct estimates of detrital export from such environments. This lack of data is at once surprising and understandable. It is surprising because coastal salt marshes are among the most intensively studied marine ecosystems, and the functional coupling of such marshes with the larger estuarine system is one of the most forcefully used arguments in the effort to conserve coastal wetlands. It is understandable because the measurement of net marsh-estuary fluxes is a difficult, laborious, and, in many ways, tedious business that requires frequent sampling over an annual cycle. Often one is trying to measure small differences between large flows, each of which is highly variable over space and time. Nevertheless, the collection of such data, as well as information on the relationship of the fluxes to environmental characteristics, such as hydraulic energy, is a crucial step in understanding the functional processes of coastal marine ecosystems. The work presented in this session represents some, though by no means all, of the recent attempts to explore the complex flows of energy and materials between coastal systems. A number of these and other studies are still underway. The findings do not always agree, and it is clear that we are still far from understanding the processes involved. This appears to be particularly true with respect to dissolved organic materials. Until a larger sample of marshes and other wetlands has been studied and the importance of site specific processes evaluated, the results presented here best serve as examples of the relationships found in particular coastal systems. It is still too early to frame reliable generalizations about the overall direction or magnitude of the fluxes that may couple coastal wetlands and offshore waters as a whole.

NUTRIENT EXCHANGES BETWEEN BRACKISH WATER MARSHES AND THE ESTUARY[1]

J. Court Stevenson, Donald R. Heinle, David A. Flemer
Robert J. Small, Randy A. Rowland and Joseph F. Ustach

Horn Point Environmental Lab, Cambridge, Maryland
and
Botany Department, College Park, Maryland 20742
(J.C.S., R.J.S., R.A.R.)
Chesapeake Biological Lab-Solomons, Maryland
(D.R.H., D.A.F., J.F.U.)

ABSTRACT: Two irregularly flooded brackish marshes of the Chesapeake Bay estuarine system were studied to detect patterns of tidal nutrient exchange and utilization. A 5.7 ha marsh on the Choptank River was monitored monthly from October 1974 to August 1975 for dissolved inorganic nitrogen and phosphorus. During these diurnal samplings, nutrient exchange with the estuary was eliminated by a tidal gate and total aquatic oxygen metabolism was measured. During winter, photosynthesis exceeded respiration, and total nitrogen increased to 80 μg-at liter^{-1} and declined through spring to 8 μg-at liter^{-1}. Phosphate varied erratically from 0.8 to 4.3 μg-at liter^{-1}. Tidal import and export of the above nutrients were also monitored monthly from January to June 1975. During winter, there appears to be a net flow of inorganic nitrogen and phosphorus to the estuary. In spring this pattern is reversed for both nutrients.

Two years tidal flow data from a larger 127 ha marsh on the Patuxent River revealed that dissolved inorganic nitrogen is also taken up in May, June and July. However, when the dissolved organic fraction is also consisered, net nitrogen flow was always to the estuary in every month sampled. It is estimated that the dissolved nitrogen flux from the marsh was 4.14 g m^{-2} yr^{-1}, or about 18-82% of

[1] Contribution No. 690, Center for Environmental and Estuarine Studies of the University of Maryland, Cambridge, Md. 21613.

the estimated nitrogen of the vegetation. Also, the net annual flux of dissolved phosphorus was 0.19 g m^{-2} yr^{-1} to the estuary or 8-23% of the phosphorus in standing crop. Data from both marshes suggest that, in contrast to previously studied high salinity marshes, brackish marshes act more as sources of nutrients to the estuary. However, these marshes may be trapping some nutrients, either from upland areas surrounding them, or possibly from infrequent massive river deposition.

INTRODUCTION

For over 300 years the extensive marshes of the upper Chesapeake Bay have been intermediaries in the coupling of man-dominated terrestrial systems and one of the largest estuarine systems in the New World. Surprisingly, until recently, little direct data has been available to evaluate the importance of extensive marsh areas in acting as buffer zones for the Chesapeake Bay ecosystem. This paucity of data is especially noticeable in terms of marsh nutrient cycling and nutrient exchanges between brackish water marshes of the Maryland portion of the upper Chesapeake Bay. Recently, there have been several studies of nutrient exchange patterns of high salinity marshes, and models of marsh nutrient cycling are in varying stages of conceptualization and quantification.

In a general review, Pomeroy (28) elaborated on the adaptive "strategy" of mineral cycling in salt marsh ecosystems vs. a variety of other high productivity systems. He contrasted the open cycling found in salt marshes, which have an excess of nutrients available in surrounding estuarine waters with the tightly closed nutrient cycling of coral reef communities which exist in nutrient depleted waters. Pomeroy attributed the excess of phosphorus in estuarine waters to clay minerals in the sediments. He estimated that a 1 m thick layer of sediment in contact with *Spartina* root systems could supply phosphorus needs of the grasses for 500 years. Other studies in high salinity marshes (40) have shown that large amounts of nitrogen can be incorporated into marsh systems through nitrogen fixation. For example, Flax Pond in Long Island has an average mean daily nitrogen fixation rate of 4.6 mg m^{-2} (41). Therefore, previous studies of sediment and atmospheric inputs into high salinity marsh systems suggest that there is a luxuriant supply of nutrients—especially phosphorus and nitrogen which often are in limiting supply in other marine systems.

In contrast, other investigators have concluded that some high salinity salt marshes might be nutrient limited and that fertilizer might increase productivity. In North Carolina, Marshall (18) reported that the biomass of *Spartina alterniflora* was greater in marshes subjected to sewage additions at Calico Creek than in control areas in a nearby creek and that productivity was enhanced in high marsh areas. Similar studies in Great Sippewisett Marsh, Massachusetts, (38) indicated that 80% - 94% of nitrogen and phosphorus inputs from sewage sludge were retained *in situ*. Further studies (39) indicated that when both urea and phosphate fertilizers were applied separately to *Spartina alterniflora* and

Spartina patens/Distichlis spicata, production was increased by nitrogen but not by phosphorus. They concluded that both low marsh and high marsh areas were nitrogen limited.

Earlier studies involving nutrient removal in brackish waters at Tinicum Marsh on the Delaware River Estuary (10) showed that 1.17 kg day^{-1} of phosphate and 2.40 kg day^{-1} of nitrogen (nitrate and ammonia) could be taken up per hectare of marsh. Interpretation of yearly nutrient exchanges is difficult in the Tinicum study, since data was obtained only during the months of August, September and October. However, if the above rates are compared with those of Valiela and Teal (39), adjusted to a daily rate basis (.35 kg of phosphorus and .43 kg of nitrogen removal per hectare), the data suggest that brackish marshes may be 3 to 4 times more efficient in nutrient removal than higher salinity marshes. Recent studies (19) of nutrient removal in the brackish Hackensack Marshes in New Jersey indicate that they also are very efficient at taking up nitrogen from polluted waters.

Studies of nitrate and nitrite in flood and ebb tides in Delaware (1) have indicated net imports of inorganic nitrogen into the marsh. In Virginia, Axelrad (2) found a similar net import of inorganic nitrogen by calculating an annual nutrient flux budget based on flow rates and monthly nitrogen concentrations. However, organic nitrogen export exceeded inorganic nitrogen import, so that net flow of dissolved nitrogen was to the estuary.

At Bissel Cove, Nixon et al. (23) analyzed nutrient patterns in conjunction with diurnal oxygen changes in order to determine whether phosphorus and nitrogen uptake and release could be correlated with marsh metabolism. During each of four sampling periods the embayment was isolated from tidal action. Strong diurnal changes in phosphate concentrations were found on two of the sampling periods. However, they concluded that diurnal nitrogen changes were difficult or impossible to predict from the 'Redfield ratio' and oxygen metabolism.

There is also some evidence that low salinity marshes may not be able to assimilate large amounts of nitrogen and phosphorus. In a study of Chesapeake Bay marshes, Bender and Correll (3) state that use of marshes as nutrient removal systems may be questionable. However, they do point out that *Spartina patens* marshes have a limited capacity for phosphorus removal. Their data indicate "that a loading of 29,000 gal of secondary sewage per day would saturate the capacity of an acre of high marsh in 45 days." Also, at Flax Pond (42) there is apparently a large export of ammonia during the summer and phosphate year round to the Long Island estuary.

The body of literature above indicates somewhat conflicting views concerning nutrient exchanges of marshes. In addition, most studies were done in higher salinity marshes, and more data is needed from lower salinity areas for interpretation of nutrient exchanges and the estuary. The present paper represents the efforts of two groups working on two Chesapeake marshes—one at Horn Point and another at Gotts' marsh—to characterize long term nutrient exchange patterns.

STUDY AREAS

The Horn Point marsh is located 4 km west of Cambridge, Maryland on the south shore of the Choptank River, the largest eastern shore sub-estuary of the Chesapeake Bay. The salinity ranges from 7 to 15 ppt and mean tidal amplitude is 0.5 m. Horn Point marsh is comparatively small, approximately 5.7 ha, but has well developed vegetation zones characteristic of the Choptank River area. Fig. 1 shows *Spartina alterniflora* (sometimes mixed with *Iva frutescens*) bordering the creek which extends to a fresh marsh dominated by *Hibiscus moscheutos*. The largest zone (comprising 21% of the surface area) consists of *Spartina patens* and *Distichlis spicata*. Smaller amounts of *Typha angustifolia, Kosteletzkya virginica, Phragmites australis* and *Baccharis halimifolia* are present in higher elevations in the marsh, while *Juncus roemerianus* is found in one isolated area. A previous study of the productivity of emergent vegetation (5) has established that the mean standing crop of the Horn Point marsh is approximately 600 g m^{-2}.

Gotts' marsh is located 2 km south of Lower Marlboro, Maryland, on the east shore of Patuxent estuary. In this area, the salinity ranges from 0 to 5 ppt, but is usually less than 5 ppt and mean tidal amplitude is 1.3 m. The total area of Gotts' marsh (Fig. 2) is comparatively large, about 127 ha. However, the area draining into the tidal slough, Middle Creek, which is the focus of this study is less, about 33 ha. The dominant vegetation on the marsh is typical of low-salinity Patuxent marshes. Dominant species were *Scirpus olneyi, Spartina cynosuroides, Typha angustifolia, Pontederia cordata,* and *Peltandra virginica*. Heinle and Flemer (13) estimated the standing crop to be between 1,000 to 1,500 g m^{-2}.

METHODS

Measurement of Diurnal Metabolism, Nutrient Patterns, and Tidal Concentrations at Horn Point

At Horn Point marsh all tidal flows enter and leave the system through a 5 m long, 2 m wide spillway. From the summer of 1974 through the summer of 1975 a gate, installed in the spillway, was closed at high tide every month for a period of 24 hours. During each sampling period it was possible to estimate the metabolism of the marsh embayment using the diurnal oxygen curve technique (25, 26). Duplicate water samples were collected for oxygen determination and fixed every three hours from a narrow pier extending 9 m into the South Pond (see Fig. 2). Simultaneously, water samples were also collected in polyethylene bottles for nutrient analysis and were immediately taken to the lab and passed through Whatman GF/C filters, using a millipore apparatus. They were analyzed immediately for ammonia (34) and the remaining water was frozen and analyzed

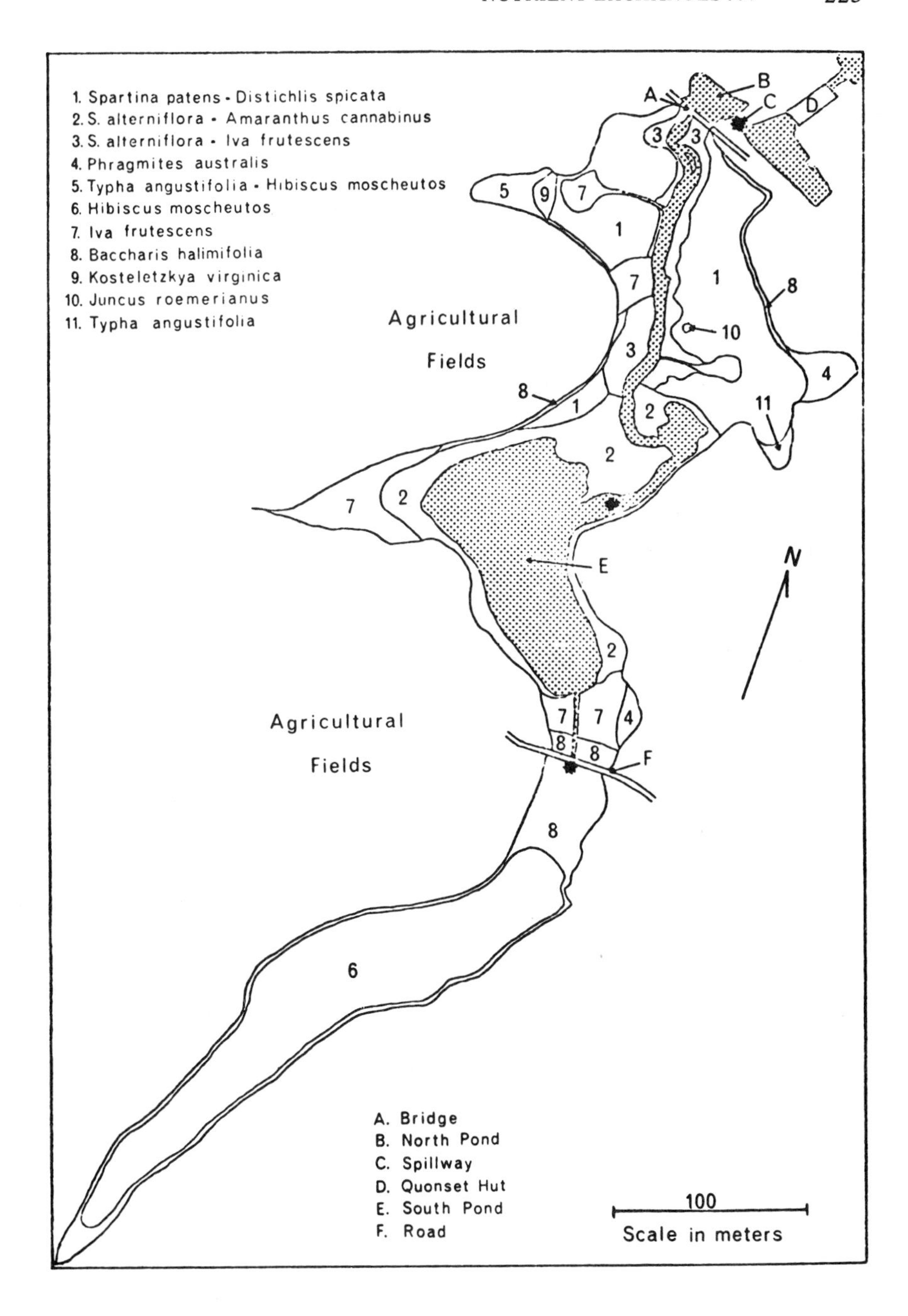

Figure 1. Vegetation map of Horn Point marsh on the Choptank River, with asterisks showing locations of sampling areas.

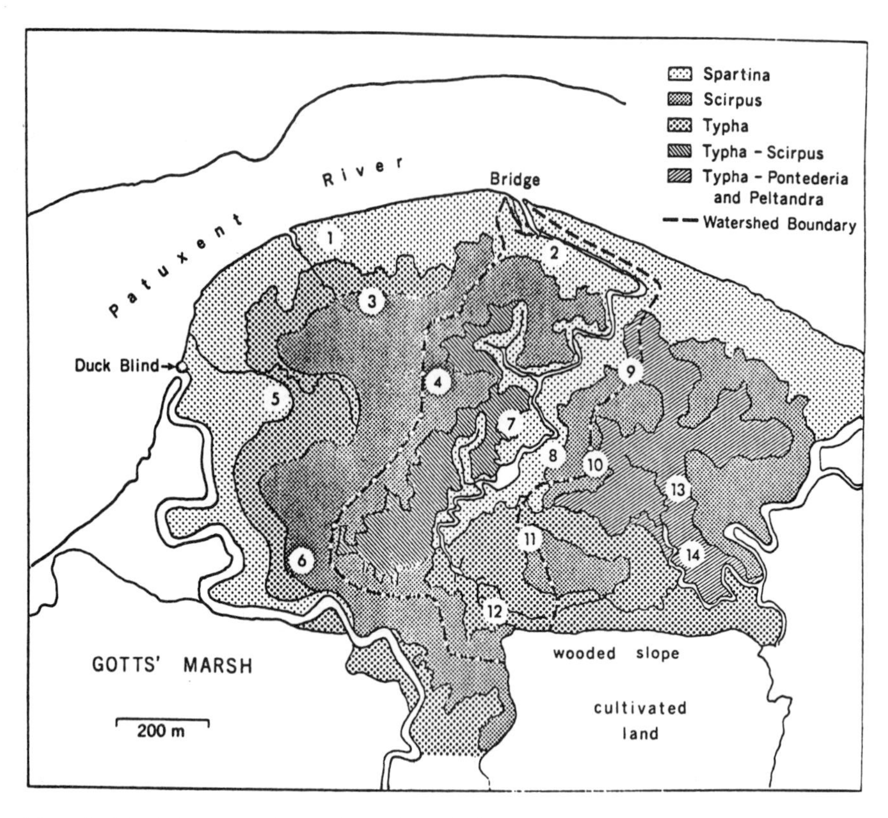

Figure 2. Vegetation map of Gotts' marsh on the Patuxent River showing watershed boundary in dotted lines. Station numbers refer to sites where biomass has been previously sampled.

later for nitrate and nitrite, organic phosphorus and orthosphosphate (35). Chlorophyll *a* was measured by using acetone extraction and the UNESCO procedure. At each sampling interval water temperature was measured with a YSI telethermometer and salinity was determined with a refractometer. Incoming solar radiation was obtained continuously during each run with a chart recorder and Epply pyranometer located on top of an instrument shack adjacent to the marsh.

Monthly diurnal metabolism was calculated from the oxygen, salinity and temperature data at each 3 hr interval using 'OXMET', a Fortran program developed by Sollins (33) and modified to run on the University of Maryland Computation Center UNIVAC 1106/1108 system. Apparent production was calculated by integrating the positive rate of oxygen change curve, and apparent respiration by integrating the negative rate of oxygen change at night. Both curves were corrected for diffusion by multiplying the oxygen diffusion constant, K, by the saturation deficit. In a previous study at Horn Point, K was determined by using the floating dome technique (8, 11). Once apparent production and apparent respiration were calculated they were added to give an estimate of gross primary production which was then compared with the diurnal nutrient curves.

In addition to the above, tidal studies at Horn Point were made monthly from January through August 1975. Concentrations of nitrate, nitrite, ammonia, orthophosphate and organic phosphorus were determined (using methods of analysis above) at ebb and flood tides. Three locations indicated by asterisks in Fig. 1 were sampled: the mouth of the marsh, the creek which drains the South Pond, and the culvert which separates the fresh marsh from the brackish South Pond.

Measurement of Nutrients and Tidal Flux at Gotts' Marsh

Tides enter and leave Gotts' marsh through Middle Creek which is about 6 m wide. A bridge was constructed near the mouth of the creek to facilitate sampling of nutrient fluxes. Nutrient flux rates were obtained monthly from June 1972 through July 1974. During each 13 hour sampling period, the following were measured: nitrate, nitrite, ammonia, soluble organic phosphorus, total phosphorus, reactive phosphorus and Chl *a*. Further details of sampling, analysis and data manipulation are given in (14).

RESULTS AND DISCUSSION

Diurnal Metabolism of South Pond at Horn Point Marsh

Table 1 shows the diurnal values for temperature, salinity and dissolved oxygen on sampling days in the South Pond. The depth of this shallow embayment averages only 15 cm below mean tide level and is therefore prone to large seasonal and diurnal changes. Salinity fluctuated from 0 to 12.6 ppt and temperature from 0 to 35 C. Solar radiation was about average (3900 Kcal m^{-2} day^{-1}) for the southeastern U.S. (32).

Dissolved oxygen levels never fell much lower than 2 g m^{-3} and increased on several days to over 14 g m^{-3}. The ratio of apparent photosynthesis to respiration (P/R) (Table 1) shows that in winter the system is autotrophic and is heterotrophic the rest of the year, except in August. Nixon and Oviatt (22) reported a similar situation at Bissel Cove when tidal action was arrested for short sampling periods. The Bissel Cove embayment never went completely anoxic and throughout the year the P/R ratio was often less than one. Furthermore, Ragotzkie (31) found in estuarine waters near Sapelo Island, Georgia, that P/R ratios generally exceeded 1 only from October to May. The rest of the year $P < R$ indicated heterotrophy of the plankton community in summer months.

Fig. 3 shows the apparent respiration and photosynthesis as well as gross primary production at Horn Point marsh throughout the 14 months of study. As

Table 1. Diurnal oxygen metabolism in South Pond at Horn Point Marsh from July 1974 to August 1975.

Month	Solar Input Ly/Day	Mean Salinity °/oo	Diurnal temp. range °C	Diurnal D.O. range g/m³	P/R Ratio
July	462.1	10.2	23-29	4.5-14.1	.88
Aug	446.4	10.8	26-31	2.0- 6.4	1.06
Sept	519.6	11.5	25-29	2.5- 8.5	.79
Oct*	367.2	6.7	5-14	6.0- 8.4	.91
Nov*	295.2	12.6	5- 8	2.0- 3.7	.58
Dec*	91.5	6.6	5- 9	3.7- 9.1	2.17
Jan*	244.2	0.0	0- 3	11.4-13.9	1.95
Feb*	553.2	2.6	1- 4	8.7-15.1	10.29
Mar*	406.8	4.3	7- 9	11.0-12.4	.79
Apr*	238.4	5.7	15-20	8.5-10.4	.83
May*	366.0	4.0	21-26	4.7-10.0	.69
Jun*	681.9	6.9	26-31	1.9-13.7	.80
July*	471.9	4.5	23-32	3.2-13.6	.89
Aug*	459.0	7.0	28-35	3.7-15.2	1.06

*Indicates months in which diurnal nutrient curves were obtained simultaneously.

expected, gross primary production was highest in summer months of both years. Declining respiration levels during winter account for P > R in winter. Apparent production levels remained moderatley high throughout the winter despite the presence of ice on the marsh in February. This indicates that the large mats of *Ectocarpus* and *Melosira* present in the marsh have high rates of photosynthesis in spite of harsh winter conditions. In contrast, Jones (17) found little phytoplankton photosynthesis occurring in winter at Horn Point marsh. Surprisingly, in August when highest water temperatures were recorded, P again exceeded R. Nixon and Oviatt (22) found that there was an exponential rise in apparent respiration with increasing water temperature to 30 C at Bissel Cove. Our data show respiration declining sharply in August of 1974 when day temperature exceeded 30 C. In 1975, water temperatures were higher earlier—exceeding 30 C in July. Again, respiration declined from the June maximum and caused P > R in August. Observations of the bottom during August indicate complete absence of higher plants and benthic algae which serve as substrates for many estuarine creatures. At the same time, however, dinoflagellate blooms of *Gymnodinium* and *Goniaulax* produce "mahogany" colored tides in the embayment. The photosynthesis from these blooms slightly offsets declining respiration rates causing the "summer P> R anomaly."

The overview in GPP patterns in Fig. 3 is that the metabolic activity is drastically accelerated in the South Pond after April. Simultaneously there are large increases in Chl *a* from .01 mg m^{-3} in March to 164 mg m^{-3} in August. The

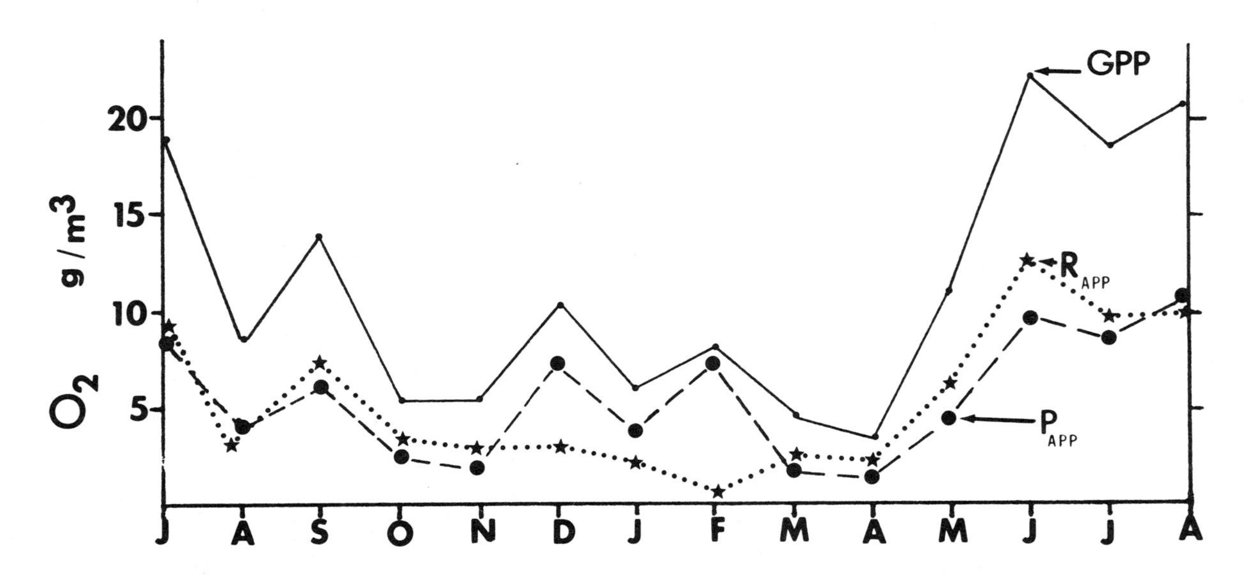

Figure 3. Metabolism in South Pond at Horn Point marsh measured monthly from June 1974 through August 1975. GPP indicates gross primary production, R app indicates apparent respiration and P app indicates apparent photosynthesis.

August Chl *a* values are among the highest reported for estuaries in the Chesapeake region except for the Potomac River, where "hypereutrophic" conditions exist and Chl *a* sometimes reaches 200 mg m^{-3} (16). Our large increase in Chl *a* can be attributed to rapid buildup of dinoflagellate populations. However, this spring buildup is not the result of P > R. Instead, the increase in phytoplankton is associated with a heterotrophic embayment–which depends on imports from detrital origin (22). At Horn Point, Jones (17) estimated that the amount of particulate nitrogen in plant tissue doubles from winter to summer, with a corresponding reduction in detrital PN in that period. Cahoon (5) reported that there is also a corresponding buildup of biomass in the marsh plants around the embayment after April with peak biomass occurring in late July.

Diurnal Concentration Changes of Dissolved N and P at Horn Point

Fig. 4 shows the change in nitrate concentrations over selected diurnal sampling periods. The diurnal changes in October and June suggest nitrate depletion during the day and slight regeneration at night. However, during the remaining months, no strong diurnal pattern was evident. Instead, erratic patterns indicated considerable environmental variation in concentrations which is a reflection of the patchiness in the South Pond. Furthermore, when the entire dissolved inorganic nitrogen fraction was considered, the same erratic pattern emerged. In view of the high system metabolism during many months of the year, a strong diurnal nitrogen variation was expected. The lack of correlation between the magnitude of nitrogen depletion and diurnal metabolism ($r = -.30$) further substantiates the findings of Nixon et al. (23) at Bissel Cove. Therefore in shallow marsh embayments of this type nitrogen fixation and sediment recycling is of greater magnitude than biological uptake and release processes on a daily basis. In lakes, sediment organic nitrogen can act as a source of inorganic nitrogen to the water column through ammonification. Byrnes et al. (4) have shown that sediment NH_4 (interstitial and exchangeable) rapidly reaches equilibrium with the overlying water and that much of the NH_4 mineralized from decaying organic matter in the sediment is soon returned to the overlying water. Since the shallow embayment at Horn Point marsh was never anoxic, NH_4 would be rapidly converted to NO_3 as it moves out of the anaerobic sediment (6). This may help account both for the high levels of nitrate (compared to ammonia and nitrite) and for part of the masking of diurnal uptake and release in the Horn Point embayment.

Unlike dissolved inorganic nitrogen, phosphorus fractions had a rather consistent diurnal depletion and regeneration pattern during high metabolism periods in the South Pond embayment. In March through August of 1975 and October of 1973, Fig. 5 shows that phosphate levels dropped consistently during daylight hours (except in April and June when almost undetectable levels were found throughout the day). During these months phosphate concentrations again increased at night. However, there is no indication of dramatic increases of phosphate similar to those that Newcombe and Lang (21) reported at night at the mouth of the Patuxent River at Solomons, Maryland. At Horn Point, during diurnal samplings, phosphorus exchange with marsh sediments may have been restricted due to absence of tidal action, whereas the Patuxent River had strong currents and tidal action during Newcombe and Lang's studies. Pomeroy et al.

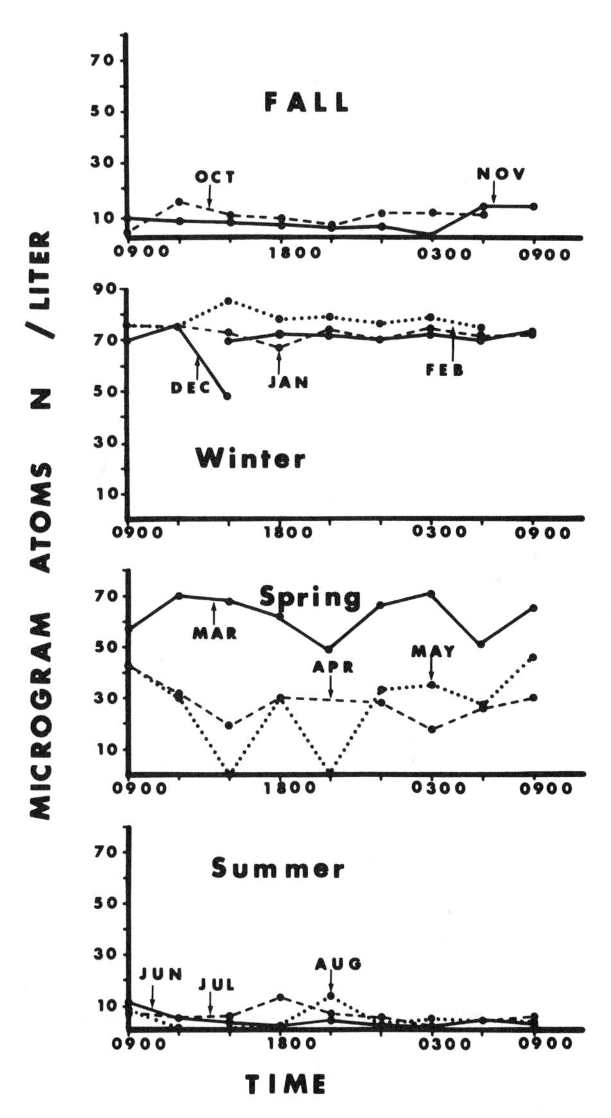

Figure 4. Diurnal dissolved nitrogen concentrations in the South Pond at Horn Point marsh from Fall 1974 through Summer 1975.

(29) found that if estuarine sediments are not disturbed, their large pool of phosphorus is isolated from overlying water, which severely limits exchange processes between compartments.

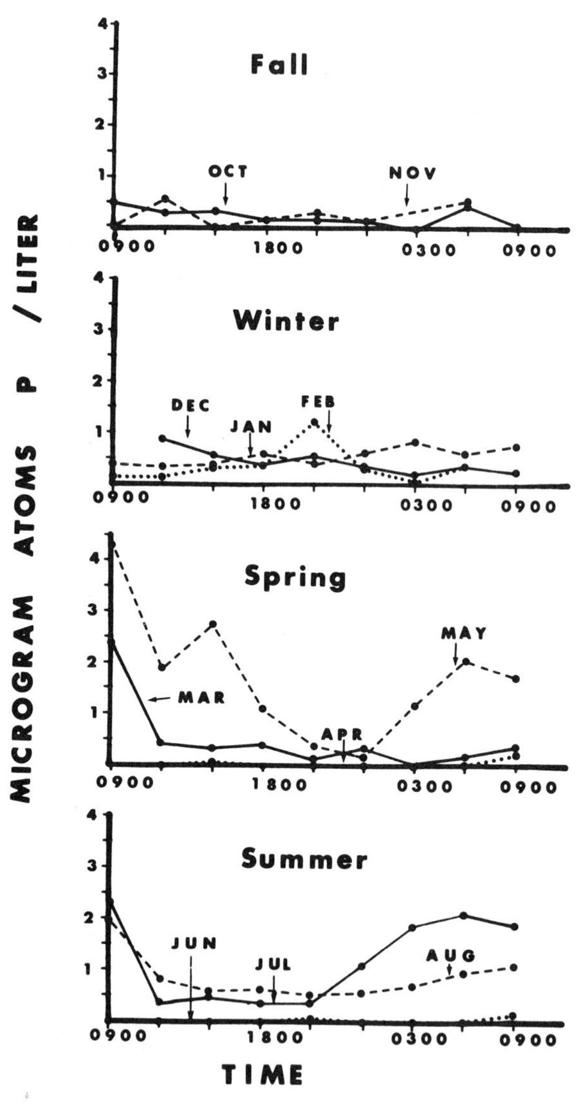

Figure 5. Diurnal dissolved phosphorus concentrations in the South Pond at Horn Point marsh from Fall 1974 through Summer 1975.

Total dissolved phosphorus (orthophosphate plus DOP) for all months dropped on the average from 3.90 μg-at liter^{-1} at 2100 h. During the night average dissolved phosphorus doubled–indicating regeneration. This shows that both orthophosphate and DOP have consistent diurnal variation patterns at Horn Point. At Bissel Cove, Nixon et al. (23) found that phosphate displayed a strong diurnal pattern on two out of four sampling days–but DOP showed no clear trend in concentration changes during diurnals. The reason for this apparent difference in DOP between the two embayments is not clear but may be due to the comparatively small number of diurnals run at Bissel Cove. In spite of this difference in DOP the overall diurnal pattern of phosphorus at Horn Point is surprisingly like the phosphate pattern at Bissel Cove–both reveal net uptake during several sampling periods. This may be related to the lack of tidal action and stirring of sediments which normally resupplies the water column with phosphorus. Therefore, it may be an artifact introduced when a marsh is "stoppered" for these types of experiments.

Seasonal Concentration Changes of Dissolved N and P at Horn Point

The seasonal changes in dissolved P and inorganic N (obtained from averaging 9 concentrations for each monthly diurnal) are very striking. Fig. 6 shows the classic nitrogen consumption pattern described for inshore marine systems by Cooper (7) in the English Channel and Harris (12) in Long Island Sound. At Horn Point, nitrate predominates in the total inorganic dissolved fraction–especially in winter months when it exceeds 75 μg-at liter^{-1}. In spring, when metabolism and Chl *a* values increased, the concentrations of nitrate declined rapidly to about 5 μg-at liter^{-1} in summer. This spring depletion pattern is also similar to that found in other Chesapeake Bay studies in the Patuxent Estuary (15) and in Virginia (27). However, the nitrate concentrations at Horn Point are much larger than maximum values in the Patuxent (3.4 μg-at liter^{-1}) or the lower Chesapeake (6.7 μg-at liter^{-1}). Horn Point concentrations seem somewhat atypical for the Chesapeake Bay, but are much more similar to those found by Aurand and Daiber (1) on the eastern side of the Delmarva peninsula. This similarity might be due to the location of these marsh watersheds in an area where intensive agriculture has gone on for some time. We do not know how much nitrate from fertilizer applications may be entering the estuarine ecosystems associated with the Delmarva area, which is sometimes called the "eastern cornbelt." High nitrogen input from farming may prevent this element from ever being an important limiting factor–as it seems to be in the marshes of North Carolina (18, 37) or Massachusetts (39). However, more studies on denitrification rates at Horn Point are needed to show conclusively whether nitrate is limiting during months in which it is in relatively short supply.

Fig. 6 shows that generally phosphorus and nitrogen have an inverse relationship on a seasonal basis. During winter DOP plus orthophosphate concentrations are about 1 μg-at liter^{-1}, but after April, they increase to about 4 μg-at liter^{-1} in

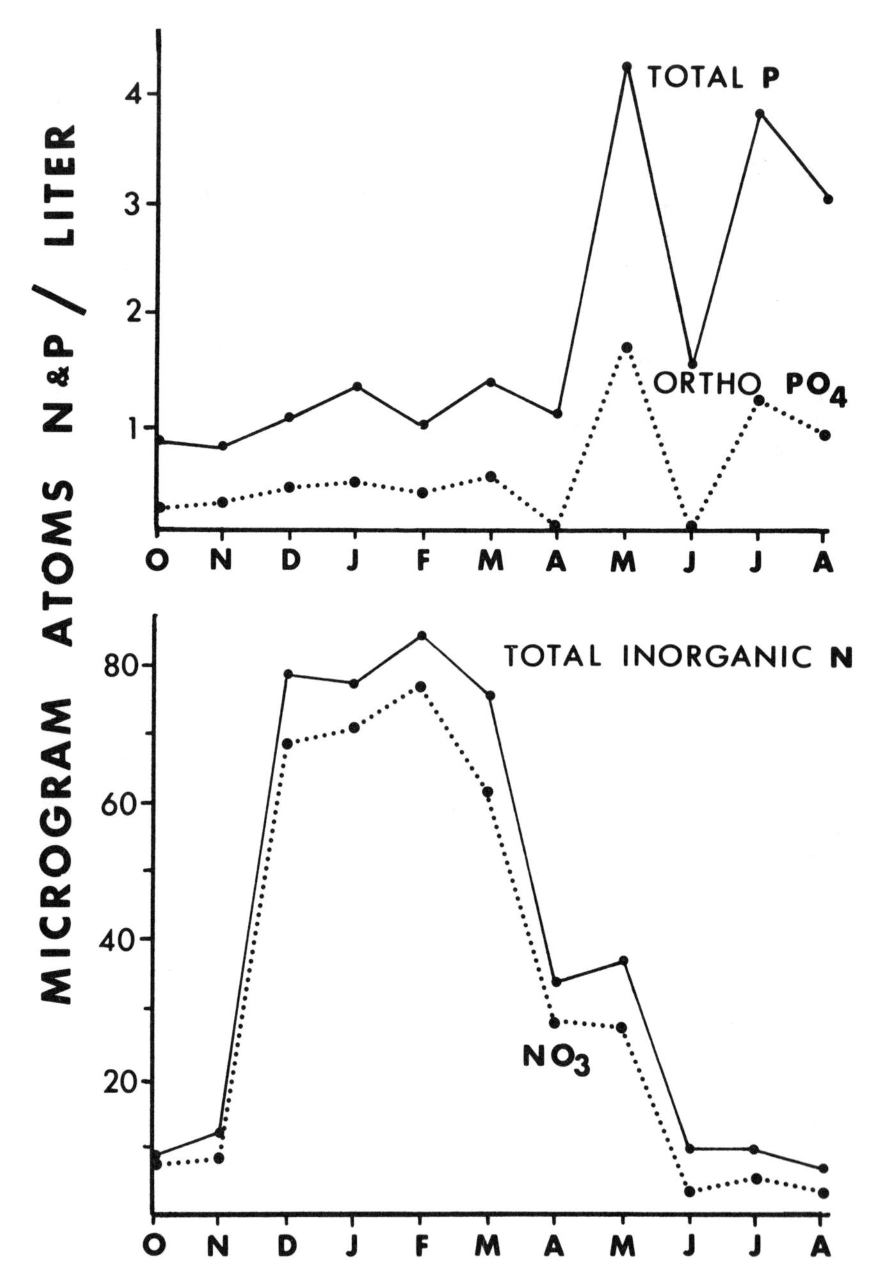

Figure 6. Monthly mean dissolved inorganic nitrogen, nitrate, orthophosphate, and total dissolved phosphorus concentrations in South Pond at Horn Point marsh from October 1974 through August 1975. Vertical axes plotted to reflect 1:16 P:N "Red-field ratio."

summer. The summer rise in phosphorus was first described in the Chesapeake Bay by Newcombe and Lang (21) and is found in many other estuaries of the world (28). Newcombe and Lang attributed increases in phosphate to a decrease in light penetration and subsequent decreases in phytoplankton utilization. However, the data at Horn Point indicate that phosphate levels rise along with the total gross primary productivity of the system. In view of the decreasing nitrogen values in late spring it is probable that phosphate is also being incorporated into protoplasm at a rapid rate. The explanation for the phosphate buildup at Horn Point marsh is not a decrease in utilization, but a very large increase of phosphorus input into the embayment. Recent investigators (36) attribute the summer maximum in the Chesapeake type estuaries to the release of orthophosphate into oxygen depleted deep water and subsequent transport to the euphotic zone by turbulent mixing. The South Pond was never anoxic during study days, so large *in situ* orthophosphate additions from sediment probably did not occur. However, the fresh water marsh is often anoxic in spring and summer, and may be a source of phosphate. During the diurnal experiments, the culvert connecting the fresh marsh with the South Pond was blocked with an inflatable plug. This may partially account for April and June values, when metabolism was very high in and around the embayment and phosphate levels diminished to undetectable levels. Thus, the uncoupling of the fresh marsh from the South Pond causes nitrate depletion in the latter ecosystem. H. T. Odum (24) elaborates on the environmental problems encountered when segments of the estuarine system are uncoupled. Normally, materials which are produced by the fresh marsh system are exported to the brackish embayment and provide high grade energy subsidies for its summer-spring productivity. Another important source of phosphate in srping is the Choptank estuary as indicated by the tidal imports during April and May.

Dissolved Inorganic N and P on Flood and Ebb Tides

Fig. 7 shows that concentrations of NO_3, NH_3 and NO_2 on ebb tides are twice as great as on flood tides during winter. However, during the period of increasing gross primary production in spring and summer there is no great difference in ebb and flood concentrations of NO_3 and NO_2. We have not yet calculated net flux of nitrogen (or phosphorus) from the Horn Point concentrations because of incomplete records of flow rates of water passing in and out of the marsh. However, precipitation and evaporation data at Horn Point suggest that the watershed surrounding the marsh delivers higher quantities of water to the embayment during winter months. This results in more water being flushed out of the marsh in winter and periods of low salinities. These higher outgoing flows plus the higher concentrations on outgoing ebb tides suggest a considerable net flux of inorganic nitrogen in winter. Fig. 7 shows that NH_3 does not follow NO_3 and NO_2 fractions closely in spring and summer, and does not decrease on the ebb tide until after June. It appears that export of NH_3 took place until the "summer P > R anomaly" discussed earlier. This may indicate that the summer dinoflagellate blooms may depend heavily on ammonia as an important nitrogen source in the Horn Point marsh.

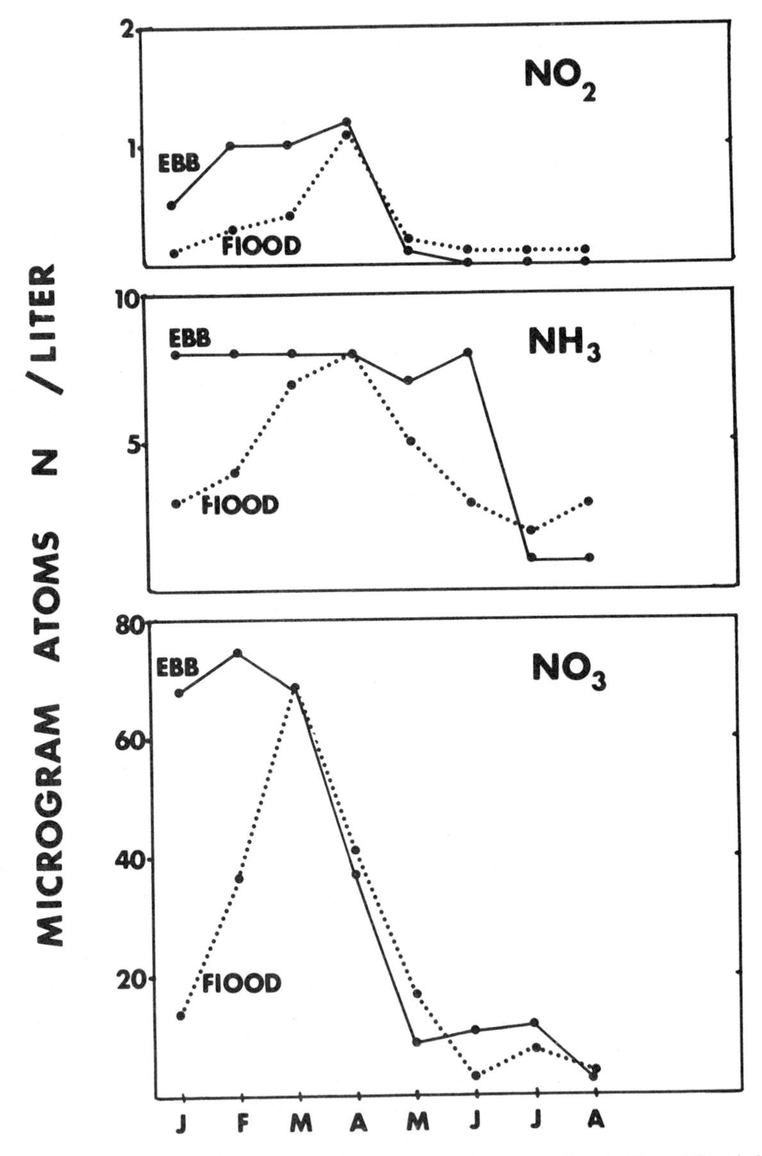

Figure 7. Dissolved inorganic nitrogen fractions sampled monthly at ebb and flood tides at entrance of Horn Point marsh from January 1975 through August 1975.

In contrast to the Horn Point marsh, Aurand and Daiber (1) reported that there was an apparent import of nitrate during the winter and export in summer from Canary Creek and Murderkill marshes. However, they point out that Gallagher's study (9) at Canary Creek showed that the greatest algal productivity was from January to May and declined in summer. The differences in productivity periods at Horn Point and the Delaware marshes may partially account for discrepancies in import-export patterns. In both ecosystems export seems to occur during periods of low metabolism. Therefore, the relative export or import of dissolved nitrogen may depend on the periodicity and magnitude of marsh metabolism throughout the year. An additional factor in the Horn Point import-export pattern is the fresh marsh surrounded by agricultural fields. Throughout the year this area has high nitrogen concentrations. In winter when metabolism is low in the brackish marsh, nitrate flushes out of the entire system in large quantities. Thus, the fresh marsh nitrate source may account for differences in nitrogen export at Horn Point versus other areas such as Gotts' marsh (2, 14).

Fig. 8 indicates the dissolved organic phosphorus (DOP) and orthophosphate concentrations in flood and ebb tides in winter and summer. These concentrations suggest export of dissolved phosphorus in winter along with dissolved inorganic nitrogen. In summer, flood and ebb concentrations of DOP are very similar and suggest little net exchange after May. However, phosphate alternates from apparent import in spring to export later during the "summer $P > R$ anomaly." This pattern suggests that during the spring, at Horn Point, there is a considerable amount of phosphorus uptake and utilization which is associated with the large increase in system metabolism. However, as metabolism stabilizes in the South Pond in summer, the marsh again resumes its export function in regard to orthophosphate. At Ware and Carter Creek, Axelrad (2) found a strong export of dissolved inorganic phosphorus almost year round. The only difference with Horn Point seems to be due to the high metabolic activity of the shallow embayment in spring.

Nitrogen and Phosphorus Fluxes at Gotts' Marsh

Fluxes of nutrients between Gotts' marsh and the Patuxent estuary have been reported in detail by Heinle and Flemer (14). They found uptake of the inorganic fractions of nitrogen during May, June, and July, but concurrent greater release of dissolved organic nitrogen to the estuary. Net flows of total dissolved nitrogen were from the estuary to the marsh on only 2 of 24 sampling dates. In contrast to the studies at Horn Point ammonia was commonly abundant at Gotts' marsh, sometimes comprising up to one-third of the total dissolved nitrogen. The annual net flow of dissolved nitrogen from the marsh to the estuary was 4.14 $g\ m^{-2}\ yr^{-1}$, about half of the nitrogen estimated to be present in the fall above-ground vegetation. Net flows of particulate nitrogen were also from the marsh to the estuary, but in much smaller amounts.

Particulate and dissolved phosphorus flowed from the marsh to the estuary at 0.138 and 0.190 g m^{-2} yr^{-1}, respectively. Heinle and Flemer (14) suggested that Gotts' marsh was probably in long term equilibrium of phosphorus with the estuary and that the observed fluxes represented year-to-year variations. Major storms might account for most of the deposition of phosphorus.

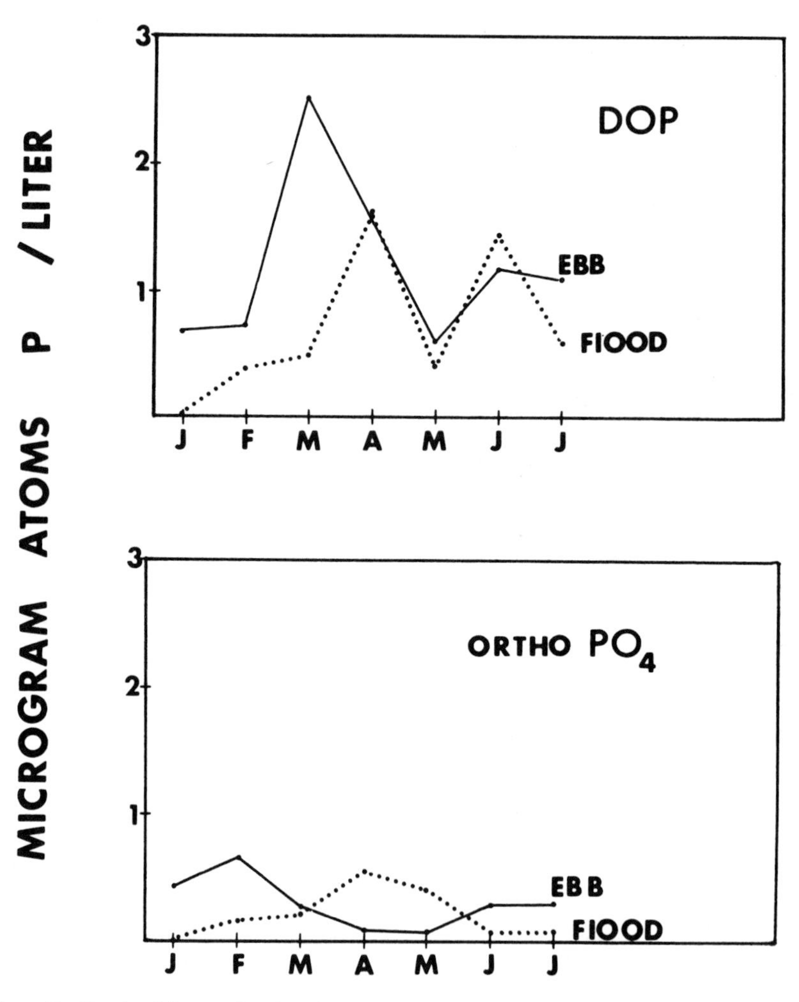

Figure 8. Dissolved inorganic phosphorus and orthophosphate sampled monthly from ebb and flood tides at entrance of Horn Point marsh from January 1975 through July 1975.

Viewed in light of other known sources, the Patuxent marshes appear to be significant sources of nitrogen and carbon to the estuary. The amounts of phosphorus contributed by the marshes were small relative to other known sources (14).

CONCLUSIONS

Comparison of these two marsh ecosystems reveals the following:

1. Although these marshes show considerable variation in nutrient exchange patterns, both appear to be exporters of nitrogen. Horn Point is characterized by sharp seasonal fluctuations, while Gotts' marsh was more stable over the study period.

2. Prediction of nutrient exchange of marshes may depend on the following factors:

A. The presence of upland sources of nutrients and sediments.

B. The magnitude and stability of nutrient flux in the estuary to which the marsh is coupled.

C. The successional age of the marsh ecosystem, which is often reflected by the ratio of open water area to marsh grass area. (Horn Point's shallow, highly metabolic embayment controls the apparent flip-flop in import-export patterns. As the marsh becomes older, the embayment is filling in with *Spartina* which may change nutrient exchange patterns.)

D. The tidal energy input which controls stirring of sediments and determines bottom composition.

E. Salinity and redox potentials, by changing species composition which especially affects nitrogen fixation and denitrification patterns.

3. Nutrient removal capacity on a yearly basis is very limited in these brackish marshes. Most studies that suggest strong nutrient removal capacity have been done in high salinity areas during the growing season, and are inappropriate models for interpretation of nutrient patterns in the Upper Chesapeake Bay.

ACKNOWLEDGMENTS

We are grateful to L. Beaven, D. Cahoon, S. Christy, M. Cole, G. Cox, R. Galloway, R. Huff, E. Hughes, T. Jones, J. Lawson, J. Manley, N. Mick, J. Morrow, R. Murtagh, D. Robison, A. Seaton, L. Simmons, L. Slacum, S. Sulkin, R. Younger, and R. Zeisel for their assistance in this study. In addition special thanks goes to Scott Nixon at the University of Rhode Island for sharing many of his innovative ideas and research design suggestions. This work was supported by the Office of Water Resources Research Center through projects B-016 Md., B-019 Md., and A-038 Md., as well as the National Science Foundation (RANN) through the Chesapeake Research Consortium. Finally we thank H. Sisler of the

University of the University of Maryland Botany Department as well as P.E. Wagner, L.E. Cronin, and P. Winn, Jr., of the Center for Environmental and Estuarine Studies for their help, support, and encouragement throughout these studies.

REFERENCES

1. Aurand, D., and F. C. Daiber. 1973. Nitrate and nitrite in the surface waters of two Delaware salt marshes. Chesapeake Sci. 14: 105-111.
2. Axelrad, D. M. 1974. Nutrient flux through the salt marsh ecosystem. Ph.D. Thesis. College of William and Mary, Williamsburg, Va. 133 p.
3. Bender, M. E., and D. L. Correll. 1974. The use of wetlands as nutrient removal systems. Virginia Institute of Marine Science Contribution 624. 12 p.
4. Byrnes, B. H., D. R. Keeney, and D. A. Graetz. 1972. Release of ammonium-N from sediments to water. Proc. 15th Conf. Great Lakes Res: 249-259.
5. Cahoon, D. R. 1975. Net productivity of emergent vegetation at Horn Point salt marsh. M.S. Thesis. Univ. of Maryland, College Park, Md. 94 p.
6. Chen, R. L., D. R. Keeney, D. A. Graetz, and A. S. Holding. 1972. Denitrification and nitrate reduction in Wisconsin lake sediments. J. Environ. Qual. 1:158-162.
7. Cooper, L. H. N. 1937. The nitrogen cycle in the sea. J. Mar. Biol. Ass. U.K. 22: 183-204.
8. Copeland, B. J., and W. R. Duffer. 1964. Use of a clear plastic dome to measure gaseous diffusion rates in natural waters. Limnol. Oceanogr. 9: 494-498.
9. Gallagher, J. L. 1971. Algal productivity and some aspects of the ecological physiology of the edaphic communities of Canary Creek tidal marsh. Ph.D. Thesis. Univ. of Delaware. 120 p.
10. Grant, R. R., and R. Patrick. 1970. Tinacum marsh as a water purifier, p. 105-123. *In* J. McCormick (ed) Two Studies of Tinicum Marsh. The Conservation Foundation, Washington, D. C.
11. Hall, C., J. Day and H. T. Odum. 1975. A new means for measuring diffusion constants of natural waters using a plastic dome. Unpubl. manuscript available from C. Hall, Ecology and Systematics, Cornell University, Ithaca, N. Y.
12. Harris, B. 1959. The nitrogen cycle in Long Island Sound. Bull. Bingham. Oceanogr. Coll. 17: 31-65.
13. Heinle, D. R., and D. A. Flemer. 1975. Carbon requirements of a population of the estuarine copepod, *Eurytemora affinis*. Mar. Biol. (Berl.) 31:235-247.
14. ——— and ———. 1976. Flows of materials between poorly flooded tidal marshes and an estuary. Mar. Biol. (Berl.) 35:359-373.
15. Herman, S. S., J. A. Mihursky and A. J. McErlean. 1968. Zooplankton and environmental characteristics of the Patuxent River estuary. Chesapeake Sci. 9: 67-82.
16. Jaworski, N. A., D. W. Lear, and O. Villa. 1972. Nutrient management in the Potomac estuary, p. 246-273. *In* G. E. Likens (ed), Amer. Soc. Limnol. Oceanogr. Special Symposium, Vol I.

17. Jones, T. W. 1975. Relative phytoplankton primary productivity and particulate organic matter in a Maryland salt marsh. M.S. Thesis. University of Maryland, College Park, Md. 79 p.
18. Marshall, D. E. 1970. Characteristics of *Spartina* marsh receiving treated municipal wastes. Dept. of Zoology and Institute of Marine Sciences, University of North Carolina, Chapel Hill, N. C. 51 p.
19. Mattson, C., R. Trattner, J. Teal and I. Valiela. 1975. Nitrogen import and uptake measured on marshes of the Hackensack River in New Jersey. A paper presented at the 38th Annual meeting of Amer. Soc. Limnol and Oceanogr. 15 p.
20. Menzel, D. W., and N. Corwin. 1965. The measurement of total phosphorus in seawater based on the liberation of organically bounded fractions by persulfate oxidation. Limnol. Oceanogr. 10: 280-282.
21. Newcombe, Curtis L., and Andrew G. Lang. 1939. The distribution of phosphates in the Chesapeake Bay. Proceedings of the American Philosophical Society 81:393-420.
22. Nixon, S., and C. Oviatt. 1973. Ecology of a New England salt marsh. Ecological Monogr. 43: 499-538.
23. ——, ——, J. Garber, and V. Lee. 1976. Diurnal metabolism and nutrient dynamics in a salt marsh embayment. Ecology. In press.
24. Odum, H. T. 1975. The coupling of marshes, estuaries, and man. The Third Biennial International Estuarine Research Conference Abstracts.
25. —— and C. M. Hoskin. 1958. Comparative studies on the metabolism of marine waters. Publ. Inst. Mar. Sci., Univ. of Texas 5: 16-46.
26. —— and R. F. Wilson. 1962. Further studies on respiration and metabolism of Texas Bays, 1958-1960. Pub. Inst. Mar. Sci., Univ. of Texas 8: 23-55.
27. Patten, B.C., R. Mulford and J. Warinner. 1963. An annual phytoplankton cycle on the lower Chesapeake Bay. Chesapeake Sci. 4: 1-20.
28. Pomeroy, L. R. 1970. The strategy of mineral cycling, p. 171-190. *In* R. F. Johnson, P. W. Frank and C. D. Michener (eds) Annual Reviews, Inc., Palo Alto, Ca.
29. ——, E. E. Smith, and C. M. Grant. 1965. The exchange of phosphate between estuarine water and sediment. Limnol. Oceanogr. 10: 167-172.
30. ——, R. J. Riemold, L. R. Shenton, and R. D. H. Jones. 1972. Nutrient flux in estuaries, p. 274-296. *In* G. E. Likens (ed.) Nutrients and Eutrophication. Amer Soc. Limnol. Oceanogr. Special Symposium, Vol. I.
31. Ragotzkie, R. A. 1959. Plankton Productivity in estuarine waters of Georgia. Publ. Inst. Mar. Sci., Univ. of Texas 6: 146-158.
32. Reifsnyder, W., and H. Lull. 1965. Radiant energy in relation to forests. U.S. Dept. of Agric., Tech. Bull. No. 1344. 111 p.
33. Sollins, P. 1970. OXMET: A program for the calculation of metabolism and productivity in equatic systems. Appendix A. to M.A. Thesis. Univ. of North Carolina at Chapel Hill. 86 p.
34. Solorzano, L. 1969. Determination of ammonia in natural waters by the phenolhypochlorite method. Limnol. Oceanogr. 14: 799-801.
35. Strickland, J. D. H., and T. R. Parsons. 1968. A practical handbook of sea water analysis. Fish Res. Bd. Canada Bull. 167. 309 p.
36. Taft, J. L., and W. Taylor. 1975. Phosphorus dynamics in some coastal plain estuaries. The Third Biennial International Estuarine Research Conference Abstracts.

37. Thayer, G. W. 1971. Phytoplankton production and the distribution of nutrients in a shallow unstratified estuarine system near Beaufort, North Carolina. Chesapeake Sci. 12: 240-253.
38. Valiela, I., J. M. Teal, and W. Sass. 1973. Nutrient retention in salt marsh plots experimentally fertilized with sewage sludge. Estuarine Coastal Mar. Sci. 1: 261-269.
39. ____ and ____. 1974. Nutrient limitation in salt marsh vegetation, p. 547-563. *In* R. J. Reimold and W. H. Queen (eds.) Ecology of Halophytes. Academic Press, New York.
40. Van Raalte, C. D., I. Valiela, E. J. Carpenter, and J. M. Teal. 1974. Inhibition of nitrogen fixation in salt marshes measured by acetylene reduction. Estuarine Coastal Mar. Sci. 2: 301-305.
41. Whitney, D. E., G. M. Woodwell, and R. W. Howarth. 1975. Nitrogen fixation in Flax Pond: A Long Island salt marsh. Limnol. Oceanogr. 20: 640-643.
42. Woodwell, G., C. Hall, D. Whitney, R. Houghton, R. Moll. 1975. The Flax Pond ecosystem: Changes in quality of coastal waters flushing a heterotrophic marsh. The Third Biennial International Estuarine Research Conference Abstracts.

NITROGEN POOLS AND FLUXES IN A GEORGIA SALT MARSH[1]

E. Haines, A. Chalmers, R. Hanson, and B. Sherr
University of Georgia Marine Institute
Sapelo Island, Georgia 31327

ABSTRACT: Preliminary data on standing stocks of nitrogen and on rates of nitrogen fixation and denitrification in *Spartina alterniflora* salt marshes on the Georgia coast have yielded a general concept of the nitrogen cycle in this system. The Georgia salt marsh is characterized by fine-textured clay soils, a tidal amplitude of two meters, and high rates of primary production throughout the year. Most combined nitrogen in the system is in the soils in forms not readily available to primary producers. The amounts of exchangeable ammonium, nitrate, and nitrite are small and change seasonally. The presence of nitrate and nitrite in the soil implies that nitrification occurs in the aerobic microzones around *Spartina* roots. The annual input to the marsh via nitrogen fixation and other quantified sources of nitrogen is as large as the seasonal accumulation of nitrogen by *Spartina* growth; the input is balanced by the loss of nitrogen via denitrification. The nitrogen input to the marsh is, however, smaller than the estimated nitrogen flux through the marsh plants. Thus, mineral regeneration must satisfy a large part of the nitrogen requirements of *Spartina* and benthic algae in the Georgia salt marsh.

INTRODUCTION

The cycling of nitrogen in marine systems is not as well understood as the cycling of other elements for which there are convenient radioisotopic tracers. In salt marshes, biogeochemical studies have focused on the fluxes of phosphorus (e.g. 21) and carbon (e.g. 7, 31, 41). The importance of nitrogen as a limiting nutrient for marine autotrophs (24) has however stimulated a number of recent studies on aspects of the salt marsh nitrogen cycle (e.g. 16, 37, 39).

Nitrogen availability can affect the biological processes of a salt marsh in several ways. The best documented instance is the limitation of primary production of marsh plants. Addition of nitrogen fertilizer has increased the standing

[1] Contribution number 308 from the University of Georgia Marine Institute, Sapelo Island.

crop of a number of plant species in marshes in Sweden (36) and at several locations along the eastern U.S. coast (4, 10, 30, 37). The shortage of nitrogen may be intensified by estuarine heterotrophs degrading carbon-rich, mineral-poor plant detritus competing with autotrophs for inorganic nitrogen (19, 33). Finally, the concentration of specific nitrogen species, i.e. ammonia, nitrate and nitrite, may inhibit or stimulate specific microbial processes in the marsh, i.e. nitrogen fixation, nitrification, and denitrification (35).

In order to elucidate the roles of nitrogen in the metabolism of salt marshes, a study of seasonal changes in the nitrogen content of various pools and of certain nitrogen flux rates in a *Spartina alterniflora* salt marsh has been initiated near Sapelo Island, Georgia. This report summarizes six months of data from the study and presents some of our emerging concepts concerning the nitrogen cycle in this marsh.

STUDY SITES AND METHODS

Two separate sampling areas were chosen in a *Spartina alterniflora* salt marsh adjacent to Sapelo Island, Georgia. These represented major differences in frequency of tidal inundation and plant productivity. One area was in streamside or low marsh, where *Spartina* culms are over 1.5 m high (11) and plant productivity is estimated to be 3711 g (dry matter) m^{-2} yr^{-1} (12). The other location was in high marsh, where *Spartina* culms are only about 0.5 m in height and chlorotic in appearance (11), and plant productivity is an estimated 1337 g m^{-2} yr^{-1} (12). The low marsh soil salinities approximate those of estuarine water (20 to 30°/oo), while high marsh soils generally have higher salinities (30 to 45°/oo) (6). Other environmental characteristics of the system include fine-textured clay soils, a tidal amplitude of about 2.0 meters, and mild seasonal changes. Additional descriptions of the Georgia salt marsh have been presented by Teal (31), Pomeroy et al. (21), Reimold et al. (23), and Christian, Bancroft, and Wiebe (6).

In each study area two sampling sites were established. Duplicate samples of aboveground plant biomass and soil cores were taken monthly in each site, in a systematic fashion to avoid sample overlap. Sampling was begun in January for the high marsh and in February for low marsh. Individual properties were analysed as follows:

Nitrogen Pools

Spartina aboveground biomass. In each site two 0.1 m^2 plots were clipped to the soil surface and the surface cleaned of most fallen litter. Samples were washed in fresh water and sorted into young culms (less than 15 cm in height for short *Spartina*, less than 25 cm in height for tall *Spartina*), older culms (remaining live shoots), and dead culms, leaves, and litter. The plants were dried to constant weight at 110°C, and subsamples ground with a Wiley mill to pass a no. 40 mesh screen.

Belowground macro-organic matter. Duplicate 7.5 cm diameter cores taken to a depth of 30 cm in each site were divided into two sections: 0 to 15 cm and 15 to 30 cm. The macro-organic matter in each section was separated into two size fractions by washing on stacked 2 mm and 1 mm sieves. No effort was made to separate live from dead roots and rhizomes. The samples were dried at 110°C and ground in a Wiley mill to pass a 40 mesh screen. The macro-organic matter was sampled bi-monthly.

Soil. Integrated 2 cc soil samples were taken from each core in the depth zones of 0 to 5 cm, 10 to 15 cm, and 25 to 30 cm. The samples were dried at 110°C and ground to a powder with a mortar and pestle.

Plant and soil percent carbon. Subsamples of 0.5 g of ground plant tissue were ashed at 520°C for 5 to 6 hours in a muffle furnace to determine percent organic matter. Dried 2 cc soil samples were weighed and ashed at 430°C for 24 hours for percent organic content. The carbon contents of selected plant and soil samples were periodically analysed with a Leco WR-12 Carbon Determinator. Conversion factors calculated from these data were used to convert percent organic matter values to percent carbon values for the remaining samples.

Plant and soil percent nitrogen. Subsamples of 20 to 100 mg of ground plant and soil samples were analysed for percent nitrogen content by the micro-Dumas technique using a Coleman Model 29 Nitrogen Analyser. The percent carbon and percent nitrogen values were used to calculate atomic C:N ratios.

Soil exchangeable ammonium. In each soil core, integrated 5 cc samples were taken in each 5 cm interval from 0 to 30 cm. Each soil sample was mixed with 20 ml of 1 N KCl solution (pH 3) (3). After a 30 minute extraction period, the sample was centrifuged and the supernatant analysed for ammonium (25).

Soil nitrate and nitrite. The supernatant was also analysed for nitrate and nitrite. Nitrate was reduced to nitrite by using a peristaltic pump to force 5 ml samples past a 1.5 m length of copper-coated 0.64 mm diameter cadmium wire in a microtubing coil (27). The nitrite was analysed colorimetrically (29).

Soil molecular nitrogen (N_2). The concentration of N_2 at 7 depths in triplicate 30 cm cores was determined as follows: 11 cc of soil were added to 19 ml of 5% formalin solution in a 50 ml flask. The gas space in the flask was flushed with helium, then the contents were vortexed and left for ten minutes to allow equilibrium partitioning of N_2 between the liquid and gas phases in the flask. The gas in the flask was analyzed for N_2 with a Carle Model 111 Analytical Gas chromatograph fitted with a 13A molecular sieve column. A blank value was obtained with the formalin solution only. An equilibrium curve was constructed so that the amount of N_2 in the liquid phase of a sample slurry could be determined by knowing only the amount in the gas phase. This curve was obtained by flushing soil and formalin solution mixtures with helium until no more N_2 could be detected in the gas phase. Known quantities of N_2 were then added to the flasks and the procedure followed as outlined above. The amount of N_2 gas in the liquid phase determined from the equilibrium curve was added to the amount measured in the gas phase, and the blank control value subtracted, to yield a total N_2 content for each soil sample.

Nitrogen Fluxes

Nitrogen fixation. Rates of fixation were estimated by acetylene reduction in three habitats in the high (short *Spartina*) marsh:

1) Marsh surface. A relatively low volume (1100 cc), large surface area (0.043 m^2) dome incubator was used for *in situ* measurement of surface algal and bacterial fixation. After clipping and removing *Spartina* plants from a 0.05 m^2 area of the marsh, the incubator was pushed down onto the soil. Acetylene and ethane (used as an internal standard) were added, yielding an atmosphere of 15% acetylene and 0.1% ethane. After an equilibrium period of 45 to 60 minutes, three gas samples were removed for analysis every hour for four hours. Concentrations of ethylene and ethane were analysed using a FID gas chromatograph with a Porpak N column. The ethylene concentration was normalized to the ethane in the incubator. The rate of ethylene production was converted to a rate of nitrogen fixation by the theoretical conversion factor of 3:1.

2) Soil depth profile. Cores 7.5 cm in diameter were taken to 30 cm. Slices 1 cm thick from selected depths were placed in small incubation chambers with a gas space of 130 cc. Acetylene and ethane were added to yield the concentrations described above, and the chambers kept in the dark. Gas samples were analysed for ethylene and ethane every two hours for eight hours. Ethylene production generally became linear after a lag period of 2-3 hours, and remained linear up to 36 hours. Control experiments using soil slices of varying thickness suggested that acetylene reduction occurred only in the top 3 mm of the 1 cm thick slices. Nitrogen fixation rates were calculated for the surface of each slice.

3) *Spartina* plants. Nitrogen fixation by epiphytes on *Spartina* was determined for the plants removed from the 0.05 m^2 plot used for the surface incubator. The plant material, both live and dead, was placed in wide mouth jars. The same gas concentrations were used as for the soil samples. The jars were incubated under full sunlight in a tub of water. Gas samples were analysed every hour for four hours and nitrogen fixation calculated from ethylene production from the linear portion of the curve.

Denitrification. The initial gas chromatographic method used to measure rates of denitrification in depth profiles and in the surface soils involved analysing N_2 accumulation in incubators flushed with helium. After several months of study, however, it was determined that this method was not sensitive enough to detect rates of N_2 release via denitrification against the background of physical diffusion of N_2, which was measured from poisoned soil samples.

The diffusion data, combined with the data on soil N_2 concentrations, were used to obtain an estimate of N_2 production rates with the equations used by de Jong and Schappert (8) to obtain CO_2 production rates from profiles of CO_2 concentrations in soils. Using the concepts of Fickian diffusion and the conservation of matter, one can determine the rates of production or consumption of a particular substance if the concentration and diffusion coefficient of the substance in the soil profile are known. That the concentration of N_2 was not

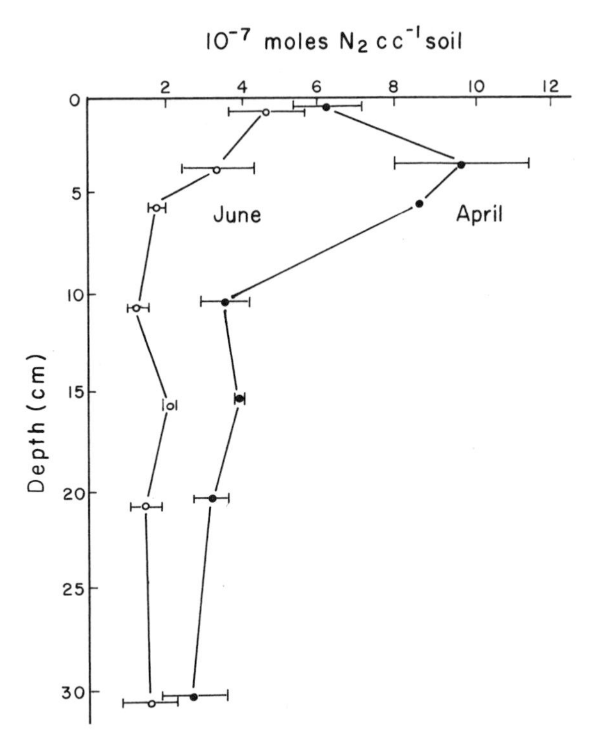

Figure 1. Depth profile of nitrogen gas content of high marsh soils in April and June. Mean ± one standard error.

constant with depth in the marsh soils (Fig. 1) indicated that the production and/or consumption of N_2 was occurring. The diffusion coefficient for nitrogen gas was empirically determined for sterile marsh soil slices. The diffusion rates determined in the initial months for known soil volumes were used to back-calculate to the concentration of N_2 in the soil. The soil N_2 concentration data were used directly. The results for December to June were pooled and extrapolated to an annual N_2 production rate.

RESULTS AND DISCUSSION

Nitrogen Pools

Soil. Over 90% of the total nitrogen in the measured pools in the marsh was soil nitrogen, which included the labile inorganic pools of exchangeable ammonium and nitrate (Table 1). However, most of the soil nitrogen is probably refractory organic nitrogen and fixed ammonium (35). The nitrogen content of the soil ranged from 0.15 to 0.50% of dry weight, or 1.5 to 5.0 mg N g^{-1}. Nitrogen content decreased with depth (Table 2). Surface soils in the low marsh had a

Table 1. Pools of nitrogen in high and low *Spartina* salt marsh; values averaged for February to July.

Pool	Six month mean value $g\ N\ m^{-2} \pm 1$ S.E.M. High marsh	Low marsh	% of total N High	Low
Soil (0-30 cm)	493 ± 33	463 ± 8	93.5%	91.6%
Macro-organic matter				
0-15 cm coarse	14.33 ± 1.97	11.39 ± 0.80		
fine	3.03 ± 0.49	0.80 ± 0.12		
15-30 cm coarse	3.80 ± 0.66	10.61 ± 1.48		
fine	3.66 ± 0.51	0.90 ± 0.27		
Total 0-30 cm	24.82	23.7	4.7%	4.7%
Spartina shoots				
Young:	0.12 ± 0.04	0.24 ± 0.12		
Old:	3.02 ± 0.59	8.04 ± 1.31		
Dead:	2.85 ± 0.26	6.79 ± 0.43		
Total	5.99	15.07	1.1%	3.0%
N_2 gas[1] 0-30 cm	3.26	3.26	0.62%	0.65%
Exchangeable NH_3	0.255	0.202	0.05%	0.04%
$NO_3 + NO_2$	0.035	0.033	0.01%	0.01%
Total in system:	527.4	505.3		

[1]From 2 months' data in high marsh.

slightly greater nitrogen content than did the high marsh soils (Table 2). The nitrogen content of Georgia salt marsh soils ranged between the reported values of 12.4 and 3.8 mg N g^{-1} for two sites in Barataria Bay, Louisiana (15), and the 0.54 mg N g^{-1} found for soil in an English marsh (16). Soil carbon-nitrogen ratios were lower than those of *Spartina* plants (Table 2). The average C:N atom

Table 2. Nitrogen content and C:N atom ratios for *Spartina* plants and soil in low and high marsh; values averaged for February to July ± 1 S.E.M.

	% N of dry weight High marsh	Low marsh	C:N atom ratio High marsh	Low marsh
Plants: shoots				
Young	1.47 ± 0.04	1.71 ± 0.07		
Mature	1.38 ± 0.04	1.51 ± 0.04	35.8 ± 1.1	32.6 ± 1.2
Dead	0.89 ± 0.03	0.92 ± 0.05	54.6 ± 2.3	54.4 ± 2.8
Soil				
0-5 cm	0.32 ± 0.02	0.44 ± 0.02	13.52 ± 1.40	10.76 ± 0.98
10-15 cm	0.27 ± 0.01	0.34 ± 0.02	13.69 ± 1.17	12.06 ± 0.94
25-30 cm	0.24 ± 0.01	0.20 ± 0.02	13.41 ± 0.92	15.26 ± 1.69

ratios of the high marsh soils were uniform with depth, while the C:N ratios of the low marsh soils increased from 0-5 cm to 25-30 cm.

Underground macro-organic matter. The nitrogen in the macro-organic matter (MOM) in the top 30 cm of soil was less than 5% of the total nitrogen (Table 1). Low and high marsh soils had different depth distributions of the MOM, as Gallagher (9) has previously observed. In the low marsh, the MOM nitrogen was uniformly distributed between the two depth zones; in the high marsh most of the MOM nitrogen was present in the top 15 cm. High marsh soils also contained a greater amount of the fine MOM fraction than did the low marsh soils. This observation suggests that the MOM may be mineralized at a faster rate in the low marsh soils, so that a fine fraction does not accumulate.

Spartina culms. The average amount of nitrogen in the high marsh *Spartina* culms, about 6 g N m^{-2}, was 1.1% of the total high marsh nitrogen, while the average 15 g N m^{-2} in the low marsh plants was 3% of the total measured nitrogen in the low marsh (Table 1). This was a result of the difference in productivity in the two marsh areas. Gallagher et al. (12) found a seasonal accumulation of 10.7 g N m^{-2} in live creekbank *Spartina*, and an accumulation of only 2.1 g N m^{-2} in live high marsh *Spartina.*

Live plants contained a higher percentage of nitrogen and had lower C:N ratios than dead plants (Table 2). The low marsh plants showed slightly greater nitrogen contents than the high marsh plants, and young culms had a slightly higher nitrogen content than older culms. The percent nitrogen contents reported here were between the 1.74 to 2.59% N reported by Burkholder (5) and the 0.7 to 1.1% N reported by Gallagher (10) for *Spartina* plants in the Sapelo salt marsh.

Soil gaseous nitrogen, exchangeable ammonium, and nitrate. The smallest of the measured nitrogen pools in the marsh were soil nitrogen gas, exchangeable ammonium, and nitrate plus nitrite, which together comprised less than 1% of the total. The N_2 gas content in high marsh soil (Fig. 1) decreased with depth. Higher concentrations of N_2 gas in the upper 10 cm of soil could be a result of transport of nitrogen gas downward by *Spartina* roots. Yoshida et al. (42) have reported a significant increase in the N_2 content of planted rice paddy soils over unplanted soils. They concluded that the increase in N_2 was a result of the transport of air by the rice culms to the roots with subsequent diffusion of N_2 into the surrounding soil. A mechanism for the transport of gas from the atmosphere to the roots has been described for rice (1, 2) and also *Spartina* (32). Depth profiles of the concentrations of exchangeable ammonium and of nitrate averaged over six months showed a uniform depth distribution of nitrate to 30 cm, and a decrease in exchangeable ammonium concentration in the top 15 cm of soil (Fig. 2). The concentrations of exchangeable ammonium were higher and more variable in the low marsh soil than in the high marsh soil.

Although nitrate and nitrite do not usually occur in water-logged soils having low redox potentials, the presence of aerobic microzones around roots may

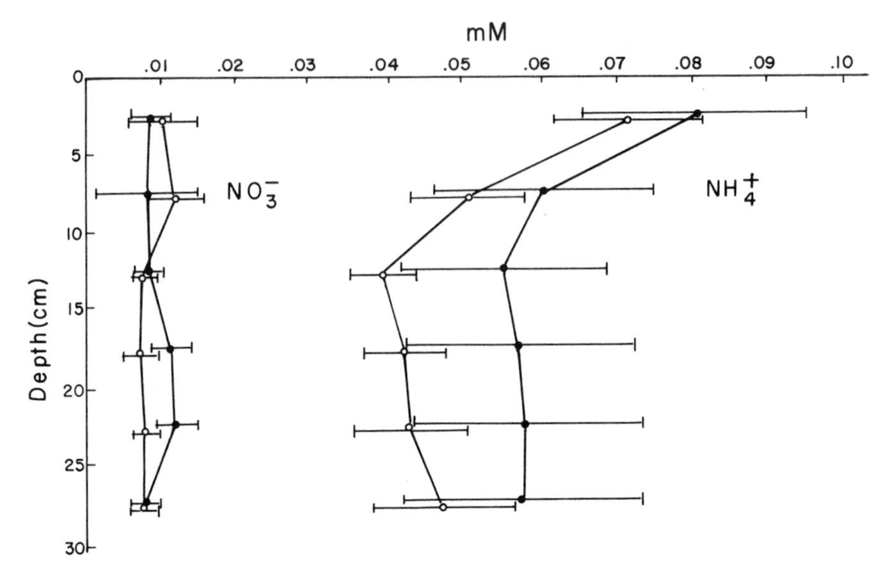

Figure 2. Depth profiles of nitrate and exchangeable ammonium concentrations in low (0) and high (●) marsh soils. Values are the mean concentration for February to July ± one standard error.

allow nitrification to proceed in an otherwise anaerobic environment (35). The measurable concentrations of nitrate we have found in Georgia marsh soils suggest that nitritification does occur around *Spartina* roots. Teal and Kanwisher (32) have estimated that the diffusion of oxygen from *Spartina* leaves to the roots may be as much as twice the respiratory requirement of the roots.

Seasonal changes in the nitrogen pools

In general, the smaller the size of the nitrogen pool in the salt marsh, the greater the relative seasonal variation. The largest pools of nitrogen in the marsh, the soil and macro-organic matter nitrogen, have shown no seasonal fluctuations. An increase in the total nitrogen content of the aboveground biomass from winter to summer has resulted from plant growth.

Ammonium and nitrate pools have shown the greatest seasonal fluctuations (Fig. 3). There was a spring peak in the content of the exchangeable ammonium pool in both marsh areas. This resembled the peak in exchangeable ammonium concentration reported by Ho and Lane (15) during the fall in Louisiana Lake sediments. Our peak could be explained by increased mineralization and nitrogen fixation rates with spring warming of the marsh soil, which may temporarily exceed the plant uptake of inorganic nitrogen. The nitrate content has decreased from winter to summer.

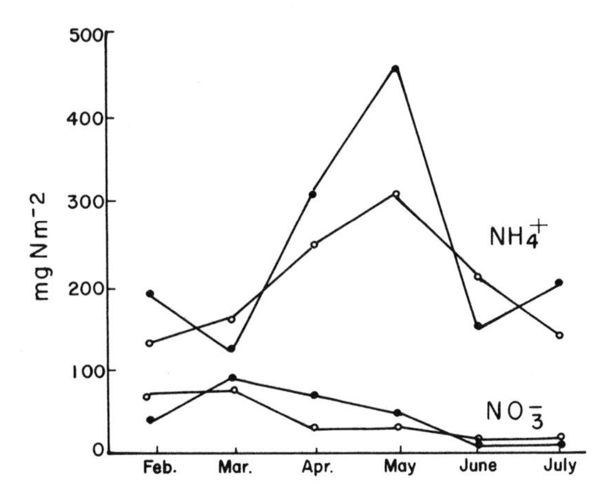

Figure 3. Seasonal contents of nitrate and exchangeable ammonium in low (0) and high (●) marsh soils to 30 cm.

Nitrogen Fluxes

Nitrogen fixation. Fixation occurred in each of the habitats studied: marsh surface, soil profile, and *Spartina* culms.

1. Marsh surface. Rates of fixation by photosynthetic and heterotrophic organisms on the marsh surface ranged from 30 to 600 μg N_2 fixed m^{-2} hr^{-1} between January and June 1975. Rates increased from January to May and dipped slightly in June. The average fixation rate corresponded to 100 mg N_2 fixed m^{-2} per month assuming fixation occurs only during daylight hours. However, measured dark fixation rates were at times as much as 10 to 20% of light fixation, so this may be an underestimate.

2. Soil profile. The integrated fixation rate for the upper 15 cm of the soil profile was 0.940 μg N_2 m^{-2} hr^{-1} in October and 705 μg N_2 m^{-2} hr^{-1} in May. Highest fixation rates were found in the 3-6 cm depth zone, within the high marsh root mat. The higher rates of soil nitrogen fixation in spring may be a result of higher soil temperatures and release of oxidizable carbon substrates into the marsh soil by *Spartina* roots.

3. *Spartina* culms. Epiphytic nitrogen fixation was primarily photosynthetic and attributable to blue-green algae and photosynthetic bacteria in the interfaces of leaf sheaths and plant stems. This habitat is probably rich in nitrogen-poor *Spartina* photosynthates and therefore ideal for nitrogen fixing organisms which require 1 gram of carbohydrate to fix 1 mg of N_2 (22). Epiphytic fixation rates ranged between 35 and 70 μg N_2 m^{-2} hr^{-1} during the period March to June. The average monthly fixation rate was 15 mg N_2 m^{-2} $month^{-1}$.

The total annual estimated input of nitrogen to the high marsh via nitrogen fixation is probably about 5.8 g N m^{-2} yr^{-1}. This rate agrees with values reported for other marine systems. Stewart (28) measured fixation rates of 1-2 g N m^{-2} yr^{-1} in brackish water and marine habitats. Van Raalte et al. (39) estimated that 1.2 g N m^{-2} was fixed for May-July in a Massachusetts marsh. Jones (16) extrapolated surface nitrogen fixation measurements in a *Spartina anglica* marsh to an annual rate of 17.1 g N m^{-2}.

Denitrification. Based on the changes in N_2 content in soil profiles, the production of N_2 in the top 30 cm of the marsh soils was estimated to be about 12 g N m^{-2} yr^{-1}. If the N_2 production was a result of denitrification alone, this flux would represent a considerable loss of nitrogen from the marsh system, and would more than balance the input of nitrogen to the marsh via nitrogen fixation. However, some of the N_2 production may result from the transport of nitrogen gas from the atmosphere to the soils by *Spartina* roots. The indirectly obtained rate presented here is similar to the denitrification rate of approximately 7 g N m^{-2} yr^{-1} which can be extrapolated from the N_2 loss rates measured for surface soils in a Massachusetts salt marsh (38).

Nitrogen Fluxes and Marsh Metabolism

We have so far quantified only a small proportion of the nitrogen fluxes occurring in the Georgia salt marsh. Our estimates of the nitrogen input to the marsh via nitrogen fixation and nitrogen loss from the marsh via denitrification are in general agreement with the rates measured in other marine systems. Another source of nitrogen for the Georgia coast is the nitrate and ammonia in precipitation. A 12 month study of the nitrogen content of rain at Sapelo Island suggests that this input is very small, only about 0.3 g N m^{-2} yr^{-1} (Haines, unpubl. data).

The most critical unknown flux at present is tidal exchange. However, some general conclusions can be drawn from existing data. The input of new inorganic nitrogen into the marsh via tidal flow is probably small. River water is the most likely source of such nitrogen nutrients in an unpolluted estuary. Windom et al. (40) have estimated that river flow can contribute only about 3.1 g N m^{-2} yr^{-1} to southeastern salt marshes. In the carbon-rich Georgia estuaries, the tidal flux of dissolved and particulate organic nitrogen, as well as inorganic nitrogen bound to clay particles, is probably much greater than inorganic nitrogen tidal fluxes.

Preliminary data on the tidal transport of nitrogen in a diked marsh at Sapelo suggest that this is the case (Haines, unpubl. data). However, the direction of net flux of organic materials between estuaries and tidal marshes is at present uncertain. The assumption that salt marshes are net exporters of dissolved and particulate organic materials (17, 18) has been challenged. Studies of the carbon flows between Flax Pond, a tidal marsh on Long Island, and Long Island Sound, have indicated either a balanced carbon budget or a net input of fixed carbon to the marsh (41). In the salt-marsh estuary adjacent to Sapelo Island, Sottile (26) found no measurable net export of dissolved organic carbon from the marsh to the estuary, and a year-long study of tidal carbon transport in a diked marsh at Sapelo suggested a small net influx of particulate materials to the marsh (Kraeuter and Hall, pers. commun.).

The total input of nitrogen via nitrogen fixation, river flow, and rain is about 9.2 g N m^{-2} yr^{-1}. This input may be completely balanced by the calculated loss of nitrogen via denitrification of 12 g N m^{-2} yr^{-1}. The gains and losses of nitrogen in the Georgia salt marsh are of the same magnitude as the seasonal accumulation of nitrogen in the salt marsh plants of 2.1 to 10.7 g N m^{-2}. However, the actual nitrogen demand by marsh autotrophs is probably greater. Reimold et al. (23) calculated average monthly rates of detritus production for *Spartina* in the salt marshes near Sapelo of 197.9 g dry matter m^{-2} for creekbank plants and 113.6 g dry matter m^{-2} for high marsh plants. Assuming the detritus is produced from standing dead material with a nitrogen content of about 0.9% (present study), this process represents a plant nitrogen loss of 12 to 21 g N m^{-2} yr^{-1}. In addition, the benthic algal production of about 200 g C m^{-2} yr^{-1} (20) would, assuming a C:N ratio of 10, require 20 g N m^{-2} yr^{-1}. The nitrogen flux through marsh autotrophs may then be as high as 32 to 41 g N m^{-2} yr^{-1}. The results of studies of material exchanges between salt marshes and estuaries suggest, as discussed previously, that there is little net loss of the plant nitrogen from the marsh.

Mineral recycling within the marsh must therefore be of greater importance than the input of new nitrogen in meeting the nitrogen requirements of the primary producers. Smaller amounts of fine macro-organic matter in low compared to high marsh soils indicate that nitrogen mineralization rates may be greater in the low marsh. Differences in mineralization rates could result in the differences in vigor and productivity between low and high marsh *Spartina* plants which can be ascribed to nitrogen availability.

On the other hand, we have preliminary data which suggest that rates of nitrogen fixation and denitrification in marsh soils are often limited by the availability of carbon substrates (Hanson, unpubl. data; Sherr, unpubl. data). These observations are supported by other evidence that heterotrophic activity in salt marsh soils is in general limited by low *in situ* concentrations of readily utilizable carbon (Christian, pers. commun.). Thus, carbon and nitrogen fluxes in the salt marsh appear to be coupled in such a way that either one may control the other.

ACKNOWLEDGMENTS

This study was partially supported by grants from the Sapelo Island Research Foundation, by Office of Water Resources Research Grant A-057-GA, and by National Science Foundation Grants GA 35806 to Dr. W. Wiebe and DES 74-21338 to Dr. W. J. Payne. We would like to thank Drs. L. R. Pomeroy, J. L. Gallagher, and B. L. Haines for their critical review of the manuscript.

LITERATURE CITED

1. Arashi, K., and N. Nitta. 1955. Studies on the lysigenous intercellular space as the ventillating system in the culm of rice and some graminaceous plants. Crop Sci. Soc. Japan Proc. 24:78-81.
2. Arikado, H. 1959. Supplementary studies on the development of ventillating systems in various plants growing on lowland and on upland. Bull. Fac. Agr. Mie University, Japan 20:1-24.
3. Bremner, J. M. 1965. Inorganic forms of nitrogen. *In* C. A. Black (ed.), Methods of Soil Analysis. Part 2. Agronomy, 9:1179-1237.
4. Broome, S. W., W. W. Woodhouse, and E. D. Seneca. 1975. The relationship of mineral nutrients to growth of *Spartina alterniflora* in North Carolina. II. The effects of N, P, and FE fertilizers. Soil Sci. Soc. Amer. Proc. 39:301-307.
5. Burkholder, P. R. 1956. Studies on the nutritive value of *Spartina* grass growing in the marsh areas of coastal Georgia. Bull. Torrey Bot. Club, 83:327-334.
6. Christian, R. R., K. Bancroft, and W. J. Wiebe. 1975. Distribution of microbial adenosine triphosphate in salt marsh sediments at Sapelo Island, Georgia. Soil Sci. 119:89-97.
7. Day, J. W., W. G. Smith, P. R. Wagner, and W. C. Stowe. 1973. Community structure and carbon budget of a salt marsh and shallow bay estuarine system in Louisiana. Pub. No. LSU-SG-72-04. Center for Wetland Resources, Louisiana State University, Baton Rouge.
8. de Jong, E., and H. J. Schappert. 1972. Calculation of soil respiration and activity from CO_2 profiles in the soil. Soil Sci. 113:328-333.
9. Gallagher, J. L. 1974. Sampling macro-organic matter profiles in salt marsh plant root zones. Soil Sci. Soc. Proc. 38:154-155.
10. _____. 1975. Effect of an ammonium nitrate pulse on the growth and elemental composition of natural stands of *Spartina alterniflora* and *Juncus roemerianus*. Amer J. Bot. 62:644-648.
11. _____, R. J. Reimold, and D. E. Thompson. 1972. Remote sensing and salt marsh productivity, p. 338-348. *In* W. J. Kosco (chairman), Proceedings of the 38th annual meeting. American Society of Photogrammetry, Falls Church, Virginia.
12. _____, _____, R.A. Linthurst, and W.J. Pfeiffer. Production, mortality, and mineral accumulation dynamics in *Spartina alterniflora* and *Juncus roemerianus* in a Georgia salt marsh. Ecology (in press).
13. Garcia, J. L. 1974. Reducation de l'oxyde nitreux dans les sols de rizieres du Senegal: mesure de l'activite denitrifiante. Soil Biol. Biochem. 6:79-84.
14. _____. 1975. La denitrification dans les sols. Bull. de L'Institut Pasteur 73:167-193.
15. Ho, C. L., and J. Lane. 1973. Interstitial water composition in Barataria Bay (Louisiana) sediment. Est. Coast Mar. Sci. 1:125-135.
16. Jones, K. 1974. Nitrogen fixation in a salt marsh. J. Ecol. 62:553-565.
17. Odum, E. P. 1961. The role of tidal marshes in estuarine production. N. Y. St. Conserv. 15:12-15.
18. _____, and A. A. de la Cruz. 1967. Particulate organic detritus in a Georgia salt marsh estuarine ecosystem, p. 383-388. *In* G. H. Lauff (ed.), Estuaries. A.A.A.S., Washington, D. C.
19. Parker, R. R., J. Sibert, and T. J. Brown. 1975. Inhibition of primary

productivity through heterotrophic competition for nitrate in a stratified estuary. J. Fish. Res. Bd. Canada 32:72-77.
20. Pomeroy, L. R. 1959. Algal productivity in salt marshes of Georgia. Limnol. Oceanogr. 4:386-397.
21. ———, L. R. Shenton, R. D. H. Jones, and R. J. Reimold. 1972. Nutrient flux in estuaries, p. 274-291. *In* G. E. Likens (ed.), Nutrients and Eutrophication. Special Symposia Vol. 1. ASLO.
22. Postgate, J. R. (ed.). 1971. The Chemistry and Biochemistry of Nitrogen Fixation. Plenum Press, New York. 326 p.
23. Reimold, R. J., J. L. Gallagher, R. A. Linthurst, and W. J. Pfeiffer. 1975. Detritus production in coastal Georgia marshes, p. 217-228. *In* L. E. Cronin (ed.), Estuarine Research, Vol. 1. Academic Press, New York.
24. Ryther, J. H., and W. M. Dunstan. 1971. Nitrogen, phosphorus, and eutrophication in the coastal marine environment. Science 171:1008-1013.
25. Solorzano, L. 1969. Determination of ammonia in natural waters by the phenol hypochlorite method. Limnol. Oceanogr. 14:799-801.
26. Sottile, W. S. 1974. Studies of microbial production and utilization of dissolved organic carbon in a Georgia salt marsh-estuarine ecosystem. Ph.D. dissertation. University of Georgia. 153 p.
27. Stainton, M. P. 1974. Simple, efficient reduction column for use in the automated determination of nitrate in water. Anal. Chem. 46:1616.
28. Stewart, W. P. D. 1967. Nitrogen turnover in marine and brackish habitats. II. Use of ^{15}N in measuring N_2 fixation in the field. Ann. Botany 31:385-407.
29. Strickland, J. D. H., and T. R. Parsons. 1972. A practical handbook of seawater analysis. Bull. 167. 2nd ed. Fish. Res. Bd. Canada, Ottawa. 310 pp.
30. Sullivan, M. J., and F. C. Daiber. 1974. Response in production of cord grass, *Spartina alterniflora*, to inorganic nitrogen and phosphorus fertilizer. Chesapeake Sci. 15:121-123.
31. Teal, J. M. 1962. Energy flow in the salt marsh ecosystem of Georgia. Ecology 43:614-624.
32. ———, and J. W. Kanwisher. 1966. Gas transport in the marsh grass, *Spartina alterniflora*. J. Exp. Bot. 17:355-361.
33. Thayer, G. W. 1974. Identity and regulation of nutrients limiting phytoplankton production in the shallow estuaries near Beaufort, N. C. Oecologia 14:75-92.
34. Todd, R. L., and J. H. Nuner. 1973. Comparison of two techniques for assessing denitrification in terrestrial ecosystems. Bull. Ecol. Res. Comm. (Stockholm) 17:277-278.
35. Tusneem, M. E., and W. H. Patrick, Jr., 1971. Nitrogen transformations in waterlogged soil. Bull. 757. Louisiana State University Agricultural Experiment Station. 75 p.
36. Tyler, G. 1967. On the effect of phosphorus and nitrogen supplied to Baltic shore-meadow vegetations. Botaniska Notiser 120:433-447.
37. Valiela, I., and J. M. Teal. 1974. Nutrient limitation in salt marsh vegetation, p. 547-563. *In* R. J. Reimold and W. H. Queen (eds.), Ecology of Halophytes. Academic Press, New York.
38. ———, S. Vince, and J. M. Teal. Assimilation of sewage by wetlands. Estuarine Research, Vol. III (in press).

39. Van Raalte, C. D., I. Valiela, E. J. Carpenter, and J. M. Teal. 1974. Inhibition of nitrogen fixation in salt marshes measured by acetylene reduction. Est. Coast. Mar. Sci. 2:301-305.
40. Windom, H. L., W. M. Dunstan, and W. S. Gardner. 1975. River input of inorganic phosphorus and nitrogen to the southeastern salt marsh estuarine environment, p. 309-313. *In* F. G. Howe, J. B. Gentry, and M. H. Smith (eds.), Mineral Cycling in Southeastern Ecosystems. U.S.E.R.D.A. Conf. 740513.
41. Woodwell, G. M., P. H. Rich, and C. A. S. Hall. 1973. Carbon in estuaries, p. 221-240. *In* G. M. Woodwell and E. V. Pecan (eds.), Carbon and the Biosphere. AEC Symposium Vol. 30.
42. Yoshida, T., Y. Takai, and D. C. Del Rosario. 1975. Molecular nitrogen content in a submerged rice field. Plant and Soil 42:653-660.

PRODUCTIVITY AND NUTRIENT EXPORT STUDIES IN A CYPRESS SWAMP AND LAKE SYSTEM IN LOUISIANA

John W. Day, Jr., Thomas J. Butler, and William H. Conner
Center for Wetland Resources
Louisiana State University
Baton Rouge, LA 70803

ABSTRACT: During a 14-month period in 1973-74, several functional properties were studied in a swamp forest and lake system in Louisiana. The study area forms the upper drainage basin of the Barataria Bay estuary. Properties measured were: a) productivity of the two swamp forest components, bottomland hardwood (BLH) and baldcypress-tupelo (CT); b) productivity of lake and bayou waters; c) hydrology; and d) carbon and nutrient export to the lower estuary.

Productivity of the BLH site was 800 g dry wt$\cdot m^{-2} \cdot yr^{-1}$ for stem biomass increase, 584 for litterfall, and 200 for understory production. Similar figures for CT were 500, 620, and 20. Total above-ground net productivity was 1584 for BLH and 1140 for CT.

Net daytime photosynthesis (NDP) for the lake was 1418 g $O_2 \cdot m^{-2} \cdot yr^{-1}$ and nighttime respiration (NR) was 1868 (P/R = 0.76). For the bayous NDP was 316 and NR was 446 (P/R = 0.71).

Water discharge from the basin is significant year-round except during the summer when evaporation equals precipitation. Annual export to the lower estuary (calculated from water discharge and materials concentrations) was 8016 metric tons of organic C, 1047 metric tons N, and 154 metric tons P. The greater part of this export occurred during the spring, corresponding to the spring peak in biological activity in the estuary.

INTRODUCTION

This paper describes an ecosystem level study of productivity and material flows in a 770 km^2 fresh water swamp and lake system in Louisiana, most of which is a continuous forest stand (about 91% forest and 9% open water). The major water body within the study area is Lac des Allemands, which is located less than thirty miles from downtown New Orleans. Thus, the area of interest comprises a large, intact, relatively undisturbed wetland system contiguous with a major metropolitan area. As we shall show, however, the swamp system is beginning to feel effects of the urban region.

The swamp basin is not an isolated system, as it forms the headwaters of the Barataria Basin, the most productive wetland-estuarine system in Louisiana. (Louisiana is the leading fishery producer in the nation and Barataria Bay accounts for 45% of this productivity.) Since the swamp zone is connected hydrologically to the lower estuary, we assumed that it played a role in maintaining estuarine productivity. Thus, we undertook this study in order to gain an understanding of functioning of the swamp system, man's effects on it, and its effect on the estuarine ecosystem. Our specific objectives were twofold, to measure overall rates of terrestrial and aquatic primary productivity in the swamp, and to measure material flows within the swamp and export to the lower estuary.

DESCRIPTION OF THE AREA

The study area is the upper freshwater basin of the Barataria Bay estuary in south Louisiana (Fig. 1). It is bounded by the natural levees of the Mississippi River on the northeast, Bayou Lafourche on the southwest, and the embankment of U.S. Highway 90 to the southeast. Prior to the construction of flood control levees, the area was an overflow swamp of the Mississippi River, and water levels were determined primarily by the river. Now the artificial levees of the Mississippi prevent riverine flooding into the basin. Presently, the only hydrologic input into the basin is precipitation which averages 156 $cm \cdot yr^{-1}$.

Elevation of the swamp is usually less than 1.5 m above sea level. The natural levees of the Mississippi River and Bayou Lafourche are the highest areas with elevations of 3 to 5 m above sea level. Small changes in elevation in the swamp result in large changes in vegetation types. In fact, an elevation change of only 15 cm in the swamp environment corresponds to one of 30 m in mountainous terrain (3). In areas of low elevation where the forest floor is innundated for several months, baldcypress (*Taxodium distichium* [L.] Rich.) and water tupelo (*Nyssa aquatica* L.) dominate while areas of higher elevation and better drainage (but still having moist soil) are dominated by bottomland hardwood species, such as red maple (*Acer rubrum* var. *drummondii* [H. & A.] Sarg.), water tupelo, boxelder (*Acer negundo* L.), and cottonwood (*Populus heterophylla* L.).

Bayou Chevreuil and Bayou Boeuf are the main streams draining into Lac des Allemands. The bayous and adjoining canals are sluggish (average velocity 14 $cm \cdot sec^{-1}$), turbid (average secchi disk depth 30 cm), acidic streams ranging from 1 to 2 m in depth and 10 to 70 m in width. The lake is a large (65 km^2), slightly alkaline, shallow (average depth 2.1 m), turbid (average secchi disk depth 41 cm), eutrophic body of water. The lake supports a commercial catfish industry (annual catch $>10^6$ kg), the upper bayou and swamp area supports a commercial crawfish industry, and the fresh marshes support a commercial fur industry. The whole area is used heavily by sports fishermen and hunters.

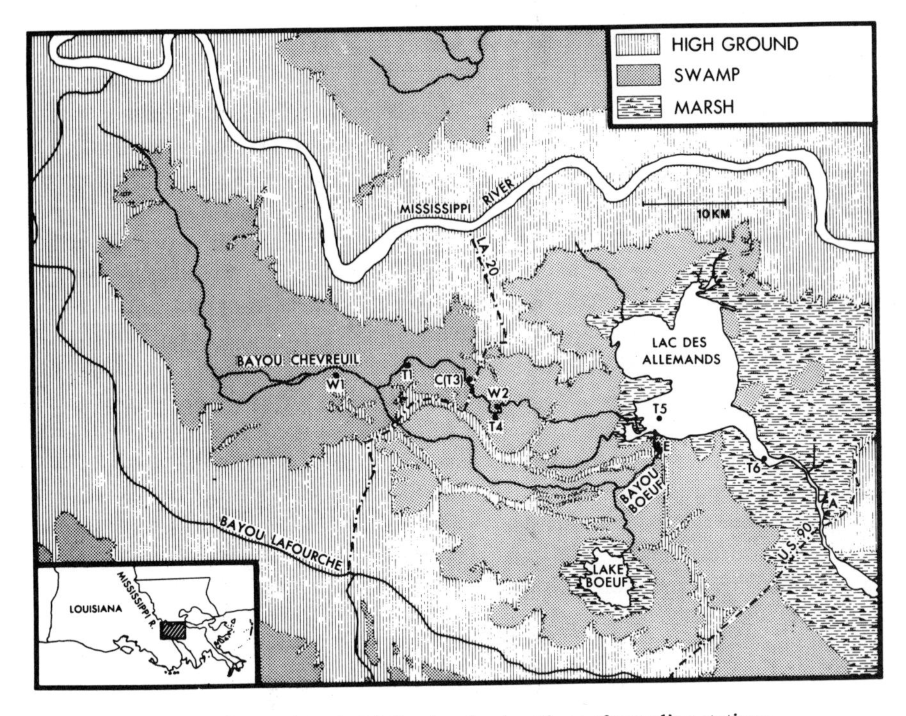

Figure 1. Lac des Allemands and vicinity showing locations of sampling stations.

MATERIALS AND METHODS

Productivity

Terrestrial Productivity

Litterfall, annual stem growth, and understory vegetation were measured at two 0.1-ha sites within the swamp basin. One site was in a typical bottomland hardwood (BLH) forest (W1, Fig. 1), and the other was in a typical cypress tupelo (CT) forest (W2).

Diameter at breast height (DBH) measurements of all the trees in each site and regression equations (23, 30) were used to calculate stem biomasses at two sampling dates. CT was measured April 20, 1974 and November 16, 1974 (210 days) and BLH was measured October 27, 1973 and November 16, 1974 (378 days). The increase between the two dates was considered annual stem growth (6). Biomass of the herbaceous understory (all new growth) was measured by harvesting four 2 × 2 m plots in each site during the first week in April. Litterfall was collected every 2 to 6 weeks from October 9, 1973 to October 12, 1974 in ten 1-m^2 litter traps randomly placed in each site. Net primary production was calculated by summing the three measured values.

Aquatic Primary Productivity

Aquatic primary productivity was measured monthly at four stations (T1, T3, T4, and T6, Fig. 1) using the diurnal oxygen curve method (24). Other parameters measured were secchi disk depth, water temperature, and weather conditions. Diffusion measurements taken at the bayou stations (T1, T3, T4) with a floating plastic dome (7) showed that diffusion was not significant when compared to biological changes ($K \leqslant 0.01\ mg \cdot m^{-3} \cdot hr^{-1}$ at 100% saturation deficit). Waves prevented the use of the dome on the lake so estimates from the literature were used (13, 14, 24, 29, 31). Estimated diffusion constants ranged from 0.1 to 0.3 mg $O_2 \cdot m^{-2} \cdot hr^{-1}$ over a 24-hr sampling period. Copeland and Duffer (7), Owens (27), and Day (8) also used variable diffusion constants for open waters, where conditions varied over the day. Since lake water was unsaturated in almost all cases, diffusion tended to lower the amount of net daytime photosynthesis (NDP) by about 10% and increase the amount of nighttime respiration (NR) by about 15%. Details of productivity calculations are included in Butler (4). Chlorophyll *a* levels were determined by using a modification of the method of Strickland and Parsons (32). Glass fiber filters used in the analysis were sonicated in 90% acetone before extraction. The extraction process took place in the dark at 0° C for 3 hours.

Hydrology and Materials Flow

Measurements of water discharge and materials concentrations were taken monthly at stations A, E, and C (T3). Because of high ground and highway embankments, water in the basin must flow past these points. Each month water samples were taken and analyzed for nitrogen and phosphorus according to the methods of Strickland and Parsons (32) and for total and dissolved organic carbon using an Oceanographic International Total Carbon System.

Total water flow at Stations A, C, and E were calculated using: a) the monthly measurements of water flow, b) daily water level readings obtained from U.S. Army Corps of Engineers (New Orleans District) gauges at Cheby, Louisiana (Station C) and des Allemands, Louisiana (Station A), and c) a hydrological model of the des Allemands basin (21). The model incorporated precipitation, evapotranspiration, and physiographic data to calculate annual discharge from bayous Boeuf, Chevreuil, and des Allemands. Because this model takes factors such as climate and hydrologic drag into consideration, it is the most accurate representation of water flows available for the basin.

The monthly current measurements were taken at the center of the channel in the surface waters using a current meter, current cross, or neutral density float. Therefore these measurements give relative flow rates. These rates along with gauge data were used to apportion the annual discharge from the model into monthly averages. Materials export was calculated by multiplying water discharge by monthly materials concentration.

RESULTS AND DISCUSSION

Terrestrial Productivity

Measured swamp forest productivity values are given in Table 1. Values for net productivity calculated from annual stem biomass increase, litterfall (Fig. 2), and measured understory biomass are minimum estimates since they do not take understory turnover or insect consumption into account.

Table 1. Productivity values for Lac des Allemands, Louisiana and surrounding swamp and bayous.

Terrestrial Productivity (g dry wt$\cdot$m$^{-2}\cdot$yr^{-1})	W1	W2
1) Stem biomass increase	800	500
2) Leaf-litter	574	620
3) Est. understory productivity	200 ± 60	20 ± 4
Total net productivity	1574	1140
Aquatic Productivity (g $O_2\cdot$m$^{-2}\cdot$yr^{-1})	**Bayou**	**Lake**
Measured		
1) Net daytime photosynthesis	296 - 523	1418
2) Nighttime respiration	433 - 623	1868
Estimated		
Gross production	762	3286
Total respiration	992	3736

Aquatic Productivity

Lac des Allemands is very productive throughout the year while the bayous are less so (Fig. 3, Table 1). Seasonally the lake is most productive from April to September, a period of extensive blue-green algal blooms (20, 10). Winter production is significant with 34% of annual NDP and 29% of NR occurring from November through March. Chlorophyll *a* concentrations averaged 49.1 mg$\cdot$m^{-3} in summer and 11.9 mg$\cdot$m^{-3} in winter.

Gross production at the bayou stations was 27% of that in the lake, probably as a result of shading from the surrounding forest and high turbidity from silt and clay particles (average secchi disk depth was 0.3 m as compared to 0.4 m for lake). This suggests that nutrients are not limiting in the bayous since concentrations of phosphorus, nitrogen, and carbon were slightly higher than in the lake (Table 2). Odum and Wilson (25) found light limiting in the case of periodically turbid Texas bays. There was no seasonal trend evident in the bayous. This, however, may be an artifact of the sampling technique. During the summer months a noticeably visible bloom of blue-green algae occurred over much of the bayou surface area in the top 5 cm. Oxygen data were collected 10-15 cm below the water surface. Since very little mixing occurs under the low flow conditions at this time, a significant amount of photosynthesis and respiration was probably

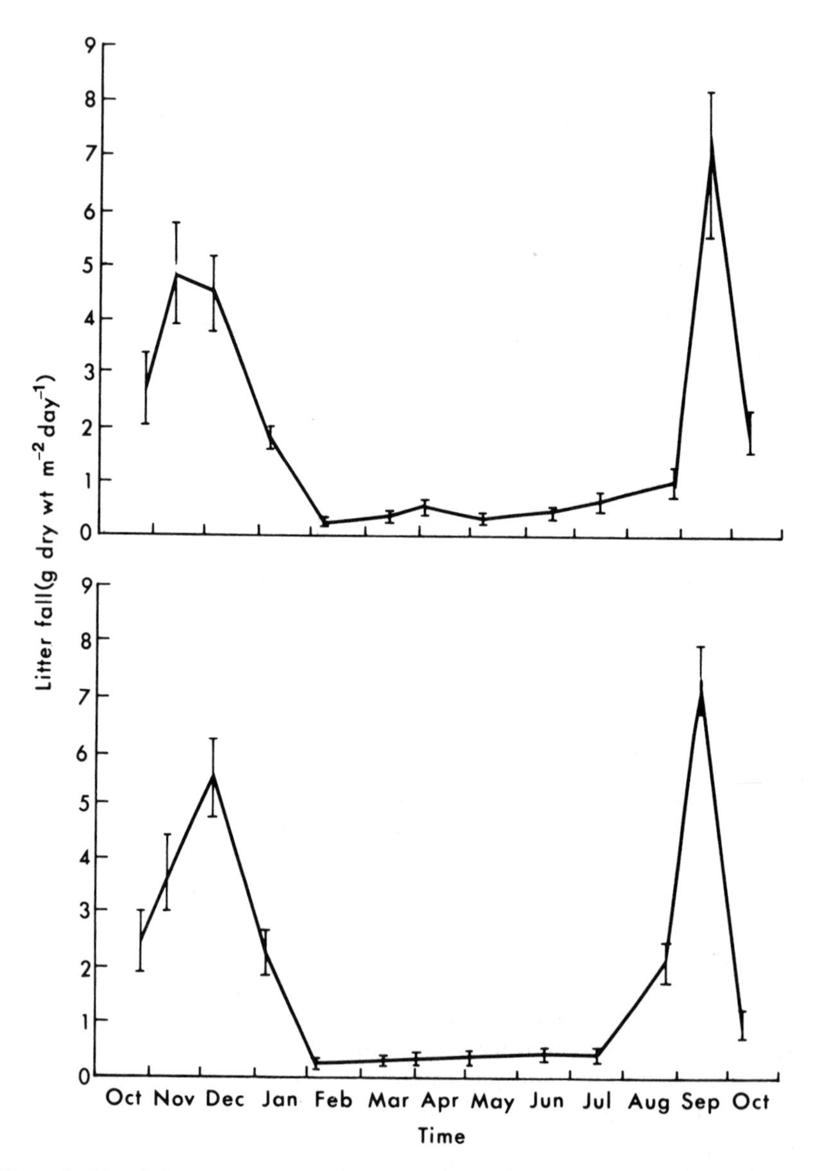

Figure 2. Litterfall at the bottomland hardwood (upper) and cypress-tupelo sites (lower). Vertical bars are ±2 S.E.

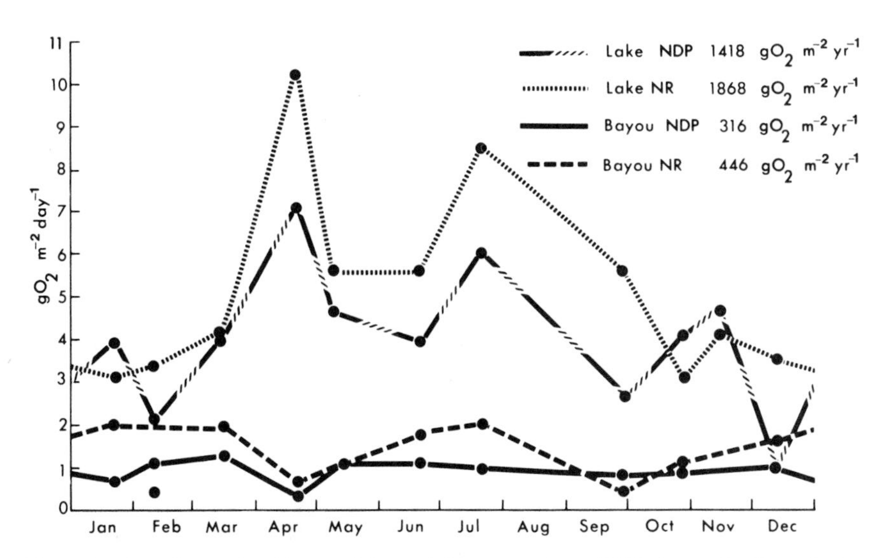

Figure 3. Aquatic productivity at lake and bayou stations. NDP is net daytime photosynthesis and NR is nighttime respiration.

unaccounted for. In addition, floating and emergent vegetation, which is abundant along the bayous, was not measured. Chlorophyll *a* data seem to indicate some seasonal trend in productivity (4). Average summer value was 21.6 mg$\cdot l^{-1}$; that for winter was 5.6 mg$\cdot l^{-1}$.

Both lake and bayou stations were heterotrophic. The dominant producers were the cyanophytes, which often have respiratory demands that exceed their photosynthesis (5), and so the major primary producers may be heterotrophic in themselves. The main source of lake water is bayou and swamp water containing high organic carbon levels, which act as an additional source of oxygen demand. These factors account for similar P/R ratios (0.76 for lake; 0.71 for bayous) but different productivities between bayou and lake.

Table 2. Annual average concentrations (±S.E.) for various chemical constituents at bayou and lake stations (in mg$\cdot l^{-1}$).

	Station A	Station C	Station E
TOC	11.6 ± 0.6	14.1 ± 1.4	12.9 ± 1.3
DOC	10.2 ± 0.5	12.2 ± 1.1	11.2 ± 0.9
Total P	0.27 ± 0.06	0.34 ± 0.07	0.20 ± 0.03
PO_4–P	0.08 ± 0.02	0.15 ± 0.02	0.12 ± 0.04
Total N	1.60 ± 0.16	2.13 ± 0.42	1.79 ± 0.35
(NO_2+NO_3)-N	0.24 ± 0.08	0.28 ± 0.11	0.14 ± 0.05
NH_4-N	0.16 ± 0.02	0.25 ± 0.03	0.19 ± 0.04
Organic-N	1.35	1.63	1.36

Hydrology

Water discharge was characterized by high rates during the spring and fall and extremely low rates during the summer (Fig. 4; [20]). The fall peak in September was the result of heavy rains accompanying the passage of Hurricane Carmen early in the month.

Extremely low discharges in the months of July and August are the result of two different phenomena. Evapotranspiration normally exceeds rainfall during the summer, so swamp drainage, the major hydrological input to the lake and bayous, is sharply reduced. Furthermore, during the summer prevailing southeasterly winds back up water in Bayou des Allemands, normally reversing the flow during the day. At night the flow is typically sluggish in a gulfward direction.

Nutrient Cycles and Materials Budgets

Bayou waters in the study area are rich in nitrogen and phosphorus (Table 2). Nutrients derived from terrestrial sources such as agriculture are a significant source for much of these nutrients. As the waters flood over the swamp floor, nutrients can be taken up by plants and incorporated into detritus. Later, some nutrients are transported downstream to the lake and marsh systems through the aquatic food chain due to direct grazing or through detrital export. Kitchens et al. (19) found this to be the case in a riverine swamp in South Carolina. The abundance of lichens, blue-green algae, and other nitrogen fixers probably

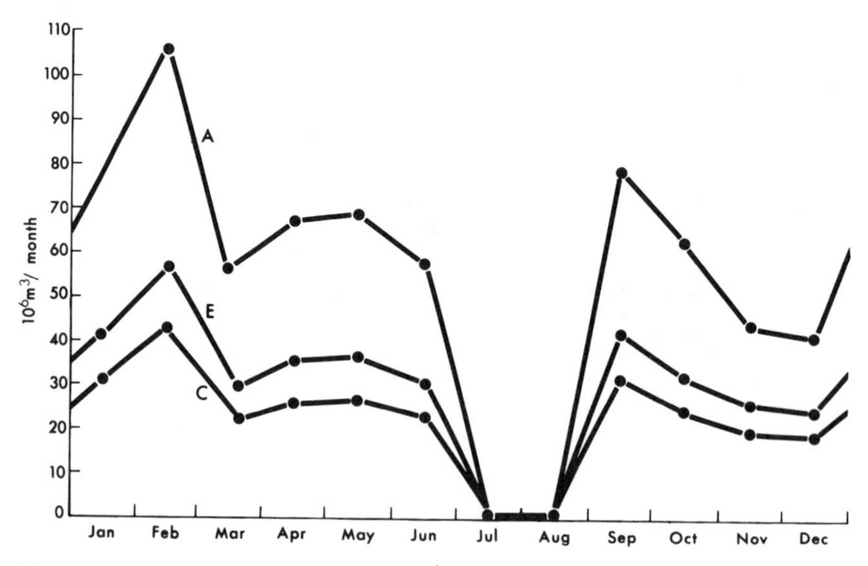

Figure 4. Monthly mean water discharge at bayous Chevreuil (C), Boeuf (E), and des Allemands (A).

represents an important pathway of nitrogen input to the swamp system. We have no quantitative estimates of this pathway yet.

Materials flow into and out of Lac des Allemands and are summarized in Figure 5. Details of these calculations are given by Butler (4). In the lake the carbon input from alloctonous sources plus the carbon fixed in the lake was nearly equal (within 10%) to that consumed, harvested, and exported. In calculating input to the lake, we took into consideration the fact that flows from bayous Chevreuil and Boeuf do not include input from areas immediately to the northwest, east, and southeast of the lake (see Fig. 1). We estimated input from these areas by assuming that it was the same (on an aerial basis) as the wetlands drained by the two bayous. Import of total nitrogen into the lake was 1.9 times greater than export while import of phosphorus was 1.8 times greater than export (Fig. 5). Activity in the sediment (nitrate stored and/or denitrified and phosphorus absorbed to the clay sediments) probably account for these differences. Engler and Patrick (12) have demonstrated that microbial NO^{-3} denitrification is a significant means of nitrate removal in submerged sediments. Patrick and Khalid (28) also found anaerobic soils more capable than aerobic soils of absorbing phosphorus at high levels of soluble phosphorus. Lac des Allemands has both high levels of phosphorus and anaerobic sediments a few centimeters from the surface. Other authors have found lakes to act as phosphorus sinks (13, 17, 22).

Therefore, the sediments probably serve as a eutrophication buffer by removal or storage of large amounts of phosphorus and nitrogen. This buffer

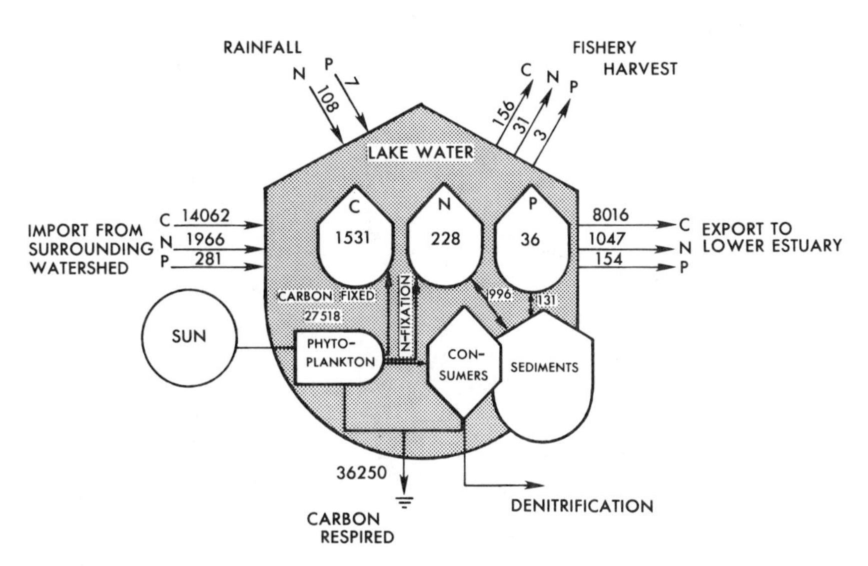

Figure 5. Summary budget of material flows in the Lac des Allemands system. Units are metric tons per year.

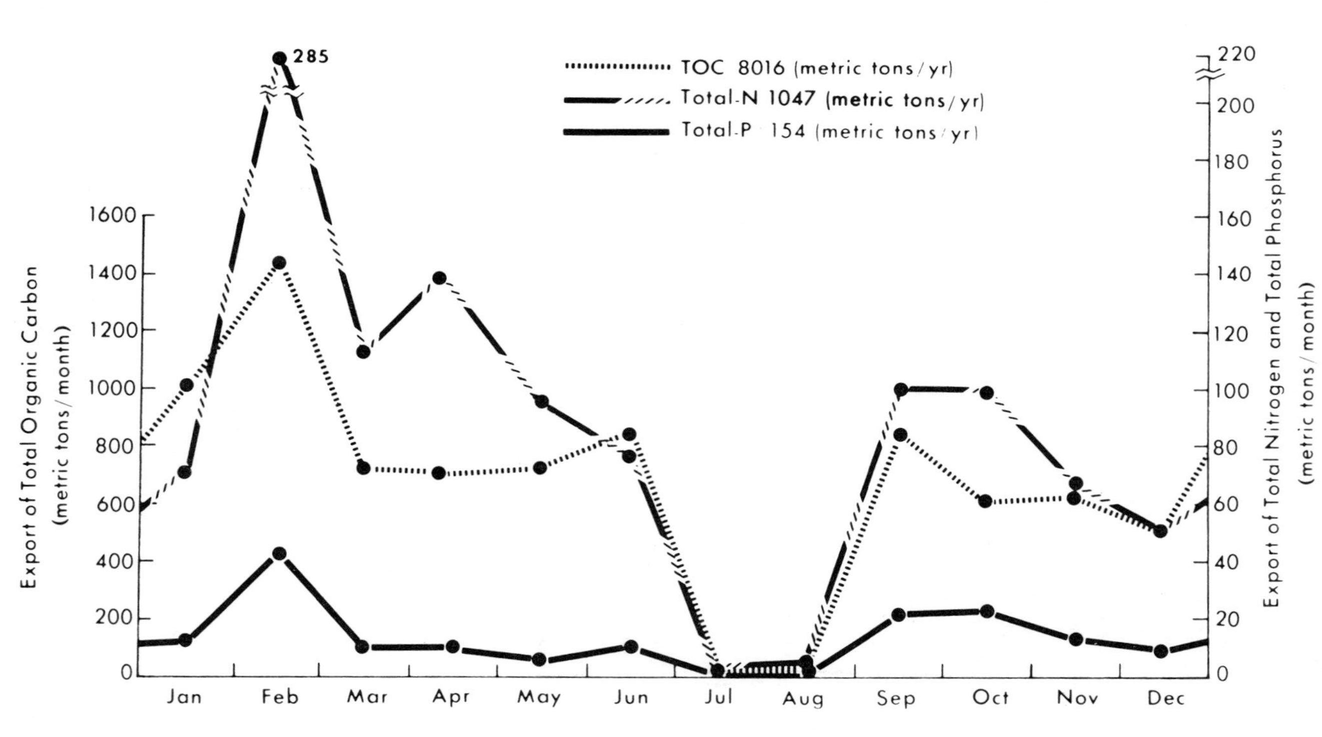

Figure 6. Materials export to the lower estuary measured at Bayou des Allemands (A).

activity slows the rate of eutrophication which is occurring in the lake. The importance of this may be in preventing even more extensive algal blooms that could result in anoxic lake conditions and major fish kills.

Export of materials from the wetland watershed into the lake (calculated by dividing total input by total watershed area) equals 19.3 g C m^{-2} of wetland yr^{-1}, 2.7 g N m^{-2} yr^{-1} and 0.39 g P m^{-2} yr^{-1}. These figures are not accurate for several reasons. Some of the input to the lake is from agricultural runoff. Parts of the swamp are isolated by low levees constructed by timber companies. Some carbon is probably recycled within the bayous. It is impossible to estimate the relative importance of these factors at this time.

Export of carbon, nitrogen, and phosphorus out of the lake (Fig. 6) represents an important source of raw materials to the lower estuary. There is a general trend of decreasing concentrations of total nitrogen, phosphorus, and organic carbon toward the brackish and saline marsh areas and the Gulf (1, 4, 15). This indicates that materials supplied by the upper watershed are being consumed by the lower system. The des Allemands system also serves as a major freshwater reservoir for maintenance of favorable salinities in the brackish and saline zones. The upper area thus contributes to the productivity of the lower estuary. The major pulse of materials to the lower system begins in February (Fig. 6) and coincides with the time of high detrital formation and the arrival of migrant species which enter the estuarine area for growth and spawning purposes (9). Another advantage of swamp detrital export in spring is that it allows for over winter enrichment of detritus by microbial action on the swamp floor. This produces a higher quality food source for the detrital-based food chains both in the swamp and the area below it. Enrichment of detritus over time has been well demonstrated (16, 18, 26).

The entire estuary from the headwater swamp to the open Gulf is linked by the water flow through it. Water flows represent an important energy source driving the system's productivity. The continual presence of water means that evapotranspiration can approach the theoretical maximum for emergent plants. Water flows serve to remove wastes, process them for recycling, supply nutrients and organic food, and aid migrating species. Maintenance of the natural hydrology which has developed this extremely rich system is thus an integral part in maintaining the high productivity of the entire wetlands.

Eutrophication

The des Allemands lake and bayou study area is a highly eutrophic aquatic environment characterized by high productivity and nutrient concentrations, and blue-green algal blooms (Tables 1, 2). The high level of primary productivity in the lake is primarily the result of large inputs of raw materials for production, mainly from the surrounding wetlands. Rainfall directly contributes a minimum of 2.4% of the total phosphorus and 5% of the total nitrogen entering the lake each year (see Fig. 5). Ten to fifteen years ago, Lac des Allemands was a clear, brown water, swamp lake (2). The increasing eutrophication of the lake seems to be caused by addition of nutrients and high BOD material due to man's activities. There is evidence that a shift over the last 10 years to larger and heavier machines for cultivation has led to increased runoff of topsoil from agricultural lands (2). Within the swamp, ongoing dredging has increased turbidity in the bayous which are a major source of lake water. Industrial phosphate waste (from the Hooker Chemical Plant) also enters the lake (20). The additional nutrient loading has resulted in the increasing eutrophication over the last 10 years of an already eutrophic lake, extending both the quantity and duration of blue-green algal blooms (2).

Cultural eutrophication appears to be occurring in a system designed to handle naturally high nutrient loads. Overloading of such systems results in destablization of the community. Fish kills and massive water hyacinth blooms are examples of excessive oscillations in the system. What long-term effects this might have are not known, but it is certain that man's activities are at least partly responsible for the eutrophication of Lac des Allemands. Here again a value judgement of whether this is good or bad is difficult to state. Fishery production of catfish has not declined over the last several years, nor has it improved (20). The recreational value of the lake for swimming and other water sports may have declined with increasing algal blooms. Aesthetically, dark "coffee colored" lakes (which was a description of Lac des Allemands 10-15 years ago) seem to be more appealing than the present more turbid condition. However, eutrophication may have increased the role of this area as a nutrient input to the lower estuary (Fig. 5).

Possible Factors Limiting Productivity

It is interesting to speculate as to what may limit productivity in the lake. Phosphorus is generally considered the limiting nutrient in algal growth in fresh-water lakes (11, 22). Lakes in Florida of similar morphology and climatic conditions are phosphorus limited (13), and there is reason to believe that on a time scale of months or years phosphorus in Lac des Allemands is the major limiting nutrient. The dominant phytoplankton are capable of fixing atmospheric nitrogen and so do not appear to be limited by that nutrient. On a shorter time scale of hours carbon may be limiting (5). From April through August, the period of most phytoplankton productivity, CO_2 is virtually absent from the water column during the day (20) indicating rapid uptake of respiratory CO_2 by the blue-greens, which are very efficient CO_2 users (22). The rate of CO_2 diffusion from the atmosphere would then limit production. Day (8) found this to be the case for heavy blooms in North Carolina brackish ponds receiving treated sewage wastes.

Hurricanes

Since hurricanes have been a factor in Louisiana coastal systems for thousands of years, it is not unreasonable to expect that they might play an ecological role in the functioning of these systems. In September of 1974, Hurricane Carmen passed approximately 75 km west of the study area bringing strong winds and torrential rains. Peak litterfall in the swamp forest was 2 months earlier than the normal mid-November period (Fig. 2) and a large pulse of organic carbon, nitrogen, and phosphorus was flushed from the area to the lower estuary after the hurricane (Fig. 6). Since detritus becomes a more nutritive food as it ages (16, 18, 26), the additional time the leaf material undergoes decomposition may mean a higher quality food. Hurricanes also may facilitate flushing of stored material which would not normally wash out, as indicated by the large export after the hurricane.

ACKNOWLEDGMENTS

This study was supported by the Louisiana Sea Grant Program, a part of the National Sea Grant Program maintained by the National Oceanic and Atmospheric Administration of the Department of Commerce. Support was also provided by Louisiana Offshore Oil Port, Inc. Appreciation is extended to the Miles Timber Company and the Bowie Lumber Company for providing information and access to the study sites and to Mr. Harvey Rodriquez, Jr., for field support.

REFERENCES

1. Allen, R.L. 1975. Aquatic primary productivity in various marsh environments in Louisiana. Unpubl. M.S. Thesis, La. State Univ., Baton Rouge. 50 p.
2. Bradley, T. Personal Communication. La. Wildlife and Fisheries Commission, New Orleans.
3. Brown, C.A. 1972. Wildflowers of Louisiana and Adjoining States. La. State Univ. Press, Baton Rouge. 247 p.
4. Butler, T.J. 1975. Aquatic metabolism and nutrient flux in a southern Louisiana swamp and lake system. M.S. Thesis, La. State Univ., Baton Rouge. 58 p.
5. Cole, G.A. 1975. Textbook of Limnology. The C.V. Mosly Co., St. Louis. 283 p.
6. Conner, W.H. 1975. Productivity and floral composition of a freshwater swamp in Louisiana. M.S. Thesis, La. State Univ., Baton Rouge. 85 p.
7. Copeland, B.J., and W.R. Duffer. 1964. Use of a clear pastic dome to measure gaseous diffusion rates in natural waters. Limnol. Oceanogr. 9(4):494-499.
8. Day, J.W., Jr. 1971. Carbon metabolism of estuarine ponds receiving treated sewage wastes. Ph.D. Dissertation, Univ. of North Carolina, Chapel Hill. 127 p.
9. ____, W.B. Smith, P. Wagner, and W. Stowe. 1973. Community structure and carbon budget in a salt marsh and shallow bay estuarine system in Louisiana. Center for Wetland Resources, La. State Univ., Baton Rouge. Publ. No. LSU-SG-72-04. 79 p.
10. ____, T. Butler, R. Allen, J.G. Gosselink, and W.C. Stowe. 1976. Flora and community metabolism of aquatic systems within the Louisiana wetlands. *In* J.G. Gosselink (ed.), Environmental Assessment of a Louisiana Offshore

Oil Port and Appertinent Pipeline and Storage Facilities. Appendix VI.4, Vol. II. Center for Wetland Resources, La. State Univ., Baton Rouge.
11. Edmondson, W.T. 1970. Phosphorus, nitrogen, and algae in Lake Washington after diversion of sewage. Science 169:690-691.
12. Engler, R.M., and W.H. Patrick. 1974. Nitrate removal from flood waters overlying flooded soils and sediments. J. Environ. Qual. 3:409-413.
13. Gayle, T. 1975. The Lake Okeechobee ecosystem and its models. M.S. Thesis, Univ. Florida, Gainesville. 65 p.
14. Haney, P.D. 1954. Theoretical principles of aeration. J. Amer. Wat. Wks. Assn. 46:355-376.
15. Happ, G. 1974. The distribution and seasonal concentration of organic carbon in a Louisiana estuary. M.S. Thesis, La. State Univ., Baton Rouge. 36 p.
16. Heald, E.J. 1971. The production of organic detritus in a south Florida estuary. Sea Grant Tech. Bull. No. 6, Univ. of Miami. 110 p.
17. Hutchinson, G.E. 1957. A treatise on limnology. Vol. I. Geography, Physics, and Chemistry. John Wiley and Sons, Inc., New york. 1015 p.
18. Kirby, C. 1971. The annual net primary production and decomposition of the salt marsh grass *Spartina alterniflora* Loisel in the Barataria Bay estuary of Louisiana. Ph.D. Dissertation, La. State Univ., Baton Rouge. 74 p.
19. Kitchens, W.M., Jr., J.M. Dean, L.H. Stevenson, and J.H. Cooper. 1974. The Santee Swamp as a nutrient "sink". *In* F. Howell, J.B. Gentry, and M. Smith (eds.), Proceedings of the Savannah River Ecology Laboratory Symposium on Mineral Cycling in the Southeastern Ecosystem. May 1-3, 1974.
20. Lantz, K. 1970. An ecological survey of factors affecting fish production in a Louisiana natural lake and river. La. Wildlife and Fisheries Comm. Bull. No. 6. 92 p.
21. Light, P., R.J. Shelmon, P.T. Culley, and N.A. Roques. 1973. Hydrologic models for the Barataria-Terrebonne area of south central Louisiana. Rept. No. 16, Hydrologic and Geologic Studies of Coastal Louisiana. Center for Wetland Resources, La. State Univ., Baton Rouge. 43 p.
22. Megard, R.O. 1972. Phytoplankton, photosynthesis, and phosphorus in Lake Minnetonka, Minnesota. Limnol. Oceanogr. 17:68-87.
23. Monk, C.D., G.I. Child, and S.A. Nicholson. 1970. Biomass, litter and leaf surface area estimates of an oak-hickory forest. Oikos 21:138-141.
24. Odum, H.T. 1956. Primary production in flowing waters. Limnol. Oceanogr. 1:102-117.
25. ——, and R.F. Wilson. 1962. Further studies on reaeration and metabolism of Texas bays, 1958-1960. Pub. Inst. Mar. Sci., Univ. of Texas 8:23-55.
26. Odum, E.P., and A.A. de la Cruz. 1967. Particulate organic detritus in a Georgia salt marsh-estuarine ecosystem, p. 383-388. *In* G.H. Lauff (ed.), Estuaries, AAAS Publ. No. 83.
27. Owens, M. 1969. Some factors involved in the use of dissolved oxygen distributions in streams to determine productivity, p. 209-224. *In* C.R. Goldman (ed.), Primary Productivity in Aquatic Environments. Univ. of California Press, Berkeley.
28. Patrick, W., Jr., and R. Khalid. 1974. Phosphate release and absorption by soils and sediments: Effect of aerobic and anaerobic conditions. Science 186:53-55.
29. Phelps, E.B. 1944. Stream Sanitation. John Wiley and Sons, New York. 276 p.

30. Schlessinger, W. Pers. Commun. Ecology and Systematics, Cornell Univ., Ithaca, New York.
31. Streeter, H.W., C.T. Wright, and R.W. Kehr. 1936. Measures of oxidation in polluted streams. III. An experimental study of atmospheric reaeration under stream flow conditions. J. Sewage. Wks. 8:282-316.
32. Strickland, J.D.H., and T.R. Parsons. 1968. A practical handbook of seawater analysis. Fish. Res. Bd. of Canada Bull. 167. 311 p.

FLUX OF ORGANIC MATTER THROUGH A SALT MARSH[1]

L. R. Pomeroy, Keith Bancroft, John Breed,
R. R. Christian, Dirk Frankenberg[2], J. R. Hall,
L. G. Maurer[3], W. J. Wiebe, R. G. Wiegert, and
R. L. Wetzel[4]
Institute of Ecology, Marine Institute, Department
of Microbiology, and Department of Zoology, The
University of Georgia, Athens, Georgia 30602

ABSTRACT: To refine our knowledge of the functional aspects of salt marshes, we have undertaken a study of the transfers of organic matter from primary producers through consumers to CO_2 in terms of quality and quantity of carbon compounds. In addition to production of particulate material, we find sources of soluble carbon compounds which may be significant. These include losses from both living and dead macrophytes, primarily *Spartina*, from algae in water and sediments, from excretion and feces of consumers, and from biological processes in the sediments. Since *Spartina* detritus is relatively indigestible, much of the flux of organic matter to detritovores must involve conversion of particulate detritus to soluble compounds and their assimilation by microorganisms, which can then be consumed by detritovores. Some dissolved material accumulates as a film on the surface of the water and is formed into organic aggregates. Several lines of evidence suggest that microorganisms in the water actively assimilate dissolved organic material during the growing season. Microorganisms in the sediments, although they reside in a large pool of organic matter, appear to be substrate limited except near the sediment-water surface.

[1] This work has been supported by grant DES72-01605-A02 from the National Science Foundation and by grants from the Sapelo Island Research Foundation. Contribution No. 298 from The University of Georgia Marine Institute.
[2] Present address: Curriculum in Marine Science, University of North Carolina, Chapel Hill, N. C.
[3] Present address: Environmental Quality Lab, General Development Engineering Company, Port Charlotte, Florida.
[4] Present address: Virginia Institute of Marine Science, Gloucester Point, Virginia.

INTRODUCTION

Studies of the energetics of the salt marsh ecosystem conducted in the 1950's and 1960's showed that it is dominated by a detritus food web. Herbivores that directly utilize the dominant plants for food appear to be a relatively minor component. The major macroscopic consumers are detritovores, a diverse group of organisms that use various feeding mechanisms to concentrate from the water or the sediments the remains of decaying vegetation, fecal pellets, and associated microorganisms that we collectively call organic detritus (8, 18, 25, 27). We think of the detritus food web in those terms, but in fact that is only part of it. The microbial portion of the food web, which involves detritus production and other transformations of organic matter, is more complex and consumes more energy and materials than the better known portion involving macroheterotrophs. Relatively little is known about microbial involvement in detritus production and consumption, and even less is known about the associated transformations of carbon compounds which must occur as *Spartina* and other primary plant materials are converted to forms that are utilized by macrodetritovores.

We have undertaken a study of the food web that emphasizes those areas of least knowledge, the transformations of organic matter and the microbial roles in those transformations. Integrated with this are studies of primary production and of the assimilation of materials by detritovores. This research, now in its third year, is concerned with the invisible but dominant microbial and biochemical aspects of the food web. Modeling of carbon flux through the system was discussed at the previous ERF conference (33).

PRODUCTION OF PARTICULATE AND SOLUBLE ORGANIC MATTER

Most research on detritus food webs has focused on production of particulate organic matter, but we also consider a number of sources of soluble organic matter in the salt marsh. Probably much of the flux of materials from producers to consumers involves one or more soluble stages. This is difficult to quantify, because there is tight coupling of the producers and consumers where soluble materials are concerned to prevent loss and dilution by the water. The fluxes of soluble material we have been able to measure are those that do leak past the first line of consumers. Therefore, they are of rather small magnitude and probably do not represent even approximately the total flux of soluble organic carbon.

Gallagher et al. (9) showed that living *Spartina* loses a few percent of the photosynthate it produces when it is washed by the rising tide. Thomas (28) found that phytoplankton in the Georgia estuaries lose 1-6% of their photosynthate as soluble material. Comparable measurements have not yet been made on the benthic algal populations in the Georgia marshes, but we might expect a similar proportionate loss. In the conversion of *Spartina* to detritus there is a significant loss of dissolved organic carbon. Gallagher and Pfeiffer (ms) found that the release of soluble organic carbon from the standing dead *Spartina* and its associated community of microorganisms and insects is twice that released from living *Spartina* per unit dry weight. Thus, with every inundation of the marsh by the tide, some soluble organic matter is washed out of both the living and the standing dead *Spartina*. There is reason to believe that this is relatively labile material that is consumed very rapidly.

There is additional loss of dissolved organic carbon by excretion from macroheterotrophs (14) and during conversion of organic compounds by microheterotrophs in the water and in the soils. Degradation of *Spartina* is an example of the latter, which is also evident in the release of dissolved organic carbon from the surface of estuarine soils[5]. We report here some measurements of that process made by John R. Hall.

Cores in sets of four were taken from the salt marsh and placed in a darkened incubator. One-hundred fifty ml of estuarine water was run gently over the surface of each core, and initial 5 ml water samples were taken from each. Filtered air was bubbled gently through the water standing over the cores. Five ml samples were withdrawn at 3, 6 and 12 hrs. Samples were filtered through glass fiber filters, and the dissolved organic carbon was measured by the method of Menzel and Vaccaro (17). Mean release rate is 6.4 mg C m^{-2} hr^{-1} on the creek banks and 5.2 on the short-*Spartina* marsh. There are few significant differences between monthly observations, and the cumulative differences between short-*Spartina* marsh and creek bank are not significantly different. There does seem to be a tendency for increased release of dissolved carbon in the summer, but it is obscured by increased variability (Table 1).

MICROBIAL LINKS IN THE FOOD WEB

The major sink for soluble organic matter appears to be microbial utilization. Macroheterotrophs, such as oysters or zooplankton may utilize soluble organic matter, but large organisms are at a distinct disadvantage when they compete with microorganisms whose high surface to volume ratio makes for rapid and efficient uptake of materials from solution. Probably most macroorganisms release more soluble organic materials in waste products than they take in from the water (13, 14). Microorganisms, however, will take dissolved materials from the water down to microgram concentrations per liter. Moreover, populations of microorganisms will develop at sites of release of dissolved organic matter, so that much uptake is closely coupled with release. During the growing season, microorganisms on the leaves of *Spartina*, as well as in the estuarine water,

[5] In this paper we refer to the areas rooted by *Spartina* and other higher plants in the intertidal zone as *soils*, since modification of sediments by the plants has produced recognizable soil profiles analogous to those of terrestrial soils. Subtidal and lower intertidal sediments, unmodified by rooted plants we call *sediments*.

Table 1. Release of dissolved organic carbon from the surface of cores of sediment collected in the salt marsh.

Date	Temp. °C	level marsh			creek bank		
		mg C m^{-2} hr^{-1}	s. d.	n	mg C m^{-2} hr^{-1}	s. d.	n
VI 19 73	28				13.8	1.2	3
VII 24 73	28				8.8	1.3	4
VIII 8 73	26	3.1	3.4	4			
XI 5 73	22				19.4	4.1	3
XII 4-5 73	17	3.5	1.0	4	3.7	1.7	4
I 15-16 74	14	3.5	1.6	4	2.6	1.9	4
II 6-10 74	16	3.6	0.7	3	1.4	0.8	4
III 18-19 74	17	1.5	0.1	4	5.1	2.0	3
IV 16-17 74	22	4.0	0.8	3	3.9	2.5	4
V 29-30 74	25	3.1	1.0	4	4.0	1.6	4
VII 9-12 74	28	10.9	4.5	4	5.2	2.6	4
VII 13 74	28	5.8	2.0	4	5.8	2.0	4
VII 29 74	29	14.8	3.6	4			
IX 13 74	28	4.0	0.4	4	3.0	0.8	4

rapidly utilize soluble organic matter as it is lost from the plants (9). Similar phenomena have been observed in both marine and fresh water (38).

In aerobic environments heterotrophic bacteria and fungi appear to contain a large number of generalists with respect to their potential carbon and energy sources (5, 31). The three major habitats for aerobic heterotrophs are *Spartina*, water, and soils. Active populations are found on both living and dead tissue of *Spartina* (Gallagher and Pfeiffer, ms). The water column contains microorganisms free in the water and attached to detritus, although it is difficult to separate these groups experimentally or observationally (32). In the soils there are two aerobic environments, the surface layers and the rhizosphere. Of all these environments, the rhizosphere is the least known at the present time. These latter two habitats contain aerobic organisms which cannot be considered generalists. Due to their proximity to the anaerobic environment several specialist groups have been found. These include methane oxidizing bacteria and *Beggiatoa.*

Anaerobic microorganisms exist primarily within the soils of the salt marsh. They are best differentiated experimentally on the basis of their specialized biochemistry (Fig. 1). Fermenting (4, 36) and dissimilatory nitrogenous oxide reducing (19) bacteria often are generalized with respect to their carbon sources. Their metabolic products are biomass, carbon dioxide, and low molecular weight fermentation products. Dissimilatory sulfate reducing bacteria require specific low molecular weight compounds, such as lactate (22). They produce biomass, carbon dioxide, and soluble end products such as acetate. Methanogenic bacteria also use a restricted group of compounds: carbon dioxide, formate, acetate, and methanol (35). Their unique end product is methane.

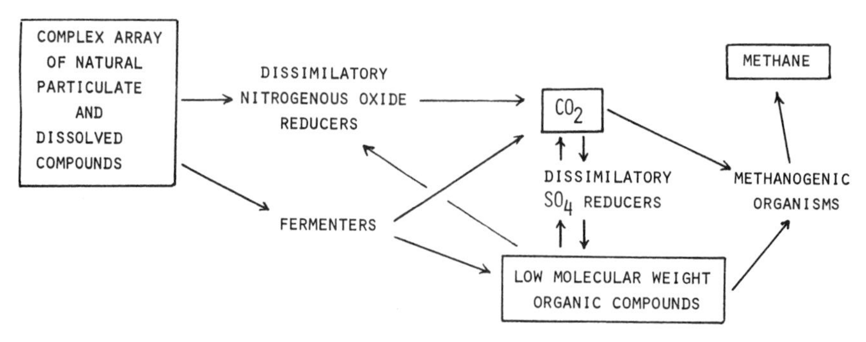

Figure 1. Pathways of anaerobic sediment community metabolism.

Heterotrophic aerobic microbial metabolism is primarily a transformation of particulate or dissolved organic matter to carbon dioxide and biomass. Anaerobic metabolism is far more complex. Intermediate dissolved and gaseous carbon compounds occur at several trophic levels of the microbial food web. All the products of anaerobic metabolism, biomass, carbon dioxide, methane, and soluble low molecular weight compounds may be used by aerobic organisms. In a compartmentalized, and in fact somewhat layered system, such as a salt marsh, microbial food webs of some complexity may exist within a small and almost entirely internalized part of the system, such as a dead *Spartina* stalk or a few cm of soil.

Much of our knowledge of the trophic dynamics of microorganisms is based on laboratory work with pure cultures. This can be translated qualitatively to field conditions by finding the organisms, their metabolic end products, and suitable substrate for growth. To evaluate quantitatively their roles in the salt marsh is a much more difficult undertaking. We have sought more generalized assays of microbial biomass and growth state that would be useful in evaluating the microbial activity in natural waters and salt marsh soils.

Initially we used the assay of ATP as an index of microbial biomass (2, 3, 6). Recently Wiebe and Bancroft (30) have used the energy charge concept of Atkinson (1) to evaluate the growth state of microorganisms in water and soils. Using a ratio of adenylates permits us to ascertain the growth state of the microbial populations, whether they are active, resting, starving, or senescent. The use of the energy charge concept in natural waters and soils is new, and we are still evaluating its potential and its pitfalls as a diagnostic tool. Preliminary measurements of energy charge in salt marsh soils indicate an increase in growth state from winter to summer.

During the part of the year when *Spartina* is both actively growing and has a significant leaf canopy (April-September in Georgia) there is a population of actively growing microorganisms in the estuarine water which does not appear to be limited by available substrates. At least, it is removing soluble materials from the water at or near its maximum uptake velocity and responds little to additional substrates, natural or defined (9). The size of this population does appear to be limited in some way, however, under ordinary circumstances. When estuarine water is put in bottles, the population begins to increase in a few hours. This may be because the population has been isolated from the detritovores in the marsh. A similar phenomenon has been reported in Sargasso Sea water (23).

There is further evidence for active microbial populations in estuarine water. ATP and oxygen uptake are highly correlated, and their ratio is comparable to that found in logarithmically growing marine bacteria *in vitro* (3). Another line of evidence is found in the response of the microbial community to antibiotics. While Yetka and Wiebe (39) showed that the addition of antibiotics to natural systems is unreliable in separating procaryotic from eucaryotic respiration, an indication of microbial activity is found in their report. Neomycin and streptomycin did reduce the community respiration within estuarine water. These results are compatible with the hypothesis that the estuarine bacteria are growing actively, because similar reductions in respiration in pure cultures are found only during early exponential growth. More direct measures of activity are needed to determine actual rates.

THE CONSUMPTION OF ORGANIC MATTER

As we begin to examine the transformations of organic matter in the salt marsh, we find a substantial amount of soluble organic matter, with a very short residence time in the water, produced and consumed principally by microorganisms. This is not the soluble organic matter that is measured ordinarily as standing stock but consists of labile compounds whose concentrations rarely exceed a few micrograms per liter. Most of the standing stock consists of relatively refractory materials of long residence time (26). The presence of this relatively large pool of refractory material obscures the labile compounds. The latter cannot be identified simply by measuring total dissolved organic carbon, for example, but must be followed by experimental work with parts of the community.

Christian and Hall (7) layered a solution of ^{14}C-labeled glycine in sterile sea water over the marsh surface, using a nine inch diameter plastic tube as a retainer. After a 4 hr incubation, loss of substrate to the soil and distribution of ^{14}C were measured. Uptake of labeled glycine was rapid, but a substantial fraction remained immobilized in the soil. After 30 days more than 50% of the ^{14}C originally incorporated in the soil as glycine remained there. An extraction regime showed that 71% of the remaining label was incorporated in lipids. Christian and Hall suggest that substrates are incorporated rapidly into the cell envelope of bacteria and removed from more labile metabolic pools. The lifetime of microorganisms in salt marsh soils is remarkably long. At the same time, removal of labile substrates from the water is rapid. This has been well documented in both estuarine and ocean water (11, 12, 34). In most cases a substantial portion of the incorporated labeled materials remain immobilized in cell constituents in water as in marsh soils. Only under unusual circumstances is most of the labeled material rapidly metabolized to CO_2 (12).

We may ask, therefore, how such primary producers as *Spartina* become available as food to detritovores and most other macroheterotrophs in the marsh community. *Spartina* is not digested readily by most detritovores (15, 29). Before *Spartina* is utilized it is digested extracellularly by microorganisms. Some of the digestion products are lost to the water, but many are incorporated into microbial protoplasm (10). Presumably this microbial protoplasm is then the digestible food source of the detritovores. We have presented some evidence to show that microorganisms are being removed from estuarine water as rapidly as they reproduce themselves (9). Whether this is also true in the sediments is less clear at the moment. Certainly, microorganisms appear to be important in the detrital food web, degrading and assimilating refractory as well as labile products of both primary and secondary production and presenting them to detritovores in small but acceptable units. At the same time microbial populations are scavenging dissolved materials from the water, making them available in part to detritovores as bacterial biomass and leaving only a small part as a residue of the most refractory compounds.

In this respect the food web of the salt marsh is much like that of the ocean. The same kinds of processes are present in both, but in the salt marsh large standing stocks and rapid fluxes occur (20, 21). Probably there are more extensive populations of metabolically active bacteria in salt marshes than there are even in the surface waters of the ocean, but when substrates are available, the responses are comparable. Therefore, the study of salt marshes is of some comparative value in understanding the function of marine ecosystems in more general terms.

SOME RESIDUAL PROBLEMS.

We can identify chemically a number of sources of dissolved organic carbon that are intrinsically associated with the so-called detritus food web. Hardly any of them are uniquely part of the salt marsh community but would be proportionately significant in sea grass, kelp, or phytoplankton communities. In fact, most of those fluxes have already been identified in those other communities. (16, 20, 24). Chemical means alone will not permit us to quantify the fluxes of soluble materials, but radiotracer studies will, especially those which utilize naturally produced materials as the experimental substrate (29). There is much more work to be done along these lines, but the means to do it are largely defined and available. Recognition that there are linkages and pathways of dissolved organic carbon all through the detritus food web presents us with a new set of problems.

There are some intriguing questions concerning the net transfer of materials between marshes and estuaries or coastal waters. Obviously, marshes are accumulations of sediments and as such are depositional systems on a geological time scale. Most marshes on the East Coast of North America are mixed and flushed by the tides about 700 times per year, so they are geochemically open and well mixed systems (the ocean, by way of contrast, mixes once per 1000 yrs). Therefore, we cannot assume that salt marshes are necessarily sources or sinks for organic materials. Woodwell et al. (37) present evidence that Flax Pond marsh on the north shore of Long Island is a net consumer of organic matter, taking in organic matter from Long Island Sound and releasing CO_2 and inorganic nutrients. We do not as yet have comparable data for Georgia marshes. With specific reference to dissolved organic material, it would appear that most of that which enters the marshes with the incoming tides is relatively refractory material. If it is retained, it must be fixed chemically by some interaction with the sediments and not assimilated biologically. We do not know how much labile dissolved material makes its way out of the marsh on the ebbing tide, because it is obscured by the refractory materials. Probably much of that produced in the marsh is utilized before it is removed by tidal circulation. If the dissolved material is assimilated by microorganisms in the marsh, several possibilities remain for the subsequent steps in the food web. The microorganisms may be consumed continuously and efficiently by detritovores, making the system internally productive but not a source of organic enrichment to the estuary. If the dissolved material is assimilated by microorganisms and then stored for relatively long time periods, gradually being converted into CO_2, a substantial loss of potential food for higher trophic levels will result. Similar questions have been raised about the roles of microorganisms in the ocean's food web (20). On the basis of the limited observations on energy charge ratio (30), the ocean does not appear to be an efficient system, but the aerobic portion of the salt-marsh estuary does appear to be an efficient converter of organic matter to microbial biomass.

These questions serve to emphasize the relative trophic importance of microorganisms in aquatic food webs. Microorganisms are biochemically and metabolically versatile. They assimilate diverse substrates, multiply very rapidly, and can activate or shut down their metabolic activity rapidly. This gives them an advantage over macroorganisms, and they exploit it. The microbial portion of the detritus food web may well prove to be both more complex and quantitatively larger than the macroorganismal portion that has received so much more attention.

LITERATURE CITED

1. Atkinson, D.E. 1971. Adenine nucleotides as stoichiometric coupling agents in metabolism and as a regulatory modifier: the adenylate energy charge. *In* Greenberg, D.M. (ed.) Metabolic Pathways. 3rd Ed. 5: 1-20.
2. Bancroft, K., E.A. Paul, and W.J. Wiebe. 1974. Extraction of adenosine triphosphate from marine sediments with boiling sodium bicarbonate. Bacteriol. Proc. 53. G203.
3. ____, J.E. Yetka, and W.J. Wiebe. 1975. Oxygen uptake, adenosine triphosphate content and bacterial growth state in estuarine waters. Abst. 75th Ann. Meeting Soc. Microbiol. p. 186.

4. Barker, H.A. 1961. Fermentations of nitrogenous oxides by microorganisms. Bacteriol. Rev. 37:409-452.
5. Burkholder, P.R., and G.H. Bornside. 1957. Decomposition of marsh grass by aerobic marine bacteria. Bull. Torrey Bot. Club. 84:366-383.
6. Christian, R.R., K. Bancroft, and W.J. Wiebe. 1975. Distribution of microbial adenosine triphosphate in salt marsh sediments at Sapelo Island, Georgia. Soil Science. 119: 89-97.
7. ——, and J.R. Hall. (In Press). Experimental trends in sediment microbial heterotrophy: radioisotopic techniques and analysis. *In* Coull, B.C. (ed.) Ecology of Marine Benthos. Univ. South Carolina Press, Columbia.
8. Darnell, R.M. 1967. Organic detritus in relation to the estuarine ecosystem, p. 376-382. *In* Lauff, G.H. (ed.) Estuaries, AAAS Publ. 83.
9. Gallagher, J.L., W.J. Pfeiffer, and L. R. Pomeroy. (In press). Release of dissolved organic carbon into tidal water by *Spartina alterniflora* leaves. Coastal and Estuarine Mar. Sci.
10. Gosselink, J.G., and C.J. Kirby. 1974. Decomposition of salt marsh grass, *Spartina alterniflora* Loisel. Limnol. Oceanogr. 19:825-832.
11. Hobbie, J.E., and C.C. Crawford. 1969. Respiration corrections for bacterial uptake of dissolved organic compounds in natural waters. Limnol. Oceanogr. 14:528-533.
12. ——, O. Holm-Hansen, T.T. Packard, L.R. Pomeroy, R.W. Sheldon, J.P. Thomas, and W.J. Wiebe. 1972. A study of the distribution and activity of microorganisms in ocean water. Limnol. Oceanogr. 17: 544-555.
13. Johannes, R.E., and K.L. Webb. 1965. Release of dissolved amino acids by marine zooplankton. Science. 150: 76-77.
14. ——, and ——. 1970. Release of dissolved organic compounds by marine and freshwater invertebrates. *In* Hood, D.W. (ed.) Organic matter in natural waters. Univ. Alaska Inst. Mar. Sci. Occ. Pap. 1: 257-273.
15. ——, and Masako Satomi. 1966. Composition and nutritive value of fecal pellets of a marine crustacean. Limnol. Oceanogr. 11: 191-197.
16. Mann, K. H. 1973. Seaweeds: their productivity and strategy for growth. Science. 182: 975-981.
17. Menzel, D.W., and R.F. Vaccaro. 1964. The measurement of dissolved and particulate organic carbon in sea water. Limnol Oceanogr. 9: 138-142.
18. Odum, E.P., and A.A. de la Cruz. 1967. Particulate organic detritus in a Georgia salt marsh-estuarine ecosystem, p. 383-388. *In* Lauff, G.H. (ed.) Estuaries. AAAS Publ. 83.
19. Payne, W.J. 1973. Reduction of nitrogenous oxides by microorganisms. Bacteriol. Rev. 37: 409-452.
20. Pomeroy, L.R. 1974. The ocean's food web: a changing paradigm. BioScience. 24: 499-504.
21. ——. 1975. Mineral cycling in marine ecosystems, p. 209-223. *In* Howell, F.G., J.B. Gentry, and M.H. Smith (eds.) Mineral Cycling in Southeastern Ecosystems. ERDA Symposium Series (Conf. 740513).
22. Postgate, J. R. 1965. Recent advances in the study of the sulfate-reducing bacteria. Bacteriol. Rev. 29: 425-441.
23. Sheldon, R.W., W.H. Sutcliffe, Jr., and A. Prakash. 1973. The production of particles in the surface waters of the ocean with particular reference to the Sargasso Sea. Limnol. Oceanogr. 18: 719-733.
24. Sieburth, J. McN., and A. Jensen. 1970. Production and transformation of extracellular organic matter from littoral marine algae: a résumé, p.

203-223. *In* Hood, D.W. (ed.) Organic Matter in Natural Waters. Univ. Alaska Inst. Mar. Sci. Occ. Pap. 1.

25. Smalley, A.E. 1960. Energy flow of a salt marsh grasshopper population. Ecology. 41:672-677.
26. Sottile, W.S. 1973. Studies of microbial production and utilization of dissolved organic carbon in a Georgia salt marsh-estuarine ecosystem. Doctoral dissertation, The University of Georgia. 153 pp.
27. Teal, J.M. 1962. Energy flow in a salt marsh ecosystem of Georgia. Ecology. 39: 185-193.
28. Thomas, J.P. 1971. Release of dissolved organic matter from natural populations of marine phytoplankton. Mar. Biol. 11: 311-323.
29. Wetzel, R.L. 1975. Macro-heterotrophic utilization of carbon in a Georgia salt marsh-estuarine ecosystem: An experimental-systems modeling approach. Doctoral dissertation, The University of Georgia.
30. Wiebe, W.J., and K. Bancroft. 1975. The use of the adenylate energy charge ratio to measure growth state of natural microbial communities. Proc. Nat'l. Acad. Sci. U.S. 72: 2112-2115.
31. ____, and J. Liston. 1972. Studies of the aerobic, non-exacting heterotrophic bacteria of the benthos, p. 281-312. *In* Pruter, A.T. and Alverson, D.L. (eds.). The Columbia River Estuary and Adjacent Ocean Waters: Bioenvironmental Studies. University of Washington Press, Seattle.
32. ____, and L.R. Pomeroy. 1972. Microorganisms and their association with aggregates and detritus in the sea: a microscopic study. Mem. Ist. Ital. Idrobiol. 29 (Suppl.): 325-352.
33. Wiegert, R.G., R.R. Christian, J.L. Gallagher, J.R. Hall, R.D.H. Jones, and R.L. Wetzel. (In Press). A preliminary ecosystem model of coastal Georgia *Spartina* marsh. *In* Recent Advances in Estuarine Research.
34. Williams, P. J. leB., and R.W. Gray. 1970. Heterotrophic utilization of dissolved organic compounds in the sea. II. Observations on the responses of heterotrophic marine populations to abrupt increases in amino acid concentration. J. Mar. Biol. Ass. U. K. 50: 871-881.
35. Wolfe, R. S. 1971. Microbial formation of methane. *In* Rose, A.H. and J.F. Wilkinson (eds.) Advancing Microbial Physiology. 6: 107-146. Academic Press.
36. Wood, W.A. 1961. Fermentation of carbohydrates and related compounds. *In* Gunsalus, I.C. and R.Y. Stanier (eds.) The Bacteria. 2: 59-150. Academic Press.
37. Woodwell, G.M., P.H. Rich, and C.A.S. Hall. 1973. Carbon in estuaries, p. 221-239. *In* Woodwell, G.M. and E.V. Pecan (eds.) Carbon and the Biosphere. USAEC Conf. - 720510.
38. Wright, R.T. 1970. Glycollic acid uptake by planktonic bacteria. *In* Hood, D.W. (ed.) Organic Matter in Natural Waters. Univ. Alaska Inst. Mar. Sci. Occ. Pap. 1: 521-536.
39. Yetka, J.E., and W.J. Wiebe. 1974. Ecological application of antibiotics as respiratory inhibitors of bacterial populations. Appl. Microbiol. 28: 1033-1039.

BENTHIC DETRITUS IN A SALTMARSH TIDAL CREEK

James C. Pickral and William E. Odum
Department of Environmental Sciences
University of Virginia
Charlottesville, Virginia 22903

ABSTRACT: A new benthic sampling device was developed to permit convenient, quantitative collection of organic detritus from submerged sediment surfaces. The apparatus was used to investigate the spatial and temporal distribution of benthic detritus in a tidal, saltmarsh creek adjacent to the York River, Virginia. Diurnal, monthly, and seasonal samples were collected and partitioned into detrital size fractions. Each size fraction was analyzed to determine the percentage of root, stem, and leaf tissue contributed by the dominant marsh plants.

The standing stock and composition of detritus varied in a characteristic manner across the width and along the length of the tidal creek. Temporal differences were also evident, especially those associated with periodic storms. The data indicate apparent relationships between the morphology of the creek, current velocities, substrate texture, and the distribution and composition of the benthic detritus.

INTRODUCTION

In recent years, it has become increasingly apparent that organic detritus plays an important nutritive role in aquatic ecosystems. Darnell (4) and Odum (19) found significant amounts of detritus in the stomach contents of a variety of estuarine organisms, including several species of fish. Laboratory feeding experiments (1, 3, 9, 21) have demonstrated that many organisms can digest and assimilate portions of the detritus which they ingest. Recent studies (7, 13, 18, 24) have indicated that bacterial populations associated with decomposing detritus are important in converting detrital material into a utilizable food source for macrodetritivores.

There is evidence that the density and structure of benthic communities are correlated with the distribution of detritus on submerged sediment surfaces. In a classical study, Newell (18) investigated the distribution of the deposit feeders

Hydrobia ulvae and *Macoma balthica* in the Thames estuary. Newell found that population densities, sediment grain size, and organic carbon concentrations were strongly interrelated. The highest densities of deposit feeders occurred in areas with fine grained sediments and high levels of organic carbon. Similar organism-substrate relationships have been reported by Longbottom (16) for the estuarine lugworm *Arenicola marina*, by Marzolf (17) for amphipods in Lake Michigan, and by Tietjen (28) for deposit feeding meiofauna in New England estuaries. Rhoads and Young (22) have suggested that distributional patterns of organic matter determine the structures of benthic communities in Buzzards Bay, Massachusetts. They found that deposit feeders were concentrated in the middle of the bay where organic matter levels were high, while suspension feeders were found around the periphery of the bay where detritus levels were low.

These studies have demonstrated that current velocities, mean grain sizes of sediments, and levels of benthic organic matter are interrelated, and that all of these factors are correlated with densities and distributions of benthic populations. However, the pertinent causal relationships have not been fully established. Several of the investigators (16, 18, 22) have suggested that hydrodynamic factors govern the distribution of benthic detritus, which in turn influences the structure and dynamics of benthic communities. This hypothesis does not preclude the possibility that hydrodynamic factors and sediment characteristics may directly and independently influence benthic communities. Complex interactions of causal factors are also possible.

Laboratory studies have demonstrated that the nutritive value of detritus depends on the origin of the detritus, the sizes of the detrital particles, and the degree of decomposition (4, 5, 11, 15). Very few studies (2, 8) have investigated the composition of detritus in benthic environments. Previous workers have determined total organic carbon either by the loss-of-weight-on-ignition method or by "wet oxidation" techniques. More detailed analyses of detritus composition would undoubtedly improve our understanding of organism-substrate relationships.

It is widely recognized that salt marshes are a major source of detritus input to estuaries and coastal waters of the Eastern United States (4, 23, 26, 30). Literature reviews organized by Keefe (14) and Wass and Wright (29) show that annual net production in these marshes is very high, especially in the *Spartina alterniflora* marshes along the Georgia coast (23, 26). Relatively little (less than five percent) of the vascular material produced in marshes is grazed by herbivores (19, 25, 26). As a result, most of the plant dies and falls to the surface of the marsh where it may decay or be removed by tidal action to decay elsewhere.

The export of detritus from marshes is of interest because this indicates how much of the net production is made available to aquatic consumers. Previous studies (4, 10, 30) indicated that a substantial portion of annual marsh plant production was exported to adjacent estuaries. Detrital export was estimated either by periodic measurements of suspended load transport or by monitoring the rate of disappearance of detritus from the marsh surface. Bedload transport of detritus was not measured, although this mechanism may contribute significantly to the total flux of detritus from marshes. In fact, there is very little data on detrital bedload transport anywhere in the literature of aquatic ecology.

This paper presents preliminary results of a continuing study of benthic detritus in a saltmarsh tidal creek. A new device for quantitative sampling of benthic detritus is described, followed by a description of laboratory techniques used to partition samples into their various constituents. Spatial and temporal patterns of detritus distribution on the creek bed are presented and interpreted. Since our study of bedload transport is still in a preliminary stage, this paper will focus on the static aspects of benthic detritus. However, the data permit certain inferences to be made concerning the dynamic aspects of the system.

STUDY SITE

Taskinas Creek is a small tributary of the York River, located about four km east of West Point, Virginia. The creek, which is a little over 2.5 km long, is approximately 10 meters wide and 3 meters deep near the mouth, and becomes progressively narrower and shallower upstream from the mouth. The creek is flanked on both sides by saltmarsh vegetation (36 hectares total). *Spartina alterniflora*, the dominant species in the saltmarsh, occupies the low elevation regions contiguous to the creek channel. The saltmeadow community, located at higher elevations behind the *S. alterniflora* zone, is dominated by *Spartina patens* and *Distichlis spicata*. *Juncus roemerianus, Spartina cynosuroides, Baccharis halimfolia, Scirpus olneyi*, and at least fifteen other minor species occur in scattered patches in the high marsh. The extreme upstream portion of the creek is flanked by a diverse freshwater community.

METHODS

The use of conventional grabs and dredges (11, 12) for sampling benthic detritus in the marsh creek proved to be cumbersome, inefficient, or time consuming. Therefore, a more convenient sampling device, the Pickral-Odum detritus sampler, was developed (Fig. 1). The central feature of the apparatus is a May's fluid transformer (A) (U.S. Patent 3664768), a suction device which functions as a large aspirator. The May's fluid transformer was previously used in a benthic organism sampler designed by Iver Brook (pers. commun.). The fluid transformer is mounted on the side of a sampling chamber, such that the suction end of the aspirator is positioned in the interior of the chamber and the output end (C) extends outside the chamber. The chamber is a box (20 cm × 20 cm × 40 cm) constructed of 1/16" aluminum sheet metal; the bottom of the sampling chamber (400 cm^2) is open. A 15 foot aluminum pipe (1 1/2") (E) mounted to the top of the chamber serves as a handle for lowering and raising the apparatus from a boat. A 3.5 hp Homelite pump located in the boat powers the fluid

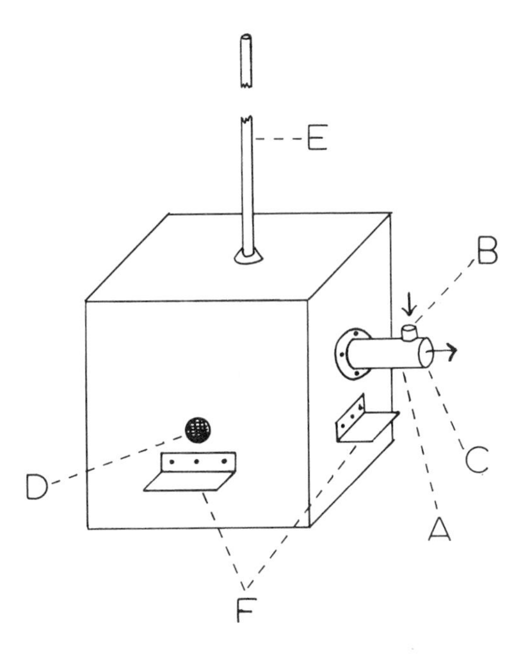

Figure 1. The Pickral-Odum benthic detritus sampler. See text for explanation.

transformer; a 1 1/2" hose connects the output of the pump to the intake of the fluid transformer (B). The intake hose of the pump, which draws water from the surface of the creek, is covered by a 0.25 mm mesh screen to prevent extraneous detritus from entering the sampling apparatus.

The detritus sampler works in the following manner. The boat is positioned directly above the site to be sampled. The detritus sampler is manually lowered by the aluminum handle to the bottom of the marsh creek. A small wooden superstructure in the boat holds the sampler in a vertical position. The open end of the chamber is pushed several inches into the mud of the creek bottom, thereby enclosing a 400 cm^2 area of the substrate surface. Flanges (F) mounted on the side of the chamber prevent excessive penetration into the mud. After the chamber has been set in place, the pump is started, and water is pumped from the surface of the creek to the submerged sampling apparatus. As water passes through the fluid transformer, suction is induced inside the chamber. Benthic particulate matter enclosed by the circumference of the sampling chamber is lifted into suspension by turbulence, drawn into the suction end of the fluid transformer, and collected in a nylon mesh bag (35 cm X 35 cm, 0.25 mm mesh) strapped to the output end of the fluid transformer on the outside of the chamber. Mesh covered holes (.25 mm mesh) in the side of the chamber (D) allow more water to enter the chamber and induce turbulence. After one minute of continuous operation, the pump is turned off, the apparatus is raised to the surface, the mesh bag containing the sample is removed and stored, and a new collection bag is strapped to the sampler.

This sampling apparatus has several favorable characteristics. It is relatively inexpensive and indestructible. It can be conveniently operated from a small boat by one person. The time required to collect and store a sample is approximately three minutes. A composite sample for several sites within a sampling area can be collected in one mesh bag without the necessity of hauling the apparatus to the surface between subsamples. Detritus particles and organisms are not damaged as they pass through the fluid transformer into the collection bag. Most importantly, the apparatus is efficient and quantitative. Essentially all particles except large pieces of tree branches are removed from the sediment surface, and essentially no particles larger than the mesh diameter of the bag escape from the apparatus. The main disadvantage of this system is its failure to retain particles smaller than the mesh size of the bag. This was not a serious drawback in the present study, since most of the benthic detritus in the creek was comprised of relatively coarse fractions. The apparatus would be impractical for sampling in depths greater than about 10 meters.

Samples of benthic detritus from Taskinas Creek were collected monthly, and sometimes more frequently, between February and August, 1975. Permanent sampling stations were established at ten locations along the length of the creek. The three stations nearest to the mouth of the creek were separated by intervals of 100 meters. Stations farther upstream were spaced at approximately 250 meter intervals. A preliminary study had suggested the feasibility of this sampling scheme.

A transect across the width of the creek was established at each of the ten sampling stations. Three samples were obtained near high slack tide along each transect, at points one-third, one-half, and two-thirds of the total distance across the creek. A rope stretched between two stakes located on opposite sides of the creek facilitated the positioning of the boat on the transect. Samples collected by the detritus sampler were stored in separate jars in five percent formalin and transported to the laboratory for subsequent analysis.

The samples collected in the field consisted of a mixture of sand grains (silica), mud pellets, pieces of mollusc shells, fragments of terrestrial vegetation, benthic fauna, and bits of decaying stems, leaves, and roots from vascular marsh plants, principally *Spartina alterniflora*. The samples were processed in the laboratory to isolate the various constituents.

Each sample was first partitioned into five different size fractions. This was accomplished by gently washing the material for five minutes on a 4.0 mm geological sieve, followed by separate five minute washings on 2.0 mm, 1.0 mm, 0.5 mm, and 0.25 mm sieves. Samples were kept thoroughly wet at all times to prevent the formation of detrital aggregates.

Table 1. Settling times (minutes) of detritus components in a 1.5 meter settling tube.

Tissue Types	2.0 mm	Size fraction 1.0 mm	0.5 mm	0.25 mm
Mud pellets	0:15	0:30	0:50	1:30
Stems	1:05	1:45	2:30	3:35
Leaves	1:40	3:20	5:40	7:20
Roots	>1:40	>3:20	>5:40	>7:20

Each size fraction was further partitioned by introducing the material into the top of a 1.5 meter settling tube (120 mm diameter). Due to differences in density, volume, and shape, the various types of particles in the original mixture settled through the tube at different rates. Sand grains, which were present only in the 0.5 mm and 0.25 mm fractions, settled most rapidly, followed by mud pellets, stem detritus, leaf detritus, and root detritus. Since the mean settling times of these particle types differed from each other considerably, it was possible to effect nearly complete physical separations by briefly opening and then closing a 10.0 mm stopcock at the bottom of the tube at appropriate intervals. Table 1 shows the empirically determined settling times for the particle types in each of the size fractions. Particles collected on the 4 mm sieve were sorted by hand because some of the larger particles would not pass through the stopcock of the settling tube.

Each particle class isolated in the settling tube was collected on a 0.063 mm sieve positioned below the stopcock. The subsample was transferred from the sieve to a tared aluminum dish, placed in a drying oven at 65°C for 48 hours, and weighed to the nearest milligram on a Mettler balance. All weights were multiplied by a conversion factor dictated by the dimensions of the sampling apparatus, and expressed as dry grams of detritus per square meter of creek bottom.

RESULTS AND DISCUSSION

Standing stock data for the period between Feburary and August, 1975, indicated characteristic patterns of benthic detritus distribution across the width and along the length of Taskinas Creek. Data obtained from the sampling transects showed a non-uniform distribution of detritus across the width of the creek at all ten sampling stations. Fig. 2 shows the standing stock values recorded for station 6 (1200 meters upstream from the mouth of the creek) on March 15, 1975. The highest standing crop of detritus at this station occurred on the left side of the channel. Slightly less detritus (32 g/m^2) was present in the middle of the creek, and considerably less (8 g/m^2) was found on the right side of the channel. The distribution of the three detritus tissue types paralleled the distribution of total detritus along the transect. Stem material was highest on the left side, lower in the middle, and lowest on the right; leaves and roots followed the

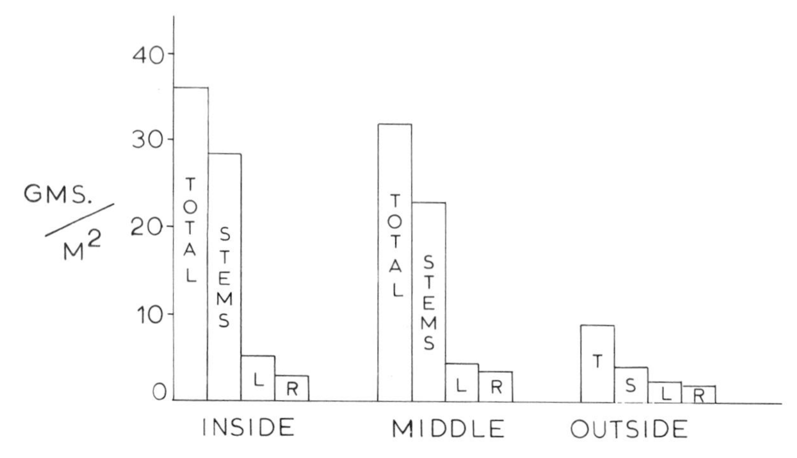

Figure 2. Standing stocks of detritus (dry grams per square meter) at the inside, middle, and outside of a meander at station 6 on March 15, 1975.

same pattern. Stem material predominated over leaf and root material at all three locations on the transect, but the percentage of stem material in the total sample was significantly lower on the right of the creek. Leaf material exceeded root material by a few grams at each of the three transect sites. Fig. 3 illustrates the distribution of detrital size fractions for the same transect on the same date. All size fractions were highest on the left side of the creek and lowest on the right side. The weight of the 0.25 mm fraction was greater than that of any other size class at each transect location, and each size class was at least slightly more abundant than all size fractions coarser than that size class. The percentage

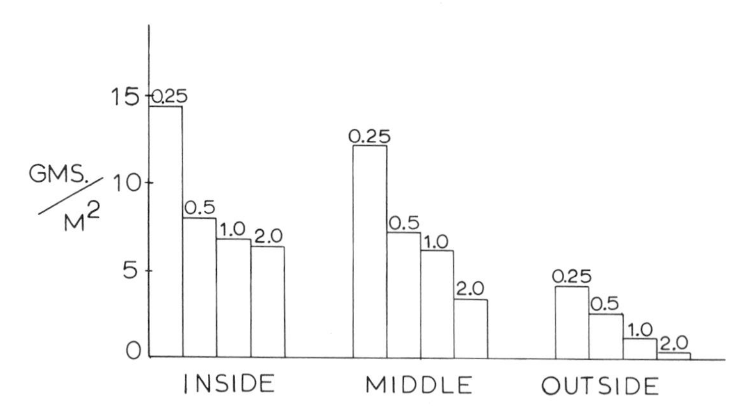

Figure 3. Standing stocks of detritus size fractions (dry grams per square meter) on the inside, middle, and outside of meander at station 6 on March 15, 1975.

of the fine fractions (0.25 mm and 0.5 mm) was highest in the sample taken from the right side of the creek.

These trends are related to the morphology and hydraulic regime of the creek at sampling station 6. The left side of the creek lies on the inside of a meander, and the right side of the creek lies on the outside of the meander. This observation, coupled with principles well established by fluvial geomorphologists (15), accounts for the nonuniform distribution of benthic detritus across the creek bed. Current velocities are relatively high on the outside of a meander where the channel is deep, and low on the inside of a meander. Consequently, fine grained sediments are eroded from the outside of a meander and deposited on the inside of a meander. Organic detritus particles, which are hydrodynamically equivalent to small sand and silt-size sediment particles, tend to accumulate in the low velocity zone on the inside of a meander. Detritus particles which have been deposited on the inside can be transported at some later time when the current velocity increases to a critical value which depends upon the density and size of the particle. The higher percentages of leaf material, root material, and fine detritus particles in samples taken from the outside of the meander reflects the relative ease with which these light weight, slow settling particles are winnowed from the detritus deposits on the insides of the meanders and moved to the outsides of the meanders where they can be more easily transported as suspended load and bedload. Detritus particles probably experience several episodes of alternating deposition and erosion before they eventually exit the creek and enter the York River.

Similar cross-sectional patterns of detrital standing stocks were recorded at the other nine stations along Taskinas Creek on March 15, and on other sampling dates throughout the study. The standing stock of total detritus, and of the three tissue types in the samples, was always higher on one side of the creek than on the other side, and the standing stock in the middle of the creek was usually intermediate. The pattern was most pronounced at stations located on sharp meanders, and less pronounced at stations on gradual bends. The "meander pattern" of detritus distribution was evident even for the two stations located on what appeared to be straight sections of the creek, thus indicating that the entire creek should be regarded as a continuous series of meanders.

The asymmetrical distribution of detritus across the bed of the marsh creek has several implications for research efforts. First, it is apparent that the standing stock of detritus at a station on the creek cannot be accurately estimated by taking one "representative" sample at that station. Secondly, the above observations concerning channel morphology, velocity fields, and detritus settling characteristics indicate that although the *standing stock* of detritus is greatest on the inside of a meander, the *transport* of detritus as bedload is greatest on the outside of a meander. Attempts to measure bedload detritus transport should focus on the outside of meanders. Thirdly, it may be hypothesized that the distribution of benthic organisms across the width of the creek parallels the nonuniform distribution of detritus. Preliminary observations indicate that this is indeed the case. We shall attempt to quantify this relationship in a future study.

Data obtained from the cross-sectional transects were used to estimate the distribution of benthic detritus along the length of Taskinas Creek. After the three samples collected from a sampling station had been individually partitioned and weighed, a graphical technique was used to compute a mean standing stock value of benthic detritus at that station. The same technique was used to compute the mean values of the several detrital size fractions and tissue types at the station. The mean standing stock of total benthic detritus and its constituents were computed in this manner for all ten sampling stations on each of the sampling dates during the study.

Fig. 4 shows the distribution of benthic detritus along the length of Taskinas Creek on June 15, 1975. Detritus levels were relatively high (40-45 grams per square meter) at the two stations farthest upstream. The standing stock steadily decreased at progressively downstream stations, and reached a minimum value at station 4, located about 400 meters upstream from the mouth of the creek. Downstream from station 4, benthic detritus levels increased sharply, reaching a maximum (50 g/m^2) a few meters upstream from the mouth. The standing stock values of stem material (Fig. 4) closely paralleled the trend observed for total standing stocks; that is, stem material comprised a nearly constant percentage of total detritus at all sampling stations. Leaf material and root material were considerably less abundant than stems at all locations, but the amount of roots and leaves was about eight percent higher in samples collected from the three stations located furthest downstream.

Although the standing stock values of benthic detritus varied widely during the seven month period, the same general longitudinal trend was observed on different sampling dates, with a few exceptions which are later noted.

The characteristic distribution of detritus along the creek bottom can be explained by the following set of observations and inferences. Taskinas Creek becomes wider and deeper at stations located progressively downstream, and these changes in channel morphology are accompanied by increases in discharge and current velocity. The capacity of the creek to transport detritus particles increases as the current velocity increases. Therefore, relatively large standing stocks of detritus can accumulate at stations where current velocities are low, and less detritus can accumulate at stations where current velocities are higher. The steady decrease in the standing stock of detritus from station 10 downstream to station 4 can be interpreted in this way.

As previously noted, benthic detritus levels increase sharply from station 4 downstream to the mouth of Taskinas Creek. This trend, which is a reversal of the trend observed upstream from station 4, and which appears to contradict the model proposed above, can be related to a special morphological feature of the creek bed. A sediment sill, composed of coarse sand derived from the York

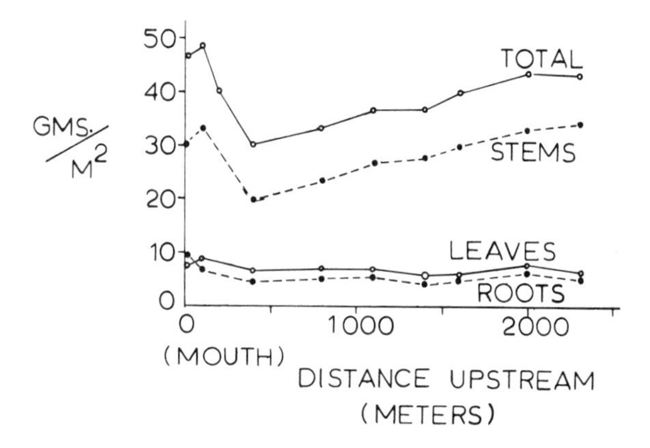

Figure 4. Mean standing stock of benthic detritus (dry grams per square meter) at sampling stations along the length of Taskinas Creek on June 15, 1975.

River, occupies the mouth of Taskinas Creek. The creek is very shallow in the region overlying the sill, but the depth increases by several meters a short distance upstream from the sill. The sill acts as a partial barrier to detritus moving downstream on the bed of the creek. Detritus accumulates in the region directly behind the sill, and some of the detritus can be alternately transported upstream and downstream for short distances during flood and ebb tides. Thus, the increasingly high levels of benthic detritus observed downstream from station 4 represent the accumulation behind the sill of detritus transported as bedload from stations upstream from station 4. The higher percentage of leaf and root material in samples collected behind the sill reflects the fact that these materials are more easily transported downstream than are the heavier stem fragments.

The interpretations of the observed differences in standing stocks of benthic detritus along the length of the marsh creek are, in part, speculative. Direct measurements of bedload transport at the various sampling stations are needed to confirm or reject our hypotheses. Laboratory flume studies are also needed to quantify bedload and suspended load transport of detritus in different current regimes.

The standing stocks of detritus at the ten sampling stations varied considerably with time. The greatest variations were observed at the three stations located furthest downstream. There was evidence that these temporal changes were related to meteorological events.

Fig. 5 demonstrates the results of a small storm which occurred during an ebb tide on June 15, 1975. Prior to the storm, the standing stocks of detritus were relatively high (greater than about 30 g/m^2) at all sampling stations along the creek. The highest levels of detritus were found a few meters upstream from the sill at the mouth of the creek. Samples taken at the same stations after the storm

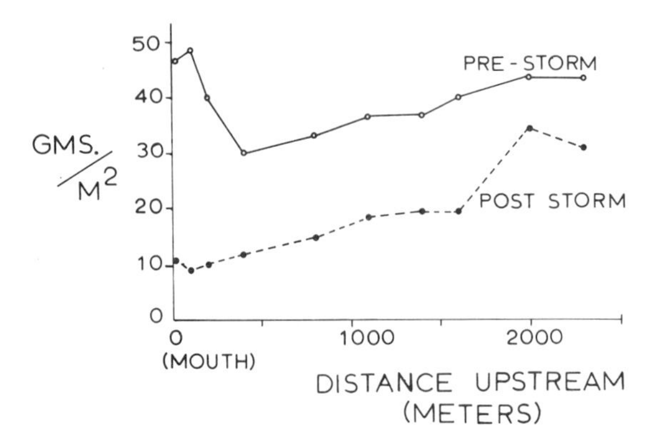

Figure 5. **Standing stocks of detritus (dry grams per square meter) at ten sampling stations along Taskinas Creek before and after a storm on June 15, 1975.**

revealed decreases in benthic detritus stocks at all stations. The most pronounced changes occurred at the downstream stations near the mouth, where the standing crop of detritus dropped to about 10 g/m^2. These changes clearly indicate that benthic detritus was scoured from the bed of the creek and exported to the York River.

The total export of detritus during the storm could not be estimated from the differences between pre-storm and post storm standing crops, because an undetermined amount of detritus entered the creek from the adjacent marsh during the storm and was transported to the York River. However, the net change in the total standing crop of detritus on the bed of the creek did indicate that *at least* 2000 kilograms of detritus was exported within a period of five hours. It is likely that total export was considerably greater than this figure, since many large wracks of floating detritus were observed moving downstream. Direct measurements of bedload, suspended load and floating material transport would be necessary to determine the total flux of detritus.

These results suggest that periodic storms are responsible for exporting large amounts of detritus from the marsh in short periods of time. Large storms may be particularly important in flushing areas of the high marsh which contribute little detritus to the estuary during normal periods. Indeed, infrequent events of extreme magnitude may account for a substantial portion of the total annual export of organic detritus from the saltmarsh.

ACKNOWLEDGMENTS

This research was supported by NSF grant GA-34100.

REFERENCES

1. Adams, S. M., and J. W. Angelovic. 1970. Assimilation of detritus and its associated bacteria by three species of estuarine animals. Chesapeake Sci. 11(4):249-254.

2. Bader, R. G. 1954. The role of organic matter in determining the distribution of bivalves in sediments. J. Mar. Res. 13:32-47.
3. Brinkhurst, R. O., K. E. Chua and N. K. Kaushik. 1972. Interspecific interactivities and selective feeding by tubificid oligochaetes. Limnol. Oceanogr. 17(1):122-133.
4. Cruz, A. A. de la 1956. A study of particulate organic detritus in a Georgia salt marsh-estuarine ecosystem. Dissertation, Univ. Georgia, 110 p.
5. ____, and W. E. Poe. 1975. Amino acids in salt marsh detritus. Limnol. Oceanogr. 20:124-127.
6. Darnell, R. M. 1961. Trophic spectrum of an estuarine community based on studies of Lake Pontchartrain, Louisiana. Ecology 43(3):553-568.
7. Fenchel, T. 1970. Studies on the decomposition of organic detritus derived from the turtle grass *Thalassia testudinum*. Limnol. Oceanogr. 15:15-20.
8. George, J. D. 1964. Organic matter available to the polychaete, *Cirriformia tentaculata* (Montagu) living in an intertidal mud. Limnol. Oceanogr. 9:453-455.
9. Hargrave. B. T. 1970. The utilization of benthic microflora by *Hyalella azteca* (Amphipoda). J. Animal Ecol. 39:427-437.
10. Heald, E. J. 1971. The production of organic detritus in a south Florida estuary. Univ. Miami Sea Grant Tech. Bull. No. 6, 110 p.
11. Holme, N. A. 1964. Methods of sampling the benthos, p. 171-260. *In* F. S. Russell (ed.), Advances in Marine Biology, 2. Academic Press, N. Y.
12. Hopkins, T. L. 1964. A survey of marine bottom samplers, p. 213-256. *In* M. Sears (ed.), Progress in Oceanography, 2. Macmillan Press.
13. Kaushik, N. K., and H. B. N. Hynes. 1968. Experimental study on the role of autumn-shed leaves in aquatic environments. J. of Ecol. 56:229-243.
14. Keefe, C. W. 1972. Marsh production: a summary of the literature. Contrib. in Mar. Sci. 16:163-181.
15. Leopold, L. B., M. G. Wolman, and J. P. Miller. 1964. Fluvial processes in geomorphology. W. H. Freeman and Co., p. 295-317.
16. Longbottom, M. R. 1970. The distribution of *Arenicola marina* with special reference to the effects of particle size and organic matter of sediments. J. Exp. Mar. Biol. Ecol. 5(2):138-156.
17. Marzolf, G. R. 1961. Substrate relations of the amphipod *Pontopore affinis*. Dissertation. Univ. Michigan. 183 p.
18. Newell, R. 1964. The role of detritus in the nutrition of two marine deposit feeders, the prosobranch, *Hydrobia ulvae* and the bivalve, *Macoma balthica*. Proc. Zool. Soc. London 144(1):25-45.
19. Odum, W. E. 1970. Pathways of energy flow in a south Florida estuary. Dissertation, Univ. Miami, 162 p.
20. ____. 1970. Utilization of direct grazing and plant detritus food chains by the striped mullet, *Mugil cephalus*, p. 222-240. *In* J. H. Steele (ed.), Marine Food Chains. Univ. Calif. Press, Berkeley.
21. Prinslow, T. E., I. Valiella and J. M. Teal. 1974. The effect of detritus and ration size on the growth of *Fundulus heteroclitus*. J. Exp. Mar. Biol. Ecol. 16:1-10.
22. Rhoads, D. C., and D. K. Young. 1970. The influence of deposit-feeding organisms on sediment stability and community trophic structure. J. Mar. Res. 28:150-178.
23. Schelske, C. L., and E. P. Odum. 1961. Mechanisms maintaining high productivity in Georgia estuaries. Proc. Gulf Carib. Fish. Inst., 14th Annual Ses., p. 75-80.

24. Seki, H., J. Skelding, and T. R. Parsons. 1968. Observations on the decomposition of a marine sediment. Limnol. Oceanogr. 13:440-447.
25. Smalley, A. E. 1959. The growth cycle of *Spartina* and its relation to the insect populations in the marsh. Proc. Salt Marsh Cong., Sapelo Island, Georgia.
26. Teal, J. M. 1962. Energy flow in the salt marsh ecosystem of Georgia. Ecology 43(4):614-624.
27. Tenore, K. R., and W. M. Dunstan. 1973. Comparison of feeding and biodeposition of three bivalves at different food levels. Mar. Biol. 21(3):190-195.
28. Tietjen. J. H. 1969. The ecology of shallow water meiofauna in two New England estuaries. Oecologia 2:251-291.
29. Wass, M. L., and T. D. Wright. 1969. Coastal wetlands of Virginia-Interim report to the Governor and General Assembly. Spec. Rpt. Appl. Mar. Sci. Ocean Eng., 10, 154 p.
30. Williams, R. B., and M. B. Murdoch. 1972. Compartmental analysis of the production of *Juncus roemerianus* in a North Carolina salt marsh. Chesapeake Sci. 13(2):69-79.

CARBON RESOURCES OF A BENTHIC SALT MARSH INVERTEBRATE *NASSARIUS OBSOLETUS* SAY (MOLLUSCA: NASSARIIDAE)[1]

Richard L. Wetzel[2]
Department of Zoology
University of Georgia

ABSTRACT: The mud snail, *Nassarius obsoletus* Say, occupies the intertidal mud flat area of the salt marsh ecosystem at Sapelo Island, Georgia, and is classed as a detritovore. The three principal carbon resources available for ingestion by the intertidal population are *Spartina alterniflora* Loisel derived carbon (plant detritus), microbial carbon (bacteria), and benthic algae (primarily pennate diatoms). Uptake studies using ^{14}C labeled substrates indicated that the structural carbohydrate fraction (crude fibre) of the plant detritus was not assimilable. Uptake and retention of ^{14}C by *N. obsoletus* occurred only on microbial and algal labeled substrates. Extrapolating the calculated short term uptake and retention rates to a daily basis for these two carbon resources, total incorporation would balance a daily body carbon loss of 5 to 6%. There was evidence from the uptake studies that *N. obsoletus* functions more as a "grazer" on the mud flat area than a strict detritovore. Label uptake was approximately 2.3 times greater on the algal substrate than the microbial and ingestion significantly reduced algal standing crop in the experimental cores. Net carbon retention efficiency for both substrates was 46%.

INTRODUCTION

The specific carbon-energy resources of macroheterotrophic organisms inhabiting the salt marsh ecosystem are poorly understood. The majority of these organisms are described as particulate, suspensory, filter or simply detrital feeders that utilize detritus as their primary carbon-energy resource. The major

[1] Contribution number 309 University of Georgia Marine Institute, Sapelo Island, Georgia. This research was supported by National Science Foundation Grant # DES 72-01605-A02.

[2] Present address: Virginia Institute of Marine Science, Gloucester Point, Virginia 23062

evidence cited is the observation that the majority of macroheterotrophs ingest detrital particles (4, 5, 6, 7, 11, 15, 17, 19, 25). However, few attempts have been made to quantify trophic relations in the detrital food web based on these observations because:

1. Ingestion is not equivalent to assimilation.
2. Detritus is a "catch-all" category not sufficiently defined for quantitative trophic studies, i.e. the major fraction as "crude fibre" has not been considered separate from associated microorganisms.
3. Classification of an organism as a "detritovore" is indicative of a feeding *strategy* but does not quantify its carbon or nutrient resource.

The role of macroheterotrophs in the flux of carbon and nutrients in the salt marsh ecosystem as well as specific information regarding their potential for regulation and control of rate processes can only be generalized from this information. Detailed information regarding their carbon-energy resources and rates of utilization are necessary for quantitatively assessing the function of these conspicuous organisms within the context of the ecosystem.

Considering only benthic macroheterotrophic invertebrates classed as detritovores of the salt marsh ecosystem, the three principal carbon resources available for ingestion and assimilation are: 1) *Spartina*-derived carbon as detritus, 2) microorganisms or microheterotrophs associated with both detrital plant fragments and the soil substrate, and 3) benthic algae (1, 9, 17, 18, 19, 20, 25, 28, 29). The results of experimental feeding studies using these three carbon resources and employing ^{14}C radiotracer techniques are presented in this report. *Nassarius obsoletus* Say, the mud snail, a common benthic salt marsh invertebrate classed as a detrital feeder, was chosen as the experimental organism for these studies and as a representative of the general benthic detritovore macrofauna.

STUDY AREA AND POPULATION SAMPLING

A population of *N. obsoletus* from the intertidal mud flat of South End Creek, Sapelo Island, Georgia was sampled. The general area is characterized by semi-diurnal tides with a mean tidal range of 2.1 m. Annual salinities range from 18 °/oo to 28 °/oo in a nearby tidal river with water and sediment temperatures ranging from 10° to 30° and 13° to 35° C respectively. The mud flat area supports a dense population of snails. Population studies conducted by J.R. Hall (pers. comm.) in the same area indicated peaks in population density in July and September for all age classes. Population biomass ranged from approximately 2.0 gC $\times$ m^{-2} in January to 11 gC $\times$ m^{-2} in July. Snails used for the experimental studies were either collected by hand or obtained in 10 cm $\times$ 10 cm core samples taken on the mud flat. All field collections were made at low tide in early afternoon as an attempt to normalize inter-experiment variability due to diel metabolic or behavioral patterns especially as they might effect changes in feeding activity.

METHODS AND MATERIALS

The three principal components, as carbon resources, of the salt marsh detrital food web, i.e. *Spartina*-derived carbon, microbial carbon, and benthic algal carbon, were specifically labeled with ^{14}C and used as feeding substrates to evaluate the uptake and retention (assimilation) of carbon by *N. obsoletus*. The labeling methods and experimental designs reported are combinations, and in most cases, modifications of the methods reported by Adams and Angelovic (1) Brock, Wiegert and Brock, (2), and Sorokin (25).

Spartina Labeling

Field labeling experiments were carried out during the summers of 1972 and 1973. Plants 30 to 50 cm in height were washed with distilled water to remove adhering mud and epiphytic growth and enclosed in plastic bags. The bags were sealed with tape just above the soil surface and incubated for approximately one hour to deplete ambient CO_2 levels. $^{14}CO_2$ was released following the initial depletion period by injecting a radioactive bicarbonate solution contained inside the bag with 1 ml, 1.0 N H_2SO_4. The plants were either pulse labeled with 10 μCi per plant for three consecutive days or given an initial dose of 25 to 30 μCi per plant. Daily incubations ran two hours for the pulse-labeled plants and three to four hours for the single dose. All plants were labeled during midafternoon and low tide. The plants were harvested at various intervals following the initial labeling (post-labeling growth period) and dried at 105° C for 48 hours. The plant material was then ground initially by Waring blender and finally by mortar and pestle with 80 to 90% passing a 0.1 mm screen. The ground and dried plant material was assayed for both translocation and biochemical fractionation of the incorporated label by specific activity analyses of leaves and stems and extracted plant tissue. The extraction procedure follows from the work of Holland and Gabbot (12) and is discussed in detail in Wetzel (28). Uniformity of label distribution (label translocation) within plant tissue was determined by the ratio of specific activity of leaves to the specific activity of stems. Biochemical homogeneity (label fractionation) was assessed by comparison of the percent reduction in specific activity over the various post labeling growth intervals by extraction of known carbon pools. When the percent change, relative to the specific activity of non-extracted tissue showed no further change with increased post labeling growth, the tissue was considered homogenously labeled with respect to the major tissue carbon pools. Both extracted and non-extracted plant tissue labeled in this manner was used as a feeding substrate.

Microbial Labeling

The microbial populations of undisturbed mud flat cores was labeled with three dissolved organic labels; glucose-^{14}C, acetate-^{14}C, and glycine-^{14}C. The acetate and glycine labeled cores were incubated 24 hours in the dark at 25 to

27°C. The glucose-^{14}C cores were incubated six to eight hours at the same temperature to allow for maximum incorporation. Following the core uptake period, the cores were used as a natural-labeled feeding substrate for the uptake and retention of microbially labeled carbon. Prior to the introduction of the experimental organisms, the cores were gently washed with 3, 50 ml aliquots of 0.22 μm membrane filtered water to remove unincorporated label and label which may have been remineralized by the sediment microbial community. Formalin injected cores served as controls.

Benthic Algae Labeling

The benthic algae were labeled by taking cores (as above) through conspicuous algal mats on the mud flat in South End Creek. The cores were taken and immediately returned to the laboratory. The cores were gently washed with 0.22 μm membrane filtered water and let stand for approximately one hour. The cores were labeled with $NaH^{14}CO_3$ solution (pH 8.0) and incubated at 25 to 27° C under incandescent lighting. Formalin injected cores served as controls. Initial concentration of isotope was 5 μCi per core. The cores were labeled for six to eight hours to allow for maximum algal incorporation. Following the uptake period, the cores were treated as before and used as a natural feeding substrate for studies with *N. obsoletus*.

Feeding Experiments

The experimental designs were of two types; 1) feeding chamber studies using labeled *Spartina alterniflora* and 2) core feeding experiments using the labeled core substrates. For each design, the experiments were run for various time intervals with each time interval replicated and with separate time controls. A total of nine feeding experiments were run using *N. obsoletus*. Table 1 summarizes the radiotracer feeding studies.

For experiments 1A and 2, finely ground plant material (non-extracted) was mixed with combusted marsh sediment (48 hours @ 500° C) to ca. 5% and 10% carbon content respectively. For the remaining experiments using labeled *Spartina*, the plant material was either offered directly to the experimental organisms or in combination with a mud slurry; Feeding Experiment (FE) 3, 8, and 9. For all experiments, 100 to 200 mg dry weight (DW) material was added to each chamber and wetted with 20 to 25 ml of membrane filtered (0.22 μm) water. The experimental organisms for each study were collected from the field the day of the experiment and held 4 to 6 hours to allow for acclimation and gut clearance. The organisms were then placed on the experimental substrates and incubated for periods ranging from 1 to 96 hours (Table 1). For the *Spartina* studies, the feeding chambers were purged with CO_2 free air during the incubation periods and the $^{14}CO_2$ collected in a phenethylamine: methanol trapping solution (30). During the incubation intervals, periodic checks were made for scrubbing efficiency of the purging system by collecting 5 to 10, 1 ml air

Table 1. Summary of feeding chamber and core feeding experiments; *N. obsoletus*.

No.	Date (mo/yr)	Temp. (°C)	Substrate	Incubation (hours)
1A	1/74	25	^{14}C-*Spartina*[1]	4-24
2	5/74	25	^{14}C-*Spartina*[1]	1- 8
3	6/74	25	^{14}C-*Spartina*[2]	24-96
8	8/74	25	^{14}C-*Spartina*[3] + mud slurry	8-72
9	8/74	25	^{14}C-*Spartina*[3] + mud slurry + glucose-^{14}C	4-24
4	7/74	25	^{14}C-acetate[4]	1- 3
5	7/74	25	^{14}C-glucose	2- 6
6	8/74	25	^{14}C-bicarbonate	2- 6
7	8/74	25	^{14}C-glycine[4]	2- 6

[1] Non-extracted *Spartina alterniflora* labeled plant material.
[2] Results loss due to scintillation counting system failure.
[3] Extracted *Spartina alterniflora* labeled plant material.
[4] Labeled cores provided by Dr. J.R. Hall, UGA Marine Institute, Sapelo Island, Ga.

samples of inlet and outlet gases. The CO_2 air samples were analysed using gas chromatography.

For experiments 1A and 9, fecal material was collected for estimation of assimilation efficiency based on specific activity. For the other experiments, the organisms were immediately killed following the incubation interval and stored frozen until analysis. The fecal samples were obtained by holding the organisms for 2 hours in finger bowls containing 25 ml 0.22 μm membrane filtered water and the fecal material collected by filtration onto tared pre-combusted glass fiber filters, washed with distilled water to remove adventitious salts, and dried at 55° C for 48 hours. Analyses for specific activity were done using the combustion analysis technique (Fig. 1). Analyses for specific activity were determined for each time interval on triplicate samples of water to estimate dissolved release substrate, and ground-dried samples of each snail. All samples were internally standardized using toluene-^{14}C.

The glucose, acetate, glycine and bicarbonate labeled cores constituted the core feeding experiments. The mud snails were added at densities occurring for the field population. The animals were incubated on the labeled core substrates for 2, 4, and 6 hours except for the first experiment (acetate label) which was run for 1, 2, and 3 hours. At the end of each time interval, the snails were immediately killed and stored frozen until analysis. Initial and final water samples were taken for each interval to check for dissolved label release. After water sampling, the cores were killed by replacing the overlying water with 10% sea water formalin solution for one hour. Surface samples (less than 0.5 cm deep)

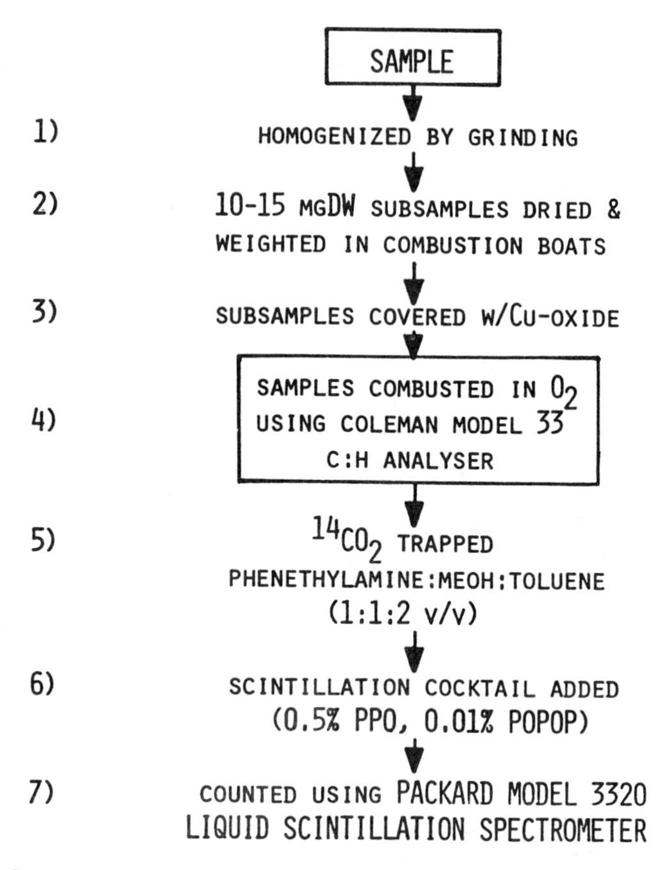

Figure 1. Specific activity combustion analysis. (30).

were randomly taken following the formalin treatment from each core for analysis of specific activity and carbon. All samples were dried at 55°C for a minimum of 48 hours prior to analysis. The animals were individually analyzed for specific activity using the combustion technique. Animal dry weights were converted to carbon based on regression analyses of carbon *vs* dry weight and dry weight *vs* ash free dry weight (AFDW). All particulate carbon determinations were done using the Coleman Model 33 C:H Analyser. Dissolved carbon was determined according to the methods of Menzel and Vaccaro (16). Gaseous carbon was determined by gas chromatography.

RESULTS

Feeding experiments 1A and 2 were designed to evaluate the uptake and net retention of non-extracted *Spartina*-^{14}C by *N. obsoletus*. Although not representative of the naturally occurring substrate, the labeled plant detritus was used to test for ingestion of the artificial substrate by offering a metabolizable-labeled source and to test for possible incorporation, either through assimilation or exchange, of dissolved label. Fig. 2 illustrates the results of analyses for $^{14}CO_2$ release, dissolved label release ($DO^{14}C$) and total label incorporation (body burden) by the experimental organisms. Table 2 summarizes the analyses for specific activity of the various labeled carbon pools.

Label uptake by *N. obsoletus* reached a maximum at 12 hours followed by a slight decline in the 24 hour incubation (Fig. 2). Increases in $^{14}CO_2$ production and dissolved label concentration reflect the decline in cummulative body burden and are probably due to metabolic losses from the organism. Control substrates showed maximum dissolved label release by the substrate during the first

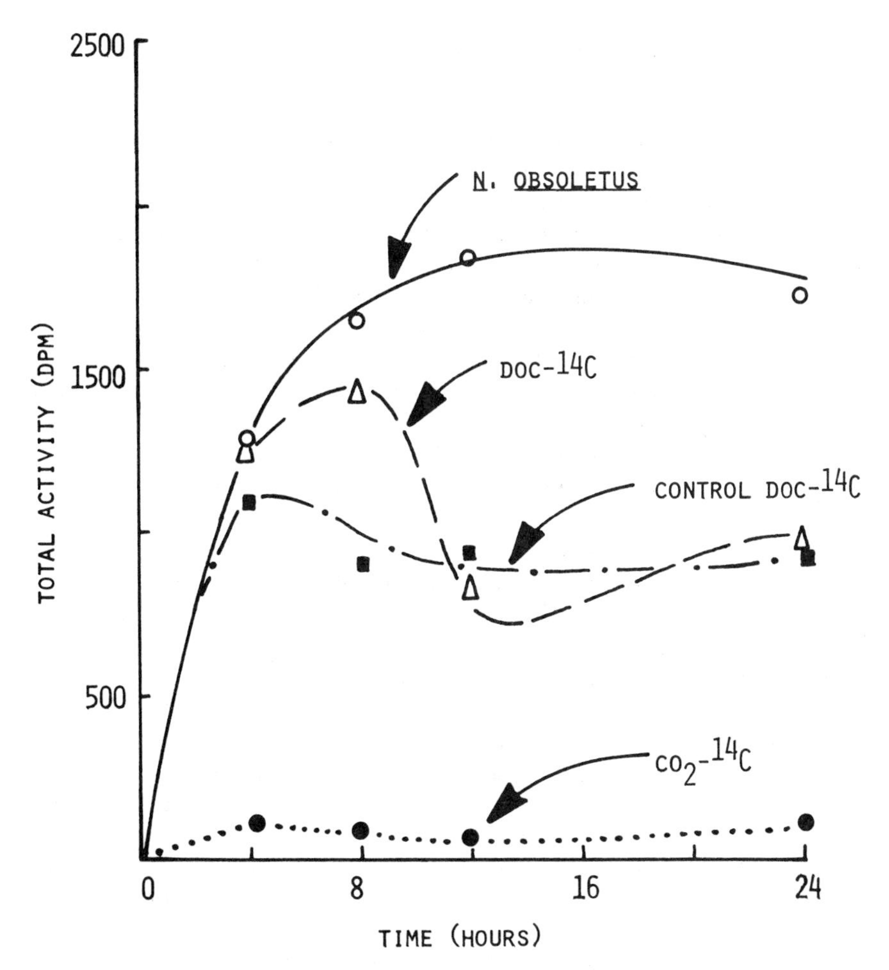

Figure 2. Specific activity of various labeled pools *Vs* incubation time; Fe 1A.

Table 2. Specific activity analyses of various labeled carbon pools; Fe 1A.

Organism	Time (hours)	Substrate dpm/mgC	H_2O dpm/mgC	$^{14}CO_2$ total	Feces dpm/mgC	*FSA/ SSA
N. obsoletus	4	1027	514	133	878	.855
	8	936	635	105	698	.746
	12	989	333	67	766	.777
	24	---	288	60	486	---
Control	4	928	562	94	---	---
	8	760	327	73	---	---
	12	938	400	60	---	---
	24	1061	402	76	---	---
Original plant material	--	1354	---	---	---	---

*Specific activity of fecal material X (specific activity of substrate)$^{-1}$.

experimental time interval (4 hours) and water specific activity remained constant for the remainder of the incubation intervals (Fig. 2). Analyses of specific activity of fecal samples collected for the time intervals indicated a 20 to 25% reduction in activity as compared to substrate activity (Table 2). The reduction in activity may not, however, have indicated a reduction by assimilation of the particulate substrate. It could be explained by dilution of substrate activity by non-substrate derived carbon, i.e., the mucous sheath characteristic of most invertebrate fecal pellets (13, 14). Although no estimate of non-substrate derived carbon in fecal pellets was determined, if the percentage lies between 10 and 20%, the assimilation efficiency based on specific activity would not be significantly different than zero, indicating that the uptake rates determined from these data are for soluble label incorporation and not due to assimilation of particulate *Spartina*-^{14}C.

The second experiment, FE 2, was of shorter duration with more frequent sampling. The substrate was offered at approximately triple the organic concentration of the first (FE 1A = 5% and FE 2 = 12 to 14% *Spartina*-^{14}C detritus). The purpose of varying both concentration and incubation time was to further evaluate label uptake and determine an adequate incubation time for subsequent experiments. The incubation intervals were 1, 2, 4, and 8 hours.

Uptake of label followed the results of the first experiment. Soluble label available for uptake was approximately three times higher, anticipated by increasing the organic concentration. Similarily, label uptake was approximately three times greater than FE 1A and appears to be concentration dependent, indicative of soluble exchange rather than particulate assimilation. Fig. 3 compares the retention rates calculated as mgC- substrate X mgC animal tissue^{-1} X hour^{-1} for the two studies.

For the last series of experiments using *Spartina* plant material (FE 8 and 9), uptake studies were run using *Spartina*-^{14}C plant material having; a) the labeled, water-soluble fractions extracted (FE 8) and b) the labeled, water-soluble and NaOH soluble (protein) fractions extracted (FE 9). The experiments were designed to measure the uptake of label by direct assimilation of labeled, crude fibre components of *Spartina* and the indirect utilization of label *via* a microbial

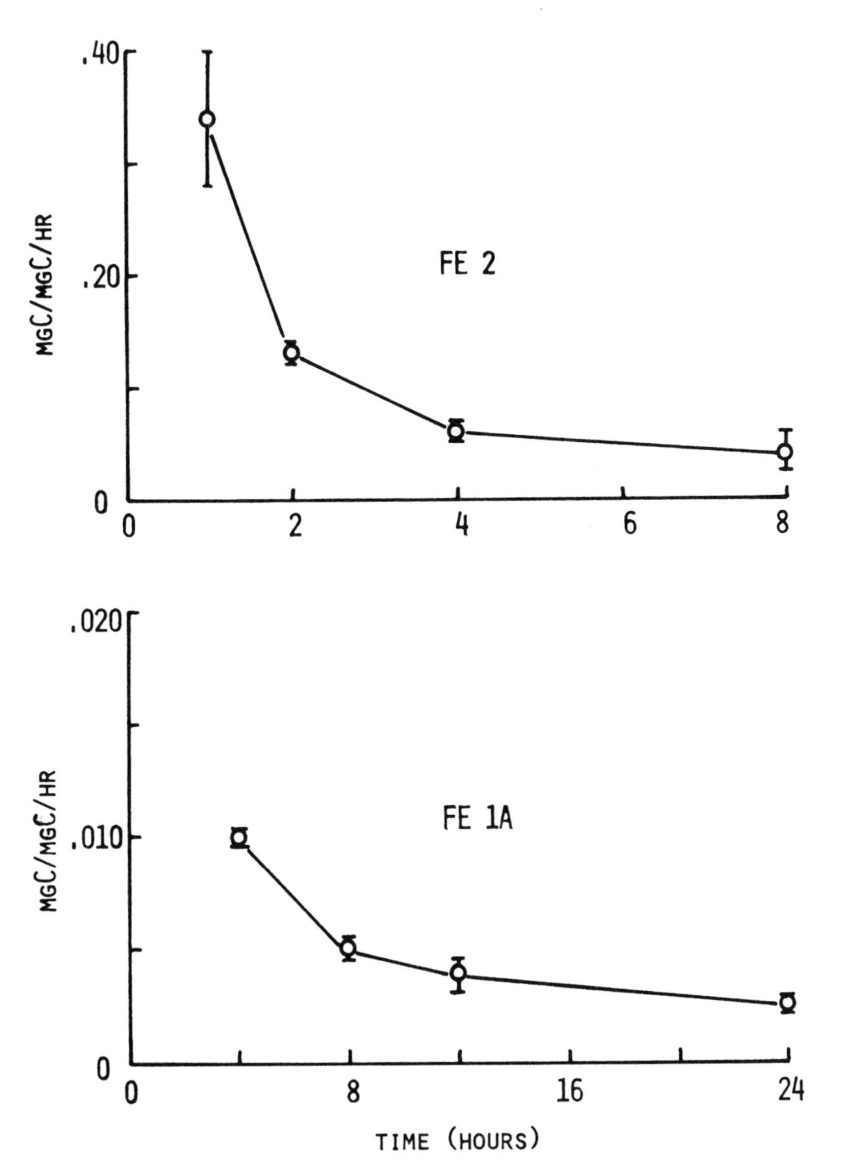

Figure 3. Comparison *N. obsoletus* uptake and retention rates feeding on non-extracted *Spartina*-^{14}C; Fe 1A and Fe 2 (± 1 S.D.).

intermediate. The microbial component added to the chambers containing the extracted plant material was a mud slurry prepared from the surface sediments of the South End Creek mud flat. The two substrates were combined (extracted plant material + mud slurry) to form the experimental feeding substrate. The uptake and retention of ^{14}C by *N. obsoletus* was followed for 8, 24, 48, and 72 hours (FE 8) and 4, 8, and 24 hours for FE 9. Results of the two long term studies are presented in Table 3.

For both experiments, there was no detectable uptake or retention of label by the snail from the extracted *Spartina* substrate alone. The only uptake and retention of label by the snail occurred on the combined substrates in experiment FE 8 which did not remove the labeled protein-fraction and included a microbial component. No uptake occurred for the completely extracted *Spartina* substrates in experiment FE 9. *Nassarius obsoletus* seems incapable of utilizing the refractory components of *Spartina* detrital material and the flux of detrital (plant) material appears totally mediated by the microbial community.

The utilization of microbial carbon by *N. obsoletus* was determined using the glucose, acetate, and glycine-^{14}C labeled core substrates. The results of ^{14}C

Table 3. Summary specific activity analyses of various labeled carbon pools; Fe 8 and Fe 9.

Substrate	Time (hr)	$^{14}CO_2$ (total)	H_2O (dpm/ml)	*N. obsoletus* (dpm/mgC)
Fe 8: H_2O soluble labeled pools extracted				
Spartina-^{14}C + Mud slurry	8	0	0	10
	24	0	0	11
	48	31	0	9
	72	204	0	14
Spartina-^{14}C	8	0	0	0
	24	0	0	0
	48	35	0	0
	72	138	0	0
Spartina-^{14}C + Mud slurry (no feeding)	8	0	0	--
	24	75	0	--
	48	129	0	--
	72	199	0	--
Fe 9: H_2O soluble and protein labeled pools extracted				
Spartina-^{14}C + Mud slurry	4	60	0	0
	8	105	0	0
	24	484	0	0
Spartina-^{14}C	4	0	0	0
	8	24	0	0
	24	338	0	0
Controls				
Fe 8 and Fe 9	4-72	0	0	--

uptake and retention are summarized in Fig. 4 for the three studies. The uptake rates were calculated from analyses of specific activity for substrate and animal tissue samples (28). Substrate specific activity was determined by carbon analyses of mud flat surface sediment samples and averaged 4.242% ± 0.411 which agrees well with reported sediment carbon values for this area (3, 10).

Maximum uptake continued for the first time interval (2 hours) based on ^{14}C body burden, indicating that a two hour incubation approximates maximum uptake rate or ingestion (Fig. 4). For the remaining incubation intervals, the specific rate of label incorporation declined and approached isotopic steady state or approximated ^{14}C retention rate for the labeled substrates. This measure

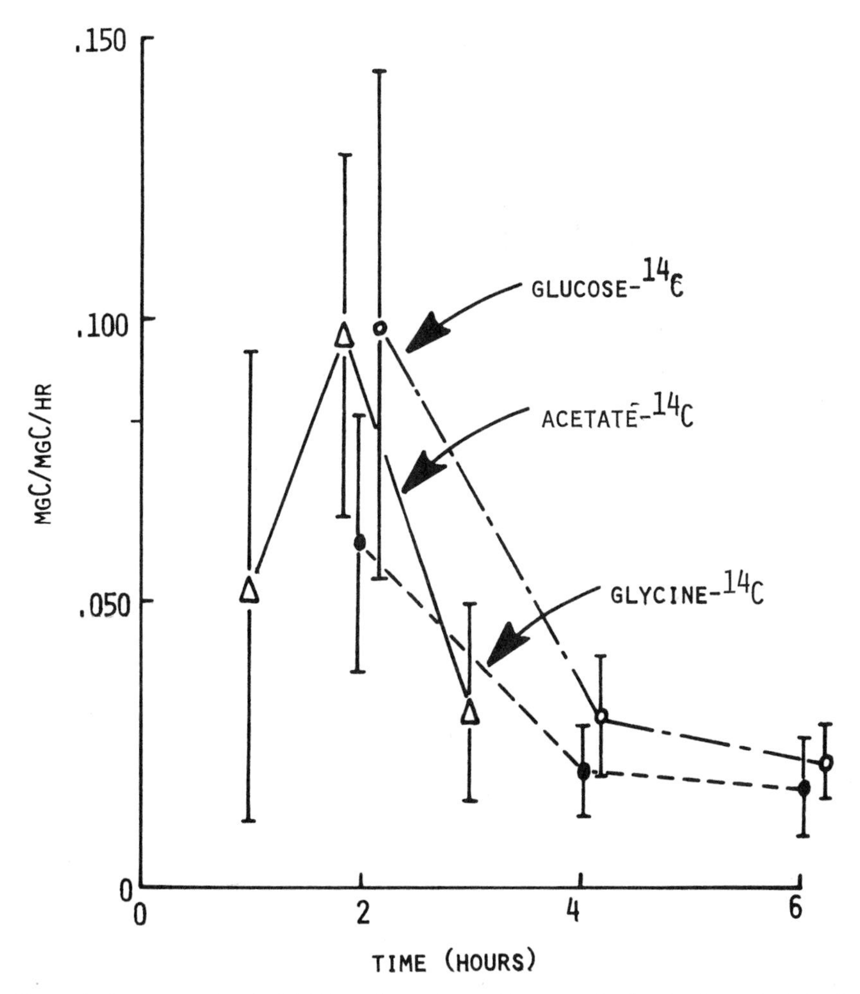

Figure 4. Summary microbial ^{14}C uptake and retention rates for *N. obsoletus*; Fe 4, Fe 5, and Fe 7 (± 1 S.D.).

(retention rate) is taken as an approximation of assimilation rate for the specifically labeled carbon source. There was no detectable release of dissolved label from any of the cores over the longest incubation period (six hours). Thus, uptake and retention of label by the snail was due solely to ingestion and assimilation of the microbial label. Analyses of control sediment samples for label absorption indicated no significant uptake or exchange and there was no detectable uptake of ^{14}C by snails fed on the control cores.

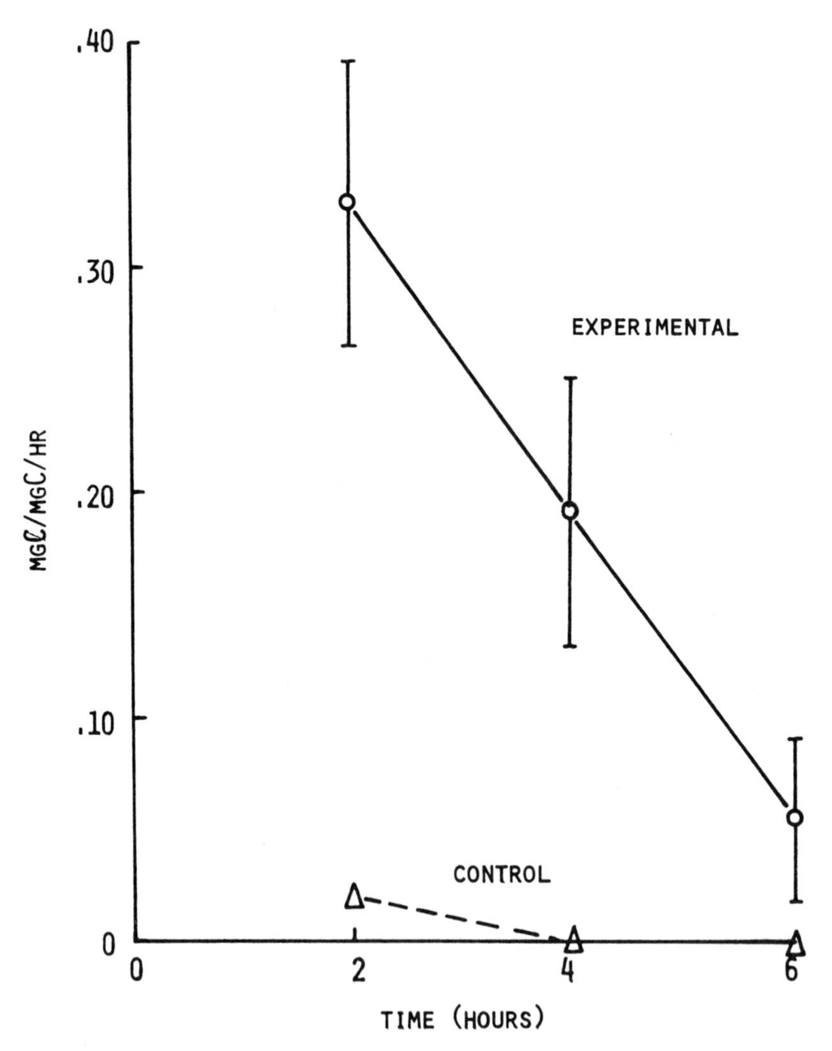

Figure 5. Uptake and retention by *N. obsoletus* on $H^{14}CO_3^-$ labeled core substrates; Fe 6 (± 1 S.D.).

The utilization of algal carbon was investigated using the same core design as for the microbial studies. Fig. 5 illustrates ^{14}C uptake and retention by the snail for the algal-labeled cores. The initial rates of uptake were significantly higher than for the microbial studies. Analyses of "grazed" and "ungrazed" core substrates indicated that there was approximately a 50% reduction in surface sediment specific activity on the "grazed" cores following the six hour incubation (grazed = 700 dpm X mgC^{-1}; ungrazed = 1175 dmp X mgC^{-1}). The snails thus appear capable of efficiently removing and incorporating the algal substrate. Formalin controls indicated no uptake or exchange of label. Non-algal utilization of the introduced label ($H^{14}CO_3$-) was assumed to be negligible.

DISCUSSION

The major input of carbon to the salt marsh and estuarine ecosystem is production by *Spartina alterniflora*. This matter/energy resource follows essentially two input pathways to the ecosystem:

1) a direct *Spartina*-grazer pathway,
2) a detrital-saprophage pathway.

Of the two, detrital production far exceeds that entering the grazer pathway (17, 23, 27). Benthic algal production, estimated at 20 to 25% of angiosperm production (9, 20), is the second major primary producer and source of fixed carbon. Unlike the detrital input, benthic algae represent a readily utilizable source of carbon and nutrients available to detrital consumers. The third major source of carbon available to benthic macrofauna is the microbial component associated with both plant fragments and the soil substrate. Using the data of Teal (27) for sediment respiration and assuming steady state conditions for the microbial community (assimilation = respiration), annual production would approximate 190 gC X yr^{-1} assuming a production ratio of 0.100 g AFDW X $kcal^{-1}$ assimilated or 0.050 gC X $kcal^{-1}$ assimilated (8, 22). The utilization of these three carbon resources by benthic feeding organisms, i.e., macroheterotrophs, is of primary concern with regard to matter/energy flux through the detrital system, as well as the possible controlling or limiting influence of these resources on individual macrohetertroph populations.

Based on the results of the experimental feeding studies, the following conclusions are drawn regarding the macroheterotrophic carbon resources:

1. The structural carbohydrate fraction of *Spartina* plant material is not directly assimilable by the mud snail, *N. obsoletus*. The material entering the salt marsh-estuarine ecosystem as detrital plant material is characterized as crude fibre-like material and primarily composed of cellulose and hemicellulose like organics (28). The direct utilization of this refractory component of the detrital input by the detritovore, *N. obsoletus*, was not supported by the experimental feeding studies using extracted, ^{14}C-labeled *Spartina*. The studies of Adams and Angelovic (1) would indicate that certain benthic feeding species are capable of assimilating detritus, by their definition. However, in these studies using labeled

eelgrass, the investigators failed to characterize, biochemically, the labeled detritus and thus cannot be certain of the assimilated fraction using ^{14}C alone. It is probable that the assimilated fraction of the labeled material was easily solublized (metabolically) materials and not characteristic of the material actually entering the food web of the eelgrass community.

2. Soluble or easily solublized components of the artifically produced plant detritus were readily incorporated by the snails. Without biochemical characterization of the labeled plant material, the results of the feeding studies using non-extracted material would have led to erroneous conclusions regarding the ability of the snail to utilize this source of detrital carbon.

3. The snails rapidly incorporated both microbial and algal labeled core substrates. Table 4 summarizes the results of the core studies extrapolated to a daily basis. Retention efficiency was calculated as the ratio between the estimated steady state uptake rate (last incubation interval) and the estimated maximum uptake rate first two hour interval). The retention rates presented in the previous section were based on specific activity of the substrate in terms of total sediment carbon. The estimate of the specific activity of the two pools (bacteria and algae) are based on the conversion values reported or derived from the works of Williams (29) for the benthic algae and Christian, et al., (3) for the microbial carbon pool. Assuming the six hour estimate to be near the steady state value, daily carbon retention rates averaged 0.0154 mgC-microbial $\times$ mgC-snail^{-1} $\times$ day^{-1} (retention efficiency = 46.6%) for the acetate, glucose, and glycine labeled core experiments and 0.0353 for the algal labeled substrate (retention efficiency = 46.4%). The retention of ^{14}C from the algal labeled substrates was 2.3 times greater than the rate for the microbial substrate. Combining the rates for the natural substrate, carbon retention could balance a daily body carbon loss of 5 to 6% for the snail, *N. obsoletus*. Benthic algae could contribute 60 to 70% of the carbon maintenance requirements (28).

Table 4. Summary microbial-^{14}C uptake and daily retention rates for *N. obsoletus* (Experiments Fe 4 to Fe 7).

Label	mgC $\times$ mgC^{-1} $\times$ day^{-1} Substrate[1]	Micro[2]	Ret. eff.[3] %
Acetate-^{14}C	0.739	0.0149	50.8
Glucose-^{14}C	0.492	0.0156	51.5
Glycine-^{14}C	0.394	0.0158	37.5
$H^{14}CO_3$	1.003	0.0353	46.4
$\bar{x}$ (bacterial label) =		0.0154	46.6
$\bar{x}$ (algae label) =		0.0353	46.4

[1] Based on substrate carbon = 0.0424 $\times$ mgDW substrate
[2] Based on bacterial label = 0.0027 $\times$ mgC sediment
[3] Based on algal carbon = 0.0027 $\times$ mgC sediments

In general, the salt marsh and estuarine ecosystem is viewed as a stable system maintained and controlled primarily by physical factors such as tides, current and temperature. The flux of matter/energy is detritally based but the interrelationships between utilizable resources and specific animal assemblages remain obscure. From the results of the feeding studies, *Spartina* detritus is not directly utilized or available as a carbon/energy resource, at least for the detritovore, *N. obsoletus*. The major carbon resources of this species are bacteria and benthic algae. If these two components of the detrital system are operating near maximum efficiency as suggested by Pomeroy et al., (21), then the role of benthic, detrital consumers in the salt marsh system becomes important, both quantitatively and as an information input to the system. Their potential role in ecosystem processes is viewed as both consumers of organic materials and nutrients and as possible regulators of microbial and/or algal community dynamics through their "grazing" activities. Little information is available regarding these hypothesized roles and further experimental research is needed to quantitate the influence of these macroheterotrophic processes on total system function.

LITERATURE CITED

1. Adams, S.M., and J.W. Angelovic. 1970. Assimilation of detritus and its associated bacteria by three species of estuarine animals. Chesapeake Sci. 11(4): 249-254.
2. Brock, M.L., R.G. Wiegert, and T.D. Brock. 1969. Feeding by *Paracoenia* and *Ephydra* (Diptera: Ephydridae) on the microorganisms of hot springs. Ecology 50(2): 192-200.
3. Christian, R.R., K. Bancroft, and W.J. Wiebe. 1975. Distribution of microbial adenosine triphosphate in salt marsh sediments at Sapelo Island, Ga. Soil Science 119: 89-97.
4. Cruz, A.A. de la. 1965. A study of particulate organic detritus in a Georgia salt marsh-estuarine ecosystem. Ph. D. dissertation, University of Georgia, 110 p.
5. Darnell, R.M. 1961. Trophic spectrum of an estuarine community based on studies of Lake Pontchartrain, Louisiana. Ecology 42(3): 553-568.
6. ____. 1964. Organic detritus in relation to secondary production in aquatic communities. Verh. Internat. Verein Limnol. 15: 462-470.
7. ____. 1967. Organic detritus in relation to the estuarine ecosystem, p. 376-382. *In* G.H. Lauff (ed.). Estuaries. AAAS Publication no. 83, Washington, D.C.
8. Doetsch, R.N., and T.M. Cook. 1973. Introduction to Bacteria and their Ecobiology. University Park Press, Baltimore, 371 p.
9. Gallagher, J.L., and F.C. Daiber. 1974. Primary production of edaphic algal communities in a Delaware salt marsh. Limnol. Oceanogr. 19: 390-395.
10. ____, F.G. Plumley, and H.F. Perkins. 1975. Measuring the carbon content of salt marsh soils. (unpubl. ms.).
11. Harrington, R.W., and E.S. Harrington. 1961. Food selection among fishes invading a high subtropical salt marsh; From onset of flooding through the progress of a mosquito brood. Ecology 42(4): 646-666.
12. Holland, D.L., and P.A. Gabbott. 1971. A microanalytical scheme for the determination of protein, carbohydrate, lipid, and RNA levels in marine invertebrate larvae. J. Mar. Biol. Assoc. U.K. 51(1): 659-668.

13. Johannes, R.E., and M. Satomi. 1967. Measuring organic matter retained by aquatic invertebrates. J. Fish. Res. Bd. Canada 24(11): 2467-2471.
14. Kraeuter, J.N., and D.S. Haven. 1970. Fecal pellets of common invertebrates of Lower York River and Lower Chesapeake Bay, Virginia. Chesapeake Sci. 11(3): 159-173.
15. Kuenzler, E.J. 1961. Structure and energy flow of a mussel population in a Georgia salt marsh. Limnol. Oceanogr. 6(2): 191-204.
16. Menzel, D.W., and R.F. Vaccaro. 1964. The measurement of dissolved organic and particulate carbon in sea water. Limnol. Oceanogr. 9(1): 138-142.
17. Odum, E.P., and A.E. Smalley. 1959. Comparison of population energy flow of a herbivorous and a deposit feeding invertebrate in a salt marsh ecosystem. Proc. Nat. Acad. Sci. 45(4): 617-622.
18. ____, and A.A. de la Cruz. 1967. Particulate organic detritus in a Georgia salt marsh-estuarine ecosystem, p. 383-388. *In* G. Lauff (ed.). Estuaries, AAAS Publ. no. 83, Washington, D.C.
19. Odum, W.E. 1968. The ecological significance of fine particle selection by the striped mullet, *Mugil cephalus*. Limnol. Oceanogr. 13(1): 92-98.
20. Pomeroy, L.R. 1959. Algal productivity in salt marshes of Georgia. Limnol. Oceanogr. 4(4): 386-397.
21. ____, K. Bancroft, J. Breed, R.R. Christian, D. Frankenberg, J.R. Hall, L.G. Maurer, W.J. Wiebe, R.G. Wiegert and R.L. Wetzel. 1976. Flux of organic matter through a salt marsh, p. 000-000. *In* M.L. Wiley (ed.), Estuarine Processes, Academic Press, N.Y.
22. Prochazka, G.J., W.J. Payne, and W.R. Mayberry. 1970. Calorific content of certain bacteria and fungi. J. Bacteriol. 104(2): 646-649.
23. Reimold, R.J., J.L. Gallagher, R.A. Linthrust, and W.J. Pfeiffer. 1975. Detritus production in coastal Georgia marshes. p. 217-228. *In* Cronin, L.E. (ed.) Estuarine Research, 1, Academic Press, New York.
24. Schleiper, C. 1972. Research Methods in Marine Biology. Univ. Washington Press, Seattle. 356 p.
25. Shanholtzer, S.F. 1973. Energy flow, food habits, and population dynamics of *Uca pugnax* in a salt marsh system. Ph.D. dissertation, University of Georgia, 91 p.
26. Sorokin, Ju. I. 1966. Carbon-14 method in the study of the nutrition of aquatic animals. Itn. Rev. ges. Hydrobiol. 51(2): 209-224.
27. Teal, J.M. 1962. Energy flow in the salt marsh ecosystem of Georgia. Ecology 43(4): 614-624.
28. Wetzel, R.L. 1975. An experimental-radiotracer study of detrital carbon utilization in a Georgia salt marsh. Ph.D. dissertation, Univ. of Gerogia Libraries, 103 p.
29. Williams, R.B. 1962. The ecology of diatom populations in a Georgia salt marsh. Ph.D. dissertation, Harvard University, 149 p.
30. Woeller, F.H. 1962. Liquid scintillation counting of $^{14}CO_2$ with phenethylamine. Anal. Biochem. 2(5): 508-511.

CONTRIBUTION OF TIDAL MARSHLANDS TO MID-ATLANTIC ESTUARINE FOOD CHAINS[1]

Donald R. Heinle, David A. Flemer[2], and Joseph F. Ustach[3]
University of Maryland
Center for Environmental and Estuarine Studies
Chesapeake Biological Laboratory
Solomons, Maryland 20688

ABSTRACT: The seasonal pulsing of flows of carbon from tidal marshes results in similarly pulsed production of zooplankton in some estuaries. Direct evidence, experimental feeding of laboratory copepods, and measurements of feeding by captured wild copepods, supports the hypothesis that detritus is rapidly incorporated into higher trophic levels.

The timing of the production of copepods based on a detrital food source is such that year-to-year variations in amount may be an important factor in the survival of larvae of anadromous fish.

INTRODUCTION

Contributions by marshes to the pool of organic matter available for consumption by organisms in estuaries have been measured at sites from Florida to Massachusetts on the Atlantic coast. Earlier studies, done in Georgia and North Carolina, suggested that a substantial part of the annual production of marsh grasses was exported to adjacent estuaries (3, 22, 27, and 28) with slightly greater quantities exported during the fall months (3). More recent measurements, on a poorly flooded tidal marsh on the Patuxent estuary in Maryland, suggest little export of organic matter from some marshes, while other marshes,

[1] Contribution No. 685, Center for Environmental and Estuarine Studies, University of Maryland.

[2] Present address: Office of Biological Studies, U.S. Fish and Wildlife Service, Department of the Interior, Washington, D.C. 20240.

[3] Present address: Department of Zoology, North Carolina State University, Raleigh, North Carolina 27607.

scoured by ice and storms, contribute all of their annual standing crop to the estuary (12). In the Patuxent, most of the detritus from marshes enters during the months of January and February, when ice occurs most often, leading to an increase in concentrations of particulate carbon during that period (5, 11). The particulate organic carbon entering the Patuxent from the marshes is roughly equal in quantity to the sum of that fixed *in situ* by algae and that contributed from upland sources by the river (12).

The upper part of Chesapeake Bay is similarly pulsed by detritus in the late winter, but that material is from upland sources (2). In the upper Chesapeake Bay, Biggs and Flemer found that detrital carbon from upland sources represents 91% of the annual carbon budget. Bissel Cove, Rhode Island, received a pulsed input of detritus following the melting of ice in the spring (19).

The involvement of ice in the formation of detritus suggested that year-to-year climatic variables might affect the production of detritus in the Patuxent (12).

Once detritus enters an estuary it rapidly loses its identity due to breakdown by physical forces and micro-organisms, perhaps enhanced by the activities of larger animals (26). The samples taken by Schubel and Schimer (23) with their clever device demonstrate the amorphous nature of estuarine particulate matter.

A variety of estuarine organisms have been found to use detritus as food, or at least to be potential users (4, 6, 15, 21, 22), including some species of fish (3, 20, 25). The mysid shrimp, *Neomysis americana*, appears to be an important estuarine detritivore (14, 25), but direct evidence for utilization of detritus is lacking for that species.

In the Patuxent estuary the pulsed input of detritus in late winter results in substantial production of the copepod *Eurytemora affinis* (11). There is direct evidence that *E. affinis* can use the micro-organisms on detrital particles as food (13). The requirements for food by *E. affinis* suggested turnover times of 8 to 83 days for the entire pool of particulate matter in the upper Patuxent during the period of high production of *E. affinis*.

The age structure and population dynamics of *E. affinis* suggest that predation during April and May is responsible for the decrease in numbers of that species (11). The timing of the spring production of *E. affinis* in the Patuxent matches the occurrence of the larvae of anadromous fish.

This paper presents additional evidence for the use of detritus by *E. affinis*, and suggests a link between climatic factors and production of detritus. A hypothesis is suggested that may eventually help explain the relationship between climatic factors and striped bass year-classes observed first by Merriman (17).

METHODS

Feeding experiments were done with freshly collected *E. affinis* from the upper Patuxent estuary and water from that location, or with animals and water from Drum Point near Solomons, Maryland, at the confluence of the Patuxent estuary with Chesapeake Bay. In some experiments mixed ages of copepods were used as they occurred in the estuary and at naturally occurring densities. In these experiments, water containing copepods and natural distributions of smaller than 63 μm particles was placed into one 180-ml bottle. Water containing only small particles was placed into a control bottle. This was done as follows. Adult copepods were pipetted into a breaker containing about 350 ml of 63 μm-filtered water from the study location. Experimental and control bottles were then filled by swirling the beaker and pouring alternately into the two bottles, being careful to put the copepods into the experimental bottle. Identical densities and size-distributions of particles in the two beakers resulted from this technique. Most of the naturally occurring particles that were smaller than copepods were less than 30 μm in diameter, so size-distributions of particles were unaffected by screening through the 63-μm filter. Specific age groups of copepods were sorted by hand with a small pipette for other experiments. Juvenile copepods were sorted into a beaker containing Millipore-filtered water from the study site, transferred to a second rinse of filtered water, and then placed into an experimental bottle partially filled with 63 μm-filtered water, with as little rinse water as possible. The volume of rinse water transferred was noted and an equivalent amount placed into the control bottle. Slight dilution of food particles resulted, but densities and size-distributions were identical in control and experimental bottles. After a period of incubation (about 24 hr) on a rotating wheel in the dark, the bottles were removed and particle densities and size-distributions determined on an electronic particle counter (Electrozone/Celloscope®, Particle Data, Inc.) coupled to a PDP 8/M computer (Digital Equipment Corp.) and a teletype with a tape punch and reader. Output from the particle counter is digital with both printed copy and punched tape. We counted 128 channels over a range of 2 to 38 μm mean spherical diameter. A computer program allowed automatic computation of filtering rates, ingestion, and electivity index in each channel, as well as total volumes of particles ingested with a "t" test for significance of filtering rates. Each bottle was analyzed in triplicate and on some occasions duplicate experiments were run. Samples were diluted to about 40,000 particles per ml or less from analysis to avoid coincidence of counts.

Visual counts of natural particles were made by allowing 500 ml of natural water preserved in 3% neutralized formalin to settle for 40 hr to one-fourth volume. Counts were made on an American Optical haemocytometer at 1,000 X oil immersion magnification with bright field optics. Five aliquots were taken from which 2 fields of 0.1 mm^3 each were counted. A No. 1 coverslip was used instead of a standard coverslip (24). Particles judged to be alive (phytoplankton with intact cells and chloroplasts evident) and nonliving particles were grouped into 2-5, 5-10, 10-20, and 20-35 μm diameter size-classes for comparison with counts by the automatic particle counter.

Turnover times of particulate matter were calculted by dividing total volumes of particles in control bottles by the volumes of particles removed by grazing in 24 hr.

Production of *E. affinis* for comparison with the data of Heinle (9, 10) was calculated from the data of Heinle and Flemer (11) by multiplying productivity (of egg-producing females) times the biomass of the entire population of *E. affinis*.

RESULTS AND DISCUSSION

Feeding experiments were done on April 3-4 and April 29-May 1, 1975, using a range of biomass of copepods that bracketed the values reported by Heinle and Flemer (11). Consumption of particulate matter was constant over a wide range of biomass of copepods in the bottles regardless of the source of the animals and water (Fig. 1).

The constancy of volumes of particles removed may be a container effect as described by Anraku (1), or may represent feeding homeostasis on the part of the copepods. More experiments are clearly needed to assess this problem.

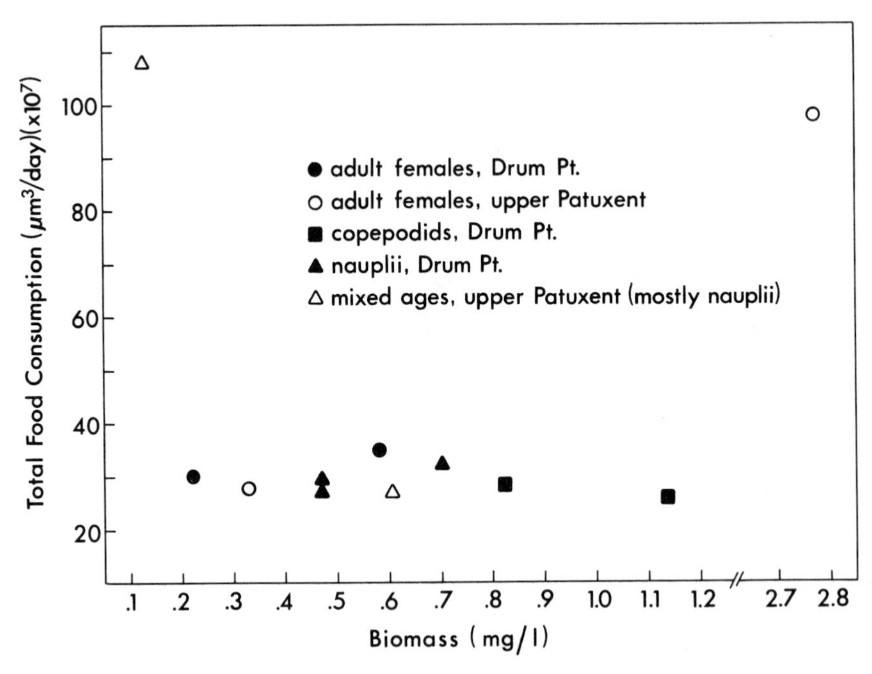

Figure 1. Consumption (μm^3 per 180-ml bottle per day) versus biomass (mg dry weight/liter) of *E. affinis* in the feeding experiments. Biomass was converted to mg/liter for comparison with the data of Heinle (10) and Heinle and Flemer (11).

The *E. affinis* in water from Drum Point generally selected particles in the way described by Wilson (29), i.e., when a biomass peak was present, filtering rates (8) were highest at the peak. The water from the upper Patuxent never had a well defined peak of volume (biomass) of particles, and size selection was not observed in those experiments. The coincidence of the data on total volumes filtered in the water from the two locations is particularly interesting in the light of the differences in size selection.

The highest value of biomass in Fig. 1 lies outside the range reported by Heinle and Flemer (11) for the months of February through May. Turnover times for detritus reported by Heinle and Flemer (11) are shown in part A of Table 1 with turnover times calculated for all the particulate matter present during their study. Part B of Table 1 shows turnover times for the particulate matter present in the feeding experiments. In water from the upper Patuxent, the *Eurytemora* removed particles at rates that suggested turnover times of 4.8 to 23 days (Table 1, part B), well within the range of 3.9 to 77 days calculated from the production of the wild population (Table 1, part A).

Table 1. Part A: Detrital turnover times (days) from Heinle and Flemer (11) and turnover times for all particulate carbon (phytoplankton plus detritus).

	March		April							May	
Station	9	10	6	7	8	9	10	12	13	6	7
Detrital Turnover Time	52	5.3	83	20	13	43	3.5	8	8	12	79
Algae plus Detritus Turnover Time	48	6.4	77	21	14	40	3.9	8	8	15	55

Part B: Turnover times (days) of suspended matter from feeding experiments and biomass of copepods present (mg per liter dry weight).

	April 3-4		April 29-May 1		
Copepods used	Adult females	Mixed ages	Adult females	Adult females	Mixed ages
Biomass	2.778	0.602	0.333	0.333	0.127
Turnover Time Upper Patuxent Water	4.79	14.1	12.2	23.2	6.54

	April 3-4			April 29-May 1		
Copepods used	Adult females	Cope-podid I	Cope-podid II	Adult females	Nauplii V & VI	Nauplii V & VI
Biomass	0.583	1.137	0.823	0.222	0.472	0.472
Turnover Time Water from Drum Point	2.84	2.76	2.82	4.08	4.35	4.27

Table 2. Size fractions of particles from the upper Patuxent estuary.

	Particles per ml				
	Size fractions (μm) of detritus				Live
Date	2-5	5-10	10-20	20-35	Phytoplankton
3/6/75	743,250	111,000	52,000	7,250	21,000
3/20/75	1,293,000	492,250	129,250	18,750	13,000
4/17/75	1,128,750	384,250	184,750	34,500	19,000

Natural particles were sized visually on three dates during March and April, 1975 (Table 2), unfortunately not the same dates as the feeding experiments. Numerically, nonliving particles were always predominant, as live phytoplankton comprised only 2.24, 0.67, and 1.08% of the total on March 6 and 20, and April 17, respectively. While most of the particles were less than 5 μm, there was usually a broad peak of particle volume in the 5 to 10 and 10 to 20 μm fractions. There were many particles smaller than 2 μm that were not counted. Most of the volume of the live phytoplankton fell within the 5 to 10 μm size-class. *Thalassiosira nana* (about 6 μm diameter) was the most abundant species on all three days and comprised approximately half the algal biomass. *Monallantus* sp. (2 X 3 μm) was second in abundance on March 20 and April 17, while an unidentified green alga (about 2 μm) was second in abundance on March 6. The same three species were among the four most abundant species on March 13 and April 24 also, when only the relative abundance of the phytoplankton was determined, suggesting a rather stable assemblage of phytoplankton during the spring period of high requirements for energy by *Eurytemora*.

One might expect that particulate carbon might range from about 3 to 20% of the total seston in the upper Patuxent (5, 13), so organic matter might comprise about 6 to 40% of the total particles counted.

The observations above strengthen the hypothesis that the requirements for food by *E. affinis* are met largely by detritus. In the upper Patuxent, about half of the total annual supply of organic carbon came from the marshes (12). Virtually all of the marsh detritus entered the estuarine system during January and February. The upper Chesapeake Bay receives a similarly timed strong seasonal input from upland sources.

Production of *E. affinis* was calculated for areas II and III of Heinle (10) who found 1.95 and 1.30 g m^{-3} respectively of dry weight production in the two areas during 1966. In 1970, production was 8.65 and 6.15 g m^{-3} dry weight in areas II and III of the Patuxent. Production of *E. affinis* was thus about 4.5 times greater during a spring that led to a strong year-class of striped bass than during one that did not. Mortalities of *E. affinis* suggest that predation is relatively light during February and March and becomes intense during April and May, causing the population to decline (11). A relatively simple food chain during April and May involving detritus, *E. affinis*, *N. americana*, and fish larvae was suggested by Heinle and Flemer (11). The fact that scouring by ice was the major means by which marsh detritus entered the estuary suggests that enrichment of the detrital food chain in the upper Patuxent, and similar estuaries, might be inversely related to the severity of winters, particularly the months of January and February. Merriman (17) observed that the occurrence of strong year-classes of striped bass, *Morone saxatilis,* was related to low winter temperatures. He observed that strong year-classes occurred only (but not always) after severe winters, while mild winters never produced a strong year-class. Merriman found a correlation of r = −0.354 (S.01) between deviations from mean winter temperatures and the "entire catch record" of striped bass, with a 2-year time lag. He also observed that *Eurytemora* and, more importantly, *Neomysis* were components in the diet of young-of-the-year striped bass, as did Ganssle (7).

Landings of striped bass (30) from the Patuxent, Chester, and Potomac rivers, upper central Chesapeake Bay, and the entire state of Maryland were correlated with an index of severity of winters, the total number of days during January and February that minimum temperatures were below freezing four years previously (Fig. 2). The correlation coefficient for the Chester River was 0.102 (N.S.). The correlation coefficients for the Patuxent and Potomac were 0.273 and 0.274, both having probability of occurrence of about 25%. The correlation for the upper central Chesapeake Bay was 0.535 (S.01) with a significant slope (t = 2.53, S.025). The total Maryland landings were also significantly correlated (not shown in Fig. 2), r = 0.446, S.05. Two of the relationships were tested using the index proposed by Merriman, i.e., deviation from long-term monthly mean temperatures. Merriman appeared to use the average deviation of the months February through May at Washington, D.C., while we used the sum of the deviations for January and February at Baltimore-Washington airport. The relationship for the Patuxent was not significant, while the correlation for the upper central Chesapeake Bay increased; r = 0.623 (S.0001), and slope b was again significant, t = 3.288 (S.005) (Fig. 3). The use of catch records from individual rivers was probably premature because it is necessary to make the unproven assumption that homing occurs. The fact that a dominant year-class is reflected in the catch for several years (16, 17) undoubtedly increased the variance in our correlations, suggesting that the trends for the Patuxent and Potomac (Fig. 2) may reflect true correlations. Merriman (17) showed that all the dominant year-classes between 1892 and 1934 originated during colder than normal winters. Koo (16) identified dominant year-classes between 1934 and 1964, and identified a 6-year cycle with three unexpectedly low year-classes, 1946 and 1952, and possibly 1928. We now know that the 1970 year-class is very strong. The coincidence between strong year-classes of striped bass and subnormal winter temperatures is universal between the years 1892 and 1970 (Table 3). In addition, three year-classes that were expected to be strong and were not (16), hatched after normal, or warmer than normal winters (1928,

1946, and 1952 in Table 3). An interesting note is that from 1928 to 1949 only 5 of 22 winters were colder than normal while from 1950 to 1971, 14 of 21 were. Catches of striped bass have been considerably higher during the latter period (16).

If detritus is the key to the food chain in the upper estuaries during the spring months, as we propose, production of detritus from upland sources must also be related to low winter temperatures. Biggs and Flemer (2) estimated that 46,200 metric tons of organic carbon entered Chesapeake Bay during February and

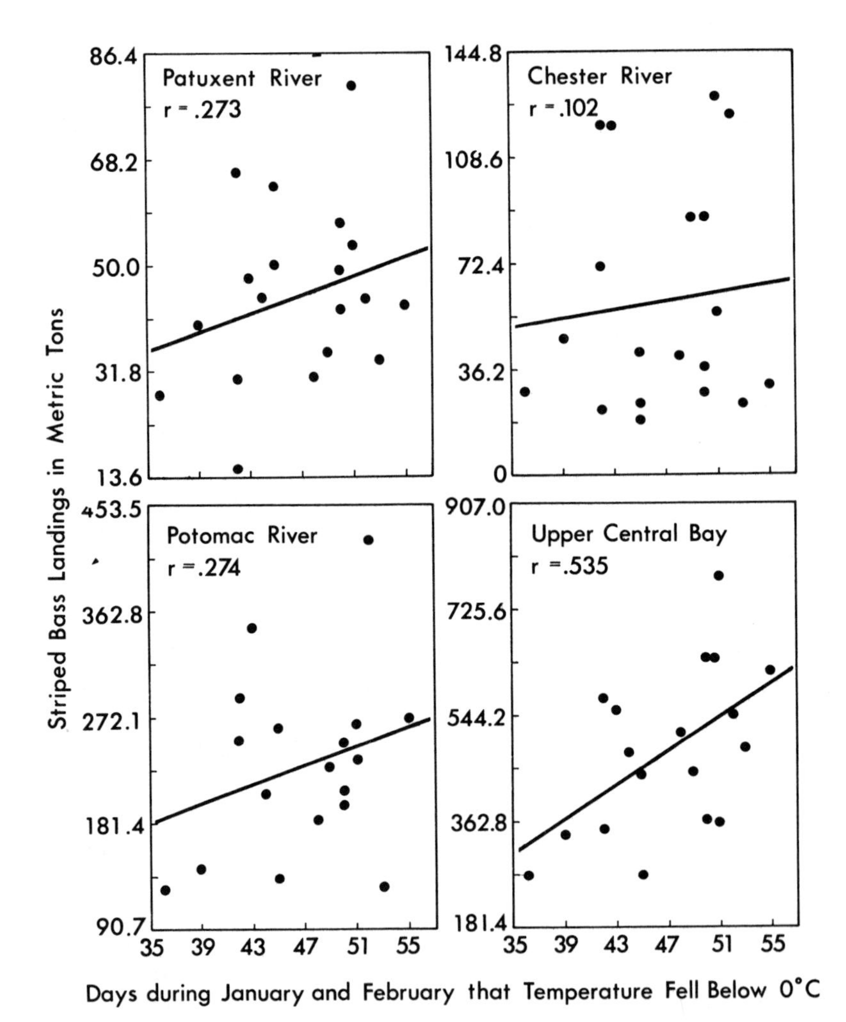

Figure 2. Regression of striped bass landings against total days during January and February that mimimum temperatures were below freezing four years previously.

March. Assuming that all of the 92.3 km^2 of tidal marshes in Maryland (18) contributed at the same rate as those in the Patuxent (12), a tenuous extrapolation at best, we calculate only 7,250 metric tons of particulate carbon from marshes during the same period.

Our hypothesized food web obviously needs additional study. The relative contributions from marshes and upland sources should be determined, the details of the detritus-microbiota-*Eurytemora-Neomysis* web should be quantified, and the effect of food supply on survival of wild striped bass young-of-the-year should be measured.

ACKNOWLEDGEMENTS

We would like to thank J.D. Allan, R.T. Huff, and S. Richman for assistance with the feeding experiments and J. Jones for the counts of particle sizes. P.W. Jones provided unpublished data on the 1973 landings of striped bass. Illustrations were prepared by F. Younger and M.J. Reber.

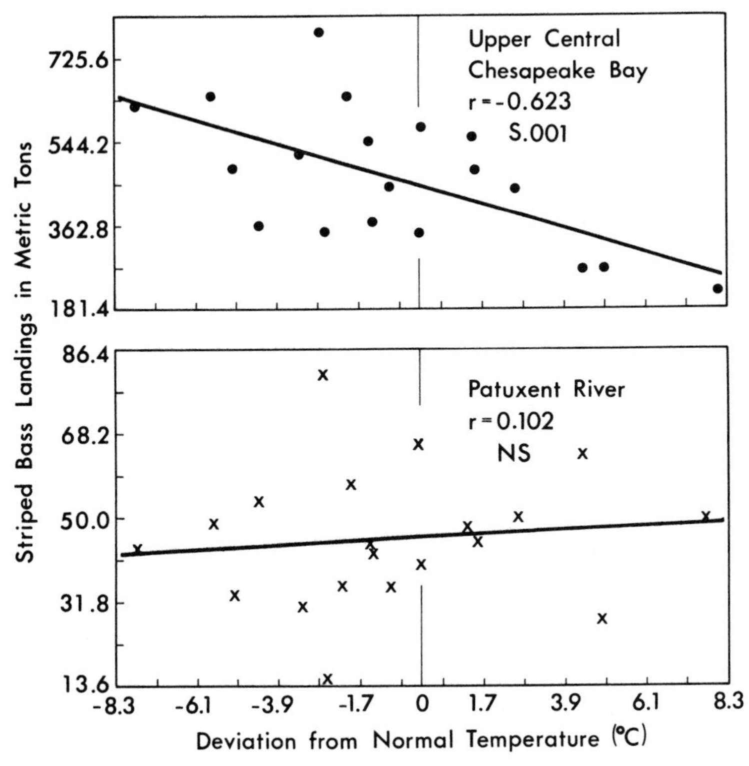

Figure 3. Regression of striped bass landings versus deviations from long-term mean temperature four years previously.

Table 3. Dominant and expected* [6-year cycle (16)], dominant year-classes and deviations from long-term mean temperatures (°C), year-classes after Merriman (17) and Koo (16) with the 1970 year-class added. Merriman's temperature deviations 1892-1934, ours 1928-1970 in parentheses.

Year-class	1892	1893	1904	1920	*1928	1934	1940	*1946	*1952	1959	1964	1970
Dev. from long-term mean temperatures	-4.2	-2.8	-6.7	-3.3	0.0 (+2.6)	-6.1 (-3.3)	(-3.2)	(+4.1)	(+4.4)	(-4.4)	(-1.9)	(-4.0)

*Expected dominant year-classes which did not lead to good catches (16).

REFERENCES

1. Anraku, M. 1964. Some technical problems encountered in quantitative studies of grazing and predation by marine planktonic copepods. J. Oceanogr. Soc. Jap. 29:19-29.
2. Biggs, R.B., and D.A. Flemer. 1972. The flux of particulate carbon in an estuary. Mar. Biol. (Berl.) 12:11-17.
3. Cruz, A.A., de la. 1965. A study of particulate organic detritus in a Georgia salt marsh estuarine ecosystem. Ph.D. dissertation, Univ. Georgia, 110 p.
4. Darnell, R.M. 1967. Organic detritus in relation to the estuarine ecosystem, p. 376-382. *In* G.H. Lauff (ed.), Estuaries. Am. Assoc. Adv. Sci., Washington, D.C.
5. Flemer, D.A., D.H. Hamilton, Carolyn W. Keefe, and J.A. Mihursky. 1970. The effects of thermal loading and water quality on estuarine primary production - Final technical report for the period August 1968 to August 1970. [Submitted to the Office of Water Resources Research, U.S. Department of Interior.] Univ. Md. Nat. Resou. Inst. Ref. No. 71-6. U.S.N.T.I.S. (#PB 209-811).
6. Frankenberg, D., and K.L. Smith, Jr. 1967. Coprohagy in marine animals. Limnol. Oceanogr. 12:443-450.
7. Ganssle, D. 1966. Fishes and decapods of San Pablo and Suisan Bays. Calif. Dept. Fish Game Fish Bull. 133:64-94.
8. Gauld, D.T. 1951. The grazing rate of planktonic copepods. J. Mar. Biol. Assoc. U. K. 29:695-706.
9. Heinle, D.R. 1969. Temperature and zooplankton. Chesapeake Sci. 10(3):186-209.
10. ____. 1969. Effects of temperature on the population dynamics of estuarine copepods. Ph.D. Thesis, Univ. Md., 132 p.
11. ____, and D.A. Flemer. 1975. Carbon requirements of a population of the estuarine copepod, *Eurytemora affinis*. Mar. Biol. (Berl.) 31:235-247.
12. ____, and ____. 1976. Flows of materials between poorly flooded tidal marshes and an estuary. Mar. Biol. (Berl.) (in press).
13. ____, ____, J. F. Ustach, R.A. Murtagh, and R.P. Harris. 1974. The role of organic debris and associated microorganisms in pelagic estuarine food chains. Univ. Md., Water Resour. Res. Cen. Tech. Rpt., 22, 54 p + Appendices. U.S.N.T.I.S. (#PB232949/AS).
14. Hopkins, T.L. 1965. Mysid shrimp abundance in surface waters of Indian River inlet, Delaware. Chesapeake Sci. 6(2):86-91.
15. Jørgensen, C.B. 1966. Biology of suspension feeding. Permagon Press, New York. 357 p.
16. Koo, T.S.Y. 1970. The striped bass fishery in the Atlantic states. Chesapeake Sci. 11:73-93.
17. Merriman, D. 1941. Studies on the striped bass (*Roccus saxatilis*) of the Atlantic coast. U.S. Fish Wildl. Serv., Fish. Bull. 35, 77 p.
18. Metzgar, R.G. 1973. Wetlands in Maryland. Md. Dept. of State Planning, Baltimore, Md.
19. Nixon, S.W., and C.A. Oviatt. 1973. Ecology of a New England salt marsh. Ecol. Monogr. 43:463-498.
20. Odum, W.E. 1970. Pathways of energy flow in a south Florida estuary. Ph.D. Dissertation, Univ. Miami, 162 p.
21. Qasim, S.Z., and V.N. Sankaranarayanan. 1972. Organic detritus of a tropical estuary. Mar. Biol. (Berl.) 15:193-199.

22. Teal, J.M. 1962. Energy flow in the salt marsh ecosystem of Georgia. Ecology 32:614-624.
23. Schubel, J.E., and E.W. Schimer. 1972. A device for collecting *in-situ* samples of suspended sediment for microscopic analysis. J. Mar. Res. 30:269-273.
24. Van Valkenburg, S.D., and D.A. Flemer. 1974. The distribution and productivity of nannoplankton in a temperate estuarine area. Estuarine Coastal Mar. Sci. 2:311-322.
25. Wass, M.L., and T.D. Wright. 1969. Coastal wetlands of Virginia–Interim report to the Governor and General Assembly. Spec. Rpt. Appl. Mar. Sci. Ocean Eng. 10, 154 p.
26. Welsh, B.L. 1973. The grass shrimp, *Palaemonetes pugio,* as a major component of a salt marsh ecosystem. Ph.D. Dissertation, Univ. Rhode Island, 76 p.
27. Williams, R.B. 1966. Annual phytoplankton production in a system of shallow temperate estuaries, p. 699-716. *In* H. Barnes, ed. Some contemporary studies in marine science. George Allen and Unwin Ltd., London.
28. ——, and M.D. Murdoch. 1969. The potential importance of *Spartina alterniflora* in conveying zinc, manganese and iron into estuarine food chains. Proc. Natl. Symp. Radio-ecol. 2:431-439.
29. Wilson, D.S. 1973. Food size selection among copepods. Ecology 54: 909-914.
30. Wilson, J.S., R.P. Morgan, III, P.W. Jones, G.B. Gray, J. Lawson, and R. Lunsford. 1974. Potomac River fisheries study, striped bass spawning stock assessment. Univ. Md. Center for Environmental and Estuarine Studies, Ref. No. 74-151, 66 p.

CIRCULATION MODELS

Convened by:
A.J. Elliott
Chesapeake Bay Institute
The Johns Hopkins University
Baltimore, Maryland 21218

The majority of estuaries along the east coast of the United States are partially-mixed, coastal plain type estuaries. These were formed by the drowning of river valleys during the most recent rise in sea level and are characterized dynamically be vertical salinity variations of 2-3°/oo and by a net circulation pattern which is directed seaward in the upper layers and landward in the lower layers. This circulation pattern is the result of the interaction between two pressure distributions: one, directed upstream, and due to the horizontal variation in salinity and the other, directed downstream, and due to the mean slope of the free surface. The interaction of these pressure forces with the tidally generated turbulence leads to the characteristic *vertical variations in salinity and velocity* which are observed in partially-mixed estuaries.

Much of the early work on the numerical modeling of estuarine systems involved vertically integrated equations and while this approach was useful for studying tidal dynamics and storm surges, it did not reveal the characteristic internal features of the estuarine circulation. This section, therefore, will discuss models which do incorporate the vertical structure, either by using two dimensional grids arranged vertically or by full three-dimensional schemes.

Perhaps the most immediate application of estuarine models is to applied problems where a model can assist in the making of management decisions. However, they do have a more fundamental use when they are used to test, by numerical experiment, theoretical ideas which relate to the dynamic and kinematic balances within an estuary. As computers become larger and faster, numerical experimentation may play an important role in our understanding of turbulent systems.

Finally, a note of caution. A numerical model can only be as good as the data used to calibrate *and independently verify it*. Therefore, modelers working with two- or three-dimensional schemes must conduct comprehensive field studies and obtain *more than one set of data*. While much effort has gone into numerical schemes, adequate physical verification has yet to be made.

A TWO-DIMENSIONAL NUMERICAL MODEL FOR THE SIMULATION OF PARTIALLY MIXED ESTUARIES

Alan Fred Blumberg[1]
Dept. of Earth and Planetary Sciences
The Johns Hopkins University
Baltimore, MD 21218

ABSTRACT: A real time numerical model is developed to describe the longitudinal and vertical distributions of velocities and salinities as well as tidal amplitudes for partially mixed estuaries. One assumes the flow to be laterally homogeneous and uses realistic estuarine bathymetry. The external inputs to the model are the salinity and tidal amplitude as a function of time at the ocean boundary and the freshwater discharge at the river boundary.

The model includes the continuity, salt and momentum balance equations, coupled by an equation of state. The elimination of the lateral momentum balance equation permits numerical solutions with little computing time. The numerical technique conserves salt, volume and momentum in the absence of dissipative effects.

Simulations show the salinity intrusion to be highly sensitive to the vertical eddy viscosity, with minor changes to the tidal amplitude. Results from the application of the model using a stability dependent eddy viscosity and eddy diffusivity to the Potomac River yield distributions comparable to field observations.

INTRODUCTION

A procedure to study and predict the dynamics and kinematics of partially mixed estuaries is important from both physical and ecological viewpoints. The purpose of this research is to present and apply a real time numerical model capable of describing the longitudinal and vertical distributions of velocities and salinities as well as tidal amplitudes for partially mixed estuaries.

[1] Present Address: Geophysical Fluid Dynamics Program, Princeton University, Princeton, NJ 08540.

GOVERNING EQUATIONS

The equations which govern the dynamics of estuaries have been simplified through a lateral (y) averaging technique by Blumberg (1) to eliminate the lateral dependence. In deriving the final form of the governing equations, equations (1) through (5), the assumptions that all variables except v are independent of y, and that the lateral average of v is zero have been used. The resulting equations using the Boussinesq approximation and the hydrostatic assumption, can be written as follows:

$$\frac{\partial}{\partial x}(uB) + \frac{\partial}{\partial z}(wB) = 0 \tag{1}$$

$$\frac{\partial}{\partial t}(B_\eta \eta) + \frac{\partial}{\partial x}\int_{-H}^{\eta}(uB)\,dz = 0 \tag{2}$$

$$\frac{\partial}{\partial t}(sB) + \frac{\partial}{\partial x}(suB) + \frac{\partial}{\partial z}(swB) - \frac{\partial}{\partial x}\left(BK_x\frac{\partial s}{\partial x}\right) - \frac{\partial}{\partial z}\left(BK_z\frac{\partial s}{\partial z}\right) = 0 \tag{3}$$

$$\begin{aligned}&\frac{\partial}{\partial t}(uB) + \frac{\partial}{\partial x}(uuB) + \frac{\partial}{\partial z}(uwB) - \frac{\partial}{\partial x}\left(BN_x\frac{\partial u}{\partial x}\right)\\&-\frac{\partial}{\partial z}\left(BN_z\frac{\partial u}{\partial z}\right) + k\,u\left|u\right|\frac{\partial B}{\partial z}\\&+ Bg\frac{\partial \eta}{\partial x} + \frac{Bg}{\rho_0}\frac{\partial}{\partial x}\int_{z'}^{0}\rho\,dz' = 0\end{aligned} \tag{4}$$

$$\rho = \rho_0(\alpha + \beta s) \tag{5}$$

where x and z are the co-ordinates pointing longitudinally up the estuary, and vertically upwards respectively; u and w, are the corresponding velocity components; B, the width of the estuary; η, the surface displacement; H, the undisturbed depth; s, the salinity; K_x, N_x, the longitudinal turbulent diffusivity and viscosity respectively; K_z, N_z, the vertical turbulent diffusivity and viscosity respectively and k, a dimensionless boundary friction coefficient. Equation (5) is an equation of state with coefficients α and β being functions of temperature.

The boundary conditions are:

1. $$\left(\frac{\partial s}{\partial z}\right)_{z=\eta} = 0, \qquad \left(\frac{\partial s}{\partial z}\right)_{z=-H} = 0 \tag{6}$$

2. $$\left(N_z\frac{\partial u}{\partial z}\right)_{z=\eta} = \text{wind stress}, \qquad \left(N_z\frac{\partial u}{\partial z}\right)_{z=-H} = \text{boundary friction}. \tag{7}$$

In order to obtain a unique solution from the governing equations, it is necessary to prescribe

$$
\begin{aligned}
&s(o, z, t) \quad \forall\, z, t \qquad && s(L, z, t) \quad \forall\, z, t \\
&\eta(o, t) \quad \forall\, t && Q(L, t) \quad \forall\, t
\end{aligned}
\tag{8}
$$

where Q is the inflow of water at the head of the estuary arising from land drainage.

VERTICAL MIXING

In this study the eddy diffusivity and eddy viscosity are taken to be:

$$
\begin{aligned}
&K_z = BK + k_1^2\, z^2 \left(1 - \frac{z}{H}\right)^2 \left|\frac{\partial u}{\partial z}\right| \left(1 - \frac{Ri}{10}\right)^{1/2} \qquad o \leq Ri \leq Ri_c = 10 \\
&N_z = K_z\ (1 + Ri) \\
&K_z = N_z = 0.0 cm^2/sec, \ Ri > Ri_c = 10 \\
&K_x = N_x = 0.0 cm^2/sec
\end{aligned}
\tag{9}
$$

where k_1, is an adjustable constant; Ri, the Richardson number which is a measure of the stability and defined as

$$
Ri \equiv \frac{-g\beta \dfrac{\partial s}{\partial z}}{\left(\dfrac{\partial u}{\partial z}\right)^2} \, . \tag{10}
$$

The critical Richardson number, Ri_c, is that value of Ri at which mixing ceases and should be determined from field measurements. The background value (BK) of 0.50 cm^2/sec avoids numerical instability when little or no turbulent mixing exists, while the coefficient k_1 (0.0354) produces the most reasonable profiles of K_z and N_z.

FINITE DIFFERENCE TECHNIQUE

The governing equations, equations (1) through (5) together with boundary conditions (6), (7) and (8) are solved by integrating (as a boundary value problem) the finite difference equations which approximate them. The staggered grid system used is shown in Fig. 1. The vertical thickness of each grid box is

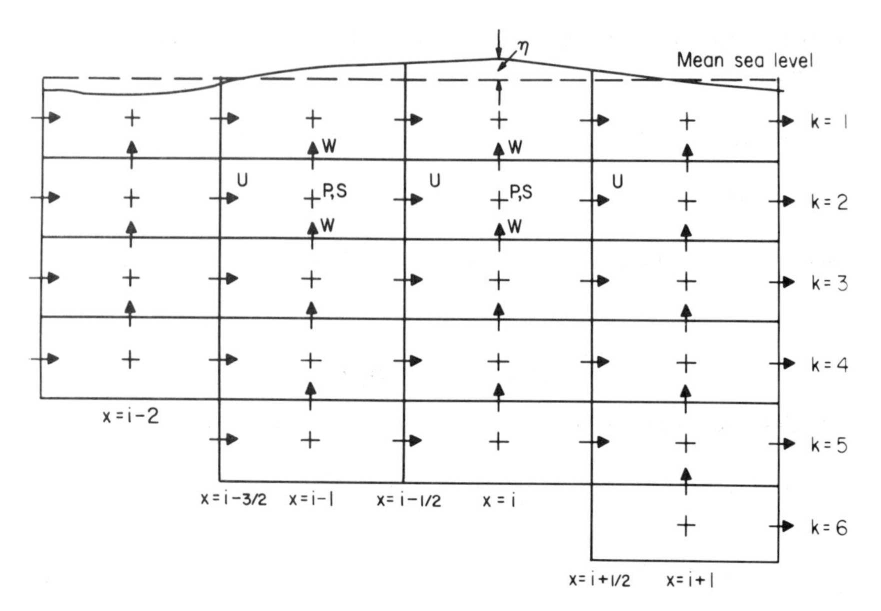

Figure 1. Finite difference grid structure.

constant except for the top one. The top grid box contains the influence of the surface gravity wave and its thickness varies in space and time.

All terms in the finite difference equations are centered differences in space and leap-frogged in time except for the diffusion and frictional terms. Those two terms are lagged one time step for computational stability. It should be pointed out that the finite difference equations, like their differential counterparts, conserve mass, salt and momentum in the asbence of dissipative effects. Also, no artifical horizontal diffusion or viscosity is introduced through this explicit finite difference scheme. The extended use of the leap-frog scheme leads to time splitting in the course of integration. In this study an Euler-backward scheme is used every 50 time steps to eliminate the time splitting and to damp gravity waves of wavelength comparable to two grid intervals (3).

It is important to insure that the specified boundary conditions are consistent with the finite difference scheme. The computational boundary conditions for the velocities are the water passing through the ocean boundary constrained to be horizontal ($w \equiv 0$), and no momentum flux imparted to the estuary by the ocean, or

$$\left(\frac{\partial uuB}{\partial x}\right)_{\text{outside}} << \left(\frac{\partial uuB}{\partial x}\right)_{\text{inside}} . \tag{11}$$

The computational boundary conditions for salinity are two fold. First, when inflow occurs, the salinity must be prescribed from field measurements. Second, during the outflow the hydrodynamic equations governing the interior of the estuary determine the salinity. It is assumed that during the outflow advective processes remove the salt from the estuary according to:

$$\frac{\partial s}{\partial t} + \frac{\partial (us)}{\partial x} = 0 \quad . \tag{12}$$

Now, the boundary value is extrapolated along the characteristic solution of the centered difference analogue of equation (12).

MODEL VERIFICATION

To test the ideas developed in this model, a configuration representing the Potomac River Estuary was chosen. The Potomac Estuary has moderate stratification and irregular bathymetry.

A large number of test cases were run to investigate various combinations of the turbulent constants, k_1 and Ri_c. Other simulations revealed that the nodal point in the tidal range is sensitive to the bathymetric schematization. In these computer runs the bathymetry presented by Cronin and Pritchard (2) was used.

The verification of the simulation of the Potomac River uses hydrographic data consisting of velocity, temperature, and salinity measurements collected during the period from November 1, 1974 through November 9, 1974. The data is not synoptic but is assumed to be representative of the conditions found in the Potomac River during this time of year. Also, the limited data is not laterally averaged. The Potomac River was not in quasi-steady state partially because of intermittent winds.

Fig. 2 depicts the measured vertical salinity profile at the ocean boundary. This profile is averaged over four tidal cycles in order to make the other salinity

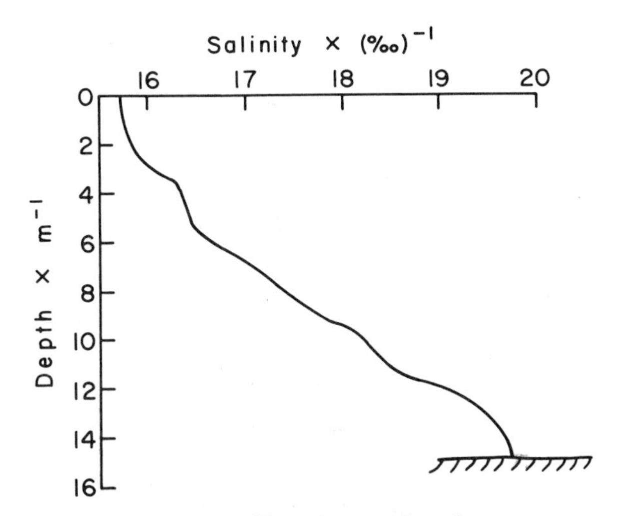

Figure 2. Measured vertical salinity profile at the ocean boundary.

measurements appear more synoptic. The free surface height at the ocean boundary consists of a pure sine wave with an amplitude equal to half the mean range at Point Lookout (18.3 cm.) and a period equal to the M_2 tidal constituent (12.42 hours). A two week mean discharge of 70 m^3/sec at the Little Falls

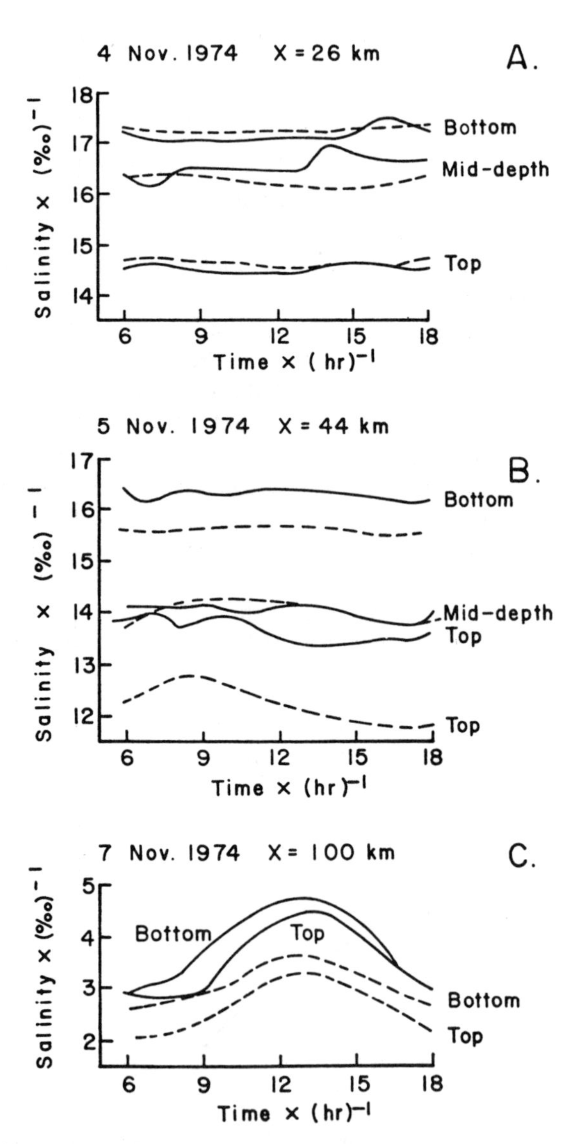

Figure 3. Comparison of observed salinity distributions (——) with computed ones (– – –) for three different sections of the Potomac River.

gauging station is the upstream boundary condition. Other parameters for the numerical simulation include,

$$
\begin{array}{llll}
\text{length of channel} & = 185 \text{ km} , & \Delta x = 4 \text{ km} & \\
\text{maximum depth} & = 15 \text{ m} , & \Delta z = 1.5 \text{ m} & (13) \\
\text{400 time steps/tidal cycle} & , & \Delta t = 111.78 \text{ sec.} & \\
\multicolumn{3}{c}{\text{Manning's n} = 0.039 \text{ cm}^{1/6} .} &
\end{array}
$$

The boundary friction parameter, k, is assumed identical to the one used in one dimensional flows and is proportional to Manning's n, the roughness. The Manning's n is not adjusted spatially as detailed tidal information is unavailable.

There are several possibilities for comparison of the numerical model with actual data and conditions. Two of these possibilities are to compare observed measurements of salinity and current velocity to their simulated counterparts. The measured surface, mid-depth and bottom salinities for three stations located 26km, 44km and 100km respectively from the mouth of the Potomac River Estuary are compared with the simulated ones in Fig. 3. The distributions found at station 26km are quite reasonable while the distributions at stations 44km and 100km from the mouth are less saline than the computed ones. The sporadic wind induced mixing could account for this discrepancy. Fig. 4 illustrates the surface and bottom current velocities (when available) for three different stations located 26km, 60km and 96km from the estuary's mouth for both the simulated and observed velocities. The numerically computed distributions always show the proper phase and in some cases, such as station 96km, the proper magnitude. Tidally averaging the computed velocities shows a distinct two layered circulation pattern with flow towards the mouth of the estuary in the upper layer and flow towards the head in the lower layer as Fig. 5 illustrates.

It should be emphasized that the data is representative and should not be used for precise comparisons. Nevertheless, the numerical model simulates the dynamic features of the Potomac River and the simulation compares well with the available field data. A 30 tidal cycle simulation takes 14 minutes of C.P.U. time on an IBM 360/195 computer.

CONCLUSION

This multilayered model, capable of simulating estuarine systems with irregular bathymetry, can easily be applied to different estuaries. The model can also be used to illustrate the response of the circulation to changes in the bathymetry, the river discharge and the wind.

ACKNOWLEDGMENTS

This work was supported primarily by the National Science Foundation under grant DES 74-08463. All the numerical computations were conducted at

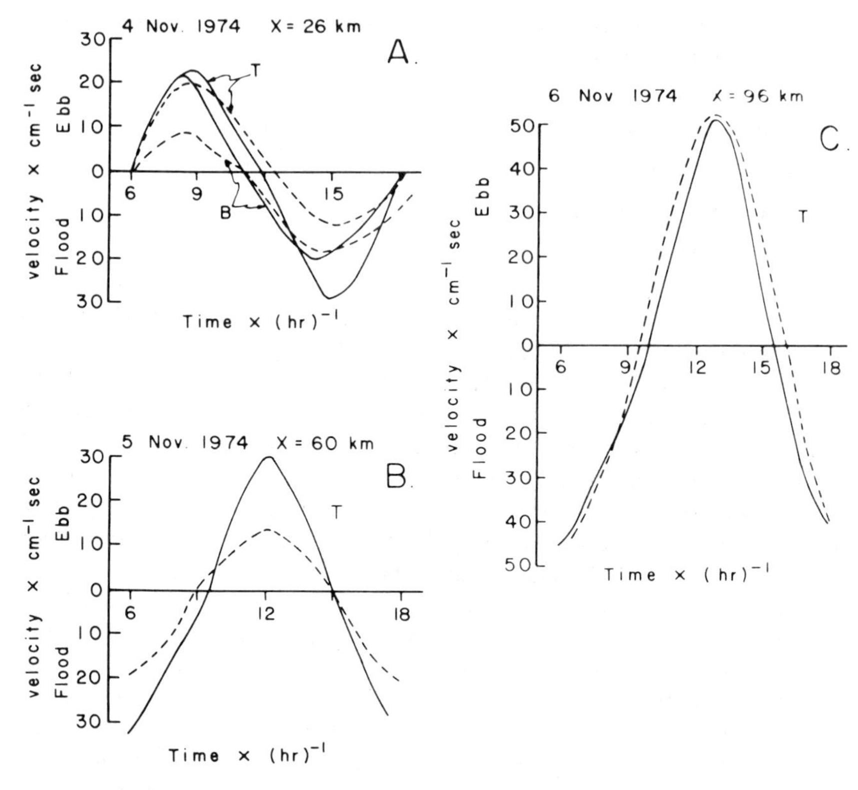

Figure 4. Comparison of observed velocity distributions (——) with computed ones (– – –) for three different sections of the Potomac River. Depth indicated by T (top) and B (bottom).

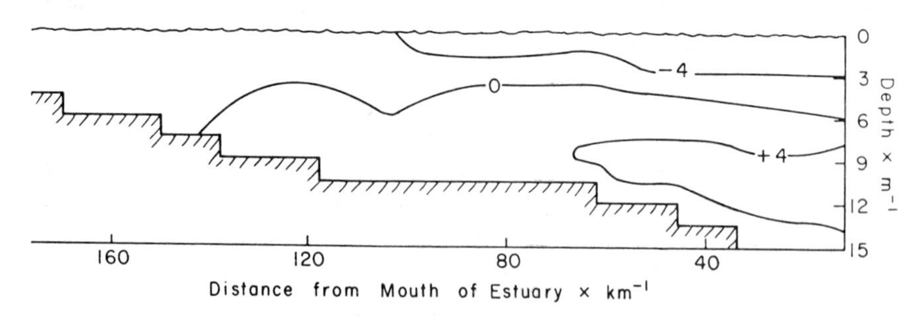

Figure 5. Numerical simulation of the Potomac River Estuary's tidally averaged longitudinal velocity distribution, isotachs in dm/sec, negative velocities indicate flow out of the estuary.

the G.F.D. Laboratory/NOAA at Princeton University. This investigation constitutes contribution No. 231 from the Chesapeake Bay Institute.

The author is indebted to Prof. F. P. Bretherton and Dr. A. J. Elliott for their comments concerning the content of this investigation.

REFERENCES

1. Blumberg, A. F. 1975. A numerical investigation into the dynamics of estuarine circulation. Chesapeake Bay Inst. Tech. Report No. 91, The Johns Hopkins University.
2. Cronin, W. B., and D. W. Pritchard, 1975. Additional statistics on the dimensions of the Chesapeake Bay and its tributaries: cross-section widths and segment volumes per meter depth. Chesapeake Bay Inst. Special Report 42, The Johns Hopkins University.
3. Haltiner, G. J. 1971. Numerical Weather Prediction. John Wiley and Sons, Inc., New York.

THE SIMULATION OF ESTUARINE CIRCULATIONS WITH A FULLY THREE-DIMENSIONAL NUMERICAL MODEL

Enrique A. Caponi[1]
Institute for Fluid Dynamics and Applied Mathematics
University of Maryland
College Park, MD 20742

ABSTRACT: The Navier-Stokes, salinity and continuity equations in the Boussinesq approximation are spatially integrated on the elementary computational cells to provide equations for the temporal rate of change of the fluxes through every cell's face and of the mean salinity in every cell. The effect of the spatial subgrid scales is lumped together with the Reynolds stresses generated by the temporal discretization procedure, and modeled by a simple Fickian relationship. The pressure field in the momentum equations is split up into a hydrostatic and a dynamic parts, the latter obtained as the solution to a finite differences Poisson equation. The three momentum equations and the salinity equation are independently updated by a forward stepping scheme. The free surface is updated by a mass conserving scheme. Required boundary conditions are river inflows and surface elevation at the sea as a function of time, as well as applied winds and atmospheric pressure. The model has been implemented in a Fortran code. It admits arbitrary coastal boundaries, openings to the sea, river inflows and bathymetry, imposed by the user through data cards. Idealized test cases are used to show that the model behaves as physically expected. A coarse application to Chesapeake Bay shows qualitatively correct results and the need to incorporate a less naive representation for the sub-grid scales.

INTRODUCTION

In recent years, several investigators have presented numerical approaches to the three-dimensional description of flows in enclosed and semi-enclosed natural

[1] Present address: Laboratorio de Hidráulica Aplicada/INCYTH, Casilla de Correos 21, Aeropuerto Ezeiza, Provincia de Buenos Aires, Argentina.

been simulated for shallow, homogenous lakes (10) with a layered formulation where convective and horizontal diffusion terms are neglected. Mixed numerical and analytical models to simulate 3-D currents without resorting to a 3-D grid system have been developed to obtain the response of the Irish Sea (7) and of coastal waters (4) to applied wind stress fields, assuming a homogeneous fluid. Tidal currents in regions with several open boundaries have been computed solving the linearized, primitive equations for a homogeneous fluid with a multilayer model which required the external specification of thermohaline currents (5).

A model specifically designed to handle as many of the variables of importance for estuarine dynamics as possible has been developed by Leendertse et al. (9). The model solves on a 3-D grid the full dynamic equations in the hydrostatic approximation, coupled to a salinity transport equation.

In the present paper, I want to report very briefly about the work performed at the University of Maryland between 1972 and 1974 in an effort to develop a different general 3-D numerical model for application to estuaries, bays and basins under the action of specific stimuli. Wind driven 3-D circulations have lakes, able to account for the effect of tides, river discharges and wind and atmospheric pressure distributions, upon the circulation of a variable density fluid contained in irregularly shaped natural basins (2). The object was to develop a tidal model able also to mimic the behavior of distorted laboratory models. Hence, the hydrostatic approximation, usually acceptable for geophysical flows, was not invoked. The following sections summarize the method used for the numerical solution [based on the Marker-and-Cell (MAC) method (6)] and present some of the available results. The reader is referred to Ref. 2 for a more extensive discussion of the mechanics of the model and its applications.

GOVERNING EQUATIONS AND SPATIAL FINITE DIFFERENCE FORMULATION

It is assumed that the laws of conservation of mass, momentum and energy can be reduced, in the Boussinesq approximation, to the volume conservation equation $\nabla \cdot v = 0$, the Navier-Stokes equation, and an advecto-diffusive equation for the temperature. Equations for other constituents (salt, DO, BOD, etc), are assumed to be also of the advecto-diffusive type, with appropriate source and sink terms. A simplified equation of state of the form

$$\rho = \rho_0 \cdot [1 + \alpha(S - S_0) - \beta(T - T_0)]$$

where ρ_0, S_0, T_0 are reference density, salinity and temperature, and α, β are experimental constants, is also adopted.

A right handed reference frame is used, with x- and y-axes in the horizontal, z-axis in the vertical, positive upwards. The physical space is divided up into a number of computational cells. For simplicity, these cells are taken of prismatic shape and uniform size (X_o, Y_o, Z_o). The differential equations are integrated over such cells, yielding back a set of conservation laws in the form of spatial finite difference equations. The reduced pressure (pressure to density ratio) is split into a hydrostatic and a dynamic part. The first one exactly balances the gravitational contribution to the vertical momentum equation, while the dynamic reduced pressure ϕ is the difference between the total and the hydrostatic part.

When the spatial and temporal coordinates are non-dimensionalized by the quantities X_o, Y_o, Z_o, τ_o (with τ_o arbitrarily chosen as 1 day), and the velocities by $U_o = X_o/\tau_o$, $V_o = Y_o/\tau_o$, $W_o = Z_o/\tau_o$, the incompressibility condition takes on the form

$$D_{i,j,k} = u_{i+1/2,j,k} - u_{i-1/2,j,k} + v_{i,j+1/2,k} - v_{i,j-1/2,k}$$

$$+ w_{i,j,k+1/2} - w_{i,j,k-1/2} = 0 \tag{2.1}$$

where u, v and w are the nondimensional average velocities through the faces of the computational cell (i, j, k) centered at $x=iX_o$, $y=jY_o$, $z=kZ_o$. Fig. 1 shows the adopted staggered grid.

The mementum equations, when similarly integrated, give the temporal rate of change of the nondimensional *volume* averages of the velocity components in terms of nondimensional *face* averages of the nonlinear and pressure terms, and nondimensional *volume* averages of the Coriolis and viscous terms. In order to

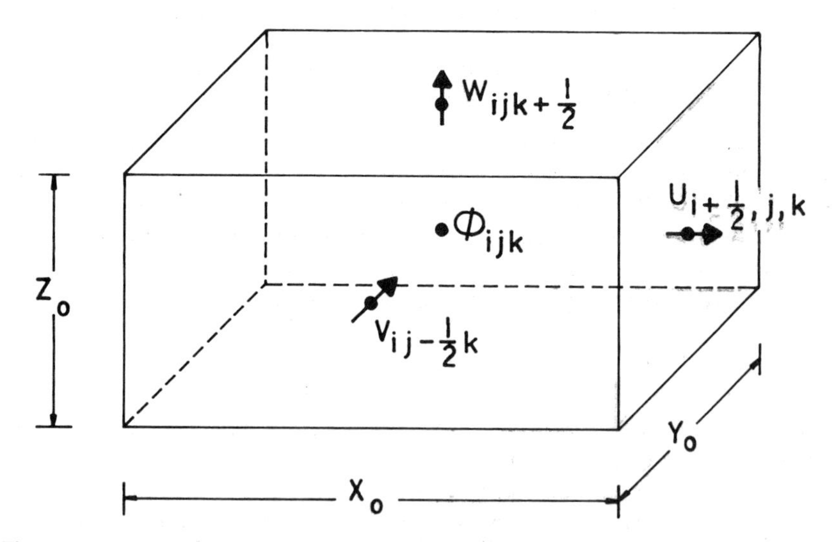

Figure 1. Location of variables in a computational cell.

reduce the number of variables to the number of equations, a set of dependent variables is chosen: the nondim. *face* averages of the velocity field, and the nondim. *volume* averages of the pressure field. When the momentum equations are expressed in terms of this new set, a number of terms, representing the effect of scales not resolved by the spatial grid upon those resolved, is generated, and they are lumped together into a dissipation term R for each equation:

$$\frac{\partial u_{i+1/2,j,k}}{\partial t} = DU_{i+1/2,j,k} + (Z_o/X_o)^2 \cdot (\phi_{i,j,k} - \phi_{i+1,j,k}) + R_u \tag{2.2}$$

$$\frac{\partial v_{i,j+1/2,k}}{\partial t} = DV_{i,j+1/2,k} + (Z_o/Y_o)^2 \cdot (\phi_{i,j,k} - \phi_{i,j+1,k}) + R_v \tag{2.3}$$

$$\frac{\partial w_{i,j,k+1/2}}{\partial t} = DW_{i,j,k+1/2} + \phi_{i,j,k} - \phi_{i,j,k+1} + R_w \tag{2.4}$$

where DU, DV, DW contain the space centered expressions for the convective, Coriolis and hydrostatic pressure terms.

The salinity equation is written as

$$\begin{aligned}\frac{\partial S_{i,j,k}}{\partial t} = {} & (uS)_{i-1/2,j,k} - (uS)_{i+1/2,j,k} \\ & + (vS)_{i,j-1/2,k} - (vS)_{i,j+1/2,k} \\ & + (wS)_{i,j,k-1/2} - (wS)_{i,j,k+1/2} + R_s\end{aligned} \tag{2.5}$$

where $S_{i,j,k}$ is the average salinity in cell (i, j, k), and a donor type scheme has been chosen for stability reasons (see below), i.e.,

$$(uS)_{i+1/2,j,k} = u_{i+1/2,j,k} \cdot \begin{cases} S_{i,j,k}, & \text{if } u_{i+1/2,j,k} > 0 \\ S_{i+1,j,k}, & \text{if } u_{i+1/2,j,k} < 0 \end{cases}$$

TEMPORAL INTEGRATION, DISSIPATION TERMS AND STABILITY

All dependent variables are filtered with a running temporal average operator. The temporal derivatives are approximated by

$$(\partial q/\partial t) \triangleq (q^{t+\delta t} - q^t) / \delta t$$

and the right hand sides of the dynamic equations (2.2) to (2.4) and (2.5) are evaluated at time t, thus defining a totally explicit scheme. The contributions due to the (temporal) sub-grid scales are incorporated to the *R* terms of the dynamic equations. As a first step towards the modeling of these terms, we have chosen:

$$(R_u)_{i+1/2,j,k} = A_1^u \cdot (u_{i+3/2,j,k} - 2 \cdot u_{i+1/2,j,k} + u_{i-1/2,j,k})$$
$$+ A_2^u \cdot (u_{i+1/2,j+1,k} - 2 \cdot u_{i+1/2,j,k} + u_{i+1/2,j-1,k})$$
$$+ A_3^u \cdot (u_{i+1/2,j,k+1} - 2 \cdot u_{i+1/2,j,k} + u_{i+1/2,j,k-1}) \tag{3.1}$$

where $A_1^u = \upsilon_1^u \tau_o / X_o^2$, $A_2^u = \upsilon_2^u \tau_o / Y_o^2$, $A_3^u = \upsilon_3^u \tau_o / Z_o^2$, with similar expressions for the corresponding term in the other momentum equations and in the salinity equation. Furthermore, we choose $\upsilon_i^u = \upsilon_i^v = \upsilon_i^w = \upsilon_i^s$ and let υ_1 and υ_2 be dependent on the horizontal grid sizes (X_o, Y_o), while taking υ_3 temporarily as a constant representing a bulk coefficient of vertical turbulent mixing.

The forward-time space-centered scheme defined above is subject to the CFL and viscous stability conditions. A heuristic analysis (8) was used for the determination of local modifications to the coefficients of eddy viscosity to avoid their artificial numerical variation. Additional terms have been added to the momentum equations in order to balance the unstable characteristics of the forward-time space-centered geostrophic equations. Test runs demonstrated that, as long as the wave CFL condition was satisfied, these modifications allowed the model to be run for the (otherwise unstable) $A_1 = A_2 = A_3 = 0$ case, without the appearance of instabilities.

The numerical viscosities generated by spatially centered convective terms in the salinity transport equation have very complicated expressions to account for. The donor cell scheme was chosen for simplicity, since it does not introduce additional stability conditions.

NUMERICAL PROCEDURE

Computational cells are labelled in the MAC tradition. BND and IN cells are located outside of the region of interest and have a common face with an interior cell. Such face represents the coastline or bottom for a BND cell, or a river mouth for an IN cell. OUT cells correspond to the opening to the sea. Cells located in the interior of the basin are flagged as EMP, SUR or FUL, depending on their being empty, containing fluid but having a common face with an EMP cell, or containing fluid but not having a common face with an EMP cell. In general, the free surface is contained in SUR cells. Cells of type BND, IN and OUT are specified by the initial data. Interior cells are examined at the beginning of each computational cycle and their EMP, SUR or FUL status is determined.

Assuming all fields known at $t-\delta t$, the right hand side of Eqs. (2.2)-(2.4) are evaluated at all locations covered with fluid, and u, v, w are explicitly updated to time t. The dynamic pressure ϕ is obtained–as in the MAC method–solving iteratively the finite difference Poisson equation obtained from Eqs. (2.2)-(2.4), subject to the divergence free condition imposed at $t+\delta t$, i.e., $D_{i,j,k}^{t+\delta t}= 0$. In this way, errors introduced because of the violation of Eq. (2.1) at every time step do not accumulate during the computation, even using a coarse convergence criterion for the solution of the Poisson equation.

The free surface is updated by computing the water inflow during δt into a computational (i, j) column from the bottom to the surface. This volume change is converted into mean surface change in column (i,j), thus assuring absolute conservation of mass in the whole system.

BOUNDARY CONDITIONS

For BND cells, normal fluxes are set equal to zero, and tangential velocities are provided by either no-slip of free-slip boundary conditions. The future modeling of the mixing coefficients in terms of mean field variables and spatial location, will allow the correct description of energy dissipation at these kinds of boundaries without the need to resort to the traditionally used Chezy type coefficients. In general, we have used free-slip conditions for w and no-slip conditions for u and v.

For IN cells, the river discharge is specified by giving the velocity values as a function of time. Values for u and v are specified in the empty region above the "free" surface by relating their jump across it to the applied x- and y-wind stresses. The w values outside the surface are consistently determined so that $D_{i,j,k} = 0$.

Velocities normal to the external faces of OUT cells are built in terms of the temporal variation of the transverse average $\overline{H}$ of the surface elevation. It is assumed that the mean velocity normal to the external face of any OUT cell equals the dynamically determined mean velocity from the interior adjacent cell into the OUT cell, plus a value Δu that is different for every OUT column that describes the same opening to the sea. An equation for the conservation of volume over every OUT (i,j) column relates the Δu value for that column to the prescribed temporal variation of $\overline{H}$. Transverse fluxes at common faces of cells belonging to consecutive OUT columns are computed dynamically, allowing the establishment of a consistent transverse slope of the surface elevation from one bank to the other (cf. Ref. 2).

Homogeneous boundary conditions for the reduced dynamic pressure ϕ are used [as in Ref. 3] for BND, IN and OUT boundaries. At SUR cells, ϕ values are interpolated or extrapolated (11).

The salinity value at IN cells may be given as a function of time (in particular, $S_{IN}=0$ for the test cases). In the case of several river inflows, though, the specification of different "reasonable" values of S at the various IN cells may force "unreasonable" interior salinity distributions and peculiar gravitational circulations. In these cases, we assume $S_{IN} = \lambda . S_{int}$, $0 < \lambda < 1$, where S_{int} is the S value in the adjacent interior cell, and λ should parameterize the effects of the river discharge, its cross section and the "mixing coefficients".

Since the normal external fluxes at OUT cells are not updated dynamically, the use of the dynamic equation for the salinity at these locations would occasion a leak of salt from the basin. Thus, the salinity value S_{OUT} is to be given by an independent prescription. We assume $S_{OUT} = (S_{int} + S_F)/2$, where S_{int} is the salinity of the interior adjacent cell, and S_F is a fixed, oceanic value.

Special treatment of the salinity equation in SUR cells ensures no salt flux through the surface, no internal computational sources and the ability to conserve homogeneity [cf. Ref. 2].

NUMERICAL EXAMPLES

Fig. 2 shows the test geometry used to obtain the response to tidal, fresh water inflow, wind and salinity gradient forcing. Horizontal eddy viscosity co-

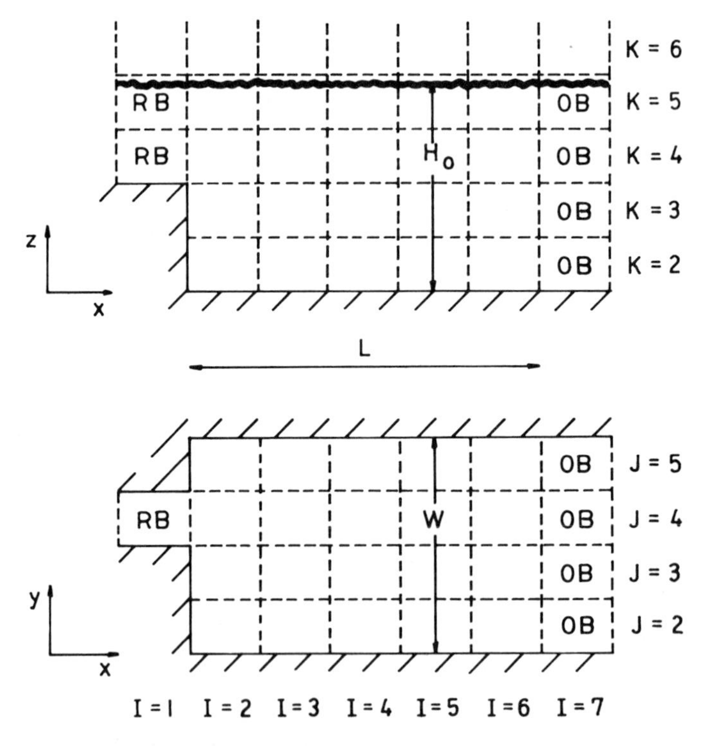

Figure 2. Test geometry and computational grid (L = 288 km, W = 96 km, H_0 = 8 m for the test cases).

efficients were chosen of $0(1\text{-}10\ m^2/s)$, according to the horizontal grid dimensions and based on experimental data (12). The vertical coefficient was arbitrarily set to $10^{-3}\ m^2/s$.

In order to test the ability of the numerical procedure employed (including the way of handling the open boundary) to develop a symmetric flow under symmetric forcing, the Coriolis parameter was set to zero, the river inflow was blocked, and a sinusoidal tidal excitation of amplitude Z_0 was applied. Examination of the velocity and surface elevation fields at arbitrary instants after 13 simulated days, showed the flow to be symmetric within the expected round-off error.

A second test case was provided by allowing a river to discharge $10^4\ m^3/s$ into the basin, and letting the system approach a steady state in the presence of Coriolis and the absence of tides. The circulation arrived at after 4 days of simulated time is shown in Fig. 3. The number of pressure iterations ITP decreases from about 20 for t=0 to 1 for t=2.5 days, and it remains at this level thereafter. ITP oscillates during the initial adjustment with a period of about 0.126 days, which is the time that it takes for a wave to travel from one lateral bank to the other. The kinetic energy and the total flux out of the system through the ocean boundary present absolute minima at about t=0.5 days and then approach constant values. Very small oscillations of period 0.126 days in the kinetic energy trace, indicate that the previously mentioned transient lateral waves carry very little energy. At t=4 days, the total outflow from the basin into the open boundary differs from the fixed river inflow by less than 0.5%.

Steady state circulations due to the application of a homogeneous wind stress field $[\tau/(\rho\upsilon_3)_{water}=0.25\ s^{-1}]$ have been obtained for upstream, downstream and

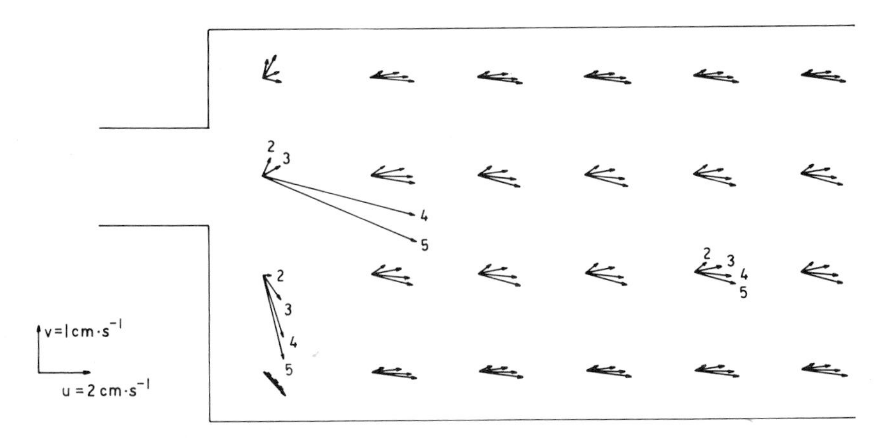

Figure 3. River driven circulation for the geometry of Fig. 2. Examples of the velocity variation with depth are shown by the values of the vertical computational index K at selected locations (K = 2 being the lowermost FUL cell, K = 5 the surface).

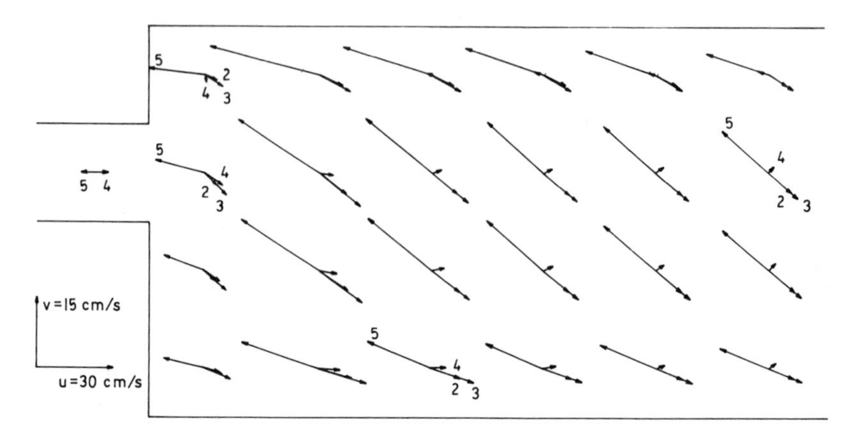

Figure 4. Upstream wind driven circulation for the geometry of Fig. 2. The meaning of numbers against selected arrows is the same as in Fig. 3.

cross winds. Initial transients are damped in about 3 days in all cases for the given parameters. Fig. 4 shows the circulation arrived at after 4 days of real time for the upstream case. Longitudinal and lateral surface slopes develop as expected.

Fig. 5 shows a typical estuarine circulation, arrived at after 10 days with no tides or wind, for S_F=30°/₀₀, S_{IN}=0, and the same fresh water discharge used above. Due to the very slow response of the salinity distribution, no steady state was established at this time: total outflow to the ocean was larger than fresh water inflow by 1.5%.

A more general application (but also restricted to constant eddy coefficients

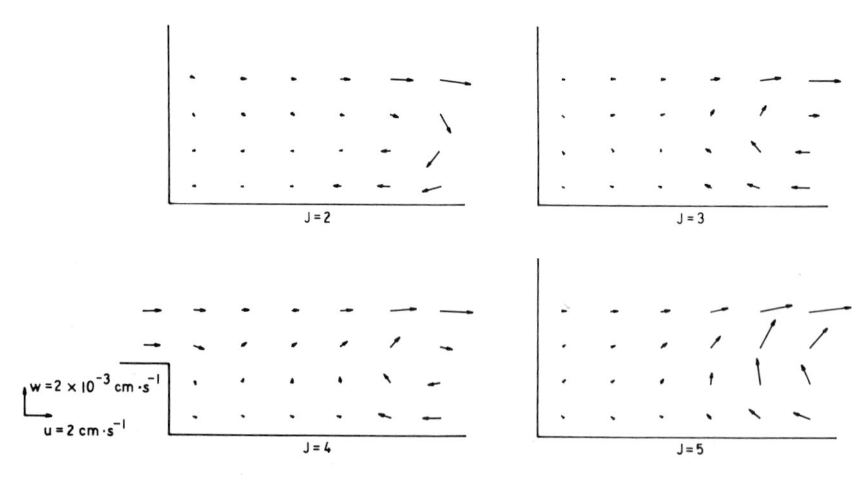

Figure 5. Density driven circulation for the geometry of Fig. 2.

$\upsilon_1 = 93\ m^2/s$, $\upsilon_2 = 57\ m^2/s$, $\upsilon_3 = 2.10^{-3}\ m^2/s$) was made on a (20 X 10 X 8) grid to a variable topography geometry with several fresh water contributors, and resembling the Chesapeake Bay (Fig. 6). Numbers inside the computational cells of

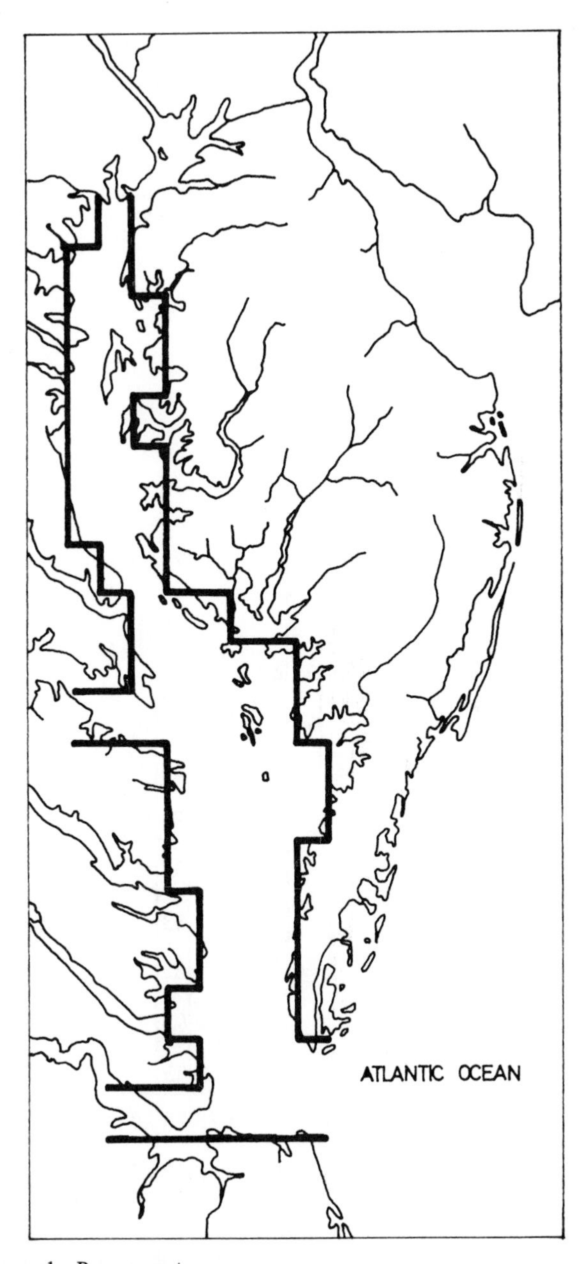

Figure 6. Chesapeake Bay geometry.

Fig. 7 indicate depth of the bottom boundary from the undisturbed surface level in terms of the vertical grid size Z_0. Computational cells were of size $X_0 = 13.7$ km, $Y_0 = 8.7$ km, $Z_0 = 3.44$ m. The initial salinity distribution was taken from Ref. (13) and surface values are shown in Fig. 8a. S_F was chosen as $31.5°/_{oo}$.

The model was allowed to evolve for a period of 4 days without tidal excitation. The resulting salinity distribution (Fig. 8b) was strongly influenced by the

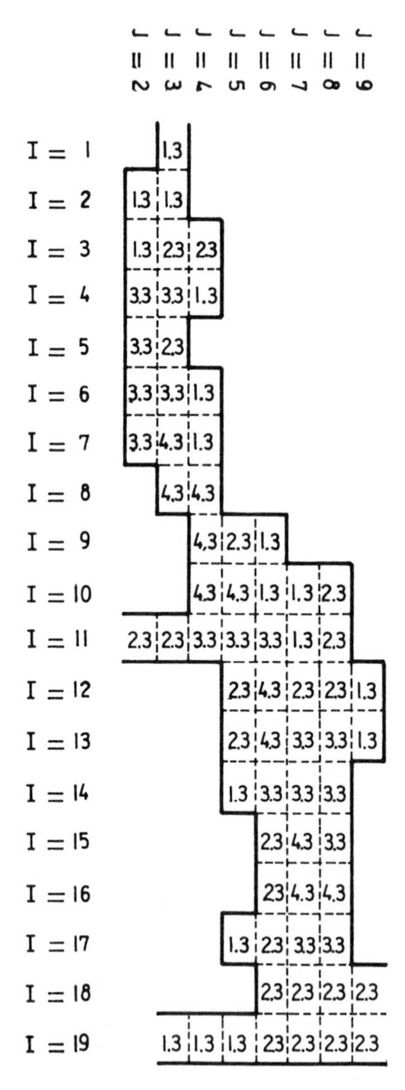

Figure 7. Computational grid for the geometry of Fig. 6. Numbers within cells indicate depth of the bottom with respect to the undistrubed surface level in units of Z_0.

local bottom topography and fresh water inflows. Salinity driven bottom flows for the lower part of the Bay were obtained as expected (cf. Fig. 9). The results do not quite represent a steady-state circulation: at the end of the four days period the net outflow through the ocean boundary still exceeded the total river inflow by almost 2%. The chosen value for υ_3 was too large to maintain the initially specified stratification.

From t=4 to t=12 days, a sinusoidal tidal excitation of amplitude 0.625 m and period 12.5 hrs was applied. The calculated tidal amplitudes and surface velocities at selected locations in the Bay were larger than available observational values (in particular, for the upper region of the Bay, the former were about twice as large as the latter). This suggests that as far as the momentum equations are concerned, the chosen vertical mixing coefficient was too small to produce

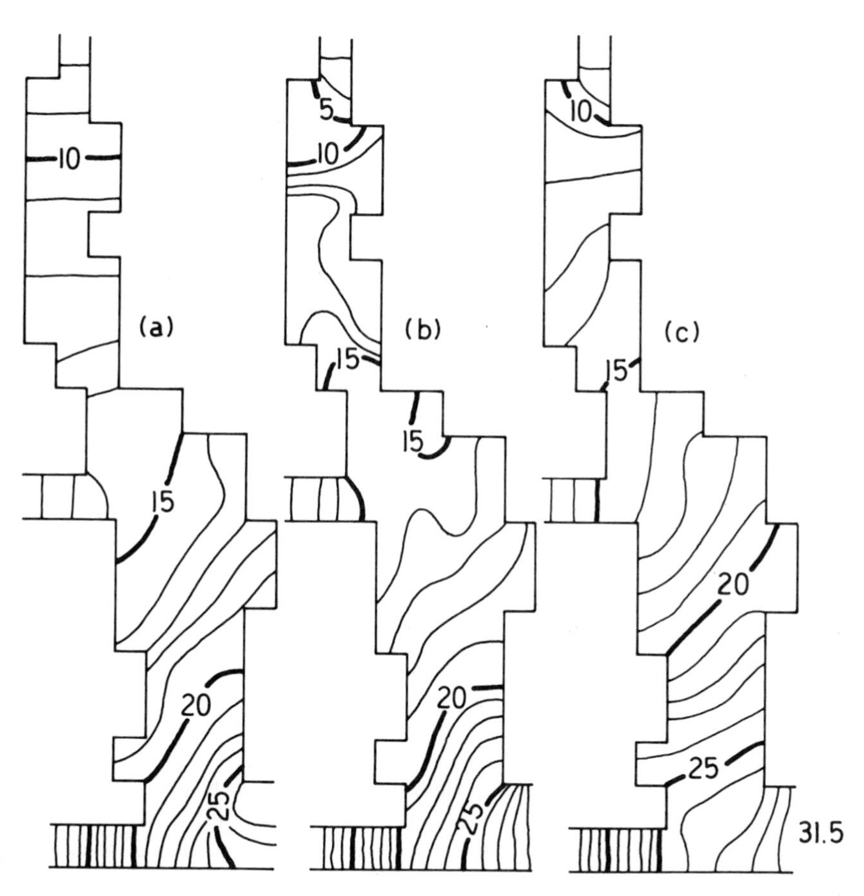

Figure 8. Surface salinity distributions: a) Initial conditions, b) After 4 days with no tides, c) At t=12 days, after 8 days of tidal action.

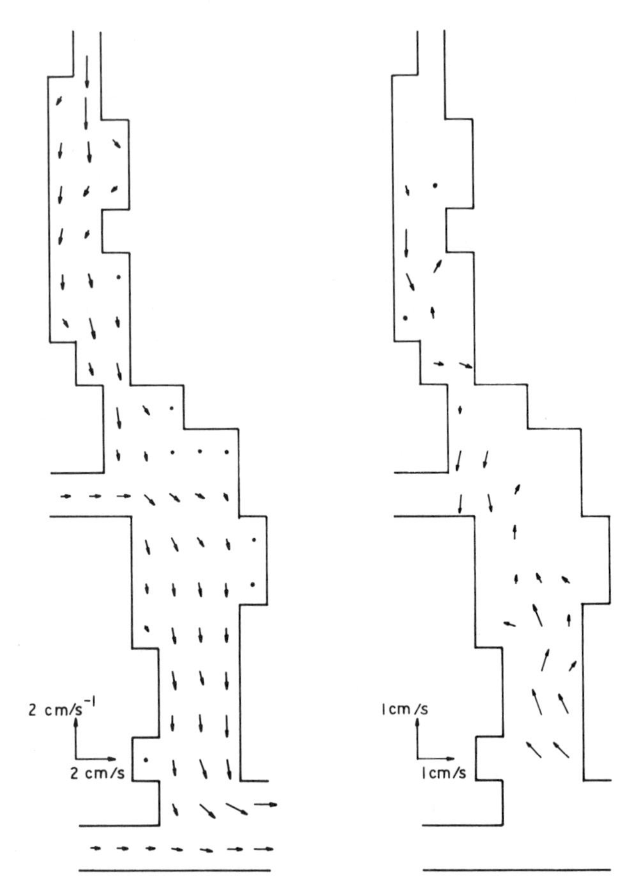

Figure 9. Circulation patterns at t=4 days (no tides): for the surface (left), for a depth of 2 Z_0 (right).

the required dissipation. The transport of salt from the ocean into the system and the enhanced effective lateral mixing within the estuary due to tidal action resulted in the salinity distribution shown in Fig. 8c.

CONCLUSIONS

Numerical experimentation has shown that it is computationally feasible to simulate estuarine flows in three spatial dimensions and time with an approach that does not invoke the hydrostatic approximation. The method is not ready for application as a management tool. Extensive research in ways of parameterizing the sub-grid motions through the several mixing coefficients involved is the primary task to be followed at this stage. The comparison between observed and computed tidal heights, velocities and vertical stratification shows that different eddy coefficients (at least in the vertical) should be used for the momentum and the salinity equations, and that—as known from experimental evidence (1)—the former must be larger than the latter.

The usefulness of a 3-D approach is restricted, not only by cost and machine considerations, but also by the ability to interpret the vast mass of information generated by it. Hence, graphics software must be adopted and made an integral part of a research or management code.

Partial results presented here do not consider the influence of temperature upon the circulation. The introduction of the corresponding equation in the present code is computationally straightforward, and will become essential when advecto-diffusive equations for biological and chemical species are incorporated.

ACKNOWLEDGEMENTS

Special thanks are due to Prof. A. J. Faller for many helpful discussions during all the stages of development of this model. This research was supported by the National Science Foundation. Computer time was supported in full by the Computer Science Center of the University of Maryland.

REFERENCES

1. Bowden, K. F. 1967. Stability effects on turbulent mixing in tidal currents. *In* Boundary Layers and Turbulence, Phys. Fluids Supplement, S278-S280.
2. Caponi, E. A. 1974. A three-dimensional model for the numerical simulation of estuaries. Ph.D. Thesis, Technical Note BN-800, Institute for Fluid Dynamics and Applied Mathematics, Univ. of Maryland, College Park, Maryland. 215 p. To be published (1976) in Adv. in Geophys. 19:189-310.
3. Easton, C. R. 1972. Homogeneous boundary conditions for pressure in the MAC method. J. Comp. Phys. 9(3):375-378.
4. Forristall, G. Z. 1974. Three-dimensional structure of storm-generated currents. J. Geophys. Res. 79(18):2721-2729.
5. Hamilton, G. D., F. R. Williams and T. Laevastu. 1973. Computation of tides and currents with Multi-layer Hydrodynamical Numerical (MHN) models with several open boundaries, p. 634-649. *In* Proc. of the 1973 Summer Comp. Simul. Conf., Montreal. Simulation Councils, La Jolla, California.
6. Harlow, F. H., and J. E. Welch. 1965. Numerical calculation of time-dependent viscous incompressible flow. Phys. Fluids 8:2182-2189. Welch, J. E., F. H. Harlow, J. P. Shannon and B. J. Daly. 1965. The MAC Method. Report LA-3425 Revised, Los Alamos Scientific Laboratory, Los Alamos, New Mexico.
7. Heaps, N. S. 1973. Three dimensional numerical model of the Irish Sea. Geophys. J. R. astr. Soc. 35(2):99-120.
8. Hirt, C. W. 1968. Heuristic stability theory for finite difference equations. J. Comp. Phys. 2(3):339-355.
9. Leendertse, J. J., R. C. Alexander and S-K. Liu. 1973. A three dimensional model for estuaries and coastal seas: Vol. I, Principles of Computation. Report R-1417-OWRR, The Rand Corporation, Santa Monica, California. 57 p.

10. Liggett, J. A. 1969. Unsteady circulation in shallow, homogeneous lakes. J. Hydraulics Div., ASCE, 95(HY4):1273-1288.
11. Nichols, B. D., and C. W. Hirt. 1971. Improved free surface boundary conditions for numerical incompressible-flow calculations. J. Comp. Phys. 8(4):434-448.
12. Okubo, A., and R. V. Ozmidov. 1970. Empirical dependence of the coefficient of horizontal turbulent diffusion in the ocean on the scale of the phenomenon in question. Izv. Atmosph. and Ocean Phys. (Engl. Ed.) 6(5)534-536.
13. Pritchard, D. W. 1969. Chemical and physical oceanography of the Bay. *In* Proc. of the Governor's Conference on Chesapeake Bay, Annapolis, Maryland.

ON THE NUMERICAL FORMULATION OF A TIME DEPENDENT MULTI-LEVEL MODEL OF AN ESTUARY, WITH PARTICULAR REFERENCE TO BOUNDARY CONDITIONS[1]

Peter Hamilton
Department of Oceanography
University of Washington
Seattle, Washington 98195

ABSTRACT: A two dimensional model of the vertical circulation of an estuary has been developed, which solves, by an explicit finite difference initial value method, the equations of continuity, salt and momentum conservation for a channel of variable width and depth, but rectangular cross section. A further semi-implicit extension of the model is given, which removes the dependancy of the time step on the depth and makes feasible long integration periods. Features of the model include a finite difference grid which allows the tidally driven free surface to move vertically through the grid points, along with accurate finite difference formulations of the surface and bottom boundary conditions. If long periods of integration are considered, the effect on the salt balance of different formulations of the vertical eddy coefficients is shown to make the choice of the mouth boundary condition on salinity ambiguous.

INTRODUCTION

The use of numerical techniques to solve the basic equations of continuity, momentum and salt conservation, in the context of an estuary, presents a challenge to the mathematical modeller. The disparity between the three dimensions (across stream, along stream and depth), the complex forcing by the external tide, freshwater flow and wind stress, and the paramaterization of turbulent viscosity and diffusion, are some of the problems not usually encountered in

[1] Contribution No. 868 from the Department of Oceanography, University of Washington, Seattle 98195.

finite difference formulations of fluid flow problems. Also in natural estuaries highly variable topography, often with tidal flats in the estuary mouth, is the case.

Until recently, numerical models of estuaries have mostly been one dimensional and have been used to investigate the propagation of the tide by explicit methods, i.e., where the time step is limited by the speed of a free gravity wave (31, 10), and also by semi-implicit methods (25), where the above restriction is overcome but iterative or matrix inversion techniques are required. There is a large literature on storm surge calculations which use a horizontal two dimensional finite difference grid, the dynamical equations being integrated with respect to the depth, (23, 17, 18), where the two grid dimensions are approximately similar and the method is explicit. Ramming (28) uses the same basic approach for a two dimensional (horizontal) model of the River Elbe in Germany, where the problem of a small grid spacing in the cross stream direction, made necessary because of the narrow channel, which would seriously restrict time step, is overcome by using a semi-implicit formulation in that direction and the usual explicit formulation in the alongstream direction. Heaps (19) has developed a three-dimensional model of the Irish sea which uses the basic two-dimensional formulation (18) but resolves the depth structure of the current into modes. Methods using fully three dimensional grids have been developed for bay-like estuaries and lakes (24, 7, 32). A model of Chesapeake Bay developed by Caponi (7) still requires that the two horizontal grid spacings be of roughly similar lengths and so is not really suitable for a narrow estuary. Similar remarks apply to the model of Leendertse and Liu (24).

One-dimensional dispersion models (33, 16, 35) have been used to investigate and predict salt intrusion lengths of estuaries, but the longitudinal dispersion coefficient contains in one bulk parameter a great deal of the physics of estuarine salt balance mechanisms. It is not surprising, therefore, that the dispersion coefficient is strongly dependent on the external forcing functions; the riverflow and the tide. The internal dynamics of an estuary have been elucidated by the similarity solutions of Hansen and Rattray (14). In particular the vertical structure of the current and salinity profiles is determined, for the steady state, by the longitudinal density gradient and strength of the turbulent mixing and Reynold stresses, characterized by eddy coefficients of diffusion and viscosity. The interdependence of the eddy coefficients and their precise dependence on details of the flow, Richardson numbers and tidal currents (the latter in many cases is the primary source of turbulent energy) is still largely a matter of speculation, though some attempts have been made at empirical formulations making use of data and some theoretical concepts (20, 4, 21).

A numerical model of the tidal estuary in which the longitudinal and vertical dimensions are considered has been developed (13) in which the method is explicit and variables are treated as continuous in the vertical, unlike most other models which consider the vertical dimension in which the quantities are integrated over separate layers and the exchange processes parameterized in terms of mean quantities of the layers (34, 24). The more direct multi-level approach allows a more precise detailed numerical treatment of the free surface and bottom boundary conditions. The method is related to marker and cell techniques developed for smaller scale two-dimensional fluid flow problems (8, 26) in which precise consideration of the boundary conditions are important in minimizing spurious instabilities. It was found for example (12) that a step-like bottom boundary produced anomalous circulations in the vertical, and a fairly elaborate procedure was developed to ensure that the finite difference scheme takes into account the true position of the bottom. The free surface is also treated similarly and as the tidal range may be more than the vertical grid spacing the number of grid points in a column covered by water will change with time and so a multi-level approach does not restrict the tidal range as a multilayer method (with fixed interfaces), where only the surface layer depth is allowed to vary and the number of layers remains fixed with time (24).

This paper is complementary to Hamilton (13), which gives the mathematical description of the model and compares numerical results with data from the Rotterdam Waterway, and Bowden and Hamilton (5), which describes a number of experiments carried out for a hypothetical channel, representative of a coastal plain estuary, to investigate the effect of changing the riverflow, tidal range, longitudinal density gradient and different formulations for the vertical eddy coefficients. This paper will give a further semi-implicit development of the model which allows longer time steps and so making long periods of integration feasible. The problem of a seaward boundary condition appropriate for long periods is also discussed.

THE MATHEMATICAL MODEL

The governing equations, using the Boussinesq approximation for a narrow channel with a rectangular cross-section, are

the equations of continuity:

$$\frac{\partial \zeta}{\partial t} + \frac{1}{b}\frac{\partial}{\partial x}\left[b\int_{-\zeta}^{h} u\,dz\right] = 0 \tag{1}$$

$$\frac{\partial}{\partial x}(bu) + b\frac{\partial w}{\partial z} = 0 \tag{2a}$$

$$w = \frac{1}{b}\frac{\partial}{\partial x}\left[b\int_{z}^{h} u\,dz\right] \tag{2b}$$

salt conservation:

$$\frac{\partial s}{\partial t} + u\frac{\partial s}{\partial x} + w\frac{\partial s}{\partial z} - \frac{1}{b}\frac{\partial}{\partial x}\left(bK_x\frac{\partial s}{\partial x}\right) - \frac{\partial}{\partial z}\left(K_z\frac{\partial s}{\partial z}\right) = 0 \quad (3)$$

momentum conservation:

$$\frac{\partial u}{\partial t} + u\frac{\partial u}{\partial x} + w\frac{\partial u}{\partial z} = -\ ag\frac{\overline{\partial s}^{z}}{\partial x} - g\frac{\partial \zeta}{\partial x} + \frac{\partial}{\partial z}\left(N_z\frac{\partial u}{\partial z}\right) \quad (4)$$

Where x, z coordinates in the plane of the undisturbed water surface (positive seawards) and vertically downwards, respectively
- t time
- ζ elevation of the water surface above the undisturbed plane
- u, w velocity components in the directions x and z respectively,
- s salinity
- h, b depth and width of the channel, respectively (functions of x only)
- N_z coefficient of vertical eddy viscosity
- K_x, K_z coefficients of horizontal and vertical eddy diffusivity, respectively
- g acceleration due to gravity.

The density ρ is related to salinity s by the linear equation of state:

$$\rho = \rho_o\ (1 + as) \quad (5)$$

where ρ_o is the density of fresh water and a is constant, numerically equal to $7{\cdot}8 \times 10^{-4}$. This is usually regarded as a satisfactory approximation because of the large horizontal changes in salinity which occur in estuaries.

In (13) the pressure term, due to the horizontal density gradients, in eq. (4);

$$\frac{\overline{\partial s}^{z}}{\partial x} = \int_{-\zeta}^{z} \frac{\partial s}{\partial x}\, dz \quad (6)$$

is approximated by $(z + \zeta)\ \partial s/\partial x$ which is satisfactory if only moderately stratified estuaries are considered and has the advantage of eliminating a number of numerical integrations which are associated with eq. (6), even though eq. (6) is the more accurate formulation. Equations (1) - (4) have been averaged across the width of the channel.

The boundary conditions for a channel of length L are:
at the free surface, the wind stress is given and there is no salt flux through the surface:

$$\tau_w = -N_z \frac{\partial u}{\partial z} \tag{7}$$

$$K_z \frac{\partial s}{\partial z} = 0 \text{ at } z = -\zeta. \tag{8}$$

At the bottom, a quadratic bottom stress law is applied and if there is a bottom slope both vertical and horizontal density fluxes are set to zero:

$$-N_z \frac{\partial u}{\partial z} = k/u_\Delta/u_\Delta \tag{9}$$

where $u_\Delta = u(x, h - \Delta, t), \Delta = 1\text{m}, k = 0{\cdot}0025.$

$$K_z \frac{\partial s}{\partial z} = 0. \tag{10a}$$

$$K_x \frac{\partial s}{\partial x} = 0 \text{ iff } \frac{dh}{dx} \neq 0 \text{ at } z = h \tag{10b}$$

The bottom stress is related to the current at 1m above the bed and is based on the assumption of a boundary layer 1m thick. The correct no flux boundary condition at the bed is of course

$$K_z \frac{\partial s}{\partial z} - K_x \frac{dh}{dx} \frac{\partial s}{\partial x} = 0 \text{ at } z = h \tag{11}$$

It was found that the numerical method was very sensitive to small errors in K_z $\partial s/\partial z$ at the boundary. Because of the difference between horizontal and vertical grid scales (10^3:1), it is easier to ensure accuracy by putting both components of flux to zero separately. In (13) and (5) only eq. (10a) is used, as the horizontal diffusion term is neglected in all experiments.

At the head of the estuary the freshwater flow is given:

$$s(0, z, t) = 0, \tag{12}$$

$$u(0, z, t) = q/((h(0) + \zeta(0,t)).\, b(0)). \tag{13}$$

where q is the riverflow, usually a constant ($m^3\ s^{-1}$).

At the mouth the elevation is given in the form of tidal input:

$$\zeta(L, t) = A(t) \tag{14}$$

If the tide is a simple semi-diurnal wave then $A(t) = -A_o \cos \sigma t$, $\sigma = 2\pi/T$, where A_o is the tidal amplitude and T the tidal period (12·42 hr). A condition is also required on salinity, which will be discussed later. Suffice it to say that in applying the model to the Rotterdam Waterway (13) the salinity at the mouth was given as a function of depth and time using data, and that in (5), for the experiments with the model, the rather artificial condition of making the salinity constant with depth and time was used.

Other boundary conditions could be considered. For example, the velocity profile could be used as a mouth boundary condition, with data taken from current meters, instead of the elevation, or the elevation condition could incorporate a radiation condition which would allow unwanted gravity waves to propagate out through the seaward boundary without reflection.

The explicit finite difference method used by (13) uses equations (1)-(4), however the semi-implicit method given as an extension of the method in (13) requires a further equation; namely the depth integrated form of the momentum equation (4):

$$\frac{\partial \bar{u}}{\partial t} + \frac{\bar{u}}{(h+\zeta)} \frac{\partial \bar{u}}{\partial x} = - ag \frac{(h+\zeta)}{2} \frac{\overline{\partial s}^{h}}{\partial x} - g (h+\zeta) \frac{\partial \zeta}{\partial x}$$

$$+ (\tau_w - k/u_\Delta/u_\Delta) \qquad (15)$$

$$\text{where } \bar{u} = \int_{-\zeta}^{h} u dz \qquad (16)$$

Note that if the velocities in the frictional term are replaced by depth mean velocities $(\bar{u}/(h+\zeta))$ and the density term is neglected, then eq. (1) and eq. (15) constitute the usual one-dimensional tidal equations (31).

NUMERICAL METHOD

The finite difference method is given in (13), so only a brief review is given here. A finite difference grid is arranged in the vertical plane to cover the profile of the estuary. The grid is staggered with salinity and velocity points being at different positions, though numbered by the same single parameter i (Fig. 1). The number of salinity grid points in a column is n and the columns are numbered by the parameter p, where

$$p = \text{int}\left(\frac{i-1}{n}\right) + 1, \qquad (17)$$

also $k = 2p$ for a salinity column and $k = 2p - 1$ for a velocity column. The grid points just beneath the water surface, i_s, and just above the bottom, i_b, the distance of i_s from the surface, d_s, and i_b from the bottom, d_b, are found from the free surface elevation, ζ, and the depth h(Fig. 1). Therefore i_s and d_s change with time. Apart from the moving free surface, the grid and the method are in essentials due to Heaps (18). The finite difference versions of the basic equations (1)-(4) for the explicit method use forward and backward time differences and central differences in space except for the horizontal advective term in eq. (3) which is a combination of upwind and central space differences. The use of an upwind scheme introduces a large apparent horizontal diffusion coefficient (30) which is a function of the horizontal current and the time step. The use of an upwind scheme was necessary because, in the presence of a strong horizontal salinity gradient, a centered method (8) produced physically unrealistic salinities near the mouth and head.

A modification introduced to improve accuracy is the use of eq. (2b) rather than eq. (2a) to calculate vertical velocities. A disadvantage of the original formulation is that due to its sequential nature a small error in the calculation of w at

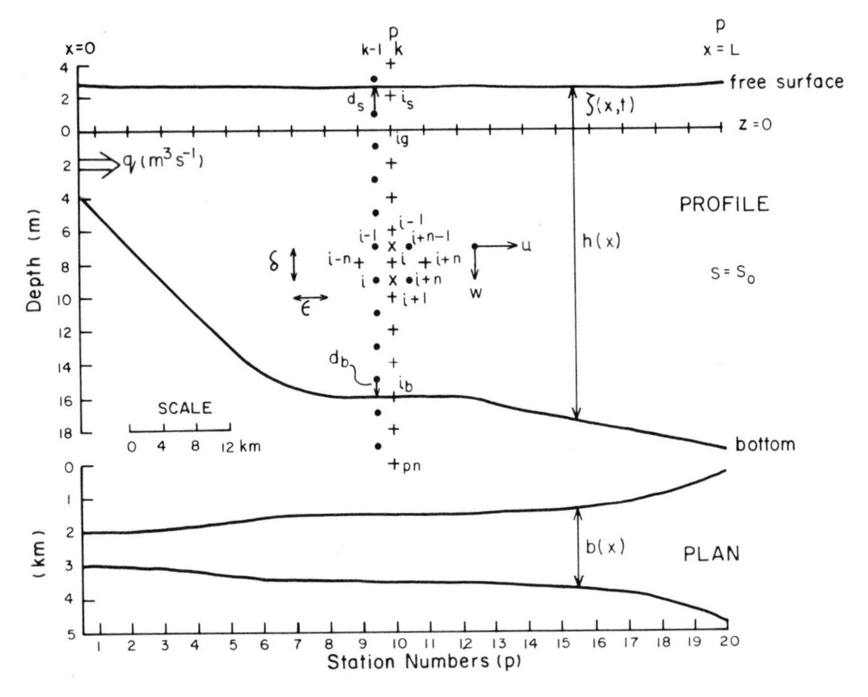

Figure 1. Geometry of the hypothetical channel used in (5), showing the quantities used in the mathematical model, including elements of the finite difference grid; i_s, i_b, i_g surface, bottom and x axis grid points respectively; + salinity, • u – velocity, X w – velocity grid points; ϵ horizontal, δ vertical grid spacings; column parameters, p and k; n, total number of grid points in a column; d_s, d_b surface and bottom grid parameters respectively.

the bottom grid point will contaminate all subsequent calculations of w in that water column. Advantages of eq. (2b) are that the surface boundary condition, $w = -\partial\zeta/\partial t$ at $z = -\zeta$, is automatically satisfied and that the integrations can be carried out as a subsidiary part of the total depth integration of u, which is calculated as a part of eq. (1) by the use of quadratic splines (c. f. § 5 in [13]) interpolated between the u grid points. Thus w is calculated at points halfway between the salinity points (Fig. 1). This also leads to an improved finite difference formulation of the vertical advection term in eq. (3).

At boundary grid points the finite difference equations are modified to account for the boundary conditions. For some of the terms in the equations extrapolation (by splines) above the surface and below the bottom is sufficient, however for the no flux condition eq. (10) and the stress condition eq. (7), this is not considered accurate enough. To include the wind stress at the surface, a series of Taylor expansions to fourth order about the velocity point $i = i_s$, making use of the fact that the first derivative at the surface is

$$\partial u/\partial z = u_o' = -\tau_w/N_z \quad \text{at } z = -\zeta, \tag{18}$$

is used to find an expression for $N_z\partial^2u/\partial z^2$.
The result is

$$\left.\frac{\partial^2 u}{\partial z^2}\right|_{i=i_s} \simeq \frac{(8\delta^2-6d_s^2)\,u_{i+1} - (\delta^2 - 3d_s^2)\,u_{i+2} - (7\delta^2 - 3d_s^2)\,u_i - 6\delta^3 u_o'}{\delta^2\,[3d_s^2 + 6d_s\delta + 2\delta^2]} \tag{19}$$

Similarly the horizontal no salt flux condition at the bottom grid point in the presence of a bottom slope eq. (10b) can be obtained by replacing δ by ϵ, the horizontal grid spacing, and d_s by $\epsilon/2$, thus

$$\left.\frac{\partial^2 s}{\partial x^2}\right|_{i=i_b} = (26\,s_{i+n} - 25\,s_i - s_{i+2n})/23\epsilon^2 \tag{20}$$

The method can be summarized as explicit; using a single time step. Equations (1)-(4) are evaluated in order, with the pressure terms in eq. (4) being at the higher time level and at each time step a new grid is defined by the position of the free surface.

The conservative form of eq. (3) is catastrophically unstable if used with this method, but this restriction can be overcome if after the calculation of a new velocity field, the salinity is recalculated using updated values of velocity and salinity before proceeding to a new cycle (36).

The main stability criterion on the explicit method is the usual dynamical condition on the propagation of gravity waves (23, 29);

$$\tau < \epsilon / (2gH)^{1/2} \tag{21}$$

where τ is the time step and H is the maximum depth. Time steps of 2 minutes, with $\epsilon = 4$km, were used for the experiments in (5) and 1 minute, with $\epsilon = 2$km, for the Rotterdam Waterway in (13). Thus if long periods of the order of weeks are required, calculations become prohibitive for these short time steps. Kwizak and Robert (21) have shown that stability criteria such as eq. (21) are determined by the explicit treatment of the velocity divergence term in the equation of continuity eq. (1) and the pressure terms in the momentum equation (4). Thus if the divergence term in eq. (1) is treated at the higher time level $t + \tau$, then the method becomes semi-implicit and the time step is then determined by the necessity of resolving the shortest wave period present, so time steps of 15 to 30 minutes become possible. In finite difference terms eq. (1) becomes [c.f. equation (23) in (13)]

$$\zeta_p^{t+\tau} = \zeta_p^t - \frac{\tau}{\epsilon}\left(\bar{u}_{p+1}^{t+\tau} - \bar{u}_p^{t+\tau}\right) \tag{22}$$

Then if eq. (15) is used to relate $\bar{u}^{t+\tau}$ to $\bar{u}^t$ and $\zeta^{t+\tau}$ (only $\partial\zeta/\partial x$ being considered at level $t + \tau$), eq. (22) can be reduced to a recurrence relation in $\zeta_p{}^{t+\tau}$, $\zeta_{p+1}^{t+\tau}$ and $\zeta_{p-1}^{t+\tau}$, which can be solved for $\zeta_p^{t+\tau}$ by matrix inversion methods. As the total number of columns, P, is of the order of 20, the matrix inversion is rapid and there is only a small increase in the computer time per time step as compared with the explicit method of (13). After the depth dependent velocity field is calculated using the finite difference form of eq. (4), then eq. (22) is used again as a check that continuity has not been violated. Therefore the equations are used in the order (2), (3), (1) and (15) semi-implicitly, (4) and (1). The two methods were compared using the hypothetical channel of (5) (Fig. 1) and the results were virtually identical. An increase in the time step from 2 to 30 minutes resulted in an approximately eight fold saving in computer time.

SUMMARY OF SOME PERTINENT RESULTS OF EXPERIMENTS PERFORMED WITH THE NUMERICAL MODEL.

Using the hypothetical channel, dimensions of which are shown in Fig. 1, a number of experiments were performed to investigate the effects of changes in tidal range, the imposed tide being semi-diurnal (eq. 14), riverflow, horizontal density gradients and the magnitude and functional form of the eddy coefficients. Initial conditions are the same for all experiments: velocities are depth independent and given by the riverflow, and salinities, also independent of depth, increase linearly from zero the head to the assumed sea water value at the mouth. The calculation was started at low water, $\zeta(x,t) = - A_o$, and run for four tidal periods, the results being taken from the fourth period.

A general formulation for the eddy coefficients used in these experiments is:

$$N_z = N_o + N_1 \left|\bar{u}\right| (1 + m\beta Ri)^q$$
$$K_z = K_o + K_1 \left|\bar{u}\right| (1 + \beta Ri)^p \tag{23}$$

where $\bar{u}$ is given by eq. (16) and the overall Richardson number by

$$Ri = ag \frac{\Delta s}{(h+\zeta)} \bigg/ \left(\frac{\bar{u}}{(h+\zeta)^2}\right)^2 \tag{24}$$

where Δs is the difference in salinity between bottom and surface. The details of the assumptions behind the formulation of eq. (23) are given in (5).

Three cases were run for the same values of tide amplitude, riverflow, and mouth salinity, with the eddy coefficients:

(a) constant: $N_o = 40, K_o = 20 \text{ cm}^2 \text{ s}^{-1}, N_1 = K_1 = 0$

(b) variable: $N_o = 5, K_o = 2{\cdot}5 \text{ cm}^2 \text{ s}^{-1}$
$N_1 = 0{\cdot}25 \times 10^{-2}, K_1 = 0{\cdot}125 \times 10^{-2}, \beta = 0$

(c) variable with Ri dependance:
$N_o = 5, K_o = 2{\cdot}5 \text{ cm}^2 \text{ s}^{-1}$
$N_1 = 0{\cdot}25 \times 10^{-2}\ K_1 = 0{\cdot}25 \times 10^{-2}$
$\beta = 1{\cdot}0, m = 7, p = -7/4, q = -1/4$

The coefficients in (c) were chosen from data on the River Mersey (4) and the magnitudes of N_1 and K_1 in (b) and (c) are based on results for a neutrally stable tidal current (3). A degree of arbitrary stratification is implicit in (a) and (b) as $K_z = ½ N_z$ and again this is typical of mean conditions in the Mersey (1). The velocity and salinity tidal curves for the three cases are shown in Fig. 2. Common features to all cases are the ebb lasting longer than the flood, a characteristic of tidal waves in estuaries, the effect of the density current circulation on the velocity shear (compare ebb and flood distributions in Fig. 2) and the reduction of stratification on the flood tide as compared with the ebb. The differences between the three cases are primarily ones of magnitude and phase and are most clearly seen in the salinity curves, which become more assymmetrical as the differences in the vertical mixing between ebb and flood increases. Fig. 3 shows the effect of the Richardson number on the eddy coefficients for case (c) as compared with (b). It would seem that case (c) with the Richardson number dependence gives the most realistic response of the salinity and velocity to the tide.

The different functional forms of the eddy coefficients also affects the overall longitudinal salt balance. Fig. 4 shows the deviation of the tidally averaged depth mean salinity, for the fourth period, from the initial linear distribution. Thus for case (a) there is a net loss of salt over the four tidal periods, whereas for case (b) there is a slight gain in salt content. Case (c) shows the largest deviations from the initial linear distribution, but the salt loss is less than for case (a). It is not the absolute magnitude of the salt loss for the three cases which is important, but the relative effects on the salt balance of the different formulations for the vertical eddy diffusion coefficients. Examination of the values of the eddy coefficients, during a tidal cycle, show that the tidally averaged depth mean salinity, S_M, at any position is related to both the magnitude of the tidal mean and the amplitude of the tidally variable components of K_z and N_z. With a Richardson

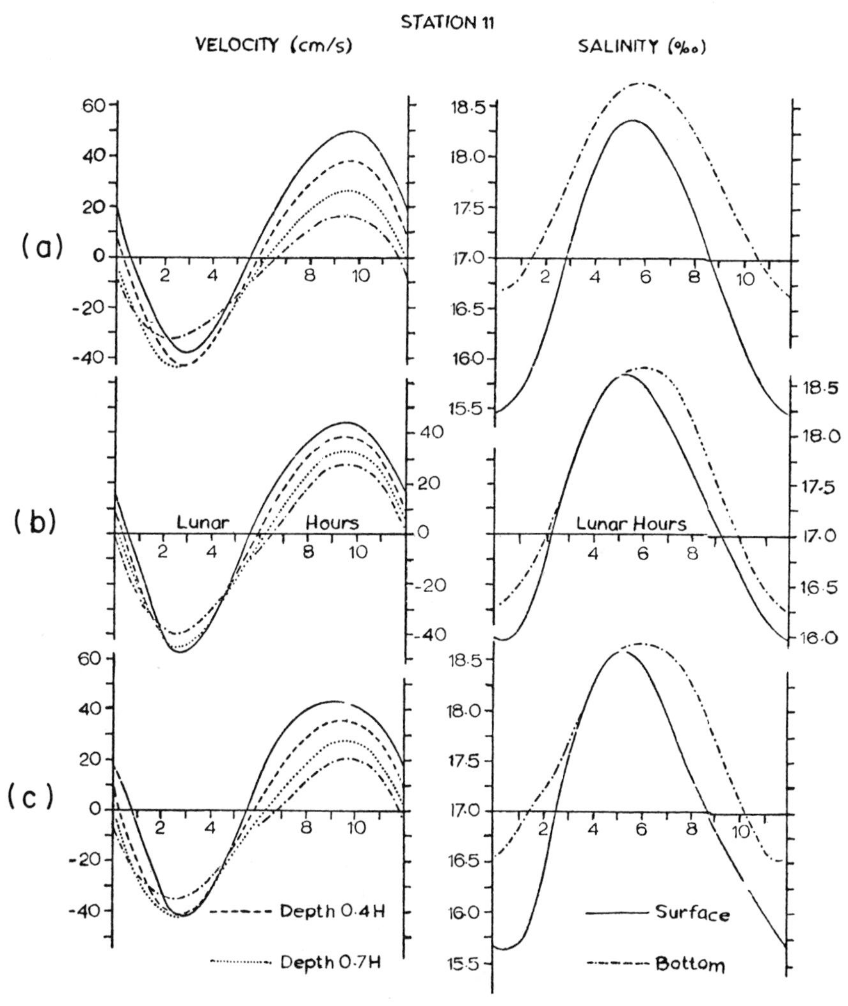

Figure 2. Velocity and salinity time curves for experiments run with three different forms of the eddy coefficients (a) constant (b) variable and (c) variable with Ri dependence. The external parameters are $A_0 = 1$m, $q = 50$ m³ s⁻¹ and mouth salinity $s_0 = 33°/oo$ [redrawn from (5)] (Note 1 Tidal period = 12 lunar hours).

number dependence, the tidally varying part of (23) can be largely suppressed by the stratification (i.e., for station numbers less than 8 for case (c) above). The physical explanation of these differences probably lies in the dispersion pro-

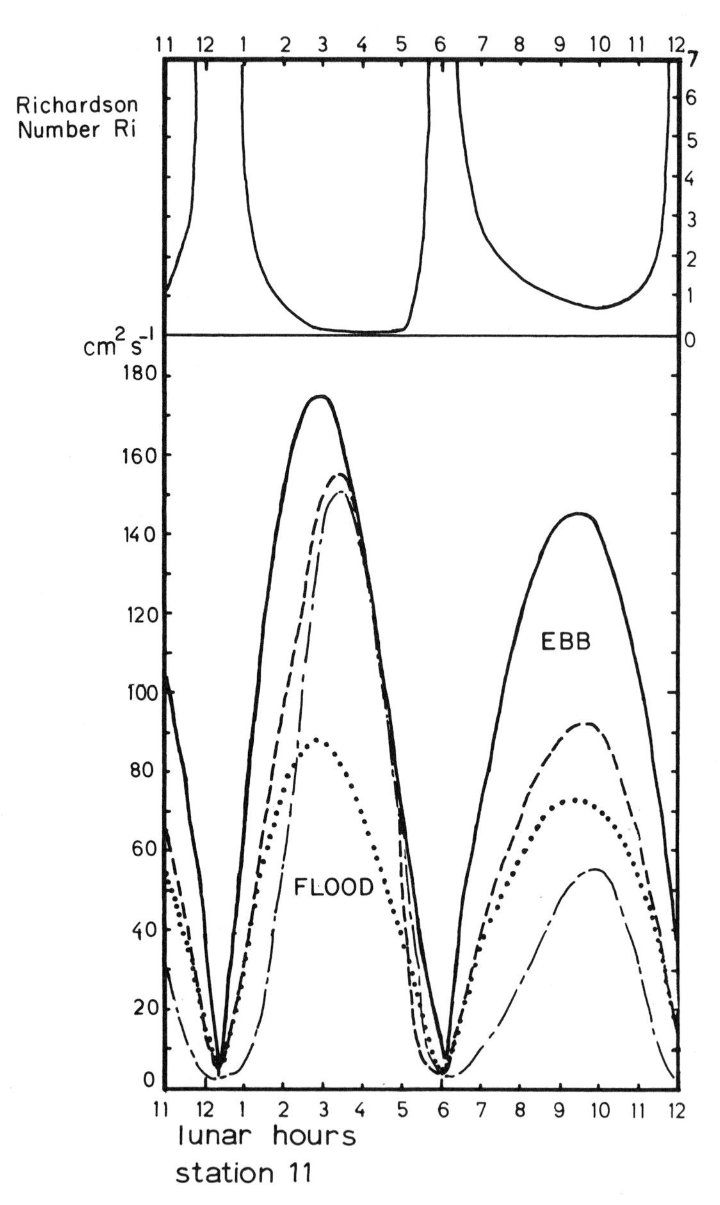

Figure 3. Variation of variable N_Z (———) and K_Z (.......) and of Ri dependent N_Z (– – – –) and K_Z (– - – -) during the tidal cycle for cases (b) and (c) in Fig. 2.

duced by shear and phase effects (2) which are difficult to delineate with a numerical model. The salt balance will be further discussed in the next section.

The experiments described in (5) were run for four tidal periods and the effect of changing the river discharge q, keeping all other parameters the same was slight. This is because for so short a period the riverflow has only a small effect on the longitudinal density gradient. Changing the horizontal density gradient by changing the sea mouth salinity produced results which agree with Hansen and Rattray's (14) analytical work. The density current circulation increased linearly with $\partial s/\partial x$ and the stratification had an approximately quadratic law dependence on the horizontal salinity gradient. Data from the Mersey (6, 4) show that $\partial s/\partial x$ increased with river discharge as $q^{1/2}$ and that stratification, Δs, was proportional to q. Thus the model results agree with nature if $\Delta s \propto q \propto (\partial s/\partial x)^2$ approximately. It should be noted in evaluating these results that the flushing time of the hypothetical channel is of the order of 50 days and in the results obtained from the Mersey q was averaged over 10 days. Thus the

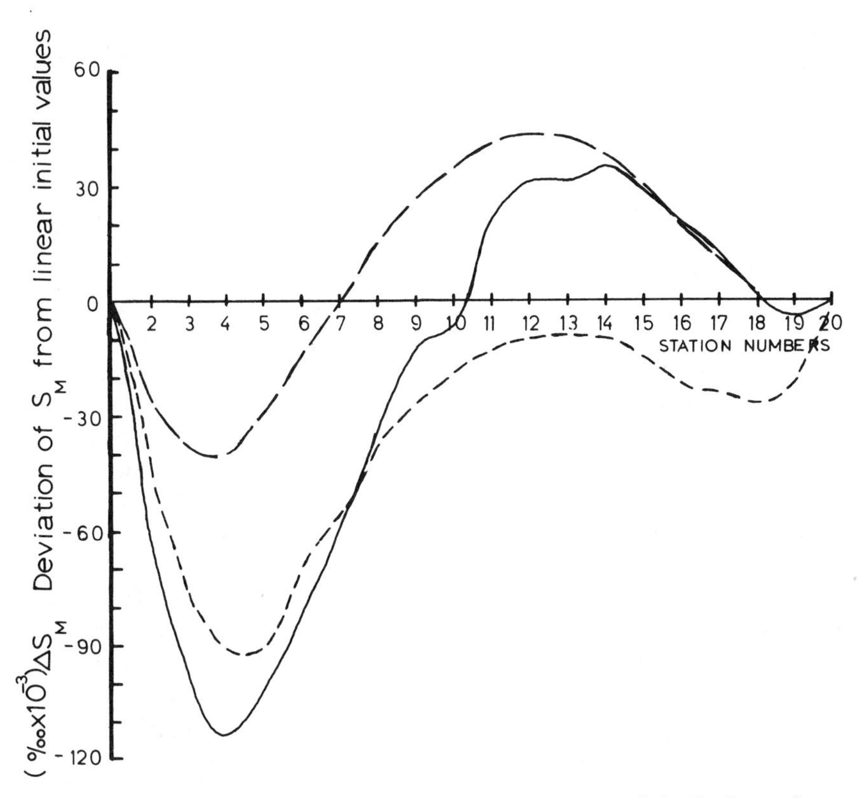

Figure 4. The deviation (ΔS_M) of the tidally averaged depth mean salinity S_M from a linear distribution, for the fourth period, with distance along the estuary. Cases (a) (– – – – – –), (b) (— — —) and (c) (________).

interaction of the density current, stratification, horizontal density gradient and riverflow, which buffers the salt content of the estuary against large changes in riverflow (14) must take place on a time scale which is a substantial fraction of the flushing time.

THE SEAWARD BOUNDARY CONDITION

On the time scale of a tidal period, the results from the explicit model, particularly the vertical distributions of current and salinity, agree with theoretical and experimental work (13). However to investigate the response, particularly of the longitudinal salt distribution, to changes in the riverflow, much longer time periods are required. The development of the semi-implicit method in section 3, makes integration periods of the order of 20 days, say, economically feasible. However for such periods, consideration of the salt balance and the seaward boundary condition is necessary.

For coastal plain estuaries of type 2 (15), into which category the Rotterdam Waterway and the hypothetical channel fall, it is known for the steady state that both advective and diffusive fluxes contribute to the salt balance (14). The upstream diffusive salt flux, however, contains contributions from correlations between the periodically varying parts of u and s (2) and also due to the variation of u and s across the estuary. The former is taken account of directly by the tidal model as is the advective contribution due to the density current (14). The latter is not, due to the lateral averaging implicit in the basic equations, and Fischer (11) considers it an important contribution to the effective longitudinal diffusion coefficient even for the Mersey where previously the vertical distributions of u and s had been thought to have the greater effect (1).

The mouth of an estuary is determined by its physical boundaries and the stratification at the mouth will be a result of interaction between the coastal sea and the estuary, which are not of course independent of each other. Perhaps the most satisfactory solution would be to construct a three dimensional model of at least a region of the sea (7, 24) which could be spliced to the two dimensional model at the estuary mouth. This would allow the boundary condition to be placed at the boundaries of the three dimensional sea model, which would be at large enough distances from the estuary mouth to be only slightly affected by the fresh water discharge from the estuary. However, there would still remain the problem of prescribing the mixing in the vicinity of the mouth. This solution would be very costly in terms of computer time, with maybe more effort being put into the sea model than the estuary proper. Ideally a seaward boundary condition should not require that the two dimensional nature of the model formulation be changed. For the hypothetical channel experiments, where the aim is to investigate the response of the estuarine circulation to a range of parameter values, the simple boundary condition is:

$$s(L, z, t) = s_o \tag{25}$$

where s_o is a constant presumed representative of the salinity of the coastal sea in the vicinity of the estuary mouth. This is the 'overmixed' case, where infinite mixing takes place at the boundary. This is clearly unphysical, though some estuaries with large tides can become almost homogeneous in their lower reaches (e.g., the Bristol Channel). However it should be noted that the estuary develops appreciable stratification and tidal variations of salinity within a few grid points of the mouth (5). Thus it may be possible to extend the channel hypothetically beyond the physical position of the mouth (many estuaries have well developed channels extending some distance into the coastal sea, e.g. the Mersey) hence the use of eq. (25) may not affect the solution in the inner reaches unduly. The stratification at the mouth could also be considered to be fixed. i.e. s_o is the function of z only. In this case the implication is that the stratification at the mouth is maintained by external means rather than being influenced by the estuarine dynamics. This may be satisfactory where the estuary opens out into a larger estuarine bay such as in the Chesapeake Bay System. If it is possible to obtain continuous time series of salinity with depth at a station near the mouth, then this data can be used as a seaward boundary condition, though data taken continuously over a long period of time are almost non-existent. The model then becomes a simulation and loses any predictive capacity beyond the range of the observed boundary conditions. Also this solution does not remove the necessity of considering the upstream diffusive component of the salt flux as the simulation of the Rotterdam Waterway indicates.

It is easy to show that the use of eq. (25) leads to a loss of salt, due to advection by the riverflow, from the estuary:

$$Q = \frac{1}{T} \int_o^T \int_{-\zeta}^{h} usdzdt = \frac{s_o q}{b} = K_x \frac{1}{T} \int_o^T \overline{\frac{\partial s}{\partial x}}^{h} dt \text{ at } x = L, \quad (26)$$

where Q is the salt flux through the mouth (positive seawards). It is difficult to use eq. (26) to calculate a value for K_x to balance this loss because there is an effective K_x already present due to the numerical scheme. The use of highly accurate dispersionless formulations for the advective terms in eq. (3) recently introduced (27) would improve this aspect of the model. One solution would be to calculate the salt loss from the numerical results comparing one tidal period with another and using a formulation similar to eq. (26) to compute an effective K_x. However it must be remembered that it is highly unlikely, taking just one tidal period, that an estuary will be in balance. There will be a net loss or gain of salt depending on the nature of the external forcing. An estuary is probably never in a steady state when short periods are considered, but is only so in some long term statistical sense. Thus an effective K_x which is designed to counteract the advective salt loss should probably only be applied at the mouth and not in the interior of the estuary, where the density current circulation and tidal effects could be assumed to be the primary salt balance mechanisms.

The above discussion is designed to show that even though the technique exists for detailed numerical modelling of an estuary with long integration periods, the highly variable nature of estuaries along with consideration of the long term salt balance makes the choice of a seaward boundary condition ambiguous. These difficulties, which can be neglected if only short periods are being considered, will probably only be resolved by detailed measurements in the field, which will need to be carried out for periods of at least several weeks.

ACKNOWLEDGMENTS

A large part of this work was carried out while I was at Liverpool University and I should like to record my thanks to Professor K. F. Bowden for suggesting the development of this type of model and for his support and interest generally. I am grateful to Dr. N. S. Heaps for a number of discussions on numerical models.

Financial support has been forthcoming from a Natural Environment Research Council Studentship, National Science Foundation grants GA 30745X, DES 74 22711 and the Coastal Upwelling Ecosystem Analysis Program (CUEA) under grant GX 33502.

REFERENCES

1. Bowden, K.F. 1963. The mixing processes in a tidal estuary. Int. J. Air Water Pollut. 7: 343-356.
2. ——. 1965. Horizontal mixing in the sea due to a shearing current. J. Fluid Mech. 21:83-95.
3. ——, L.A. Fairbairn and P. Hughes. 1959. The distribution of shearing stresses in a tidal current. Geophys. J. Roy. astr. Soc. 2: 288-305.
4. ——, and R.M. Gilligan. 1971. Characteristic features of estuarine circulation as represented in the Mersey estuary. Limnol. Oceanogr. 16: 490-502.
5. ——, and P. Hamilton. 1975. Some experiments with a numerical model of circulation and mixing in a tidal estuary. Estuarine Coastal Mar. Sci. 3: 281-301.
6. ——, and S.H. Sharaf El Din. 1966. Circulation, salinity and river discharge in the Mersey estuary. Geophys. J. Roy. astr. Soc. 10: 383-399.
7. Caponi, E.A. 1974. A three dimensional model for the numerical simulation of estuaries. Tech. Note BN-800, Institute for Fluid Dynamics and Applied Mathematics, University of Maryland, 215 p.
8. Chan, R. K-C, and R.L. Street. 1970. Computer study of finite amplitude water waves. J. Comp. Phys. 6: 68-94.
9. Crowley, W.P. 1968. Numerical advection experiments. Mon. Weath. Rev. 96: 1-11.
10. Dronkers, J.J. 1964. Tidal computations in rivers and coastal waters. John Wiley and Sons, New York, 518 p.
11. Fischer, H.B. 1972. Mass transport mechanisms in partially stratified estuaries. J. Fluid Mech. 53:671-688.

12. Hamilton, P. 1973. A numerical model of the vertical structure of tidal estuaries. Ph.D. thesis, University of Liverpool, 164 p.
13. ____. 1975. A numerical model of the vertical circulation of tidal estuaries and its application to the Rotterdam Waterway. Geophys. J. Roy. astr. Soc. 40: 1-21.
14. Hansen, D.V., and M. Rattray, Jr. 1965. Gravitational circulation in straits and estuaries. J. Mar. Res. 23: 104-122.
15. ____, and ____. 1966. New dimensions in estuary classification. Limnol. Oceanogr. 11: 319-326.
16. Harleman, D.R.F. 1971. One dimensional models, p. 34-39. *In* Ward, G.H. Jr., and W.H. Esprey, Jr. (eds), Estuarine Modelling: An Assessment, TRACOR, Inc., Austin, Texas.
17. Heaps, N.S. 1967. Storm Surges. Oceanogr. Mar. Biol. Ann. Rev. 5: 11-47.
18. ____. 1969. A two dimensional numerical sea model. Phil. Trans. Roy. Soc. A 265: 93-137.
19. ____. 1974. Development of a three dimensional numerical model of the Irish Sea. Rapp. P. - v. Reun. Cons. int. Explor. Mer. 167: 147-162.
20. Kent, R.E., and D.W. Pritchard. 1959. A test of mixing length theories in a coastal plain estuary. J. Mar. Res. 18: 62-72.
21. Kullenberg, G. 1971. Vertical diffusion in shallow waters. Tellus 23: 129-135.
22. Kwizak, M., and A. Robert. 1971. A semi-implicit scheme for grid point atmospheric models of the primitive equations. Mon. Weath. Rev. 99: 32-36.
23. Leendertse, J.J. 1967. Aspects of a computational model for long period water wave propagation. Memor. Rand. Corp., Santa Monica, Calif. RM-5294-PR.
24. ____, and S-K Liu, 1975. A three dimensional model for estuaries and coastal seas. Vol. II. Aspects of computation. Memor. Rand. Corp., Santa Monica, Calif. R-1764-OWRT.
25. McDowell, D.M., and D. Prandle. 1972. Mathematical model of the River Hoogly. Journal of the Waterways and Harbors Division, ASCE, 98 WW2: 226-242.
26. Nichols, B.D., and C.W. Hirt. 1971. Improved free surface boundary conditions for numerical incompressible flow claculations. J. Comp. Phys. 8: 434-448.
27. Pedersen, L.B., and L.P. Prahm. 1974. A method for the numerical solution of the advection equation. Tellus 26: 594-602.
28. Ramming, H.G. 1972. Hydrodynamical dispersion of seston in the River Elbe. Memoires Sociètè Royale des Sciences de Liége. 6e serie, tome II: 181-208.
29. Richtmyer, R.D., and K.W. Morton. 1967. Difference methods for initial value problems. Interscience Publ., New York, 405 p.
30. Roach, P.J. 1972. On artificial viscosity. J. Comp. Phys. 10: 169-184.
31. Rossiter, J.R. 1961. Interaction between tide and surge in the Thames. Geophys. J. Roy. Astr. Soc. 6: 29-53.
32. Simons, T.J. 1973. Development of three dimensional numerical models of the Great Lakes. Canada Centre for Inland Waters Sci. Ser. No. 12, 26 pp.
33. Stigter, C., and J. Siemons. 1967. Calculation of longitudinal salt distribution in estuaries as a function of time. Publ. no. 52, Delft Hydraulics Laboratory, 35 p.

34. Vreugdenhil, C.B. 1970. Computation of gravity currents in estuaries. Publ. no. 86, Delft Hydraulics Laboratory, 108 p.
35. Williams, D.J.A., and J.R. West. 1973. A one dimensional representation of mixing in the Tay estuary. Water Pollut. Res. Tech. Pap. 13: 118-125. Department of the Environment, London. HMSO.
36. Young, J.A., and C.W. Hirt. 1972. Numerical calculation of internal wave motion (2D stratified flows) J. Fluid Mech. 56: 256-276.

FORMULATION AND CLOSURE OF A MODEL OF TIDAL-MEAN CIRCULATION IN A STRATIFIED ESTUARY

George H. Ward, Jr.
Manager, Engineering Programs
Espey, Huston and Associates, Inc.
3010 South Lamar
Austin, Texas 78704

ABSTRACT: We consider the formulation of the lateral-mean tidal-mean momentum and salt balances in an estuary with rectilinear cross-section. Explicit treatment of the density current, typical of tidal-mean estuarine circulation, requires solution of the coupled equations of momentum and salinity. Central to this is the specification of the vertical eddy fluxes. Analysis of data from the Mersey and James estuaries indicates that the entirety of the vertical structure of the salt flux is determined by the local gravitational stability and the proximity to the vertical boundaries of the system, the absolute magnitude of the flux being established by the bulk hydrodynamic characteristics of the estuary. The structure of the momentum flux is further dependent upon the Richardson number ratio Rf/Ri, for which a functional form is suggested. An idealized analytical solution is given and numerical solution of the more general equations is discussed.

1. TIDAL-MEAN DYNAMICS

We consider an estuary of the drowned-river-valley type, characterized by a prominent longitudinal dimension in which the zone of saline intrusion is of the order of a few hundred kilometers and in which mechanical mixing is sufficient to eliminate wedge-type stratification. These features are typical of many of the estuaries along the east coast of the United States, and it is this type of estuary whose dynamics received the earliest and, perhaps, most thorough attention in oceanography.

In this type of estuary. as in most geofluid systems, there are many scales of motion extant simultaneously, ranging from the second-millimeter fluctuations of fine-scale turbulence, through the semi-diurnal oscillations of the tidal wave, up to shifts in the thermohaline structure extending over the entire estuarine reach on a time scale of days. From the standpoint of the large-scale distribution of waterborne properties, such as salinity, temperature and many indicator constituents for water quality, it is the tidal-mean circulation of the estuary that is most important. This scale of motion is exposed by performing a time average of the governing equations over one or more tidal cycles. In estuaries of the type considered here, the tidal-mean circulation is predominantly density driven. Unless there is significant cross-channel variation in physiography, this circulation is two-dimensional, in the longitudinal-vertical plane, so a further integration across the channel is justified, reformulating the dynamic equations in terms of lateral-mean properties.

To obtain a simple paradigm of the dynamics of this sort of system, we adopt the following assumptions:

(i) neglect of molecular viscosity and diffusivity,
(ii) Boussinesq approximation,
(iii) incompressibility,
(iv) hydrostasy,
(v) linear relation between density and salinity, viz. $\rho = \rho_o + \Lambda s$ (for ρ in g m^{-3} and s in °/oo, $\rho_o = 10^{-6}$, $\Lambda = 8.2 \times 10^{-10}$),
(vi) neglect of coriolis accelerations,
(vii) zero fluxes of mass and momentum through the lateral boundaries,
(viii) a rectilinear estuary channel, i.e., constant breadth and depth,
(ix) nil surface stress, i.e., zero wind,
(x) neglect of atmospheric pressure gradients.

The last five may be easily relaxed, and are used in this discussion merely to simplify the arithmetic. Under these conditions, the governing equations become

$$\rho_o \left(\frac{\partial u}{\partial t} + u\frac{\partial u}{\partial x} + w\frac{\partial u}{\partial z}\right) = -\frac{\partial p}{\partial x} + \frac{\partial}{\partial x}\tau_{xx} + \frac{\partial}{\partial z}\tau_{xz} \tag{1}$$

$$\frac{\partial p}{\partial z} + \rho g = 0 \tag{2}$$

$$\frac{\partial u}{\partial x} + \frac{\partial w}{\partial z} = 0 \tag{3}$$

$$\frac{\partial s}{\partial t} + u\frac{\partial s}{\partial x} + w\frac{\partial s}{\partial z} = \frac{\partial}{\partial x}\sigma_x + \frac{\partial}{\partial z}\sigma_z \tag{4}$$

All variables are lateral and tidal averaged, x is positive downstream and z positive upwards with z = 0 coinciding with the estuary bed. Here τ_{xx}, τ_{xz} and σ_x,

σ_z denote the residual or "nonadvective" momentum and salt fluxes respectively. These consist of the tidal-mean lateral-mean eddy fluxes plus fluxes due to secondary circulations which induce correlations between the lateral and/or tidal structures of velocity components and salinity. Provided independent expressions for the fluxes τ_{xx}, τ_{xz}, σ_x and σ_z can be established, the system of equations (1) - (4) constitutes a complete set in the variables u, w, s and p. (An additional term, $-u_T \partial u_T/\partial x$ where $u_T \equiv \max \left\{u - \frac{1}{T}\int_t^{t+T} u dt\right\}$, should appear on the righthand side of (1). This term, though not negligible, is usually of lower order and is omitted here for simplicity. When retained, it is generally evaluated from independent knowledge of the tidal current amplitude variation along the estuary, therefore is of the character of a small specified forcing function.) Detailed discussion of these equations may be found in Dyer [5].

The longitudinal pressure gradient may be evaluated from (2) as

$$\frac{\partial p}{\partial x} \cong g\int_z^h \frac{\partial \rho}{\partial x}\, dz + \rho g \frac{\partial h}{\partial x} \tag{5}$$

where h denotes the tidal-mean water surface level. It is this term which is responsible for the characteristic circulation of drowned river valley estuaries: counterflowing layers in the saline-intrusion reach, with the lower layer flowing upstream. The upstream flow is a density current driven by the longitudinal salinity gradient.

From experiential studies in real estuaries (e.g., Pritchard [16]), evidence has accumulated that suggests the dominant terms of (1) to be the pressure gradient acceleration and the residual xz-flux term. Indeed, except under extremely dynamic conditions, the momentum structure appears to be antitryptic, i.e., the pressure gradient is balanced by the residual stress gradient, so that the equation of motion (1) simplifies to

$$\frac{\partial p}{\partial x} = \frac{\partial}{\partial z}\tau_{xz} \tag{6}$$

Similarly, the term $\partial\sigma_x/\partial x$ in the salt-balance equation (4) may be omitted, but the advective terms must be retained.

The residual fluxes τ_{xz} and σ_z must be specified in order to close the system. These are traditionally parameterized in terms of the mean properties of the flow by the introduction of "eddy" diffusivities, viz.

$$\tau_{xz} = \rho_o N_z \quad \partial u/\partial z \tag{7}$$

$$\sigma_z = K_z \, \partial s/\partial z \tag{8}$$

It should be remarked that, because τ_{xz} and σ_z are the residual fluxes in a tidal-mean lateral-mean model, and u and s are the tidal-mean lateral-mean current and salinity, the diffusivities N_z and K_z cannot rigorously be considered to be merely the means of the intratidal "turbulent" diffusivities. The same sort of dynamic reasoning applies, however, as demonstrated in the following section, and, further, N_z and K_z approximate the mean intratidal coefficients if the tidal variation of the shear $\partial u/\partial z$ and stratification $\partial s/\partial z$ is small. (This approximation is degraded by significant covariance in the tidal or lateral structure of velocity and salinity, see Dyer [5] and Fisher [7].)

Boundary conditions on salinity are straightforward: the salinity profiles at the upstream and downstream boundaries of the system and the salt fluxes at surface and bottom (which we assume to be zero). Use of the antitryptic balance equation (6) obviates the need for longitudinal boundary conditions on u, although we must use the integral-mean condition

$$\int_0^h u\,dz = hv_f \tag{9}$$

to close the system for h. Here $v_f = Q/A$ is the throughflow velocity, Q denoting the inflow to the head of the estuary and A the cross-sectional area. At the surface $w = 0$ and $\tau_{xz} = 0$ and at or near the bed $u = w = 0$.

We must further specify the bed stress τ_o. The intratidal current profile is logarithmic near the bed, hence the tidal-mean profile must also be logarithmic, see [23],

$$u = \frac{\tau_o}{|\tau_o|} \sqrt{\frac{|\tau_o|}{\rho_o}} \; \frac{1}{\kappa_T} \log \frac{z}{z_o} \tag{10}$$

in which $\kappa_T \equiv 2\kappa(u_T/T|u|)^{1/2}$, u_T is the tidal current amplitude, and κ is the von Karman constant. The roughness length z_o is the height above the bed at which the tidal-mean current vanishes; $z_o \ll h$ and is well approximated by the intratidal roughness length, the level at which the ensemble-mean current vanishes. The value of z_o may be estimated from the Chèzy coefficient C according to

$$C = \frac{2\sqrt{g}}{\kappa} \left(\frac{\tanh^{-1} \sqrt{1 - \frac{z_o}{h}}}{\sqrt{1 - \frac{z_o}{h}}} - 1 \right) \tag{11}$$

see reference [23]. Typically $z_o \cong 10^{-3}$m.

From (7) the logarithmic profile is seen to be equivalent to a near-bed profile of N_z given by

$$N_z = \kappa_T z \sqrt{\frac{|\tau_0|}{\rho_0}} \tag{12}$$

i.e., linear in z.

2. SPECIFICATION OF DIFFUSIVITIES

Due to the importance of the vertical fluxes of salt and momentum σ_z and τ_{xz}, the closure of the equations by specification of these fluxes is central to the model formulation. Unfortunately, no adequate theoretical model currently exists for their formulation, but rather they are quantified by a mix of empiricism and physical reasoning.

The distinction between the diffusivity for momentum N_z and that for salt (i.e., mass) K_z is important. Although for homogeneous waterbodies it appears that $K_z = N_z$, as the vertical density stratification increases, the ratio $N_z : K_z$ becomes greater than unity. A partial explanation is that density stratification permits the existence of internal waves which can transport momentum (through the pressure-velocity covariance) without transporting mass, thus offering an additional transfer mechanism for momentum.

The vertical structure of the residual fluxes is a critical element of their formulation. Two factors that strongly influence the vertical structure are the presence of confining boundaries, viz. the surface and bottom, and the density stratification. The former has been analyzed in terms of a "mixing-length" theory, relating the flux to the rms vertical excursions of water parcels. For the case of a fluid confined between horizontal planes, Montgomery [14] suggests the mixing length to be proportional to $z(h-z)/h$, and the usual manipulations (e.g., Sutton, [21], Kent and Pritchard [12]) show K_z to be proportional to the dimensionless quadratic factor $z(h-z)h^{-2}$ or its powers. It is convenient therefore to separate this boundary effect from the other elements of the structure of K_z by defining

$$k_z \equiv K_z/d_z \tag{13}$$

where $d_z \equiv [z(h-z)h^{-2}]^m$. We consider only the cases m = 1 and m = 2.

The influence of vertical stratification is more complex. The kinetic energy in the eddying motion is given by

$$\frac{\partial}{\partial t}\frac{1}{2}(\overline{u'^2} + \overline{v'^2} + \overline{w'^2}) = -\frac{\partial}{\partial z}\overline{w'\left[\frac{p'}{\rho} + \frac{1}{2}(u'^2 + v'^2 + w'^2)\right]}$$
$$-\overline{u'w'}\frac{\partial \bar{u}}{\partial z} - \frac{g}{\rho}\overline{\rho' w'} + \Sigma \tag{14}$$

where overbars denote a space-time or ensemble mean, the primes local departures from that mean, and Σ the addition of turbulent energy from external sources, e.g., bottom roughness elements, free surface perturbations, ship traffic. The first term on the right is an internal redistribution of turbulent energy and need not be further considered. The term $-\overline{u'w'}\frac{\partial \bar{u}}{\partial z}$ represents the working of the Reynolds stress on the vertical shear of the mean current, and is (usually) a net source of turbulent energy whose supply is the energy of the mean current. The term $-g\,\overline{\rho' w'}/\rho$ represents the effect of density stratification, and is a sink of kinetic energy when the density is stably stratified ($\partial\rho/\partial z < 0$). Because of vertical motions along the density gradient, part of the energy of the eddying motion must be invested in increasing the potential energy in the water column; hence for a fixed rate of energy supply, the greater the density stratification, the smaller the intensity of the turbulence.

How do we parameterize this effect? Since Taylor's original analysis [22], it seems to have become traditional to employ the gradient Richardson number

$$\mathrm{Ri} \equiv -\frac{g\,\partial\rho/\rho\partial z}{(\partial u/\partial z)^2} \tag{15}$$

This emerges as a fundamental parameter in the theory of the instability of shearing stratified flow. In fact, the necessary condition for instability, $\mathrm{Ri} < \frac{1}{4}$, is often adopted as an index for the initiation of turbulence. It is not unnatural therefore to seek a dependence of the eddy flux upon Ri, as, for example, Pritchard [17] has done. However, there are reasons to reject the relevancy of the shear, and to consider solely the gravitational stability of the fluid, which we measure by the square Brunt-Väisällä frequency

$$N^2 \equiv -g\ \partial\rho/\rho\partial z \tag{16}$$

The model of tidal-mean mass diffusivity derived by Kent and Pritchard [12] and Pritchard [17] from their data from the James estuary is given by

$$K_z = \eta\ \frac{2hu_T}{\pi}\ \frac{z^2\,(h-z)^2}{h^4}\ (1 + \beta\,\mathrm{Ri})^{-2} \tag{17}$$

with $\eta = 8.6 \times 10^{-3}$ and $\beta = .276$. Here the boundary effect is quartic $d_z = z^2 (h-z)^2\ h^{-4}$. The stability effect is represented as a function of Ri, but Pritchard defined Ri as $N^2/(2u_T/\pi h)^2$ so that in his work the stability factor really only varied as N^2. This expression has been employed in the numerical models of Fisher et al. [8] and Overland [15]. Pritchard found the predicted diffusivities from (17) to be too low, particularly near the surface; he attributed this to the additional mixing effect of surface waves and added a term to (17) which decayed as the sinh of depth.

To test the efficacy of the various parameterizations of vertical structure, data from the James and the Mersey have been employed. The former were taken from the 1950 current and salinity measurements of Pritchard and Kent [18]. The latter were taken from the work of Bowden and his collaborators, e.g., ref. [2]. A particular boundary function was selected, either quadratic (m=1) or quartic (m=2), and $k_Z = K_Z/d_Z$ computed at each depth. The linear correlation of log k_Z against both log Ri and log N^2 was then tested. The James data set is superior in quality and extent to that of the Mersey; the computed explained variations (square of the linear correlation coefficient) for the James are given in Table 1. It is apparent that the combination of a quadratic boundary factor with an N^2 stability dependency is remarkably successful in predicting the structure of the diffusivity. A similar conclusion resulted from the Mersey data, although the explained variations were somewhat lower. It is interesting to note that with a quadratic boundary factor, rather than a quartic, the need for the near-surface "wave term" of Pritchard is obviated.

The disposition of the data is illustrated by Figs. 1 and 2; note in particular the organization of the data in Fig. 2, in which the quadratic boundary factor and the N^2 - dependence are plotted. (The points in parentheses in Fig. 1 are at levels near the lower extremum in the velocity profile, where calculation of Ri by (15) is noisy.) In Fig. 2, data from the Kattegat [11] are also plotted. Although the Kattegat is not an estuary, it is an instance of a counterflowing circulation with pronounced horizontal and vertical salinity gradients, so is pertinent to the present discussion.

On the basis of the behavior of this data, the following postulates are proposed to delineate the structure of K_Z:

(i) The boundary effect is represented by a factor of form $z(h-z)h^{-2}$.

(ii) The effects of density stratification are accounted for by the single parameter of vertical stability, e.g., the Brunt-Väisällä frequency N.

(iii) For sufficiently large N^2, the effect of stability is given by $k_Z \propto N^{-2}$. (The least-squares lines of Fig. 2 have slopes nearly −1. As $N^2 \to 0$, however, the value of k_Z must approach asymptotically the homogeneous value, therefore we expect a dependency like $(1 + \mu N^2)^{-1}$ with $\mu N^2 << 1$ for the range of N^2 considered here.)

Table 1. Explained variation of log k_Z by log N^2 and log Ri for quadratic and quartic boundary factors. James Estuary data, ref. [18].

Predictor variable:	d_Z quartic (m=2) Ri	N^2	d_Z quadratic (m=1) Ri	N^2
18-23 June 1950	.19	.89	.25	.98
26 June - 7 Jul 1950	.20	.84	.45	.98
17-21 Jul 1950	.14	.86	.37	.98

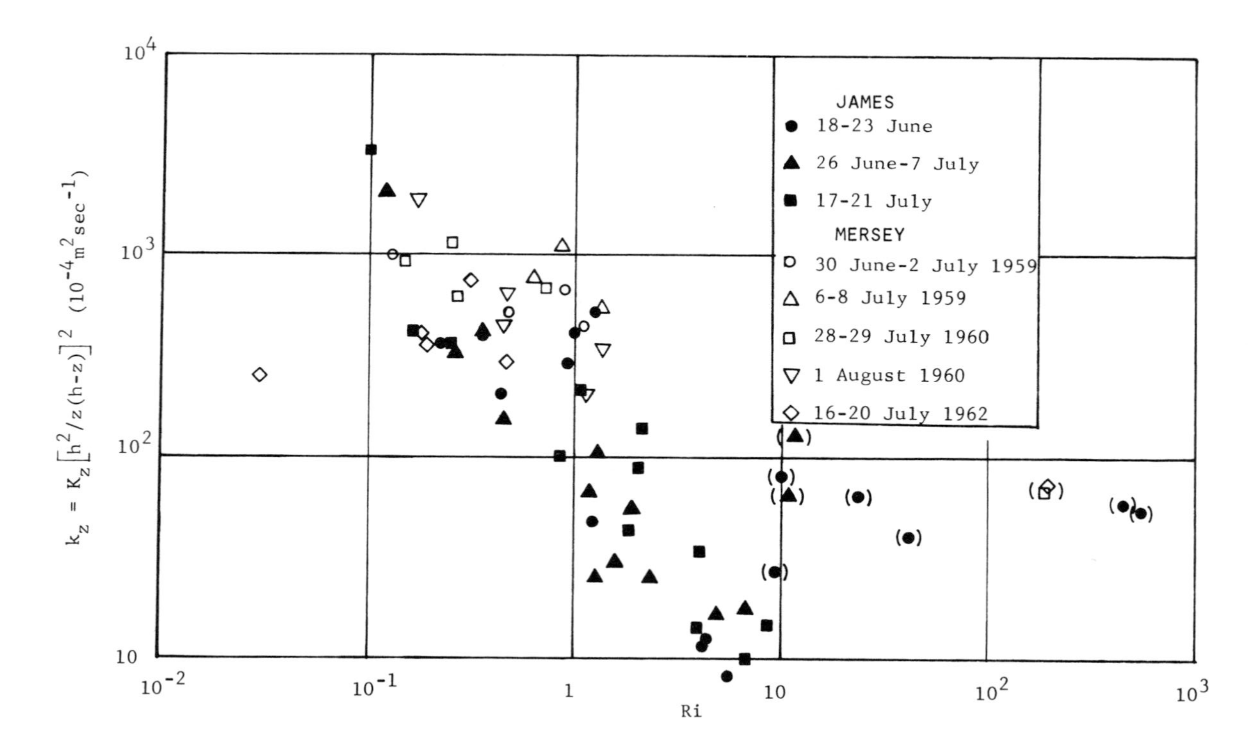

Figure 1. k_z versus Ri for quartic boundary effect.

(iv) The entirety of the vertical structure of K_Z is determined by the combined influences of the boundary effect d_Z and the stability N^2. Further quantification of K_Z is determined only by the bulk hydrographic parameters of the estuary.

The last postulate is motivated by the "sorting" of the data of Fig. 2 into separate k_Z–N^2 clusters for each waterbody.

On the basis of these postulates and further analysis of the James and Mersey data, the proposed model for mass diffusivity is

$$K_Z = \frac{z(h-z)\,[\mu_1 u_T h + \mu_3 (u_s + v_f) h^{-1}]}{h^2 (1 + \mu_2 N^2)} \tag{18}$$

where u_s is the mean current in the upstream flowing layer. Rough values for the

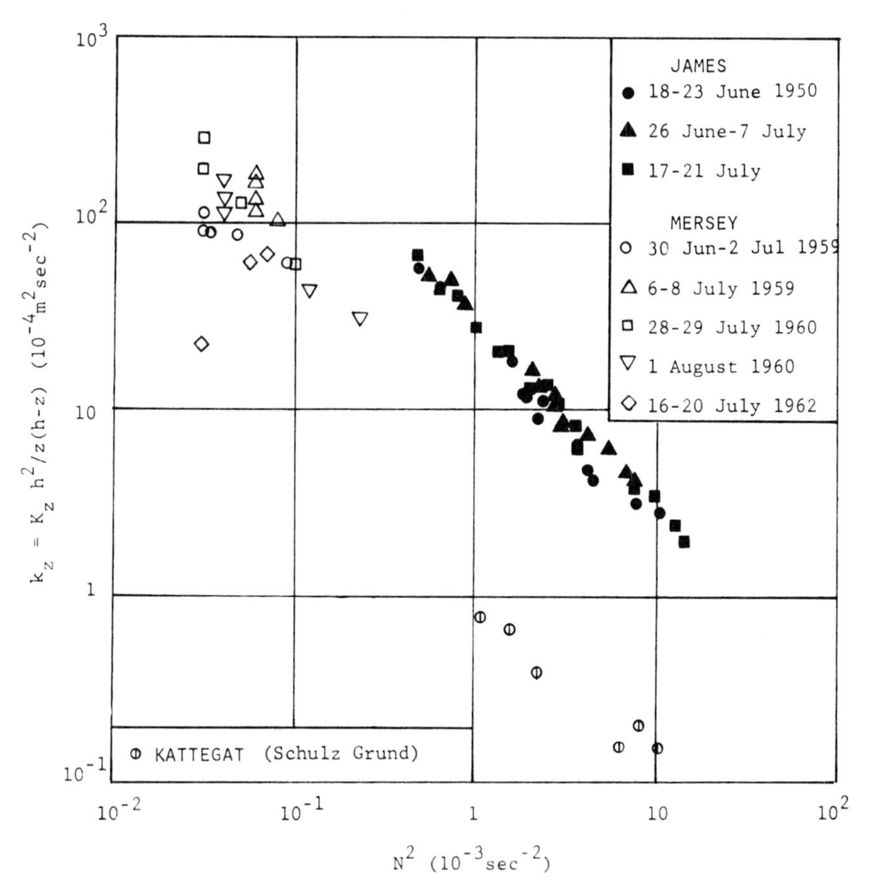

Figure 2. k_Z versus N^2 for quadratic boundary effect.

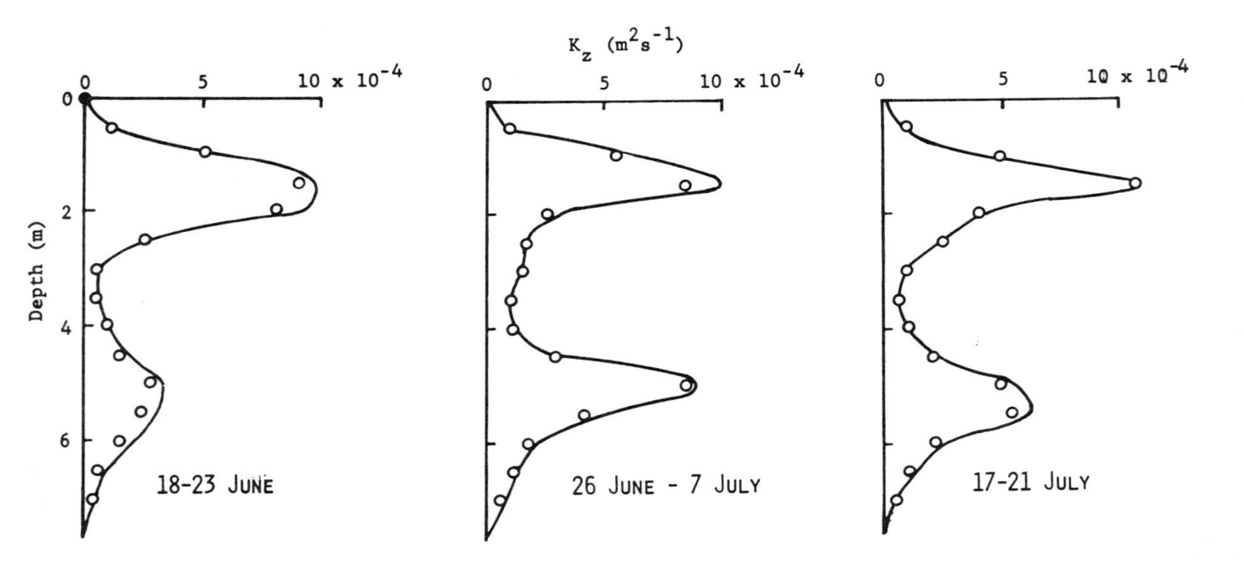

Figure 3. Observed and predicted eddy diffusivities for James Estuary.

constants are $\mu_1 = 6.4 \times 10^{-3}$, $\mu_2 = 3 \times 10^5 s^2$, $\mu_3 = 10^2 m^2$. Details of the inference of this form and its evaluation are given in ref. [23]. The bulk-hydrographic factor, in brackets in (18), is spongy and needs additional study.

To investigate the structure of the momentum diffusivity N_z we define the flux Richardson number

$$Rf \equiv - \frac{g\,\overline{\rho' w'}}{\rho\,\overline{u' w'}\,\frac{\partial u}{\partial z}} \tag{19}$$

This measures the relative importance of the buoyancy loss of turbulent energy to the shear production of energy, cf. (14). In the absence of other sources and sinks of turbulent energy Rf>1 if and only if there is a decay of turbulent energy. *A priori* we would expect Rf to be much less than unity in such an intensely turbulent system as an estuary; in fact, values around .5–.8 appear to be typical, which implies the existence of strong external sources of turbulence. From (7) and (8)

$$N_z = Ri\, K_z/Rf \tag{20}$$

where Ri and Rf are evaluated with tidal-mean quantities. An analysis of the James and Mersey data suggests a dependency of the form

$$Rf = \mu_4 \tanh(\mu_5\, Ri) + \mu_6 \left(\frac{\partial u}{\partial z}\right)^{-2} \tag{21}$$

where $\mu_4 = .5$, $\mu_5 = 6$, $\mu_6 = 5 \times 10^{-4}$, see ref. [23]. From (18), (20) and (21), the value of N_z may be calculated.

In figs. 3 and 4 are displayed the empirical values of K_z and N_z from the James data and the computed values using (18), and (20) and (21). The vertical structure is evidently well replicated.

3. COMMENTS ON THE MODEL AND ITS SOLUTION

The complete hydrodynamic model consists of the momentum equations (2) and (6), the continuity equation (3), the salt-balance equation (4), and the expressions for the diffusivities (18), (20), and (21), together with the various defining relations and boundary conditions. Despite all of the simplifications introduced, the complete system is not soluble analytically. Even its numerical solution is difficult due to the mathematical properties of the model. For example, the partial differential equations are both strongly coupled and non-linear. The former becomes especially apparent if one traces the functional dependencies among u, s, K_z and N_z.

If it is assumed that N^2 and the ratio Ri/Rf are independent of depth, then $N_z = N_o\, z(h-z)h^{-2}$ and the momentum equation (6) can be integrated directly.

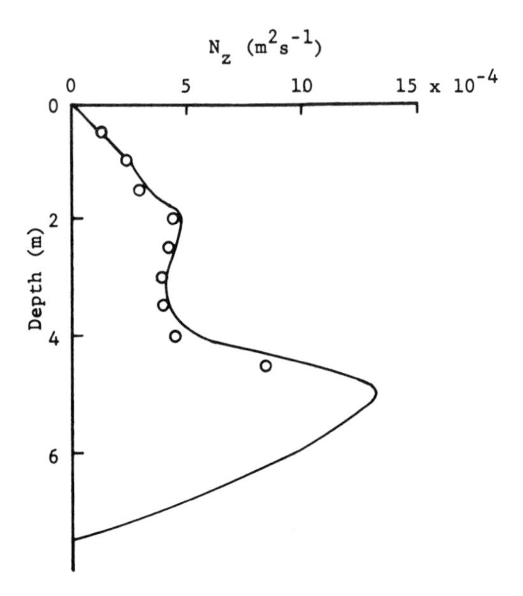

Figure 4. Observed and predicted eddy viscosities for James Estuary, 18-23 June 1950.

The current profile is then given by

$$u(z) = \frac{gh^2}{N_o}\left[\frac{1}{2}\frac{\partial \log \rho}{\partial x}(z-z_o) - \frac{\frac{h}{4}\frac{\partial \log \rho}{\partial x} - \frac{N_o v_f}{gh^2}}{\log \frac{h}{z_o} - 1}\log \frac{z}{z_o}\right] \tag{22}$$

In spite of the oversimplification of a constant N^2, the shape of u(z) given by (22) is in surprising agreement with those obtained in the James and Mersey: a shallow near-bed logarithmic layer and a near-linear variation of u with depth elsewhere.

Numerical solutions of these types of equations have been implemented by Overland [15], who uses Pritchard's formula (17), and by Ward [23]. (The approach of Fisher et al. [8] avoids the problems of the salt balance by using an empirical representation of the longitudinal salinity profile.) In both cases the approach was the method of finite differences. The former solved the time-dependent equations, advancing the solution forward until an equilibrium was attained. The latter solved the steady-state equations directly by a numerical quadrature of the momentum equation coupled with successive overrelaxation of the salinity field.

Related numerical models of vertical-longitudinal circulations have been developed by Hamilton [9] and Blumberg [1]. These models are similar in that longitudinal pressure gradient accelerations due to density and vertical eddy fluxes of momentum and salt are included in the formulation, but differ in the important respect that the equations are not tidal averaged but rather treat the complete intratidal time variation. These represent an improvement over the one-dimensional treatment of tidal hydrodynamics (e.g. Harleman and Lee [10]), and fairly good replications of observed intratidal stage, current and salinity variations are obtained. However, resolution of the less intense, intertidal circulation in an intratidal model demands a high degree of accuracy in the numerical solution. We note, for example, that while Hamilton [9] obtained qualitatively good tidal variation of salinity in the Rotterdam Waterway, the upstream intrusion of salinity is not particularly well replicated. On the other hand, a scale separation by averaging, which yields the tidal-mean equations, places a greater demand upon parameterization of the residual fluxes. This, of course, is not a new problem in geofluid dynamics, see for example Bolin et al. [2].

Numerical solution of the model equation, whether tidal-averaged or intratidal, requires consideration of the usual problems of numerical approximation, viz. stability, both linear and nonlinear, conditioning of the coefficient matrices, and accuracy. In general, estuarine modeling presents no novelty, and the standard methods in geofluid dynamics apply, e.g. Kreiss and Oliger [13], Saul'yev [20], Roache [19], etc. One exceptional problem appears to be the high degree of nonlinear coupling, notably the dependence of the eddy coefficients upon the density and velocity fields. In the work of both Overland [15] and Ward [23], numerical interaction through the coupled terms was a serious difficulty, in that feedback degradation of the solution was encountered which required special measures to control. Apparently, the same sort of difficulty was met by Bowden and Hamilton [4] and Blumberg [1]: the former found it necessary to remove vertical dependence in the eddy coefficients (and, therefore, local dependence upon fluid properties), and the latter bounded the range of the diffusivity by constants both above and below to keep the solution well-behaved.

The viability of modeling the tidal-mean vertical-longitudinal circulation in estuaries by numerical integration has been demonstrated by the above studies. From the standpoint of formulating the basic physics of this circulation, specification of the vertical residual fluxes (or, equivalently, the eddy diffusivities for mass and momentum) is critical, for which a more satisfactory theoretical basis is needed. Some injection of empiricism (such as presented in Section 2 of this paper) is probably unavoidable, at least for the immediate future, but a more fundamental approach that will sharpen and confine that empiricism is essential.

REFERENCES

1. Blumberg, A. F. 1975. A numerical investigation into the dynamics of estuarine circulation. Tech. Rep. 91, Chesapeake Bay Institute.
2. Bolin, B., et al. 1972. Parameterization of sub-grid scale processes. GARP Publ. Ser. No. 8, World Meteorological Organization, Geneva.

3. Bowden, K. F. 1960. Circulation and mixing in the Mersey estuary. Int. Assn. Sci. Hydro., Comm. of Surface Waters, Publ. No. 51: 352-360.
4. ——, and P. Hamilton. 1975. Some experiments with a numerical model of circulation and mixing in a tidal estuary. Estuarine Coastal Mar. Sci., 3: 281-301.
5. Dyer, K. R. 1973. Estuaries: A Physical Introduction. John Wiley & Sons, London. 140 pp.
6. ——. 1974. The salt balance in stratified estuaries. Estuarine and Coastal Marine Science, 2: 273-281.
7. Fischer, H. B. 1975. Numerical models of estuarine circulation and mixing. Numerical Models of Ocean Circulation, National Academy of Sciences, Washington, 10-17.
8. Fischer, J. S., J. D. Ditmars and A. T. Ippen. 1972. Mathematical simulation of tidal time-averages of salinity and velocity profiles in estuaries. Rep. No. MITSG 72-11, Massachusetts Institute of Technology, Cambridge.
9. Hamilton, P. 1975. A numerical model of the vertical circulation of tidal estuaries and its application to the Rotterdam Waterway. Geophys. J. R. Astr. Soc., 40: 1-21.
10. Harleman, D. R. F., and C. H. Lee. 1969. The computation of tides and currents in estuaries and canals. Tech. Bull. No. 16, Comm. Tidal Hydraulics, U. S. Army Corps of Engineers.
11. Jacobsen, J. P. 1913. Beitrag zur Hydrographie der Dänischen Gewässer. Medd. Komm. f. Havundersøgelser, Ser. Hydrografi, Copenhagen.
12. Kent, R. E., and D. W. Pritchard. 1959. A test of mixing length theories in a coastal plain estuary. J. Mar. Res., 18(1): 62-72.
13. Kreiss, H., and J. Oliger. 1973. Methods for the approximate solution of time dependent problems. GARP Publication Series No. 10, World Meteorological Organization, Geneva.
14. Montgomery, R. B. 1943. Generalization for cylinders of Prandtl's linear assumption for mixing length. Ann. N. Y. Acad. Sci. 44 (Art. 1): 89-103.
15. Overland, J. E. 1973. A model of salt intrusion in a partially mixed estuary. TR 73-1, New York Institute of Ocean Resources.
16. Pritchard, D. W. 1956. The dynamic structure of a coastal plain estuary. J. Mar. Res. 15(1): 33-42.
17. ——. 1960. The movement and mixing of contaminants in tidal estuaries. Proc., First Conf. on Waste Disposal in the Marine Environment. Pergamon Press, London.
18. ——, and R. E. Kent. 1953. The reduction and analysis of data from the James River operation Oyster Spat. Tech. Rep. VI, Ref. 53-12, Chesapeake Bay Inst.
19. Roache, P. J. 1972. Computational fluid dynamics. Albuquerque, Hermosa Publishers.
20. Saul'yev, V. K. 1964. Integration of equations of parabolic type by the method of nets. Oxford, Pergamon Press, Ltd.
21. Sutton, O. G. 1953. Micrometeorology. McGraw-Hill Book Company, New York. 333 p.
22. Taylor, G. I. 1931. Internal waves and turbulence in a fluid of variable density. Rapp. Proces-Verbaux Con. Perm. Int'l. pour d'Exploration de la Mer 76: 35-42.
23. Ward, G. H. 1973. Hydrodynamics and temperature structure of the Neches estuary, Vol. II. Doc. No. T73-AU-9510, Tracor, Inc., Austin, Texas.

TOOLS AND METHODS

Convened by:
Jerome Williams
Environmental Sciences Department
U.S. Naval Academy
Annapolis, Maryland 21402

An attempt has been made in this section to present a few examples of tools and methods representative of the large strides made in this general area within the last few years. Kjelson and Colby, for example, suggest the use of a specially designed net for use in estimating fish populations. Even when the well known otter trawl is used, there is still some doubt as to the reproducibility of the data obtained. McKay describes the new hydraulic model of Chesapeake Bay, showing the many differing studies that may be performed, but also indicating a few that even this newest model facility cannot handle.

Klemas is perhaps the most optimistic of the authors in this group, spelling out the myriad of measurements that can be made by remote sensing. New ideas continue to be developed as new portions of the electromagnetic spectrum become available. But even he sounds a pessimistic note when he indicates the continuing need for "ground truth." It appears there will always be a need for fixed surface instrumentation in conjunction with remote sensing.

In summary, this group of papers raises more questions than it answers. It has always been so and probably always will be.

REMOTE SENSING OF COASTAL WETLAND VEGETATION AND ESTUARINE WATER PROPERTIES

Vytautas Klemas
College of Marine Studies
University of Delaware
Newark, Delaware 19711

ABSTRACT: The advantages and limitations of remote sensing techniques for collecting synoptic data over large coastal and estuarine areas are reviewed with emphasis on the need for a proper balance between remotely sensed data and "ground truth". Specific applications include mapping wetland vegetation and coastal land use; monitoring natural and man-induced changes in the coastal zone; charting current circulation, including the movement and dispersion of known water pollutants; and determining the type and concentration of suspended matter in coastal waters. The photo-interpretation of aircraft and satellite imagery with the aid of "ground truth" is illustrated, employing both direct visual and automated computer techniques. For some applications, it is shown that an integrated boat-aircraft-satellite approach can produce better results or cost less, than the deployment of large numbers of boats or field teams without remote sensor support.

INTRODUCTION

Economic pressures to extract oil, to increase the harvest of food and to find new or maintain existing waste disposal sites are creating a need to understand the environment of large estuarine and coastal areas, including the entire Continental Shelf. The excessive amount of boat time and cost of ground crews required to collect synoptic data over such regions is causing investigators to look for more cost-effective means of performing this task. One technique which appears promising, involves the use of remote sensing, including standard aerial cameras and other sensors operating beyond the normal visual range of photographic films. The physical and technical aspects of imaging with remote sensors were reviewed by Colwell et al. (9, 10). Among the advantages enumerated are:

1. Wide area coverage, including regions with difficult access.
2. High resolution.
3. High cartographic accuracy with precision cameras.
4. Improved discrimination with multispectral sensors, including spectral bands outside the visible region.
5. Rapid, automated interpretation of imagery using optical and digital enhancement techniques.
6. Improved transmission, storage and update of the data in digital form.

The objective of this paper is to make estuarine and coastal investigators aware of the advantages and limitations of remote sensing techniques, including integrated boat-aircraft-satellite systems, for collecting synoptic data over large coastal areas.

MAPPING WETLAND BOUNDARIES AND PLANT SPECIES

The commitments to environmentally sound coastal land management that have been generated in federal and state governments over the past few years have produced a demand for accurate and complete bodies of scientific data on which to base policy decisions. Inventories of wetlands are now specifically required by individual state laws, such as the New Jersey Wetlands Act of 1970, the Maryland Wetlands Act of 1970, and the Delaware Wetlands Act of 1973. Further incentive to coastal states to inventory and manage their coastal resources was provided by the Coastal Zone Act of 1972. Since plant community composition appears to be a good indicator of relative marsh value and also of the wetlands-uplands boundary, wetlands vegetation is currently being mapped using various techniques, most of them involving remote sensing (3, 28, 44).

Coastal wetlands of the type found along the East and Gulf coasts of the United States are well suited to remote sensing techniques. The uniform flatness of marsh topography eliminates variations in reflectance due to sloping surfaces and shadows. Marsh plant communities are composed of a few abundant species, thus simplifying photointerpretation. Environmental changes generally take place over large horizontal distances in the marsh. Therefore, zones of relatively uniform vegetation are usually large enough to be discernible, even on very high-altitude imagery. Finally, the major plant species are different enough in their morphologies to have distinct reflectance characteristics, particularly in the near-infrared portion of the spectrum. The net result is that aerial photographs can be used to make a detailed wetlands map showing vegetation growth patterns which are related to local environmental factors (2, 29, 44).

Most of the wetlands mapping is being performed at scales of 1:2,400 and 1:24,000. A scale of 1:2,400 is usually employed to define the "legal" wetlands-uplands boundary and inventory and the plant species composition. The maps contain considerable detail and at that scale can be readily related to local zoning or taxation maps. The maps are generally prepared by direct photointerpretation of color and color-infrared prints and transparencies obtained from low-flying aircraft. In the case of the Delaware wetlands mapping the photographs were taken from an aircraft altitude of 6,000 feet with six-inch focal length cameras, producing nine-inch original photographs at a scale of 1:12,000. The nine-inch color transparencies were then used to make 5x black-and-white enlargements on stable "Mylar" film material at a scale of 1:2,400. The lines separating the plant species and the wetlands from the uplands were next drawn in on the enlargements by photointerpreters using the original color and color-infrared prints and supporting data from ground surveys. Since these maps will constitute a legal definition of the wetlands boundaries, highest map accuracy standards must be maintained and accuracy limitations well known.

On the other hand a scale of 1:24,000 is being used by many investigators and agencies to map land cover and land use for planning coastal development and managing coastal resources. These maps are somewhat easier and less expensive to prepare than the detailed wetlands maps for several reasons. The scale is smaller, requiring considerably fewer maps to cover the same area. For instance, to map Delaware's 115,000 acres of wetlands at a scale of 1:2,400 required 360 maps, whereas fifteen maps sufficed at a scale of 1:24,000. Since base maps already existed at a scale of 1:24,000, such as USGS topographic maps, only overlays of the vegetation species were prepared, eliminating the need for expensive geometric corrections and ground grid controls. Fig. 1 shows a typical overlay map, at a scale of 1:24,000, of ten plant species in the wetland region around Taylor's Bridge, Delaware. While primary and secondary species were identified by visual photointerpretation, percentages of minor species in each of the rectangular areas were obtained by automated computer techniques using the General Electric Multispectral Data Processing System (GEMS), a hybrid analog/digital system permitting man-machine interaction at nearly real-time rates (29). A modified color TV camera scans the color transparency and produces three video signals representing the red, blue and green spectral components of the image. The output from the camera is displayed on a color monitor. The human interpreter selects the area of interest and then employs various available electronic data processing techniques to identify the spectral characteristics of the area, search the scene for areas with similar spectral signatures, and compute the percentage of the total scene occupied by these areas. The fact that most coastal plant species differ in their spectral signatures, i.e. the amount of light they reflect at various wavelengths, forms the basis for their discrimination by remote sensors using multispectral techniques.

A relationship between spectral reflection characteristics and productivity of certain marsh plant species has been found, making it feasible to remotely map marsh productivity (44). However, large amounts of "ground truth" are required and the reliability of the technique leaves much to be desired.

MONITORING LAND-USE AND COASTLINE CHANGES

Monitoring natural changes, such as beach erosion, or man-made changes, such as land-use, requires repetitive photographic coverage of the coastal zone either by aircraft or satellites (41). Fortunately, aircraft from the U.S. Coast and

Geodetic Survey or from the U.S. Department of Agriculture, Agricultural Stabilization and Conservation Service (ASCS) have photographed many coastal regions at least once per decade since 1938. Figs. 2 and 3 show black-and-white

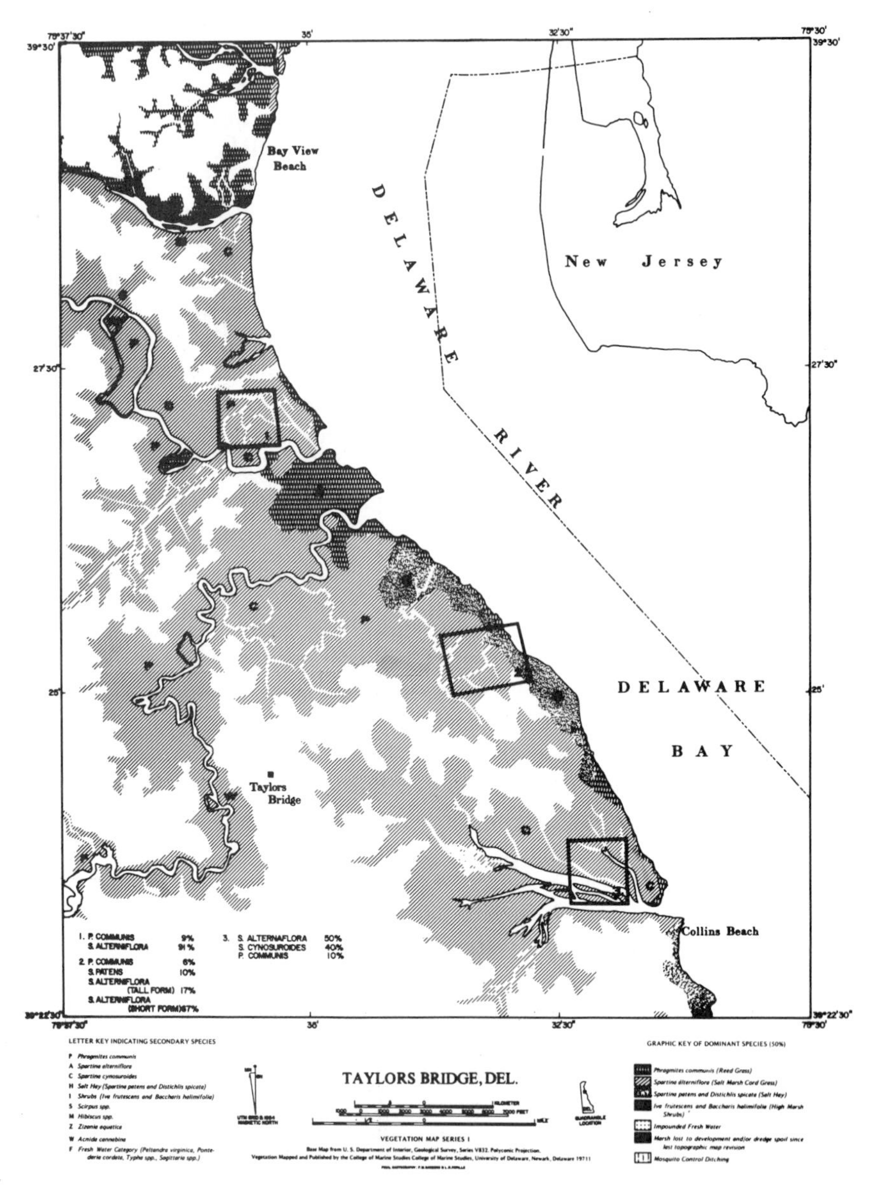

Figure 1. Overlay map showing ten species of marsh vegetation, ditching, impounded water, and marsh lost to development. Fifteen such maps covering Delaware's wetlands have been prepared using multispectral analysis of RB-57 imagery.

aerial photographs of the Indian River Inlet area in Delaware, obtained in 1938 and 1968 respectively. Reshaping of the inlet and construction of the jetty is causing accretion of sand south of the inlet and erosion north of it, with imminent danger to the highway above it. To map the coastline change accurately one could use the procedure described in (47) consisting of selecting stable reference points on the aerial photographs taken in different years and measuring the distance between these points on the transient beach. The measurements obtained are then multiplied by the scale of the aerial photographs to produce ground distances, and the change in location of the beach over the period of the time lapse. As a final step one can attempt to relate the volume of material eroded to the linear distances of beach erosion perpendicular to the beach (18, 25, 47).

Figure 2. Aerial photograph of the Indian River Inlet area of the Delaware coast, at a scale of 1:20,000 in 1938. (USDA-ASCS).

Color, color-infrared or black-and-white photographs such as shown in Figs. 2 and 3 can be used to map land use change. The most effective way to accomplish that is to map the land use for each year represented by the photographs on an overlay superimposed on a base map. To compensate for scale differences between the photographs and the base map one can use a Zoom Transfer Scope (ZTS) or similar viewing system. For instance, the Bausch and Lomb ZTS enables the user to view both an aerial photograph and a topographical map of the same area. Simplified controls allow the matching of differences in scales and provide other optical corrections so that the two images appear superimposed. Information from the photograph may then be compared or traced onto the map.

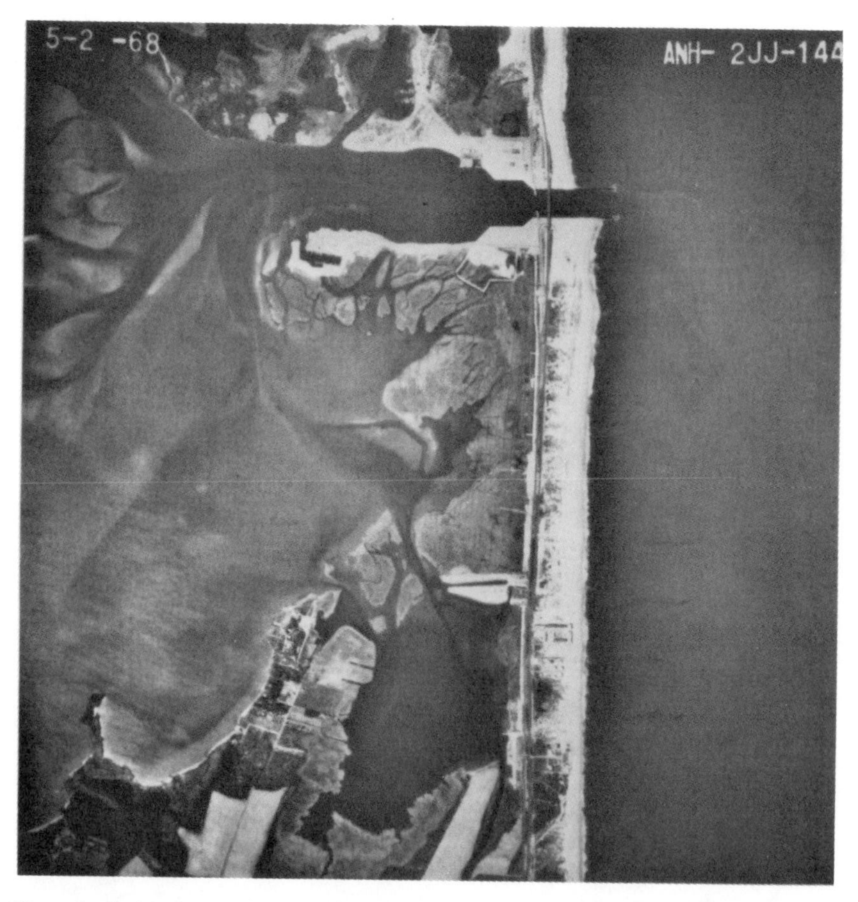

Figure 3. Aerial photograph of the Indian River Inlet area of the Delaware coast, at a scale of 1:20,000 in 1968. (USDA-ASCS).

Table 1. Vegetation and land-use categories.

1. Forest land.
2. *Phragmites communis* (Reed grass).
3. *Spartina patens* and *Distichlis spicata* (Salt marsh hay and spike grass).
4. *Spartina alterniflora* (Salt marsh cord grass).
5. Cropland.
6. Plowed cropland.
7. Sand and bare sandy soil.
8. Mud and asphalt.
9. Saline deep water.
10. Sediment laden and shallow saline water.

Satellites, such as NASA's LANDSAT-1, offer wider and more regular coverage than aircraft (16, 28). For instance, LANDSAT-1 passes over the Delaware Bay test site every 18 days, and even if on the average two out of three passes are obscured by cloud cover, a successful pass every 54 days is more than sufficient to detect changes in coastal land-use. From an altitude of 920 km, the satellite uses a four-channel Multispectral Scanner (MSS) and a Return-Beam Vidicon Camera to image an area of 185 x 198 km in each frame. The location and bandwidths of the four MSS channels are band 4 (0.5-0.6 μm); band 5 (0.6-0.7 μm); band 6 (0.7-0.8 μm); and band 7(0.8-1.1 μm).

Several investigators have successfully used LANDSAT-1 to monitor coastal land-use and its changes (14, 15, 16, 29). The large amount of data generated from repetitive coverage and the digital tape format of the satellite data make it attractive to analyze it with computers using multispectral techniques (28, 45). For instance, digital LANDSAT-1 MSS scanner data and SKYLAB photographs have been used in an attempt to inventory and monitor significant natural and man-made cover types in Delaware's coastal zone. Automatic classification of LANDSAT data yielded classification accuracies of over 83 per cent for all categories shown in Table 1 (28). The classification accuracy of several important categories is shown in Table 2.

Visual interpretation of Skylab Earth Terrain Camera photographs distinguished a minimum of 10 categories with classification accuracies ranging from

Table 2. Classification accuracy table derived by comparison of LANDSAT thematic data with NASA-RB-57 aircraft photography.

Category	Forest	*S. alternaflora*	*S. patens*	Water	Agriculture
Forest	89.9%	0.0%	4.5%	0.0%	5.6%
Spartina alterniflora	0.0%	93.7%	5.7%	.6%	0.0%
Spartina patens	0.0%	7.7%	87.0%	2.2%	3.0%
Water	0.0%	2.6%	3.9%	93.5%	0.0%
Agriculture	3.5%	0.3%	2.1%	0.0%	94.1%

75% to 99% (28). A land-use map derived from SKYLAB imagery is shown in Fig. 4. Note that the scale is 1:125,000. Maps derived from satellite or spacecraft imagery generally have scales smaller than 1:100,000. Thus, to obtain wide area coverage from satellite altitudes, one must give up the detailed resolution attainable from aircraft imagery. The size of the smallest resolvable object at high contrast is about 80 meters for LANDSAT, 20 meters for SKYLAB and less than 1 meter for most mapping aircraft altitudes.

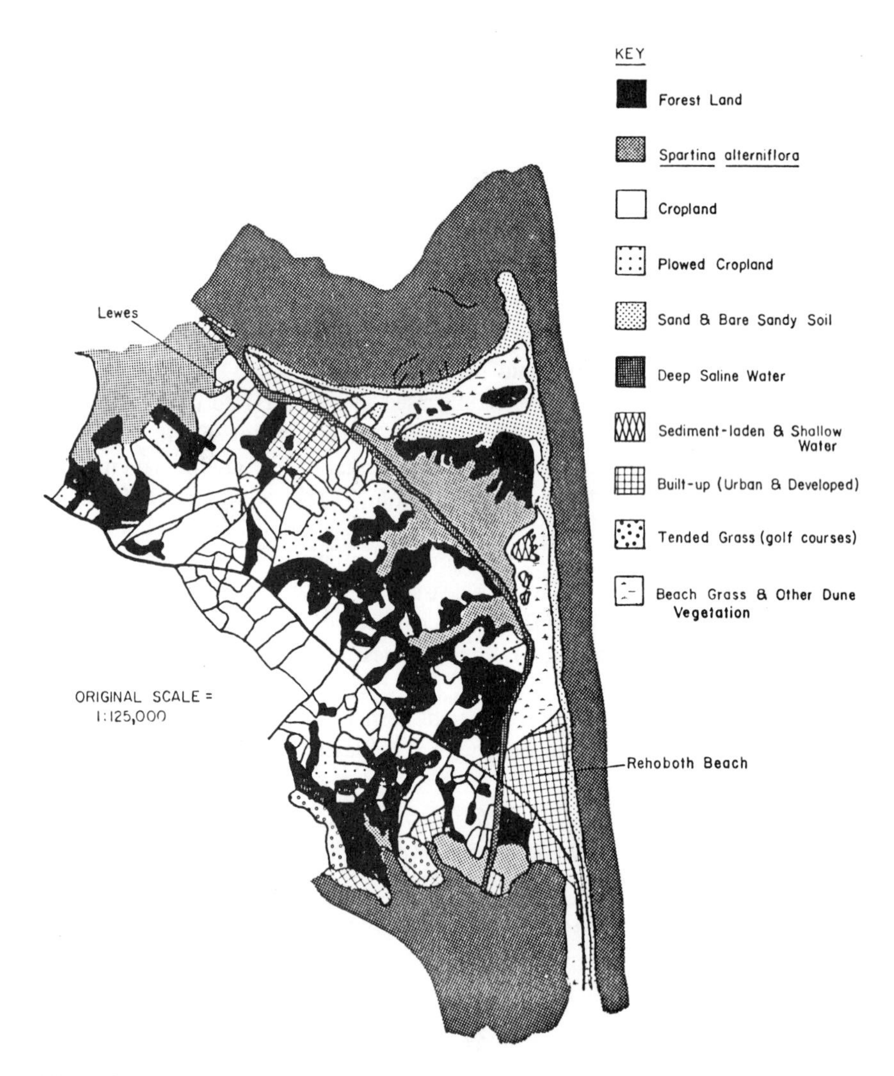

Figure 4. Land-use map derived from Skylab/EREP image by visual photo-interpretation.

Various land-use classification schemes have been proposed by individuals and agencies (1). Most investigators are adopting the Federal Land-Use Classification System for the upper levels and modifying the lower level categories to suit the needs of their application and geographic region. Once a user has selected the classification categories, he can instruct a computer to perform "supervised" or "unsupervised" classification of the imagery. Supervised classification begins with identifying certain sets of resolution cells within a scene that represents known classes of categories on the ground. These groups of cells are known as training sets. The spectral responses in the spectral channels of each training set provides the information needed to identify the remaining cells in the image. The "decision rules" that are used to identify the class of each cell are defined by the user (12).

Non-supervised classification randomly selects resolution cells within the scene. The spectral characteristics of these random points eventually provide the statistics for classifying remaining cells. Simply stated, a sample cell is either placed in a cluster with other sample cells of similar spectral responses, or it forms the core of a new cluster. These sample-derived clusters provide the statistics used to classify the remainder of the scene. However, each computer-aided data processing and interpretation approach still requires that a human interpreter be "in the loop" to verify the final results.

Both visual and computer-aided interpretation of imagery requires some "ground truth" data, i.e. a minimum amount of information about the area being imaged. As a result, remote sensing techniques do not eliminate the need for ground surveys, but only decrease significantly the amount of field data required.

CHARTING CURRENT CIRCULATION AND POLLUTANT DISPERSION

Passive remote sensing systems, such as described in this paper, sense either reflected solar radiation or radiated thermal infrared energy. The solar radiation wavelengths are attenuated exponentially by water, whereas the thermal infrared ones are limited to a very thin surface layer of water. Only the surface layers are sensed and as the sensor "looks" at deeper layers of the water column, the signal-to-noise ratio decreases in an exponential manner. Thus, very sophisticated sensors are required to monitor subsurface processes. In the ocean, where scattering and absorption effects are of the same order of magnitude, the penetration depth of detectable visible light may exceed 50 meters (20, 51). In tidal estuaries and coastal waters, scattering by suspended matter becomes severe, resulting in measured penetration depths and Secchi depths of the order of a few meters. As a result, it is difficult to use remote sensors to map bottom contours in coastal waters. Laser profilers with the help of intense light beams can penetrate to a detectable depth several times the Secchi depth, but that is still insufficient to chart the depth contours in most estuaries (5, 19). Remote sensors, because of detector dynamic range and sensitivity, are generally limited to a narrow band in the visible region since energy at other wavelengths, such as the ultra-violet or infrared, is more strongly absorbed by the water. Therefore, multi-spectral remote sensing and analysis over wide renges of the electromagnetic spectrum cannot be applied as readily to sensing substances in water as it has been over land. The attenuation coefficients over a portion of the visible spectrum for distilled, oceanic and coastal waters are plotted in Fig. 5. Note that as one goes from deep ocean to coastal conditions, the attenuation not only increases, but the wavelength for best penetration also shifts from the blue towards the green and red.

Water features tend to change more rapidly than those on land, especially in tidal estuaries. For instance, at the mouth of Delaware Bay water samples from ships or helicopters must be obtained within 20 minutes of a satellite overpass to be valid as "ground truth" for suspended sediment mapping, compared to an acceptable delay of several weeks for vegetation inventories. Despite these problems, remote sensing techniques are being applied to attempt to chart the current circulation and dispersion of known pollutants; to map the concentration of suspended matter and thickness of certain films; and to determine the identity of unknown slicks and suspended matter.

Current circulation patterns have been studied remotely by time lapse photography of current drogues, tracer dyes, natural tracers, such as suspended sediment, or thermal gradients (11, 23, 31, 36, 46). Surface water movement studies utilizing fluorescent tracer dyes have been conducted on most of the major surface water bodies and near coastal waters of the United States. These studies have been conducted to determine the dynamic characteristics of the water bodies with the primary objective being to trace the current flow rate and direction and the rate of dispersion (49). Systems used to monitor tracer dyes are visual inspection, photographic recording, grab sample collection and laboratory analyses, continuous field sampling and recording with flow-through fluorometers, field measurements with a submersible pulsed light fluorometer, and remote measurements of dye concentrations from an aerial platform (13). Rhodamine B and Rhodamine WT are two of the more commonly used dyes, having specific gravities at room temperature of 1.12 and 1.19, and maximum emission wavelengths of 0.579 microns and 0.582 microns, for solutions of 40% to 20%, respectively.

Rhodamine WT dye can be tracked for several hours by low altitude aircraft carrying color cameras (31). After several hours, however, the dye is diluted to concentrations that can be discerned only by aircraft cameras with special optical filters that are optimized for the spectrum of each dye used. This is particularly true for dye experiments in coastal waters. The emission spectrum of Rhodamine WT and the spectral transmission of suitable filters are shown in Fig. 6. The Wratten 73 is a band-pass filter with its transmission band closely matched to the spectral emission maximum of Rhodamine WT. Field tests, however, showed that the Wratten 25A filter was more effective in enhancing the dye patches (31). This result could be partly explained by Eliason and Foote's (13) observation that dyes at high concentrations are self-absorbing, which causes the effective peak of the dye fluorescence line to shift to longer wavelengths. Therefore, the dye was tracked from aircraft that carried cameras

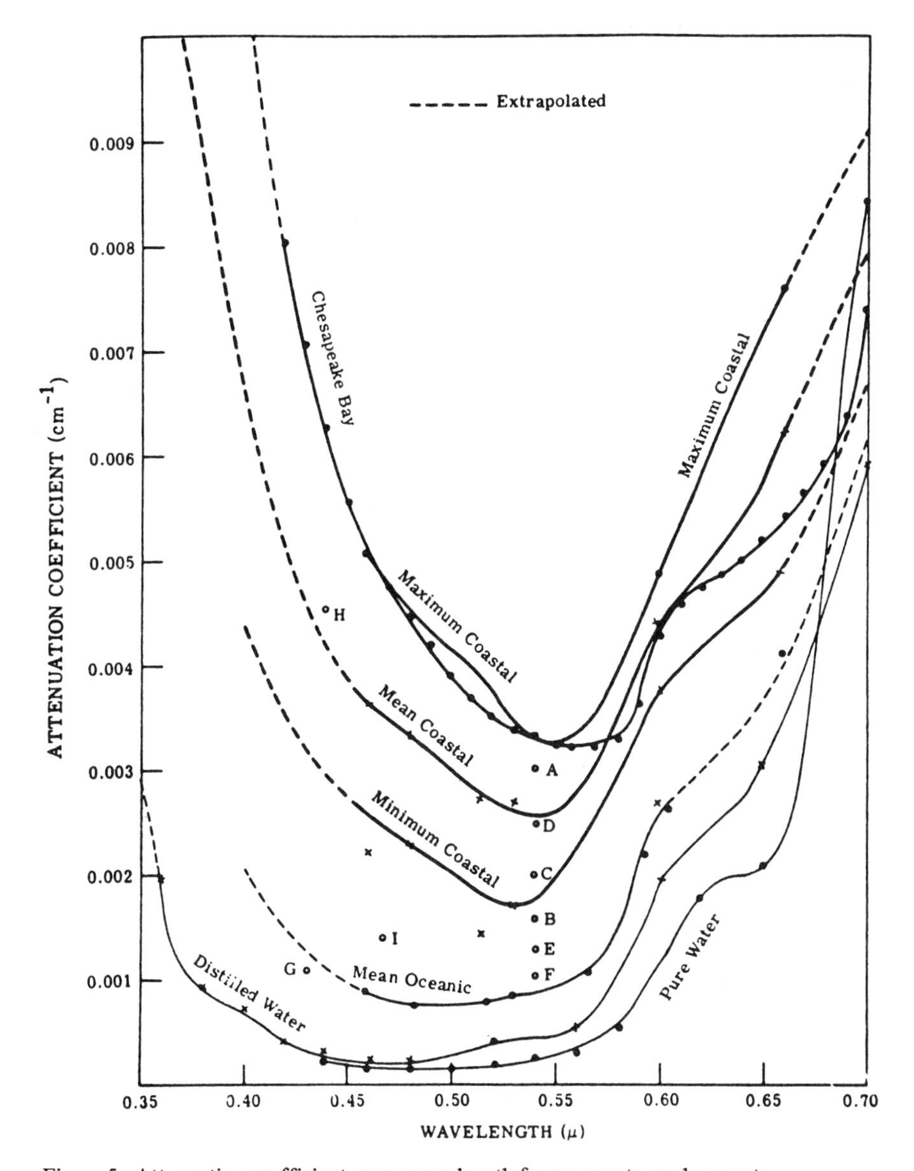

Figure 5. Attenuation coefficient versus wavelength for pure water and sea water.

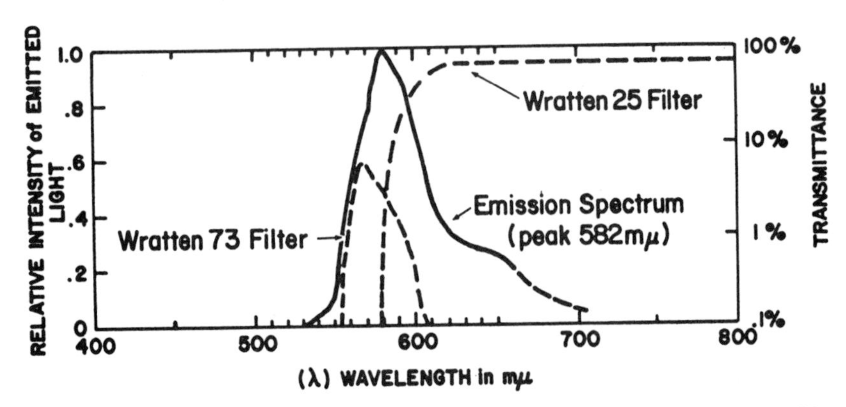

Figure 6. Emission spectrum of Rhodamine WT dye and transmittance of camera filters used in the dye study.

containing Kodachrome-X film with Wratten 1A Skylight filters and Tri-X film with Wratten 25A filters. Aircraft altitudes ranged from 300 to 1,000 m.

Small current drogues have been developed which can be dropped and tracked from low-flying aircraft. Their basic design does not differ significantly from that of drogues used by various investigators during the past few decades (40). These small drogues are deployed whenever a detailed charting of current circulation over a relatively small area, such as four square miles, is desired. The drogues usually consist of a styrofoam float and line to which is attached a current trap consisting of a stainless steel biplane. The length of the line determines at what depth currents will be monitored. The floats are color-coded to distinguish their movement and mark the depth of the biplanes. Packs with dyes of two different colors can be attached to the float and the biplane (30). The movement of the dye and drogues is tracked by sequential aerial photography, using fixed markers on shore or on buoys as reference points to calibrate the scale and direction of drogue movement. The results of a combined dye and drogue experiment at the mouth of Delaware Bay are illustrated in Fig. 7. As shown, subsurface currents differed significantly from surface currents during both the ebb and the flood tidal cycles (31).

Satellites, such as LANDSAT-1, have been used to obtain a synoptic view of current circulation over large coastal areas (22, 27, 36, 43). Since in turbid coastal regions suspended sediment acts as a natural tracer, cost is minimized by eliminating the need for expensive injections of large volumes of dye such as Rhodamine-B. Fig. 8 shows the LANDSAT-1 MSS band 5 image and predicted tidal currents of a satellite overpass on February 13, 1973, about one hour after maximum ebb at the mouth of Delaware Bay. The intensity variations throughout the bay are caused by suspended sediment, and not bottom contours, since the actual water depth in most areas was at least three times the Secchi depth. Strong sediment transport out of the bay in the upper portion of the water column is clearly visible, with some of the plumes extending up to 30 km out of the bay. The northward curvature of small sediment plumes along New Jersey's coast clearly indicates that the direction of the nearshore current at that time was towards the north. The wind velocity at the time of the satellite overpass

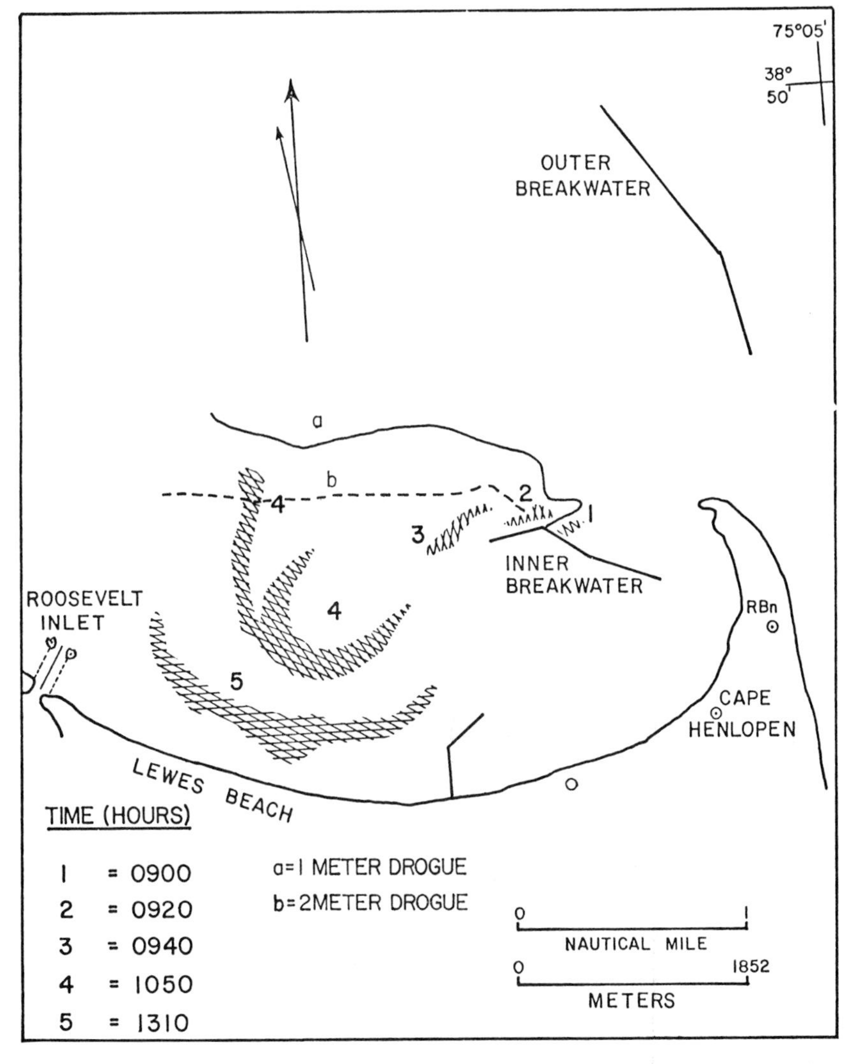

Figure 7. Dye patterns and drogue tracks during circulation studies in the Lewes harbor breakwater area (flood tide).

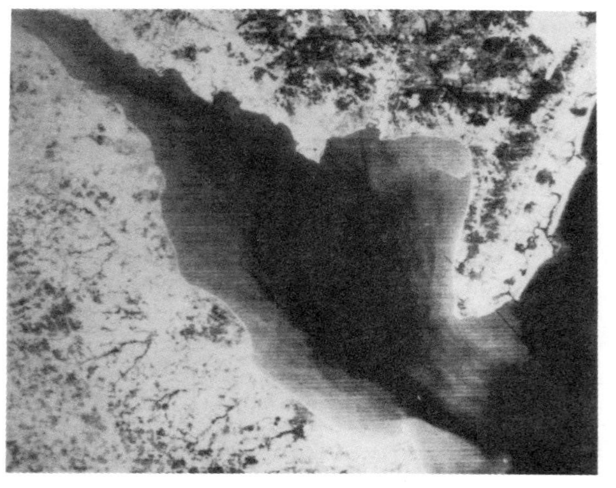

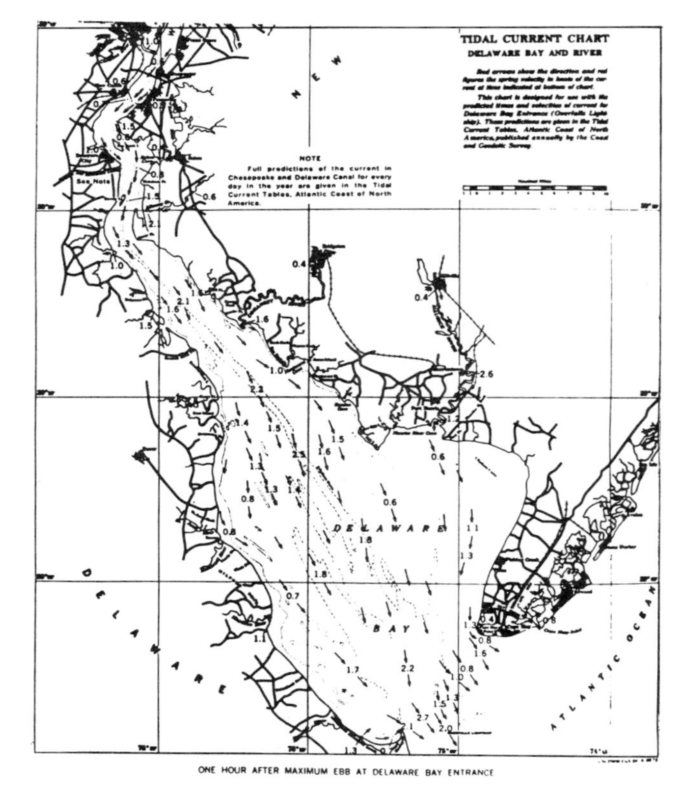

Figure 8. Predicted tidal currents and ERTS-1 MSS Band 5 image of Delaware Bay taken on February 13, 1973 (I.D. NO. 1205-15141).

was about 13 km per hour from the west-northwest, reinforcing the tidal current movement out of the bay.

The suspended sediment and current circulation patterns in Fig. 8 can be significantly enhanced not only by careful print development, but also by multispectral enhancement techniques such as color density slicing and color additive viewing. Color density slicing breaks up gradual grey tone variations into digital steps and converts each grey tone step into a distinct color, helping the eye recognize subtle grey tone changes. This is effective because the average human eye can distinguish several hundred color hues while it can only discriminate about a dozen grey scale levels. Color additive viewing involves the exact superposition of transparencies of the same scene obtained with different filter film combinations. Both enhancement techniques have been used with some success to improve the contrast of water features (27, 33, 38, 53).

One of the principal shortcomings of satellite imaging of coastal currents has been the inability to determine current magnitude and to penetrate beyond the upper few meters of the water column. These objections have been overcome by complementing satellite observations with drogues tracking currents at various selected depths (30). One type of drogue used was developed by ITT-Electro Physics Laboratories and emits a radio signal which is tracked from shore. The drogue consists of a plastic pipe less than two inches in diameter, with all of its electronics and antenna totally enclosed within the pipe (Fig. 9). It is also provided with a water temperature sensor. A current trap (biplane) is attached to the bottom of the drogue and can operate at a variety of depths from about one meter to a hundred meters. The intended radiated power of the drogues is such that the position of each drogue can be fixed by triangulation from shore with a mobile antenna over a range in excess of 300 km, with an accuracy approaching ±0.5 degree. By combining the satellites' wide coverage with aircraft or shore stations capable of tracking the expendable drogues, a cost-effective, integrated system has been devised for monitoring currents over large areas, various depths and under severe environmental conditions (30).

Aircraft and satellites, supported by water sampling conducted from boats, have also been used to study the movement and dispersion of various pollutants in estuaries and on the Continental Shelf (4, 32, 42, 50). Approximately forty nautical miles off the Delaware coast is located the disposal site for waste discharged from a plant processing titanium dioxide. The discharge is a greenish-brown, 15% to 20% acid liquid which consists primarily of iron chlorides and sulfates. The barge which transports this waste has a 1,000,000 gallon capacity and makes at least three trips to the disposal site per month. The frequency of this dumping made it possible for the LANDSAT-1 satellite to image the acid plume in various stages of degradation, ranging from minutes to days after dumping (30). Nine photographs were found which show water discoloration. The dump pattern and the time difference between the dump and photograph give strong indications that the discolorations are the acid plume. Careful examination of an overpass on January 25, 1973, disclosed a fishhook-shaped plume caused by a barge disposing acid wastes about 40 miles southeast of Cape Henlopen. The plume shows up more strongly in the green band than in the red band. Enlarged enhancements of the acid waste plumes, prepared from the LANDSAT MSS digital tapes (Fig. 10) aided considerably in studies of the

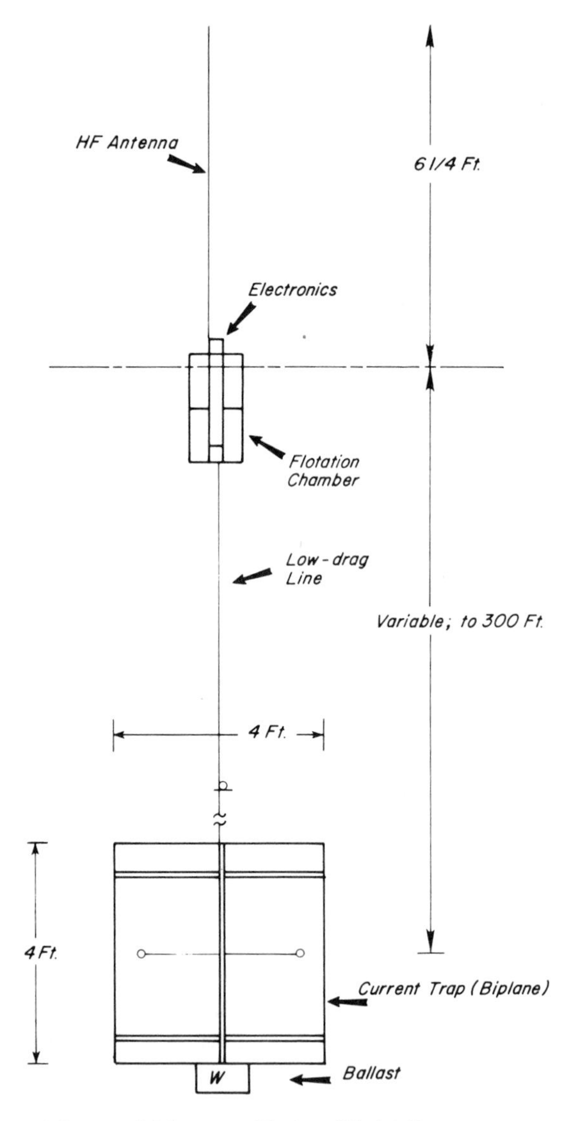

Figure 9. Deep current drogue with low parasitic drag (Model 3).

Figure 10. Enlarged digital enhancement of acid waste plume imaged by Landsat-1 on January 25, 1973.

dispersion of the waste plume. Currently acid dumps are being coordinated with LANDSAT overpasses in order to determine the dispersion and movement of the waste materials along the Continental Shelf. Sludge disposal plumes in the ocean off the Delaware coast and in the New York Bight have also been detected in LANDSAT-1 imagery (38).

Thermal scanners on various platforms are frequently employed to chart current circulation patterns in the ocean and to study specific thermal effluent plumes (46, 48). The accuracy of such thermal maps is of the order of $\pm 1^\circ C$ without "ground truth" and about $\pm 0.2^\circ C$ with calibration provided by surface water temperature measurements.

DETERMINING THE CONCENTRATION AND IDENTITY OF WATER POLLUTANTS

It is far more difficult to remotely sense the concentration and identify of an unknown pollutant than to monitor the movement and dispersion of a known substance. Oil is one example. Oil slicks have been tracked successfully with remote sensors employing the ultraviolet, visible, infrared and microwave regions of the electromagnetic spectrum (6, 26). Both passive and active sensors are being used, such as film cameras and ultraviolet lasers, respectively. The least expensive means of tracking oil slicks is from a single engine aircraft with a camera using color film or a sensitive black-and-white film, with an ultraviolet filter (e.g. Tri-X film with Kodak Wratten 18A filter). However, except for partially successful attempts under controlled conditions, no reliable technique has yet been developed for remotely determining oil slick thickness, the concentration of emulsified oil, or the type of oil in a slick.

At the beginning of this paper, I pointed out that one of the most effective ways to identify and discriminate certain vegetation and land-use types was by their spectral signatures. This technique is more difficult to apply to substances in water, due to complex mixing, multiple scattering and absorption processes, especially in coastal waters (17, 20, 39). For instance, approximate chlorophyl concentrations have been mapped remotely near upwelling regions of the oceans, but whenever the chlorophyl is mixed with suspended sediment in estuarine or near-shore areas, its assessment becomes extremely complicated (4, 7, 8). When different types of pollutants are injected into the ocean, their classification clusters in spectral signature space tend to merge. This difficulty to differentiate spectral signatures tends to make remote sensing of water substances heavily dependent on a well-coordinated ground truth acquisition program including water sampling at various depths from boats or helicopters (33, 52).

The suspended sediment concentration map in Fig. 11 illustrates what can be accomplished with proper ground truth data. The image radiance was extracted from computer compatible digital tapes comprising a LANDSAT MSS band 5 image of the Delaware Bay similar to the one shown in Fig. 8. The image

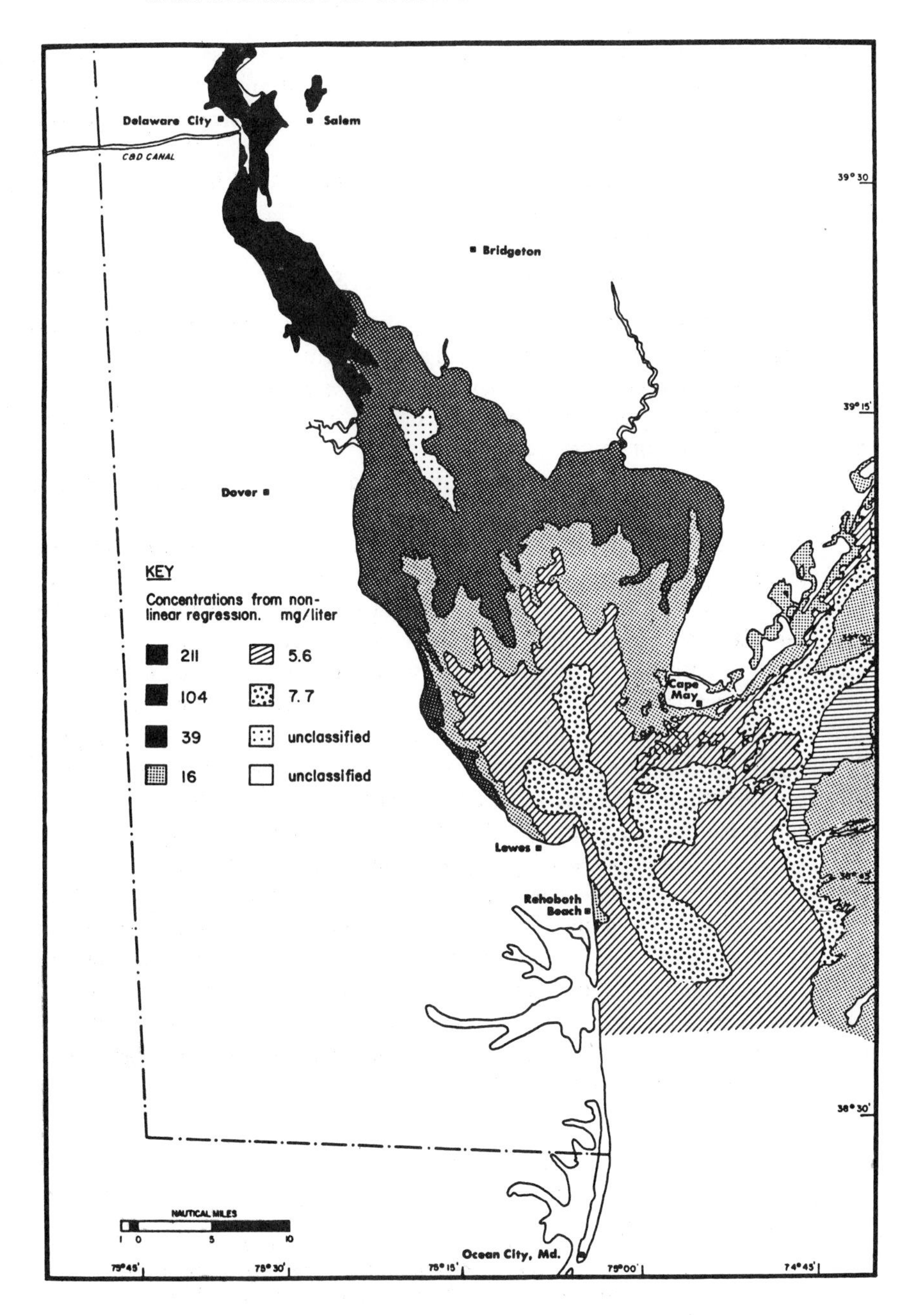

Figure 11. Sediment distribution map.

radiance was then correlated with data obtained from water sample analyses of suspended sediment concentration (27). Since a high degree of correlation was obtained, the map in Fig. 11 was prepared, showing the suspended sediment concentration in the upper one meter of the entire Delaware Bay area. Other investigators are combining several of the LANDSAT MSS bands to map suspended sediment concentrations (21, 52).

SUMMARY AND CONCLUSIONS

Remote sensing techniques have been applied with varying degrees of success to accomplish the following in the coastal and estuarine areas:

—mapping wetland boundaries, plant species diversity and productivity.

—monitoring man-made and natural changes in the coastal zone, such as land-use change and shoreline erosion.

—charting current circulation and pollutant dispersion, including slicks and suspended matter.

—determining the identity and concentration of certain natural and man-made pollutants.

Imaging suspended matter or other subsurface features in water is more difficult than mapping surface slicks or land-use cover because of complex mixing, scattering and absorption processes in the water column. The failure of remote sensors to penetrate beyond a few meters into turbid coastal waters makes it difficult to map bottom contours or track near-bottom sediment transport. Wavelengths outside the visible region are more strongly absorbed in the water column, diminishing the value of ultraviolet and near infrared bands, which are used quite effectively for surface slick and land-cover discrimination.

Closely coordinated ground truth collection programs are required for most remote sensing efforts, especially for the assessment of marsh productivity, the identification of water pollutants and the mapping of their concentration. In general, the acquisition of data from aircraft or satellites has not eliminated the need for data collection from ships and by ground survey teams; however, well coordinated remote sensing efforts have significantly decreased the number of samples that have to be collected on the ground, resulting in cost savings on ship time and ground personnel.

REFERENCES

1. Anderson, J. R. 1971. Land-use classification schemes. Photogrammetric Engineering 37(4):379-387.
2. Anderson, R. R., L. Alsid, and V. Carter. 1975. Applicability of Skylab orbital photography to coastal wetland mapping. Proc. of the American Society of Photogrammetry, 41st Annual Meeting, p. 371-377.
3. ——, F. J. Wobber. 1973. Wetlands mapping in New Jersey. Photogrammetric Engineering 39(4):353-358.
4. Bressette, W. E., and D. E. Lear. 1973. The use of near infrared reflected

sunlight for biodegradable pollution monitoring. Proc. of E.P.A. Second Environmental Quality Sensors Conference, Las Vegas, Nevada, 2:69-89.

5. Brown, W. L., F. C. Polcyn, and S. R. Stewart. 1971. A method for calculating water depth, attenuation coefficients and bottom reflectance characteristics. Proc. of the Seventh Int. Symp. on Remote Sensing of Environment, (1971):663-682.
6. Catoe, C. E. 1972. The applicability of remote sensing techniques for oil slick detection. Fourth Annual Offshore Technology Conference, Houston, OTC 1606: I 887-901.
7. Clark, D. K., W. S. Glidden, L. V. Strees, and J. B. Zaitzeff. 1974. Computer derived coastal waters classifications via spectral signatures. Proc. of Ninth Int. Symp. on Remote Sensing of Environment, 9:1213-1239.
8. Clarke, G. L., G. C. Ewing, and C. J. Lorenzen. 1969. Remote measurement of ocean color as an index of biological productivity. Proc. of Sixth Int. Symp. on Remote Sensing of the Environment, University of Michigan, Ann Arbor, 2:991-1001.
9. Colwell, R. N. 1963. Basic matter and energy relationships involved in remote reconnaissance. Photogrammetric Engineering 29 (5):761-799.
10. ____. 1966. Uses and limitations of multi-spectral remote sensing. Proc. of Fourth Symp. on Remote Sensing of Environment, Ann Arbor, Michigan, (1966):71-100.
11. Dolan, R. 1973. Coastal processes. Photogrammetric Engineering 39(2):255-260.
12. Egan, W. G., and M. E. Hair. 1973. Automated delineation of wetlands in photographic remote sensing. Proc. of Seventh Int. Symp. on Remote Sensing of Environment, Ann Arbor, Michigan, (1971):2231-2251.
13. Eliason, J. R., M. J. Doyle, and H. P. Foote. 1971. Surface water movement studies utilizing a tracer dye imaging system. Proc. of Seventh Int. Symp. on Remote Sensing of Environment, Ann Arbor, Michigan, 7:731-749.
14. Erb, R. B. 1974. The ERTS-1 Investigation (ER-600): ERTS-1 Coastal/Estuarine Analysis, NASA Report TMX-58118.
15. Estes, J. E., and L. W. Senger. 1972. Remote sensing and detection of regional change. Proc. Eighth Int. Symp. on Remote Sensing of Environment, Ann Arbor, Michigan, p. 317-324.
16. Feinberg, E. B., R. L. Mairs, J. A. Stitt, and R. S. Yunghans. 1973. Impact of ERTS-1 images on the management of New Jersey's coastal zone. Proc. Third ERTS Symp., NASA-GSFC, Washington, D.C.
17. Gordon, H. R., and W. R. McCluney. 1975. Estimation of the depth of sunlight penetration in the sea for remote sensing. Applied Optics 14(2):413-416.
18. Herbich, J. B., and F. L. Hales. 1971. Remote sensing techniques used in determining changes in coastlines. Proc. Third Annual Off-shore Technology Conference, Houston, II:319-334.
19. Hickman, G. D., and J. E. Hogg. 1969. Application of an airborne pulsed laser for near-shore bathymetric measurements. Remote Sensing of Environment 1:47-58.
20. Jerlow, N. G., and E. S. Nielsen. 1974. Optical Aspects of Oceanography. Academic Press, New York.
21. Johnson, R. W. 1975. Quantitative sediment mapping from remotely sensed multispectral data. Proc. of Fourth Ann. Remote Sensing of Earth Resources Conference, Tullahoma, Tennessee.

22. Keene, D. F., and W. G. Pearcy. 1973. High altitude photographs of the Oregon coast. Photogrammetric Engineering 39(2):163-168.
23. Keller, M. 1963. Tidal current surveys by photogrammetric methods. Photogrammetric Engineering 29(5):824-832.
24. Kennedy, J. M., and E. G. Wermund. 1971. Oil spills, IR and microwave. Photogrammetric Engineering 37(12):1235-1242.
25. Kidson, C., and M. M. M. Manton. 1973. Assessment of coastal change with the aid of photogrammetric and computer-aided techniques. Estuarine Coastal Mar. Sci. 1:271-283.
26. Klemas, V. 1971. Detecting of oil on water: A comparison of known techniques. Proc. of Joint Conference on Sensing of Environmental Pollutants, Palo Alto, California.
27. ——, D. Bartlett, W. Philpot, and R. Rogers. 1974. Coastal and estuarine studies with ERTS-1 and Skylab. Remote Sensing of Environment 3:153-174.
28. ——, D. Bartlett, and R. Rogers. 1975. Coastal zone classification from satellite imagery. Photogrammetric Engineering and Remote Sensing 4;(3):499-512.
29. ——, O. Crichton, F. Daiber, and A. Fornes. 1974. Inventory of Delaware's wetlands. Photogrammetric Engineering 15(4):433-440.
30. ——, G. Davis, G. Tornatore, H. Wang, and W. Whelan. 1975. Monitoring estuarine circulation and ocean waste dispersion using an integrated satellite-aircraft-drogue approach. Proc. Int. Conf. on Environmental Sensing and Assessment, Las Vegas, Nevada.
31. ——, P. Kinner, W. Leatham, D. Maurer, and W. Treasure. 1974. Dye and drogue studies of spoil disposal and oil dispersion. J. Water Poll. Cont. Fed. 46(8):2026-2034.
32. Klooster, S. A., and J. P. Scherz. 1974. Water quality by photographic analysis. Photogrammetric Engineering 40(8):927-934.
33. Kritikos, H., H. Smith, and L. Yorinks. 1974. Suspended solids analysis using ERTS-1 data. Remote Sensing of the Environment 3:69-78
34. Lillesand, T. M., J. L. Clapp, and F. L. Scarpace. 1975. Water quality in mixing zones. Photogrammetric Engineering and Remote Sensing 41(3):285-297.
35. Mairs, R. L., and D. K. Clark. 1973. Remote sensing of estuarine circulation dynamics. Photogrammetric Engineering 39(9):927-938.
36. Maul, G. A. 1974. Applications of ERTS data to oceanography and marine environment. Proc. of COSPAR Symp. on Earth Survey Problems, Akademie-Verlag, Berlin, p. 335-347.
37. ——. 1975. A new technique for observing mid-latitude ocean currents from space. Proc. Am. Soc. of Photogrammetry, 41st Annual Meeting, p. 713-716.
38. ——, R. L. Charnell, and R. H. Qualset. 1974. Computer enhancement of ERTS-1 images for ocean radiances. Remote Sensing of Environment 3:237-254.
39. McCluney, W. R. 1974. Ocean color spectrum calculation. Applied Optics 13(10):2422-2429.
40. Monahan, E. C., and E. A. Monahan. 1973. Trends in drogue design. Liminol. Oceanogr. 18:981-985.
41. Nichols, M., and M. Kelly. 1972. Time sensing and analysis of coastal water.

Proc. of Eighth Int. Symp. on Remote Sensing of Environment, Ann Arbor, Michigan, p. 969-981.

42. Piech, F. R., and J. E. Waler. 1971. Aerial color analyses of water quality. Proc. of ASCE Nat. Water Resources Eng. Meeting, Phoenix, Arizona.
43. Pirie, D. M., and D. D. Steller. 1973. California coastal processes study. Proc. Third ERTS-1 Sym., NASA-GSFC, Washington, D.C., 3:1413-1446.
44. Reimold, R. J., J. L. Gallagher, and D. E. Thompson. 1973. Remote sensing of tidal marsh. Photogrammetric Engineering 39(5):477-489.
45. Rogers, R. H., and L. E. Reed. 1974. Automated land-use mapping from spacecraft data. Proc. of National ACSM-ASP Convention, St. Louis, Missouri.
46. Scarpace, F. L., and T. Green. 1973. Dynamic surface temperature structure of thermal plumes. Water Resources Research 9(1):138-152.
47. Stafford, D. B., and J. Langfelder. 1971. Air photo survey of coastal erosion. Photogrammetric Engineering 37(6):565-576.
48. Szekielda, K. -H. 1973. The validity of ocean surface temperatures monitored from satellites. J. Cons. Int. Explor. Mer. 35(1):78-86.
49. Teleki, P. 1975. Data acquisition methods for coastal currents. Ocean Engineering III Conference, Newark, Delaware.
50. Wezernak, C. T. 1974. The use of remote sensing in limnological studies. Proc. Ninth Int. Symp. on Remote Sensing of Environment, Ann Arbor, Michigan, (1974):963-979.
51. Williams, J. 1970. Optical properties of the sea. United States Naval Institute, Annapolis, Maryland.
52. Williamson, A. N., and W. E. Grabau. 1973. Sediment concentration mapping in tidal estuaries. Proc. Third ERTS-1 Symp., NASA-GSFC, Washington, D.C., 3L1347-1386.
53. Zaitzeff, J. B., and J. W. Sherman, III. 1968. Ocean applications of remote sensing. Proc. of Fifth Int. Symp. on Remote Sensing of Environment, 5:497-527.

THE HYDRAULIC MODEL OF CHESAPEAKE BAY[1]

James H. McKay, Jr.
U.S. Army Corps of Engineers
P.O. Box 1715
Baltimore, Maryland 21203

ABSTRACT: The Hydraulic Model of Chesapeake Bay is located at Matapeake, Maryland. It is a fixed bed geometrically distorted model constructed to a horizontal scale of 1 to 1000 and a vertical scale of 1 to 100, a distortion ratio of 10. The model will be operated using salt water introduced into the model ocean and fresh water flowing into the system through model tributaries. Linear geometric scales in conjunction with model laws determine hydraulic similitude between the model and its prototype. One year of hydrologic record in nature can be reproduced on the model in 3.65 days. The model will be used in the study of many different estuarine problems, including:

1. The effects on salt water intrusion that are due to modifications of the physical or hydraulic regimen of the estuary.
2. Diffusion, dispersion, and flushing of wastes.
3. The effects of power plant cooling water discharges.
4. Tidal flooding by storm surges.

The Hydraulic Model of Chesapeake Bay will be a powerful addition to the tools available for analysis of estuarine physical problems by scientists, engineers, and planners working on Chesapeake Bay.

Chesapeake Bay, one of the world's larger and more productive estuaries, lies within the Atlantic Coastal Plain. Its tributary arms are the drowned valley system of the Susquehanna River resulting from the most recent rise of sea level. As such, the bay is at most only 10,000 years old. Chesapeake Bay, its estuarine

[1] The author gratefully acknowledges information used in the preparation of this paper supplied by personnel from the Hydraulics Laboratory, U.S. Army Engineers Waterways Experiment Station, Vicksburg, Mississippi.

tributaries, the freshwater rivers and streams that empty into it, as well as the lands draining to it, comprise a highly complex hydrologic system. The estuarine system has a water surface area of approximately 11,400 km^2, a shoreline length of some 11,700 km, and varies in width from 6 to 50 km. The bay receives freshwater inflow from a drainage area of approximately 166,000 sq km through over 50 rivers of varying hydraulic and geochemical properties. The Susquehanna River alone supplies over half of the freshwater contributed to the bay, exerting profound hydraulic and ecological effects on the system. Salinities range from 33 parts per thousand inside the mouth of the bay to near zero at the head of tide. Average maximum tidal currents in the mid channel of the bay range between one and 3.5 kilometers per hour. The mean tidal fluctuation is small, generally between 0.3 and 0.6 meters.

Accelerating urban and economic development, as well as a generally expanding level of personal income, have generated both industrial and recreational demands on the bay's aquatic resources. It not only meets the demands of a $40 million commercial seafood industry, but also has the capacity to carry 135 million tons of yearly commerce. Over 125,000 pleasure boats call its waters home. Large numbers of tourists enjoy many different forms of outdoor recreation as well as cultural and historical aspects of the tidewater community.

The realization of increasing pressures on Chesapeake Bay led to the legislation authorizing the Chesapeake Bay Study. Section 312 of the River and Harbor Act of 1965 directed the Corps of Engineers to undertake a complete investigation of water use and control in the Chesapeake Bay Basin. Further, it directed that a hydraulic model of Chesapeake Bay be constructed in the State of Maryland to assist in accomplishing this study.

In response to this legislation the Chesapeake Bay Study was established and organized in a form that was capable of using to its best advantage the large reservoir of existing administrative, scientific, and engineering talent in the governmental and educational institutions in the immediate area.

An elaborate coordination mechanism was established to handle the resource study work. The study organization developed an Advisory Group to provide broad study guidance, a Steering Committee to advise on technical matters, and five task groups to perform basic study work. Constant liaison, work review, and agency interaction are being maintained between the participating study entities through the above mechanism. In addition, in order to provide interaction between public organizations, conservation groups and the Chesapeake Bay Study, a public participation program has been initiated.

The Chesapeake Bay Study Group has recently published the *Chesapeake Bay Existing Conditions Report*. Knowledge and experience gained during the preparation of this report reinforced the need for comprehensive planning to guarantee the future ecological integrity of Chesapeake Bay. Continued fundamental and applied research will be required to assist in the formulation of rational guidelines supplementing action plans. It became abundantly clear that the hydraulic model of Chesapeake Bay will contribute much to the development of knowledge concerning internal flow conditions within the estuary and how they will be affected in the future by works of man.

The use of scale models as an aid in the resolution of hydraulic problems is by no means a new concept. It is an evolutionary outgrowth of three centuries of applied and theoretical engineering studies. It is known that some early hydraulic investigators assembled rudimentary models simulating natural phenomena. It was not until the last decades of the nineteenth century, however, that hydraulic modeling techniques were used in the solution of practical estuarine engineering problems. Then, as now, many of the flow conditions encountered in estuarine areas were not subject to rigorous mathematical analysis. It became necessary to build models to observe or predict hydraulic phenomena that, then as now, could not easily or economically be investigated in the prototype.

Professor Osborne Reynolds built the first estuarine hydraulic model to study the interaction between shoaling and the construction of controversial training works on the Mersey River Estuary near Liverpool in England. As a result of Reynold's work, the proposed work on the Mersey River was extensively revised. Reynolds also, in that early period, called attention to the fact that hydraulic models had potential for use in pollution studies.

The design, construction, and effective use of a modern estuarine hydraulic model similar to that of Chesapeake Bay is a significant engineering project that requires the mobilization and coordination of many technical disciplines. Called upon for the Chesapeake Bay Model Project were the skills of structural and hydraulic designers, hydrologists and oceanographers, coastal and hydrographic surveyors, as well as the talents of other technical specialities such as electronic, instrumentation, and computer technicians.

Developing the Chesapeake Bay Hydraulic Model complex constituted a two pronged design and construction effort by the U.S. Army Engineers Waterways Experiment Station (WES) and the Baltimore District Office. Concurrent with, but distinct from the design and construction work, other agencies and institutions were collecting prototype field data for the adjustment and verification of the model.

The hydraulic model shelter, including the model water supply and treatment system was designed by Whitman, Requardt and Associates of Baltimore, Maryland. The work was supervised by the Baltimore District, with significant design input by WES.

The hydraulic model was designed at WES and is presently under construction. Included in the design of the hydraulic model was the sizing and layout of a complex water distribution system supplying saltwater to generate tides in the model ocean and freshwater to simulate inflow into the system from tributary drainage areas. A very significant portion of model design included the plotting of masonite templates to which concrete is molded to form the topographic features of the model. Twenty-six miles of templates will be installed during model construction. The entire Chesapeake Bay System is being modeled to the head of tide both in the bay proper and its tributary waters and to the +6.0 meter contour on the overbank areas. Model construction is being done by government labor under the supervision of experienced construction men from WES. Fig. 1A is a photograph of the model under construction. Fig. 1B shows a

Figure 1. A. View within the shelter showing pre-cut model templates set to grade and backfilled with sand. B. 'Sketcher' drawing contours on concrete in preparation for molding.

"sketcher" detailing topographic features between templates for model molders. Model construction is scheduled to be complete in the spring of 1976.

As previously mentioned, other agencies and institutions were collecting prototype data during a four year period to be used to adjust the hydraulic model in order to verify that the model does, in fact, act like the prototype. Fig. 2 is a map showing locations of tidal elevation gaging stations and transects at which current velocity and salinity data were taken both at the surface and at various specified depths. Tidal elevation data were recorded at 72 stations established by the National Ocean Survey. In turn, The Johns Hopkins University, the University of Maryland, and the Virginia Institute of Marine Science cooperated in supplying personnel and boats to occupy specified stations in order to collect current velocity and salinity data. In all, data was collected at 105 transects containing 704 measuring points for periods of from three to five days. All of the prototype data were submitted to WES where a data package to be used to verify the model is being assembled.

As designed and built, the hydraulic model of Chesapeake Bay is a fixed bed distorted scale model. Its horizontal scale is 1:1,000 and its vertical scale is 1:100; a scale distortion ratio of 10. Many years of experience at WES have shown that these are the smallest practical scales to which an estuarine model can be built and still accurately reproduce the lateral and vertical distribution of current velocities and salinity. The model is approximately 315 meters long and 225 meters wide. Its total wetted area when flooded to mean low water is about 15,400 sq. meters. Construction of the model is scheduled to be completed in the spring of 1976.

It must be emphasized that upon completion of construction that the hydraulic model complex will be a finely tuned system. Fig. 3 shows the relationship of the hydraulic model and its appurtenances to the shelter. Included in Fig. 3 is:

1. The model shelter, required to shelter the model from the elements during operation. Wind generated waves as well as debris carried into the model could stop model operation.

2. The water treatment plant is needed because of the high mineral content of the groundwater. This plant is capable of supplying water to a community of 10,000 people.

3. The elevated water tower to store treated water to be used for both the model and domestic purposes.

4. The elevated water supply sump that provides saltwater to generate tides in the model ocean.

5. The degraded saltwater return sump, where saltwater returns as the model tide ebbs, has its salinity adjusted, and is then pumped up into the elevated water supply sump for reuse.

6. The lixator, a tank where brine used to adjust water salinity is made and stored until used.

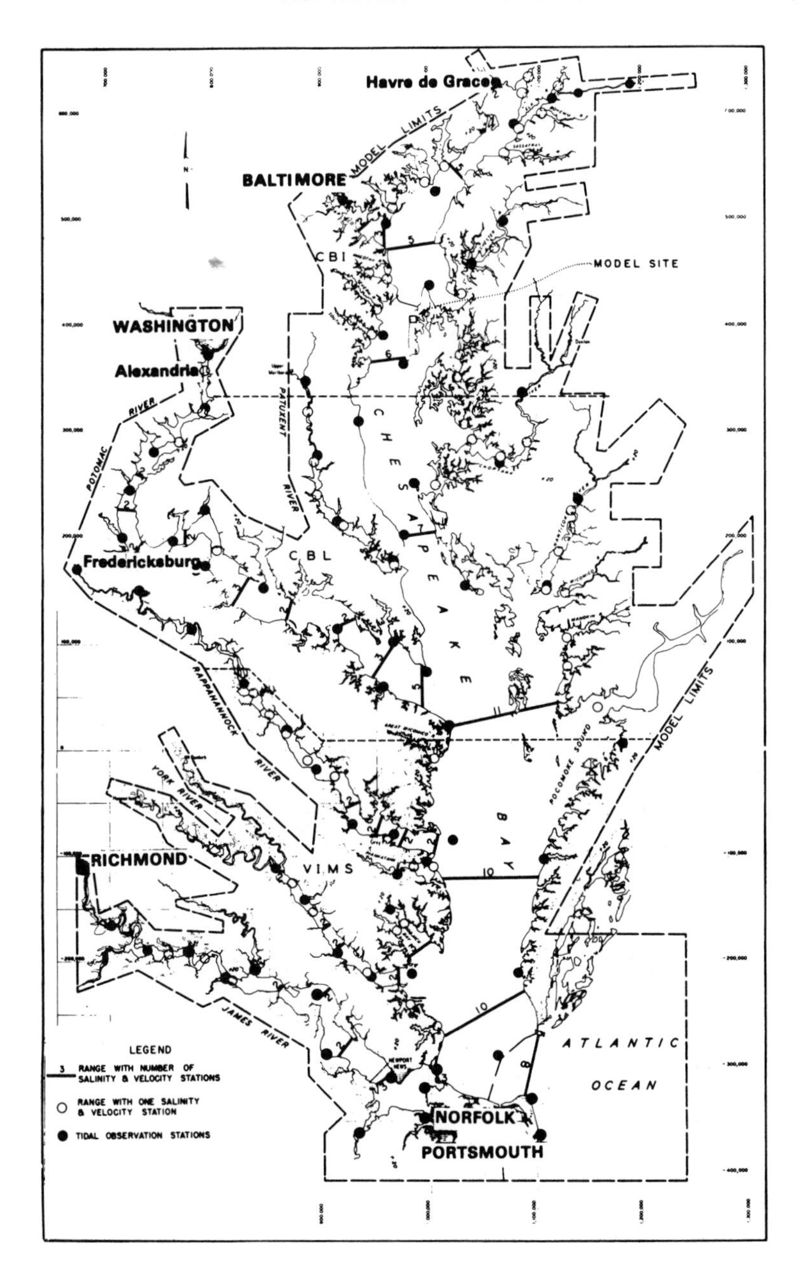

Figure 2. Location map of field data stations occupied to gather prototype data for adjustment and verification of the model.

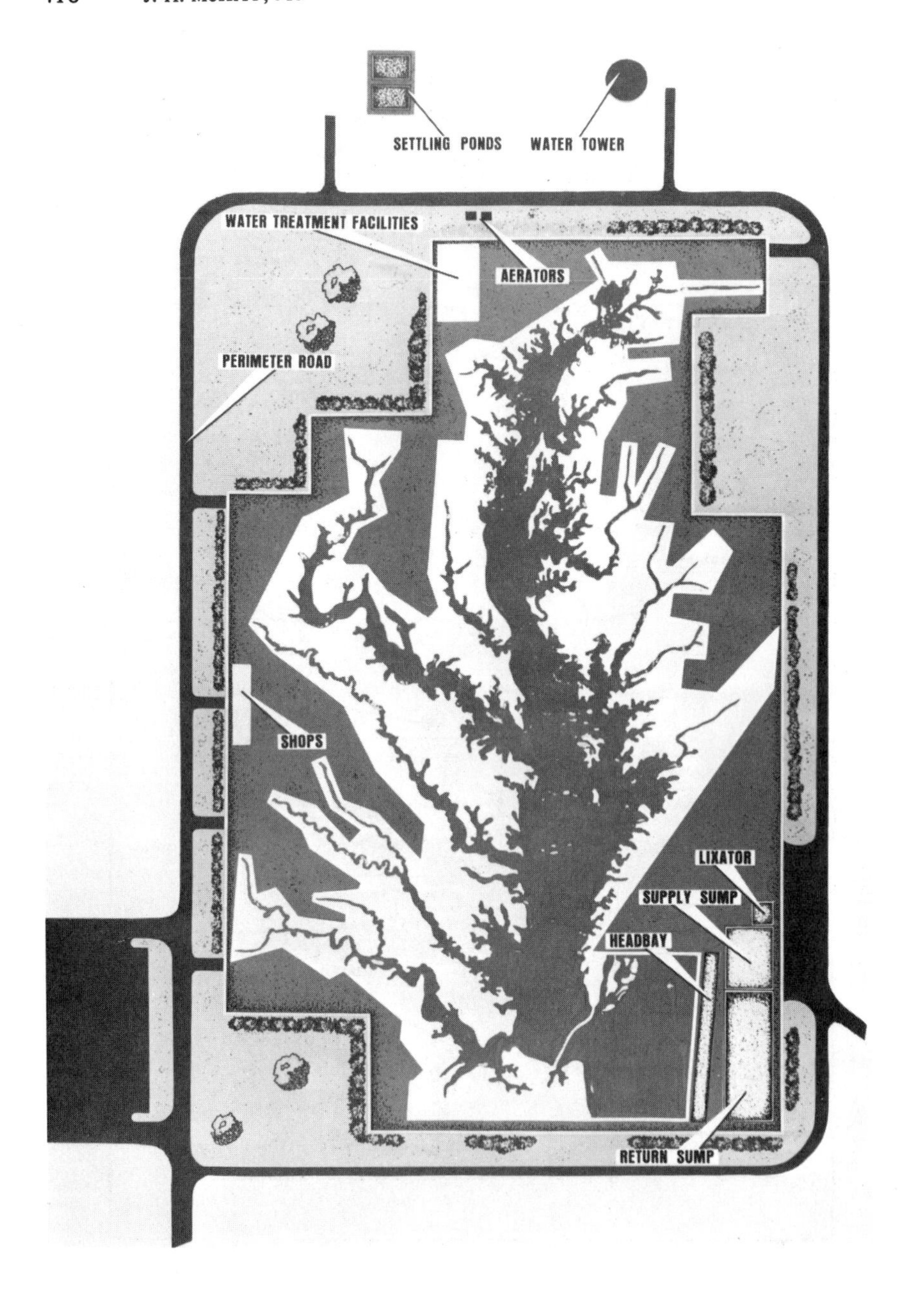

Figure 3. Schematic diagram of the layout of the Hydarulic Model and its appurtenances in the shelter.

Every component of this system must function perfectly for the model to properly operate. A digital control unit will moderate the generation of tides in the model ocean as well as operate the valves that control the flow of freshwater into the model through its tributary rivers and streams.

The model adjustment and verification phase will begin immediately upon completion of construction and is scheduled to be completed within one year. The objective of the verification of the Chesapeake Bay Model is to adjust the model and to verify that, in fact, model hydraulic and salinity phenomena are in acceptable agreement with the prototype. This process is both time consuming and tedious, and also of very great importance in that the validity of a hydraulic model investigation is totally dependent on the ability of the model to reproduce prototype hydraulic phenomena within reasonable limits of accuracy.

Because of the previously mentioned scale distortion, slopes within the model are ten times those in the prototype, making the model hydraulically more efficient than the prototype. To compensate for this increased efficiency additional roughness in the form of metal strips is being installed, extending from the surface of the model vertically through the water mass. Without this additional flow resistance the hydraulic model could not reproduce the lateral and vertical distribution of current velocity and salinity.

Briefly, adjustment and verification work will be accomplished in two phases: (1) a hydraulic verification that will establish that tidal elevations and current velocities are in reasonable agreement with the prototype, and (2) a salinity verification, ensuring that salinity conditions in the model reflect those in the prototype. The basis for this work will be the previously mentioned prototype data package being prepared at WES.

Initially, using freshwater, a specified tide is generated in the model ocean. At the same time appropriately scaled inflows are reproduced in all of the model tributaries. Model roughness strips are progressively adjusted by hand until prototype tidal elevations are acceptably reproduced throughout the model within the time phases in which they occurred. When it has been determined that average tidal volumes and discharges are being satisfactorially reproduced the model is brought into agreement with that in the prototype through further adjustment of model roughness. The next step in the adjustment process consists of operating the model with salt-water in the model ocean and further refining the model roughness distribution in order to achieve an accurate reproduction of both lateral and vertical current distribution. The final step in the adjustment and verification of the Chesapeake Bay Model is the proper adjustment of both salinity in the model ocean and the location and quantity of freshwater inflow into the model to establish the longitudinal, lateral, and vertical distribution of salinity throughout the model. Although the Chesapeake Bay Model has not yet been verified, its operational capability can be anticipated from the results of previous model verification studies. Fig. 4 shows the relationship between previously constructed models and their respective prototype estuarine water bodies

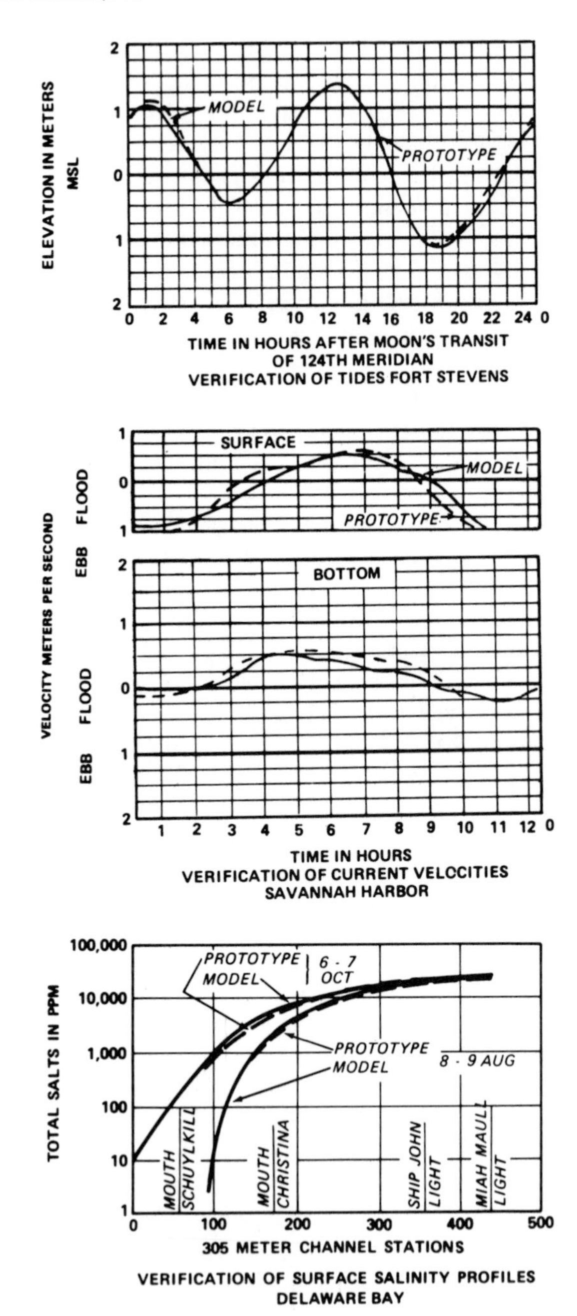

Figure 4. Verification curves for existing models.

for tidal elevation, salinity, and current velocity. It should be noted that the various models closely simulate their prototype estuaries for the specified physical properties.

The foregoing description of model verification procedures for the Chesapeake Bay Model is necessarily brief and represents only the basic hydraulic and salinity verification. There are, depending on studies to be done, further verifications that may be necessary. These include: (1) a fixed bed shoaling verification that assures acceptable reproduction of prototype shoaling characteristics, (2) dye dispersion verification for waste water dispersion studies, and (3) storm surge verification to ensure that the model will accurately reproduce water surface elevation and flow patterns resulting from a range of possible storm surges.

There are six basic measurements that can be made on estuarine hydraulic models. These include water surface elevation, salinity, current velocity, dye concentration from dye dispersion studies, temperature, and sediment distribution (considered a qualitative measure). Properly collected and reduced, this data can be used to describe the physical impact on an estuarine water body of the works of man. In turn, biological stress can often be predicted from the knowledge of changing hydraulic conditions. A partial listing of the types of problems addressed during studies on other models at the U.S. Army Engineers Waterways Experiment Station include: (1) investigations of the changes in water surface elevations, current velocities, salinity distribution, flushing rates, and waste dispersion characteristics due to the geometrical modification of an estuarine water body resulting from the construction of such facilities as navigation channels and port facilities; (2) studies of the distribution of sediment as affecting the alignment and maintenance costs of navigation channels; (3) investigation of the hydraulics of storm surges and the planning and design of protective works; (4) the dispersion characteristics and the area of influence of waste water discharges, including heated discharges of power plant cooling water; (5) studies concerning the feasibility of using the upper portions of estuaries as sources of water supplies; (6) investigations of the impact on estuarine salinity regimes due to upstream modification of freshwater inflows resulting from the construction of reservoirs or increased consumptive losses due to intensive industrial development; and (7) provision of basic data for the adjustment and verification of other models, both physical and analytical.

There are, however, some phenomena that will not be able to be reproduced in the Chesapeake Bay Model. Shoreline erosion and bottom scour cannot be reproduced in a fixed bed (concrete) model, waste water dispersion studies that are done on the model using conservative dyes cannot directly yield values of non-conservative constituents. Physical model results must be treated analytically before thay can be compared to field conditions. In studies of the effects of heated discharges, near field dispersion cannot be readily simulated in a distorted model because of the scale effects on the discharge jet.

Estuarine hydarulic models have been used for many years in the study of physical processes in the marine environment. They are extremely useful tools in studies leading to a more complete understanding of estuarine phenomena as well as providing a technique for predicting the effects on a specific water body of both structural and geometric change. As such, they are an important tool for both planning and designing works in the estuary. Through their use it is possible to evaluate a number of alternative problem solutions rapidly and economically. To obtain maximum benefits from these studies, however, it is necessary to employ skilled hydraulic engineers thoroughly familiar with both the uses and limitations of the models involved.

Preliminary planning for the formulation of the first year of hydraulic studies on the Chesapeake Bay Model has recently been completed. The primary purpose of this initial effort is to develop a study program that is both responsive to problems of immediate importance and at the same time ensure that from the very beginning of operation maximum economical use is made of the model. The formulation of this preliminary study plan involved an extensive analysis of the environmental, economic, and social aspects of a series of current problems in order to establish a priority listing of their importance. The study program that evolved is oriented towards the analysis of the effects of some of the works of man on the Chesapeake Bay estuarine environment. Included in the first year's work will be:

1. The Low Freshwater Inflow Study. This investigation is designed to study the effects on the salinity regime of the Chesapeake Bay System that will result from significantly decreased freshwater inflows due to drought conditions or due to upstream construction such as reservoirs or to increased consumptive losses.

2. The Baltimore Harbor Study. This work will be undertaken to define the effects on the estuarine system due to increasing the depth of Baltimore Harbor navigation channels to 15 meters. Conditions to be investigated include rates of harbor flushing, waste dispersion patterns, salinity distribution, and dredge material disposal.

3. The Potomac River Estuary Water Supply and Waste Water Dispersion Study. This study will explore the ramifications of using the Potomac River Estuary as a supplemental source of water supply for Washington, D.C. One of the concerns generated by using the estuary as a source of water supply is the possibility of recycling waste water into the public water supply during periods of low freshwater inflow and the possibility of changing the salinity levels and current patterns in the estuary.

4. The James and Elizabeth Rivers Water Quality Study. If at the completion of the above three studies there is both sufficient time and funds remaining, this study will be done. The James and Elizabeth rivers are subject to considerable stress from both municipal and industrial waste discharges. The purpose of this

study is to define existing waste water dispersion patterns of treatment plant discharges in the area of concern.

The hydraulic study program is scheduled to begin in the spring of 1977 and will be completed within one year. Although the presently authorized Chesapeake Bay Study has funding sufficient for only a one year program of hydraulic investigations, it is anticipated that with both future funding and expanded use, the Chesapeake Bay Model will have a long productive life and will play an increasingly important role in future investigations concerning the formulation of rational plans of development for the bay system.

THE EVALUATION AND USE OF GEAR EFFICIENCIES IN THE ESTIMATION OF ESTUARINE FISH ABUNDANCE

Martin A. Kjelson and David R. Colby
Atlantic Estuarine Fisheries Center
National Marine Fisheries Service
Beaufort, North Carolina 28516

ABSTRACT: Accurate estimates of fish abundance are necessary for models of many estuarine processes, but they have generally been unavailable because fish sampling methods are biased. A major source of bias is a fish's avoidance of sampling gear. A brief discussion of theoretical and qualitative information concerning net avoidance and a simple gear efficiency model is presented. Results of quantitative studies on the efficiency of sampling gear have potential for correcting bias due to net avoidance. Discussion of gear efficiency studies for a plankton net, beam trawls, portable drop-net, haul seines and otter trawls are provided and indicate that such research is feasible and can yield useful information. Our ongoing investigation on the efficiency of a 6.1 m otter trawl for *Lagodon rhomboides* and *Leiostomus xanthurus* in a North Carolina estuary during 1975 indicates that trawl abundance estimates for these species were only 9 to 51% of the true values. General recommendations are given to assist in limiting the problem of fish sampling bias.

INTRODUCTION

Accurate estimates of fish abundance are necessary for estuarine trophic dynamic and management models, yet such estimates generally have been unavailable because fish sampling methods are biased. A major source of bias is the selective nature of fishing gear resulting from mesh selection and fish avoidance. Mesh selection refers to a process which results in the retention of larger fish and escapement of smaller fish through a given size of mesh. Avoidance simply refers to a process wherein fish dodge an approaching sampling gear and thereby elude capture. Mesh selection is relatively well understood and has been overcome through the use of various mesh sizes. Conversely, there is very little information concerning the problem of avoidance.

The objectives of this paper are: 1) to briefly discuss the theoretical and empirical attempts to evaluate avoidance, 2) to provide an example of ongoing research concerning the efficiency of an otter trawl in sampling juvenile estuarine fishes, and 3) to recommend actions designed to limit the problem of fish sampling bias. We will assume throughout the remainder of the paper that bias resulting from mesh selection has been overcome and that gear efficiency therefore is a result of avoidance. Gear efficiency information is necessary to adjust catch data to provide more accurate estimates of fish density. Gear efficiency will be defined as the probability that a given fishing gear will collect a given fish present in the water volume sampled. Emphasis will be given to active sampling gears such as trawls and seines as opposed to more passive devices such as gill nets or traps.

THE EVALUATION OF AVOIDANCE

Evaluations of avoidance have been both theoretical and empirical. Important theoretical efforts have included Barkley's (3) model for the design and evaluation of a towed net where he demonstrated, in the context of a specific problem, that an optimal net size will exist for which avoidance will be minimized. Barkley (4) further related the reaction distance and escape speed of an organism to the radius of the mouth of the net and the towing speed. In particular, an equation was derived that allows calculation of a minimum probability of capture. Such approaches as well as other theoretical approaches discussed by Clutter and Anraku (5) can be powerful aids in the analysis and interpretation of catch data and the choice of sampling gear.

Empirical evaluations of avoidance have compared catches when towing speeds, light intensity and net size were varied (5). Generally, more and larger fish are taken at faster towing speeds and at night. An increase in gear size usually imposes a reduction in towing speed, and therefore most towed gears now in use represent a compromise between the desirability of a higher towing velocity and that of a larger net. A comprehensive review of the relationship of fish behavior to fishing techniques is provided in the proceedings of the FAO conference held in Bergen, Norway in 1967 (9).

Although past theoretical efforts and comparative studies have provided a greater understanding of the factors influencing the gear avoidance of sampling gear by fish, few studies have quantified gear efficiencies. Nevertheless, the need of gear efficiencies for fish sampling gears is well documented (1, 5, 10, 16) but it is apparent from the paucity of efficiency data that it is not easily obtained.

Ideally, information on the relative ability of a given gear to capture individuals of a given species of fish would be in the form of a multiple dimensional model where all variable characteristics of the individual fish and environmental conditions are included. A simpler, two-dimensional efficiency model is provided in Fig. 1A to clarify the idea of gear efficiency. It provides for sampling bias due to both mesh selection and avoidance. In this model, the efficiency of the gear is

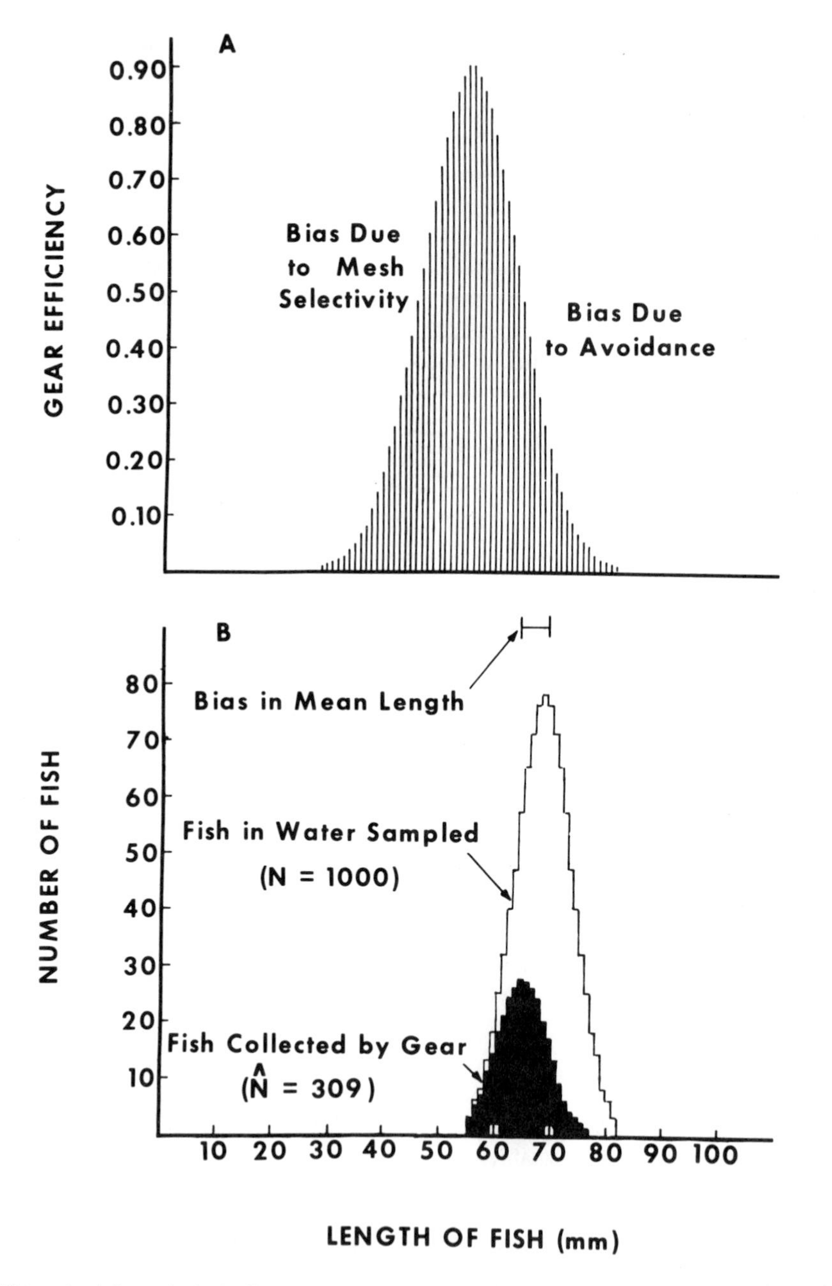

Figure 1. A hypothetical efficiency model for a fish sampling gear. A. Gear efficiencies for fish of different sizes. B. Comparsion of the number of fish collected by a gear to those present in the water sampled.

entirely dependent upon the size of the individuals, with the smaller fish passing through the meshes and the larger individuals avoiding the trawl. For convenience, we have hypothesized a symetrical distribution of gear efficiencies, but this is probably atypical.

Fig. 1B describes the size distribution and numbers of fish collected by the gear under a given condition where there is a population of 1000 individuals in the water sampled by the gear. In this hypothetical example the size distribution of the fish in the path of the trawl is such that all of the fish are larger than the size class for which the gear is most effective. Thus, loss of fish through the meshes is a less serious source of bias than avoidance. The fish collected by the gear are clearly a biased sample of the 1000 fish in the water strained by the gear. The average size of the fish captured is 64.4 mm, which is an estimate having a bias of 5 mm associated with it. However, our estimate of total numbers (and biomass) is even more seriously biased because we captured 309 individuals, whereas 1000 fish were in the water strained by the gear. We calculated the number of individuals of each size that will be collected (Fig. 1B) by simply multiplying the number of fish of each size in the water strained, by the probability that a fish of that size will be captured (Fig. 1A). Conversely, if the catch and efficiency model are known, the "true" abundance of fish of each size can be calculated by dividing the catch of each size by the corresponding gear efficiency.

A great deal of effort is required to reliably measure the efficiency of active gears and few studies have been reported (Table 1). Gear efficiencies for the 1 m plankton net (15) were based on comparison of its catch to that made by a plankton purse seine assumed to be unbiased in its estimation of larval fish abundance. Gear efficiencies of the 2 m beam trawl (14) and 16 m^2 portable drop net (12) were measured by taking test samples of fish restricted to an enclosed body of water. The efficiencies of all three haul seines (8, 13 and Kjelson, unpubl.) were gained by releasing fish directly into the path of the approaching net. Jones' refined study (11) of the Granton (otter) trawl was made using acoustic transponding tags and high resolution sector scanning sonar to observe the continual location of individual tagged fish as the trawl is pulled toward it. Lastly, Edwards' (7) estimates of the "36 Yankee" otter trawl were obtained from more qualitative information based upon a combination of general ecological and behavioral evidence and gear observations.

The above studies, although few in number, indicate that research on the efficiency of sampling gear is feasible. Two studies (7, 12) indicate that a given gear may exhibit very different efficiencies for different species. Secondly, only half of the efficiencies provided are greater than 0.50 (Table 1), emphasizing that most raw catch data is considerably inaccurate. Finally, a more detailed review of the above studies indicates that there is a major need for replication in order to assess the reliability of estimated gear efficiencies.

Table 1. Examples of gear efficiencies for several gears and fish species.

Gear	Species	Size	Gear efficiency	Reference
1 m plankton net	Anchovy	10 mm	0.07	15
2 m beam trawl	Plaice	14-20 cm	0.28	14
16 m² portable drop-net	Atlantic croaker	50-70 mm	0.52	12
	Atlantic menhaden	100-130 mm	0.05	12
21 m haul seine	Pinfish	30-57 mm	0.60	Kjelson Unpubl.
	Spot	45-70 mm	0.72	Kjelson Unpubl.
25 m haul seine	Tomtate	–	0.53	8
350 m long-haul seine	Pelagic fishes	>12 cm	0.41	13
	Semi-demersal fishes	>12 cm	0.47	13
Granton (otter) trawl	Plaice	>40 cm	0.65	11
"36" Yankee (otter) trawl	Alewife	>20 cm	0.90	7
	Barndoor Skate	>40 cm	0.10	7
	Cod	>30 cm	0.80	7
	Pollock	>30 cm	0.50	7

SAMPLING EFFICIENCY OF AN OTTER TRAWL

A brief discussion of *ongoing* research concerning the selectivity of a 6.1 m otter trawl for pinfish, *Lagodon rhomboides,* and spot, *Leiostomus xanthurus*, exemplifies some of the practical problems involved in attempting to measure gear efficiency under semi-natural conditions in shallow estuarine environments characteristic of many estuarine regions on the Atlantic and Gulf coasts.

The research site is a 400 X 125 m tidal embayment near Beaufort, N.C., with a depth of 1-2.5 m. It is surrounded on three sides by *Spartina* marsh; the open end of the bay was completely blocked by a 13 mm mesh net. The trawl consisted of 19 mm bar mesh wings, a 6 mm bar mesh tailbag, and a 6.1 m footrope with tickler chain attached. The average width of the trawl during operation was 5 m while the average height was 1 m. Secchi readings ranged from 52 to 100 cm. Fish were collected at an adjacent sampling location and transported to the test site for marking and release. Fish were marked by clipping a small portion of a single lobe of the caudal fin. To eliminate loss of fish due to mesh selection, only fish of sufficient size to be held by the net were marked. Recovery either was attempted immediately upon release or after a waiting period of two to three days. An experimental (recovery) tow was made at a speed of 1.6 knots along the longitudinal axis in each of three zones dividing the bay lengthwise. An estimate of average density for clipped fish was calculated from the trawl catches and from the total area sampled in the three tows. The "true" density of marked fish was calculated by dividing the total number of released fish, less those estimated lost by mortality and escapement, by 50,000, the total area of the embayment in m^2. Mortality over the 2-3 day "waiting period" was estimated from tank experiments, while loss by escapement was estimated from the catch of a single tow made in the 25,000 m^2 embayment immediately adjacent to the enclosure. Each gear efficiency was calculated by dividing the trawl estimate of mean density by the corresponding "true" density.

The results of seven tests using pinfish and six using spot are summarized in Table 2. The average difference in the two values for gear efficiencies from each of the five replicated experiments was 0.15.

The trawl apparently was more efficient for larger pinfish than for smaller pinfish or for spot. Only the data for adult spot suggested that efficiency of the trawl varied for day and night tests. A greater proportion of both juvenile pinfish and spot were recovered if recovery operations occurred immediately following release rather than after a delay of 2-3 days.

We must interpret these gear efficiencies cautiously because we were not able to satisfactorily evaluate several assumptions of the test procedure. Estimates of mortality were: juvenile spot, 39%; juvenile pinfish, 3%; adult spot, 16%; and adult pinfish, 10%. The relationship of tank mortality to actual field mortality is unknown. The effect of herding by trawl warps and doors upon the catch is unknown. [See Alverson and Pereyra (2) for discussion of this topic]. Escapement from the "enclosed" test site may have occurred at extremely high tides, when border areas of dense *Spartina* marsh grass were flooded, but we consider this to be a minor problem. Minimum estimated losses due to escapement, based upon recovery of clipped fish immediately outside the blocking net, were 1.5%

Table 2. Gear efficiencies of a 6.1 m otter trawl for pinfish and spot.

Size class	Time of day	Species	
		Pinfish	Spot
Juvenile[1] (38-85 mm)	day	0.69*(0.64,0.74)	0.35*(0.44,0.25)
Juvenile[2] (38-80 mm)	day	0.21*(0.17,0.25)	0.23
	night	0.23*(0.34,0.12)	0.16*(0.09,0.23)
Adult[2] (90-157 mm)	day	0.51	0.30
	night	0.50	0.09

[1] Immediate recovery
[2] Recovery delayed 2-3 days
*Mean value resulting from 2 experiments (individual values in parentheses)

for spot and 1.1% for pinfish. The average coefficient of variation for the three sample tows was 55% (range 14%-175%). We assumed that the fish were available to the gear at all times, but future studies of their distribution patterns will enable us to evaluate this assumption.

The use of identical gear to collect fish, as well as for experimental recovery, may cause problems in measuring gear efficiencies, since fish exposed to capture once may respond differently to a second trawl attack. In addition, the selectivity of the gear used to collect fish for gear testing may mean that the test population is unrepresentative of the estuarine population. When possible, several methods of collection should be used to collect specimens for gear testing.

The stimulus of capture and handling appears to increase vulnerability (Table 2) and therefore, it is desirable to provide a time interval between establishment of the test population and the gear-testing operation. A break between the two operations also allows the test fish to acclimate to the test environment and thus more adequately represent wild fish populations.

It is apparent from this study that there is considerable inaccuracy in estimates of pinfish and spot density gained from 6 m otter trawl studies. For example, trawl estimates of juvenile pinfish abundance would be low by a factor of approximately five based upon the gear efficiency of 0.21 (Table 2). Even for the immediate recovery tests, using test fish that had recently been handled and released into unfamiliar surroundings, the gear efficiencies averaged only 0.52 (Table 2). The distortion resulting from using such biased estimates in models of most estuarine processes is apparent.

GENERAL RECOMMENDATIONS

The above discussion has emphasized one aspect of the bias of fish sampling methods. Although such bias may never be completely eliminated, a unified approach by estuarine ecologists involved in fish sampling activities can do much to limit this problem. The following recommendations are designed to limit the problem of fish sampling bias.

Fishery biologists should continue their attempts to evaluate the bias involved in their respective sampling gears and the bias and limitations of sampling gears and sample design should be reported and discussed more frequently so that other investigators using that data will not draw incorrect inferences.

The standardization of estuarine fish sampling gears and their operating procedures would facilitate collaborative investigations of gear selectivity and comparison of specific results of different investigations. The standard use of bongo gear for ichthyoplankton sampling on a world-wide basis is one example of such action. Otter trawls, surface trawls and haul seines have been used frequently in sampling estuarine fish populations, yet design and operation of these three gears has varied greatly from one study to another.

The development of new and improved technologies directed at improving fish abundance estimates must continue. Acoustical and other remote sensing devices have potential for the estimation of estuarine fish populations, but their application in shallow turbid estuaries requires extensive development and field experimentation. Here, as for classical gears, a major problem involves the evaluation of sampling bias. There also is a need for continued development of fishing gear specifically for sampling, such as purse seines for sampling adult pelagic fishes in open water habitats and for improved gear with higher sampling efficiencies. Specialized sampling gear is particularly important in achieving the potential to sample whole fish communities in a restricted habitat site at one time, a requirement if we are to understand the functional relationships between populations.

CONCLUSIONS

As a rule estimates of estuarine fish abundance are seriously biased. It appears sensible to measure the sampling efficiency of a gear for a species of fish and to then use the resulting information for adjusting estimates of fish abundance. However, the diversity that characterizes the fish communities, habitats and environmental conditions of most estuaries, implies that a great deal of effort must be devoted to such gear calibration before we will be in a position to accurately describe estuarine fish populations. The magnitude of the task before estuarine fish ecologists in turn underscores the desirability of gear standardization and collaborative effort among interested institutions.

LITERATURE CITED

1. Allen, G.H., A.D. DeLacy and D.W. Gotshall. 1960. Quantitative sampling of marine fishes–a problem in fish behavior and fishing gear, p. 448-511. *In* E.A. Pearson (ed.), Waste disposal in the marine environment. Pergamon Press.
2. Alverson, D.L., and W.T. Pereyra. 1969. Demersal fish explorations in the Northeastern Pacific Ocean–an evaluation of exploratory fishing methods and analytical approaches to stock size and yield forecasts. J. Fish. Res. Board Can. 26(8):1985-2001.
3. Barkley, R.A. 1964. The theoretical effectiveness of towed-net samplers as related to sampler size and to swimming speed of organisms. J. Cons. Cons. Int. Explor. Mer. 29(2):146-157.
4. ——. 1972. Selectivity of towed-net samplers. U.S. Natl. Mar. Fish. Serv. Fish. Bull. 70(5):799-820.
5. Clutter, R.I., and M. Anraku. 1968. Avoidance of samplers, p. 57-76. *In* Zooplankton Sampling. UNESCO, Monogr. Oceanogr. Methodol.
6. Edwards, R., and J.H. Steele. 1968. The ecology of O-group plaice and common dabs at Loch Ewe I. Population and food. J. Exp. Mar. Biol. Ecol. 2:215-238.
7. Edwards, R.L. 1968. Fishery resources of the North Atlantic area, p. 52-60. *In* D. Gilbert (ed.) The Future of the Fishing Industry in the U.S. College of Fish. Univ. Wash. Publ.

8. Edwards, R.R.C. 1973. Production ecology of two Caribbean marine ecosystems I. Physical environment and fauna. Estuarine Coastal Mar. Sci. 1:303-318.
9. Food and Agriculture Organization. 1968. Proceedings of the conference on fish behavior in relation to fishing techniques and tactics. *In* A. Ben-Tuvia and W. Dickson (eds.) FAO Fisheries Report No. 62. Vol. 1-3. Rome.
10. Grosslein, M.D. 1971. Some observations on accuracy of abundance indices derived from research vessel surveys. Int. Comm. Northwest Atl. Fish. Rebd. Part III. p. 249-265.
11. Jones, F.R.H. 1974. Objectives and problems related to research into fish behavior. *In* F.R.H. Jones (ed.), Sea Fisheries Research. John Wiley Sons, New York. 510 p.
12. Kjelson, M.A., and G.N. Johnson. 1973. Description and evaluation of a portable drop-net for sampling nekton populations. Proc. 27th Annu. Conf. Southeast Assoc. Game Fish. Comm. p. 653-662.
13. ——, and ——. 1975. Description and evaluation of long-haul seine for sampling fish populations in offshore estuarine habitats. Proc. 28th Annu. Conf. Southeast Assoc. Game Fish. Comm. 1974, p. 171-178.
14. Kuipers, B. 1975. On the efficiency of a two-metre beam trawl for juvenile plaice (*Pleuronectes platessa*). Neth. J. Sea. Res. 9(1):69-85.
15. Murphy, G.I., and R.I. Clutter. 1972. Sampling anchovy larvae with a plankton purse seine. U.S. Natl. Mar. Fish. Serv. Fish Bull. 70(3):789-798.
16. Pearcy, W.G. (ed.). 1975. Workshop on problems of assessing populations of nekton. Santa Barbara, Calif. Feb. 25-27, 1975. Office of Naval Research, Final Report ACR 211, 30 p.

INDEX

A 6
B 7
C 8
D 9
E 0
F 1
G 2
H 3
I 4
J 5

ECH